Lecture Notes in Mathematics

1934

Editors:
J.-M. Morel, Cachan
F. Takens, Groningen
B. Teissier, Paris

Catherine Donati-Martin · Michel Émery ·
Alain Rouault · Christophe Stricker (Eds.)

Séminaire de Probabilités XLI

 Springer

Editors

Catherine Donati-Martin

Laboratoire de Probabilités
  et Modèles Aléatoires
Université Pierre et Marie Curie
Boîte courrier 188
4, place Jussieu
75252 Paris cedex 05, France
e-mail: donati@ccr.jussieu.fr

Alain Rouault

Laboratoire de Mathématiques
  Bâtiment Fermat
Université Versailles-Saint-Quentin
45, avenue des Etats-Unis
78035 Versailles cedex, France
e-mail: rouault@fermat.math.uvsq.fr

Michel Émery

Institut de Recherche
  Mathématique Avancée
Université Louis Pasteur
7, rue René Descartes
67084 Strasbourg cedex, France
e-mail: emery@math.u-strasbg.fr

Christophe Stricker

UFR Sciences et techniques
Université de Besançon
16, route de Gray
25030 Besançon cedex, France
e-mail: christophe.stricker@
  univ-fcomte.fr

ISBN: 978-3-540-77912-4   e-ISBN: 978-3-540-77913-1
DOI: 10.1007/978-3-540-77913-1

Lecture Notes in Mathematics ISSN print edition: 0075-8434
                ISSN electronic edition: 1617-9692

Library of Congress Control Number: 2008921482

Mathematics Subject Classification (2000): 15A52, 60Gxx, 60Hxx, 60Jxx, 82B20, 91B28

Typesetting by the editors and EDV-Beratung Frank Herweg using a Springer LATEX macro package

Cover design: WMXDesign GmbH, Heidelberg

Printed on acid-free paper

9 8 7 6 5 4 3 2 1

springer.com

Preface

As usual, some of the contributions to this 41st Séminaire de Probabilités were exposed during the Journées de Probabilités (held in Nancy in 2005 and in Luminy in 2006). The other ones come from spontaneous submissions or were solicited by the editors. We hope that the whole volume is a good sample of the main streams of current research on probability and stochastic processes, in particular those active in France.

The last two volumes of the Séminaire (vol. XXXIX, LNM 1874 and vol. XL, LNM 1899) have sustained long delays, ascribable to the editors and to the publisher. We have to admit that in extreme cases, the lapse between submission and publication has exceeded three years. This backlog propagates from one volume to another: the submission dates of the articles collected in this volume XLI range from May 2005 to May 2007, and publication is expected in winter or spring of 2008; we apologize to contributors and readers of the Séminaire for this inconvenience. With Springer's help, the situation is now improving. The next volume, vol. XLII, is started with no backlog at all. This means that we are now back to the normal state, where the batch of articles sent to Springer for a new volume contains only contributions having reached their final form less than one year before.

Authors can help us shorten the processing of their contribution and speed up the whole volume, by formatting their manuscript with Springer's own LaTeX environment (see the Note to Contributors on page 463) and by avoiding personal macros or style files.

7 November 2007

C. Donati-Martin
M. Émery
A. Rouault
C. Stricker

Contents

Spectral gap inequality for a colored disordered lattice gas

Azzouz Dermoune[1] **and Philippe Heinrich**[2]

Laboratoire Paul Painlevé, Université Lille 1, Bât. M2
59655 Villeneuve d'Ascq Cedex, France
[1] *e-mail: azzouz.dermoune@univ-lille1.fr*
[2] *e-mail: philippe.heinrich@univ-lille1.fr*

Summary. We establish a spectral gap property related to a model called colored disordered lattice gas. The main result is stated for an auxiliary Markov generator which, thanks to the general strategy developped in the work of Caputo [1], produces a uniform Poincaré inequality with respect to the original dynamics of the model.

Key words: simple exclusion process, disordered systems, lattice gas dynamics, Markov process, spectral gap.

1 A simple exclusion colored process with site disorder

1.1 Background

Consider the d-dimensional lattice $\mathbb{Z}^d$ with canonical basis $\mathscr{E}$. A bond of $\mathbb{Z}^d$ is a non oriented couple of sites $\{x, x + e\}$ where $x \in \mathbb{Z}^d$ and $e \in \mathscr{E}$. To each site x, we assign a disorder, that is a random variable α_x, and a particle configuration η_x defined by

$$\eta_x = \begin{cases} +1 \text{ if there is a } blue \text{ particle at } x, \\ 0 \text{ if there is } no \text{ particle at } x, \\ -1 \text{ if there is a } white \text{ particle at } x. \end{cases}$$

As in [6], we assume that the α_x's are i.i.d. and bounded by some constant B. We set for simplicity

$$\eta_x^+ = \mathbf{1}_{\{\eta_x=1\}}, \quad \eta_x^- = \mathbf{1}_{\{\eta_x=-1\}}, \tag{1}$$

so that $\eta_x = \eta_x^+ - \eta_x^-$ and $|\eta_x| = \eta_x^+ + \eta_x^-$.

Dynamics in a volume Λ.

Consider only a finite set $\Lambda \subset \mathbb{Z}^d$ of sites. A (particles) configuration is the collection $\{\eta_x, x \in \Lambda\}$ which is simply denoted by η. If $\{x, y\}$ is a pair of sites, we denote by $\eta^{x,y}$ the configuration derived from η by permuting η_x with η_y. Namely, $\eta_x^{x,y} = \eta_y$, $\eta_y^{x,y} = \eta_x$ and the rest is unchanged. The dynamics of the particles is given by a Markov process $\{\eta(t), t \in \mathbb{R}^+\}$ and can then be described as follows: a particle at x waits an exponential time and attempts to jump to a neighbour site $x \pm e$. If this site is occupied then the jump is aborted, otherwise it is realized with a probabilistic rate

$$
\begin{aligned}
c_{x,x+e}^{\alpha}(\eta) &= \lim_{t \to 0} \frac{\mathbf{P}\left(\eta(s+t) = \eta^{x,x+e} \middle| \eta(s) = \eta\right)}{t} \\
&= f_e\left(\alpha_x, |\eta_x|, \alpha_{x+e}, |\eta_{x+e}|\right)
\end{aligned} \tag{2}
$$

for all $s \geq 0$ and where f_e is a bounded function on $(\mathbb{R} \times \{0,1\})^2$ satisfying the following conditions:

1. $f_e(a, s, a', s') = f_e(a', s', a, s)$ (*symmetry* condition),
2. $ss' \neq 0 \Rightarrow f_e(a, s, a', s') = 0$ (*exclusion* condition),
3. $ss' = 0 \Rightarrow f_e(a, s, a', s') \geq \delta > 0$ (*uniform bound* condition),
4. $f_e(a, s, a', s') = f_e(a, s', a', s) \exp\left(-(s'-s)(a'-a)\right)$ (needed for the *detailed balance* condition).

Markov generator. These conditions allow us to define a disordered Markov generator $\mathscr{L}$. Disordered means depending on the random collection $\alpha = \{\alpha_x, x \in \Lambda\}$ which is also called disorder. The mentioned generator $\mathscr{L} := \mathscr{L}_\Lambda^\alpha$ is given for bounded functions f on $\{-1, 0, 1\}^\Lambda$ by

$$
\begin{aligned}
\mathscr{L}f(\eta) &= \lim_{t \to 0} \frac{\mathbf{E}\left[f\left(\eta(s+t)\right) - f\left(\eta(s)\right) \middle| \eta(s) = \eta\right]}{t} \\
&= \sum_{\substack{(x,e) \in \Lambda \times \mathscr{E} \\ \{x, x+e\} \subset \Lambda}} c_{x,x+e}^{\alpha}(\eta)\left[f(\eta^{x,x+e}) - f(\eta)\right]
\end{aligned} \tag{3}
$$

for all $s \geq 0$. We have thus a Markov process $\{\eta(t), t \in \mathbb{R}^+\}$, with state space $\{-1, 0, 1\}^\Lambda$ and generator $\mathscr{L}$, which induces a colour-blind one, namely $\{|\eta(t)|, t \in \mathbb{R}^+\}$, studied in the work [6] of FAGGIONATO and MARTINELLI.

Grand canonical and canonical measures. Given a disorder α and real numbers λ^+, λ^-, we set $e^\lambda = e^{\lambda^+} + e^{\lambda^-}$ and we consider the product Gibbs measure $\mu = \mu_\Lambda^{\alpha, \lambda^+, \lambda^-}$ on $\{-1, 0, 1\}^\Lambda$ defined by

$$
\mu(\eta_x^\epsilon) = \frac{\exp(\alpha_x + \lambda^\epsilon)}{1 + \exp(\alpha_x + \lambda)} \quad (x \in \Lambda, \, \epsilon \in \{-, +\}) \tag{4}
$$

where η_x^ϵ is a Bernoulli variable as in (1). Thus we have for all $\eta \in \{-1, 0, 1\}^\Lambda$,

$$\mu(\{\eta\}) = Z^{-1} \exp\left(-H(\eta)\right)$$

where Z is a normalizing constant and with Hamiltonian

$$H(\eta) = -\sum_{x \in \Lambda} \left[\alpha_x |\eta_x| + \lambda^+ \eta_x^+ + \lambda^- \eta_x^-\right].$$

The probability measure μ is called *grand canonical measure* on Λ with disorder configuration α and chemical potential couple (λ^+, λ^-).

Besides, to count blue and white particles in Λ, we introduce the number of ϵ-particles for $\epsilon \in \{-, +\}$,

$$N_\Lambda^\epsilon(\eta) = \sum_{x \in \Lambda} \eta_x^\epsilon.$$

For a couple $(m^+, m^-) \in \{0, 1/|\Lambda|, 2/|\Lambda|, \dots, 1\}^2$ such that $m^+ + m^- \leq 1$, we define also integers $N^\epsilon = m^\epsilon |\Lambda|$, and the corresponding *canonical measure*

$$\nu(\,\cdot\,) = \mu\left(\,\cdot\, \mid N_\Lambda^+ = N^+, N_\Lambda^- = N^-\right),$$

which in turn does not depend on (λ^+, λ^-). We showed in [2] that for such a couple (m^+, m^-) and for almost all α, there exists a unique couple (λ^+, λ^-) depending on $(\Lambda, m^+, m^-, \alpha)$ such that

$$\forall \epsilon \in \{-, +\} \quad \mu(N_\Lambda^\epsilon) = N^\epsilon. \tag{5}$$

Motivation. Previous works [2,3] were a first step to get the hydrodynamic limit of this colored disordered simple exclusion process similar to [6]. The aim of our work is to add another step towards this hydrodynamic limit. More precisely, we want to get a uniform Poincaré inequality. Namely, we want to show that there exists a universal constant $c > 0$ such that for every $f \in L^2(\nu)$ and every (Λ, m^+, m^-),

$$\nu(f; f) \leq c\mathscr{D}(f)$$

where $\nu(f; f)$ stands for the variance of f w.r.t. ν and $\mathscr{D}$ is the Dirichlet form defined by

$$\mathscr{D}(f) = \frac{1}{|\Lambda|} \sum_{x, y \in \Lambda} \nu\left(\left|f(\eta^{x,y}) - f(\eta)\right|^2\right).$$

Of course, $|\Lambda|$ denotes the cardinality of Λ. In this paper, we establish a spectral gap property **(SGP)** which implies, as shown in [1], the uniform Poincaré inequality.

2 The main result: spectral gap property

Set for every $f \in L^2(\nu)$,

$$\mathscr{P}f = \frac{1}{|\Lambda|} \sum_{y \in \Lambda} \nu(f \mid \eta_y). \tag{6}$$

Note that this linear operator $\mathscr{P}$ on $L^2(\nu)$ preserves positivity, and is of norm less or equal to one, and satisfies $\mathscr{P}1 = 1$. These properties ensure that $\mathscr{P} - I$ is a Markov generator. Moreover, note that $\mathscr{P} - I$ has reversible (and thus invariant) measure ν since $\nu(f(\mathscr{P} - I)g) = \nu(g(\mathscr{P} - I)f)$ for all $f, g \in L^2(\nu)$.

We are now able to state our main result, called as in [1] (SGP). Introduce the following assumptions:

(H1) $m^+ \geq m^- \geq |\Lambda|^{-1+\tau_1}$ for some $\tau_1 \in (0,1)$ and $\dfrac{m^+}{m^-} = O\left(\ln |\Lambda|\right)$,

(H2) $m^2 \leq |\Lambda|^{-1-\tau_2}$ for some $\tau_2 > 0$ where we set $m = m^+ + m^-$.

Theorem 1 (SGP). *Assume that (H1) or (H2) holds. There exist numbers $c \in (0, \infty)$ and $\tau \in (0, 1/8)$ such that for $f \in L^2(\nu)$ with $\nu(f) = 0$ and $|\Lambda| \geq 3$,*

$$\nu\left(f(I - \mathscr{P})f\right) \geq \frac{|\Lambda| - 2}{|\Lambda| - 1}\left[1 - c|\Lambda|^{-1-\tau}\right]\nu(f^2). \tag{7}$$

In the work of Caputo ([1]), a similar result is proved with $\mu'(\cdot) = \mu \circ |\cdot|^{-1}$ and $\nu'(\cdot) = \mu'(\cdot \mid N_\Lambda = N)$ where $N_\Lambda = N_\Lambda^+ + N_\Lambda^-$, instead of μ and ν.

The sequel is devoted to the proof of Theorem 1. We will assume that m is at most $1/2$. This is not a restriction since, if $m > 1/2$ we could work with $m' = 1 - m$. Moreover, as it is shown in Caputo [1], it is sufficient to prove the Theorem for $f = \sum_{x \in \Lambda} f_x^+ \overline{\eta_x^+} + f_x^- \overline{\eta_x^-}$, where $\overline{\eta_x^\epsilon} = \eta_x^\epsilon - \nu(\eta_x^\epsilon)$ and f_x^+, f_x^- are real numbers.

Remark. Theorem1 should be true in the whole domain m^+, m^-, $m := m^+ + m^- \in \{\frac{1}{|\Lambda|}, ..., \frac{|\Lambda|-1}{|\Lambda|}\}$. But our hypothesis (H1),(H2) excludes many cases, due to the requirement on the ratio $\frac{m^+}{m^-}$. Let us explain how this requirement appears in our technic. The key of our proof is the estimate of correlation coefficient of $(\eta_x^\epsilon, \eta_y^{\epsilon'})$ via a Fourier representation and Gaussian estimates. Such estimates need a control of the ratio $\frac{m^+}{m^-}$ by a function $f(|\Lambda|) = o(|\Lambda|^\gamma)$ for some $\gamma \in (0,1)$ and $f(|\Lambda|) \to +\infty$ as $|\Lambda| \to +\infty$. Hence, the choice $f(|\Lambda|) = \ln(|\Lambda|)$ is natural in our technic. The extension of our result to the whole domain needs a new trick and remains open. The projection on the monocolour system (see [4]) could be a track.

3   A scheme of proof of Theorem 1

We will adopt the following notations and conventions:

- c will denote positive constants, not necessarily the same at each occurrence.
- $\preccurlyeq$ means less than or equal to up to a positive absolute constant, $\succcurlyeq$ means more than or equal to up to a positive absolute constant, and $\asymp$ means that $\preccurlyeq$ and $\succcurlyeq$ hold together.
- δ_x^y will denote the Kronecker symbol that takes value 1 if $x = y$ and 0 if $x \neq y$. It will also be convenient and rather natural to set $\epsilon\epsilon' = 2\delta_\epsilon^{\epsilon'} - 1$ for $\epsilon, \epsilon' \in \{+, -\}$.

3.1 Estimation of the affine combination $c_1\nu(\eta_x^+) + c_2\nu(\eta_x^-) + c_3$

We shall show that for all $(c_1, c_2, c_3) \in \mathbb{R}^3$ and $(x, y) \in \Lambda^2$,

$$c_1\nu(\eta_x^+) + c_2\nu(\eta_x^-) + c_3 \asymp c_1\nu(\eta_y^+) + c_2\nu(\eta_y^-) + c_3, \tag{8}$$

from which we will get by summing over $y \in \Lambda$,

$$c_1\nu(\eta_x^+) + c_2\nu(\eta_x^-) + c_3 \asymp c_1 m^+ + c_2 m^- + c_3. \tag{9}$$

As a by-product and for a further use, we have

$$1 - \nu(|\eta_x|) \asymp 1 - m \tag{10}$$
$$\nu(\eta_x^\epsilon; \eta_x^\epsilon) \asymp m^\epsilon(1 - m^\epsilon), \tag{11}$$

where we set $\nu(f; g) = \nu(fg) - \nu(f)\nu(g)$.

Proof of (8)

It is similar to [1] and works as following. For simplicity we only detail the particular case where $c_1 = 1$ and $c_2 = c_3 = 0$. Assume $x \neq y$ and introduce the following subset of configurations:

$$E_{j^+ j^-} = \left\{ \eta \in \{-1, 0, 1\}^\Lambda : \sum_{z \neq x, y} \eta_z^+ = N^+ - j^+, \ \sum_{z \neq x, y} \eta_z^- = N^- - j^- \right\}.$$

Then, by definition of ν, and independence w.r.t. μ

$$\frac{\nu(\eta_x^+)}{\nu(\eta_y^+)} = \frac{\mu(\eta_x^+)}{\mu(\eta_y^+)} \frac{\left[(1 - \mu(|\eta_y|))\mu(E_{10}) + \mu(\eta_y^+)\mu(E_{20}) + \mu(\eta_y^-)\mu(E_{11}) \right]}{\left[(1 - \mu(|\eta_x|))\mu(E_{10}) + \mu(\eta_x^+)\mu(E_{20}) + \mu(\eta_x^-)\mu(E_{11}) \right]}, \tag{12}$$

and (8) will follow from (12) in this particular case if we can prove that

$$\mu(\eta_x^\epsilon) \asymp \mu(\eta_y^\epsilon), \quad 1 - \mu(|\eta_x|) \asymp 1 - \mu(|\eta_y|).$$

But it's enough to notice from (4) that we have

$$\frac{\mu(\eta^\epsilon_x)}{\mu(\eta^\epsilon_y)} = e^{\alpha_x - \alpha_y} \frac{1 + e^{\alpha_y + \lambda}}{1 + e^{\alpha_x + \lambda}} \le e^{\alpha_x - \alpha_y} \left(1 + e^{\alpha_y - \alpha_x}\right) \le 2e^{2B}$$

$$\frac{\mu(1 - \eta^\epsilon_x)}{\mu(1 - \eta^\epsilon_y)} = \frac{1 + e^{\alpha_x + \lambda^{-\epsilon}}}{1 + e^{\alpha_y + \lambda^{-\epsilon}}} \frac{1 + e^{\alpha_y + \lambda}}{1 + e^{\alpha_x + \lambda}} \le 1 + e^{|\alpha_y - \alpha_x|} \le 1 + e^{2B}$$

$$\frac{\mu(1 - |\eta_x|)}{\mu(1 - |\eta_y|)} = \frac{1 + e^{\alpha_y + \lambda}}{1 + e^{\alpha_x + \lambda}} \le 1 + e^{\alpha_y - \alpha_x} \le 1 + e^{2B}. \quad \square$$

Here $-\epsilon$ is the opposite sign of ϵ.

3.2 Matrix representations of $\nu(f^2)$ and $\nu(f\mathscr{P}f)$

Our aim here is to obtain handable expressions of $\nu(f^2)$ and $\nu(f\mathscr{P}f)$. Let for $x, y \in \Lambda$ and $\epsilon, \epsilon' \in \{+, -\}$

$$C^{\epsilon\epsilon'}_{xy} = \nu(\eta^\epsilon_x; \eta^{\epsilon'}_y), \quad R^{\epsilon\epsilon'}_{xy} = \frac{C^{\epsilon\epsilon'}_{xy}}{\sqrt{C^{\epsilon\epsilon}_{xx} C^{\epsilon'\epsilon'}_{yy}}}.$$

$C = \left(C^{\epsilon\epsilon'}_{xy}\right)$ is a $2|\Lambda|$ by $2|\Lambda|$ covariance matrix and $R = \left(R^{\epsilon\epsilon'}_{xy}\right)$ the corresponding correlation matrix.

Let $(f^\epsilon_x, x \in \Lambda, \epsilon = +, -) \in \mathbf{R}^{2|\Lambda|}$. We will identify $f = \sum_{x \in \Lambda} f^+_x \overline{\eta^+_x} + f^-_x \overline{\eta^-_x}$, where $\overline{\eta^\epsilon_x} = \eta^\epsilon_x - \nu(\eta^\epsilon_x)$, with the row $\left((f^+_x)_{x \in \Lambda}, (f^-_x)_{x \in \Lambda}\right)$. Hence we can write

$$\nu(f^2) = \sum_{x,y} \sum_{\epsilon,\epsilon'} f^\epsilon_x f^{\epsilon'}_y \nu(\overline{\eta^\epsilon_x}\overline{\eta^{\epsilon'}_y}) = fCf^T. \tag{13}$$

Besides, note that we have

$$\nu(f\mathscr{P}f) = \frac{1}{|\Lambda|} \sum_{z \in \Lambda} \nu\big(f\nu(f|\eta_z)\big) = \frac{1}{|\Lambda|} \sum_z \sum_{x,y} \sum_{\epsilon,\epsilon'} f^\epsilon_x f^{\epsilon'}_y \nu\big(\overline{\eta^\epsilon_x} \nu(\overline{\eta^{\epsilon'}_x}|\eta_z)\big). \tag{14}$$

Since η^+_y, η^-_y are Bernoulli variables satisfying $\eta^+_y \eta^-_y = 0$, we can write

$$\nu\left(\overline{\eta^\epsilon_x}|\eta_y\right) = a^{\epsilon+}_{xy} \overline{\eta^+_y} + a^{\epsilon-}_{xy} \overline{\eta^-_y}, \tag{15}$$

for some numbers $a^{\epsilon+}_{xy}$ and $a^{\epsilon-}_{xy}$. Multiply (15) by $\overline{\eta^+_y}$ or $\overline{\eta^-_y}$ and integrate, it gives

$$\begin{pmatrix} a^{\epsilon+}_{xy} \\ a^{\epsilon-}_{xy} \end{pmatrix} = \frac{1}{\det(C_{yy})} \begin{pmatrix} C^{--}_{yy} & -C^{+-}_{yy} \\ -C^{+-}_{yy} & C^{++}_{yy} \end{pmatrix} \begin{pmatrix} C^{\epsilon+}_{xy} \\ C^{\epsilon-}_{xy} \end{pmatrix}, \tag{16}$$

where C_{yy} denotes the 2 by 2 matrix $\left(C^{\epsilon\epsilon'}_{yy}\right)_{\epsilon,\epsilon' \in \{-,+\}}$. Note that, whatever y, $\det(C_{yy})$ can not vanish unless $\nu(|\eta_y|) = 1$. But, by (10), this would imply

$m = 1$. Inspecting (16), we notice that the matrix $A = (a_{xy}^{\epsilon\epsilon'})$ satisfies

$$A = CD, \tag{17}$$

where $D = (D_{xy}^{\epsilon\epsilon'})$ is the $2|\Lambda|$ by $2|\Lambda|$ symmetric matrix defined by

$$D_{xx}^{\epsilon\epsilon'} = \frac{(2\delta_\epsilon^{\epsilon'} - 1)C_{xx}^{\epsilon\epsilon'}}{\det(C_{xx})}, \quad D_{xy}^{\epsilon\epsilon'} = 0 \text{ for } x \neq y, \tag{18}$$

where $\delta_\epsilon^{\epsilon'}$ denotes the Kronecker symbol. From (14), (15), (17) and (18) we get

$$
\begin{aligned}
\nu(f\mathscr{P}f) &= \frac{1}{|\Lambda|} \sum_{x,y} \sum_{\epsilon,\epsilon'} f_x^\epsilon f_y^{\epsilon'} \left[\sum_z a_{xz}^{\epsilon+} \nu\left(\overline{\eta_z^+ \eta_y^{\epsilon'}}\right) + a_{xz}^{\epsilon-} \nu\left(\overline{\eta_z^- \eta_y^{\epsilon'}}\right) \right] \\
&= \frac{1}{|\Lambda|} \sum_{x,y} \sum_{\epsilon,\epsilon'} f_x^\epsilon f_y^{\epsilon'} (AC)_{xy}^{\epsilon\epsilon'} \\
&= \frac{1}{|\Lambda|} \sum_{x,y} \sum_{\epsilon,\epsilon'} f_x^\epsilon f_y^{\epsilon'} (CDC)_{xy}^{\epsilon\epsilon'} \\
&= \frac{1}{|\Lambda|} fCDCf^T.
\end{aligned}
\tag{19}
$$

Change of variables in $\nu(f^2)$ and $\nu(f\mathscr{P}f)$

Now, we shall find a non negative matrix Q and a row g such that

$$\nu(f^2) = gQg^T, \quad \nu(f\mathscr{P}f) = \frac{1}{|\Lambda|} gQ^2 g^T, \tag{20}$$

so that if we set furthermore $h = Q^{1/2}g$ and $\Gamma = I - Q$, (20) yields

$$\nu(f^2) = hh^T, \quad \nu(f(I - \mathscr{P})f) = \frac{|\Lambda| - 1}{|\Lambda|} h\left(I + \frac{\Gamma}{|\Lambda| - 1}\right) h^T. \tag{21}$$

We seek for a $2|\Lambda|$ by $2|\Lambda|$ upper triangular matrix U such that $D = U^T U$. Note from (18) that D consists of four diagonal submatrices of equal size so that U must satisfy the following "block" identities:

$$U^{-+} = 0, \quad U^{++}U^{++} = D^{++}, \quad U^{+-}U^{++} = D^{+-},$$
$$U^{+-}U^{+-} + U^{--}U^{--} = D^{--}.$$

This implies that $U_{xy}^{\epsilon\epsilon'} = 0$ if $x \neq y$ and

$$U_{xx}^{++} = \sqrt{D_{xx}^{++}}, \quad U_{xx}^{+-} = \frac{D_{xx}^{+-}}{\sqrt{D_{xx}^{++}}}, \quad U_{xx}^{--} = \sqrt{D_{xx}^{--} - \frac{|D_{xx}^{+-}|^2}{D_{xx}^{++}}}. \tag{22}$$

The matrix U is easily seen to be well defined and invertible since D has a dominating diagonal. Set $g = fU^{-1}$ and $Q = UCU^T$ so that (13) and (19) yield (20) and thus (21).

Expression of Γ by using the correlation matrix R

Note first that

$$\det\left(C_{xx}\right) = C_{xx}^{--}C_{xx}^{++} - |C_{xx}^{+-}|^2 = C_{xx}^{--}C_{xx}^{++}\left[1 - |R_{xx}^{+-}|^2\right] > 0,$$

where positivity follows from Cauchy-Schwarz's inequality. For further computations, keep in mind that

$$1 - |R_{xx}^{+-}|^2 = \frac{1 - \nu(\eta_x^+) - \nu(\eta_x^-)}{\left(1 - \nu(\eta_x^+)\right)\left(1 - \nu(\eta_x^-)\right)} \asymp \frac{1 - m^+ - m^-}{(1 - m^+)(1 - m^-)} \geq \frac{1}{2}.$$

Combining with (18), we get

$$D_{xx}^{++} = \frac{1}{C_{xx}^{++}\left[1 - |R_{xx}^{+-}|^2\right]},$$

$$D_{xx}^{--} = \frac{1}{C_{xx}^{--}\left[1 - |R_{xx}^{+-}|^2\right]},$$

$$D_{xx}^{+-} = \frac{-R_{xx}^{+-}}{\sqrt{C_{xx}^{++}C_{xx}^{--}}\left[1 - |R_{xx}^{+-}|^2\right]},$$

so that by (22),

$$U_{xx}^{++} = \frac{1}{\sqrt{C_{xx}^{++}}\sqrt{1 - |R_{xx}^{+-}|^2}}, \quad U_{xx}^{+-} = \frac{-R_{xx}^{+-}}{\sqrt{C_{xx}^{--}}\sqrt{1 - |R_{xx}^{+-}|^2}}, \quad U_{xx}^{--} = \frac{1}{\sqrt{C_{xx}^{--}}}. \tag{23}$$

Since $Q = UCU^T$ and $U_{xy}^{\epsilon\epsilon'} = 0$ for $x \neq y$, we have

$$Q_{xy}^{\epsilon\epsilon'} = C_{xy}^{++}U_{xx}^{\epsilon+}U_{yy}^{\epsilon'+} + C_{xy}^{+-}U_{xx}^{\epsilon+}U_{yy}^{\epsilon'-} + C_{xy}^{-+}U_{xx}^{\epsilon-}U_{yy}^{\epsilon'+} + C_{xy}^{--}U_{xx}^{\epsilon-}U_{yy}^{\epsilon'-},$$

and since $\Gamma = I - Q$, we deduce for all $x, y \in \Lambda$

$$\Gamma_{xy}^{++} = \delta_x^y - \frac{R_{xy}^{++} - R_{xy}^{+-}R_{yy}^{+-} - R_{xx}^{+-}R_{xy}^{-+} + R_{xx}^{+-}R_{xy}^{--}R_{yy}^{+-}}{\sqrt{1 - |R_{xx}^{+-}|^2}\sqrt{1 - |R_{yy}^{+-}|^2}}, \tag{24}$$

$$\Gamma_{xy}^{+-} = -\frac{R_{xy}^{+-} - R_{xx}^{+-}R_{xy}^{--}}{\sqrt{1 - |R_{xx}^{+-}|^2}}, \tag{25}$$

$$\Gamma_{xy}^{--} = \delta_x^y - R_{xy}^{--}, \tag{26}$$

where δ_x^y denotes the Kronecker symbol. In particular, note that $\Gamma_{xx}^{\epsilon\epsilon'} = 0$ whatever $\epsilon, \epsilon' \in \{+, -\}$.

3.3 Reduction to a key lemma

Suppose we have proved that for some $\tau \in (0, 1/8)$ and $c(\tau) > 0$,

$$\Gamma \geq -c(\tau)|\Lambda|^{-\tau}, \tag{27}$$

in the matrix sense. Then it is immediate to see that (27) and (21) imply (7) and we are done. As in the work Caputo [1], (27) will in turn come from the following.

Lemma 1. *Assume that **(H1)** or **(H2)** holds. There exist $c > 0$ and $\tau \in (0, 1/8)$ such that for all ω, Λ, for $x \neq y$,*

$$\left| \Gamma_{xy}^{\epsilon\epsilon'} - \frac{\beta_{xy}^{\epsilon\epsilon'} \ln |\Lambda|}{|\Lambda|} \right| \leq c|\Lambda|^{-1-\tau}, \tag{28}$$

where $(\beta_{xy}^{\epsilon\epsilon'})$ is a non-negative $2|\Lambda|$ by $2|\Lambda|$ matrix depending on (ω, Λ) and where $\beta_{xx}^{\epsilon\epsilon}$ are non-negative and uniformly bounded by c.

Let us show how Lemma 1 implies (27). Indeed, it follows from (28) that

$$\langle v, \Gamma v \rangle = \sum_{\substack{x \neq y \\ \epsilon, \epsilon'}} \Gamma_{xy}^{\epsilon\epsilon'} v_x^\epsilon v_y^{\epsilon'}$$

$$= \sum_{\substack{x \neq y \\ \epsilon, \epsilon'}} v_x^\epsilon v_y^{\epsilon'} \left[\frac{\beta_{xy}^{\epsilon\epsilon'} \ln |\Lambda|}{|\Lambda|} + \left(\Gamma_{xy}^{\epsilon\epsilon'} - \frac{\beta_{xy}^{\epsilon\epsilon'} \ln |\Lambda|}{|\Lambda|} \right) \right]$$

$$\geq -\sum_{x, \epsilon, \epsilon'} v_x^\epsilon v_x^{\epsilon'} \frac{\beta_{xx}^{\epsilon\epsilon'} \ln |\Lambda|}{|\Lambda|} - c|\Lambda|^{-1-\tau} \sum_{\substack{x \neq y \\ \epsilon, \epsilon'}} |v_x^\epsilon v_y^{\epsilon'}|$$

$$\geq -\sum_x \left(\sum_\epsilon |v_x^\epsilon| \right)^2 \frac{c^2 \ln |\Lambda|}{|\Lambda|} - c|\Lambda|^{-1-\tau} \left(\sum_{x, \epsilon} |v_x^\epsilon| \right)^2$$

$$\geq -2 \sum_{x, \epsilon} |v_x^\epsilon|^2 \frac{c^2 \ln |\Lambda|}{|\Lambda|} - c|\Lambda|^{-1-\tau} \left(\sum_{x, \epsilon} |v_x^\epsilon| \right)^2$$

$$\geq \left(-\frac{2c^2 \ln |\Lambda|}{|\Lambda|} - c|\Lambda|^{-\tau} \right) \sum_{x, \epsilon} |v_x^\epsilon|^2$$

$$\geq -c(\tau)|\Lambda|^{-\tau} \langle v, v \rangle,$$

for some positive constant $c(\tau)$ depending only on τ, and this is exactly (27).

3.4 Proof under assumption (H1)

We assume $m^- \geq |\Lambda|^{-1+\tau_1}$ for some $\tau_1 \in (0, 1/2)$ and $\dfrac{m^+}{m^-} \preccurlyeq \ln |\Lambda|$. The proof uses mainly Gaussian estimates and expansion of characteristic functions. The aim is to expand for distinct x, y the covariances

$$C_{xy}^{\epsilon\epsilon'} = \nu\big(\eta_x^\epsilon \eta_y^{\epsilon'}\big) - \nu\big(\eta_x^\epsilon\big)\nu\big(\eta_y^{\epsilon'}\big).$$

A Fourier representation of $C_{xy}^{\epsilon\epsilon'}$

By definition of ν and μ, we have

$$\nu\big(\eta_x^\epsilon \eta_y^{\epsilon'}\big) = \mu(\eta_x^\epsilon)\mu(\eta_y^{\epsilon'})$$

$$\cdot \frac{\mu\left(N_{\Lambda\setminus\{x,y\}}^+ = N^+ - \delta_\epsilon^+ - \delta_{\epsilon'}^+, N_{\Lambda\setminus\{x,y\}}^- = N^- - \delta_\epsilon^- - \delta_{\epsilon'}^-\right)}{\mu\left(N_\Lambda^+ = N^+, N_\Lambda^- = N^-\right)},$$

$$\nu\big(\eta_x^\epsilon\big) = \mu(\eta_x^\epsilon)\frac{\mu\left(N_{\Lambda\setminus\{x\}}^+ = N^+ - \delta_\epsilon^+, N_{\Lambda\setminus\{x\}}^- = N^- - \delta_\epsilon^-\right)}{\mu\left(N_\Lambda^+ = N^+, N_\Lambda^- = N^-\right)}, \qquad (29)$$

$$\nu\big(\eta_y^{\epsilon'}\big) = \mu(\eta_y^{\epsilon'})\frac{\mu\left(N_{\Lambda\setminus\{y\}}^+ = N^+ - \delta_{\epsilon'}^+, N_{\Lambda\setminus\{y\}}^- = N^- - \delta_{\epsilon'}^-\right)}{\mu\left(N_\Lambda^+ = N^+, N_\Lambda^- = N^-\right)}.$$

Let M_Λ be the covariance matrix of the random vector $(N_\Lambda^+, N_\Lambda^-)$ under μ and set

$$d_\Lambda = [\det(M_\Lambda)]^{1/4}. \qquad (30)$$

Set also for any $\xi = (\xi^+, \xi^-) \in \mathbb{R}^2$,

$$\widetilde{v}_z(\xi) = \mu\left(\exp\left\{i\frac{\xi \cdot \big(\eta_z^+ - \mu(\eta_z^+), \eta_z^- - \mu(\eta_z^-)\big)}{d_\Lambda}\right\}\right),$$

and to shorten notations, let $d\xi = d\xi^+ d\xi^-$ and introduce for any $\Delta \subset \Lambda$ and any map $\xi \mapsto f(\xi)$.

$$\widetilde{v}_\Delta(\xi) = \prod_{z \in \Delta} \widetilde{v}_z(\xi),$$

$$I(f) = \iint_{[-\pi d_\Lambda, \pi d_\Lambda]^2} f(\xi)\widetilde{v}_\Lambda(\xi)\, d\xi.$$

Using then Fourier's transform, we notice that for any integers j^+, j^-,

$$\mu\big(N_{\Lambda\setminus\Delta}^+ = N^+ - j^+, N_{\Lambda\setminus\Delta}^- = N^- - j^-\big) =$$

$$\frac{1}{4\pi^2 d_\Lambda^2} I\left(\frac{1}{\widetilde{v}_\Delta(\xi)}\exp\left\{i\frac{\xi \cdot \big(j^+ - \mu(N_\Delta^+), j^- - \mu(N_\Delta^-)\big)}{d_\Lambda}\right\}\right). \qquad (31)$$

Define then

$$g_x^\epsilon(\xi) = \frac{1}{\widetilde{v}_x(\xi)} \exp\left\{ i \frac{\xi \cdot (\delta_\epsilon^+ - \mu(\eta_x^+), \delta_\epsilon^- - \mu(\eta_x^-))}{d_\Lambda} \right\}, \tag{32}$$

and apply (31) for $\Delta = \{x, y\}$, $\Delta = \{x\}$ and $\Delta = \{y\}$. It is easily seen using (29) and (32) that

$$C_{xy}^{\epsilon\epsilon'} = \mu(\eta_x^\epsilon)\mu(\eta_y^{\epsilon'})I(g_x^\epsilon; g_y^{\epsilon'}), \tag{33}$$

with

$$I(f; g) = \frac{I(1)I(fg) - I(f)I(g)}{I(1)^2}.$$

Gaussian estimate of $\widetilde{v}_\Delta$

Let M_Δ be the covariance matrix of (N_Δ^+, N_Δ^-) under μ and denote by $\theta_\Delta^{(1)} \leq \theta_\Delta^{(2)}$ its eigenvalues. It is proved in [3] that

$$|\Delta|(1 - m)\min(m^+, m^-) \preccurlyeq \theta_\Delta^{(1)} \leq \theta_\Delta^{(2)} \preccurlyeq |\Delta|\max(m^+, m^-), \tag{34}$$

and

$$|\xi| \leq \pi d_\Lambda \implies |\widetilde{v}_\Delta(\xi)| \leq \exp\left(-\frac{2\xi M_\Delta \xi^T}{\pi^2 d_\Lambda^2}\right) \leq \exp\left(-\frac{2\theta_\Delta^{(1)}|\xi|^2}{\pi^2 d_\Lambda^2}\right). \tag{35}$$

Expansion of $\widetilde{v}_z(\xi)$. By Taylor's formula, we have

$$e^{it} = 1 + it - \frac{t^2}{2} - it^3 \int_0^1 \frac{(1-u)^2}{2} e^{iut} du.$$

Applying this formula for $t = \dfrac{\xi \cdot (\eta_z^+ - \mu(\eta_z^+), \eta_z^- - \mu(\eta_z^-))}{d_\Lambda}$ and integrating over μ, we note that

$$\mu(t) = 0, \quad \mu(t^2) = \frac{\xi M_z \xi^T}{d_\Lambda^2}, \quad \mu(t^3) \leq \frac{2|\xi|}{d_\Lambda}\mu(t^2) \leq \frac{2|\xi|^3 \theta_z^{(2)}}{d_\Lambda^3},$$

and get

$$\widetilde{v}_z(\xi) = 1 - \frac{\xi M_z \xi^T}{2 d_\Lambda^2} + O\left(\frac{|\xi|^3 \theta_z^{(2)}}{d_\Lambda^3}\right). \tag{36}$$

Expansion of $\widetilde{v}_\Delta(\xi)$. Let c be positive and small enough such that

$$|\xi| \leq c \ln |\Lambda| \implies \forall z, \frac{\xi M_z \xi^T}{2 d_\Lambda^2} \in [0, 2).$$

This is possible since we have

$$\xi M_z \xi^T \preccurlyeq \theta_z^{(2)} |\xi|^2 \preccurlyeq m^- \ln |\Lambda| |\xi|^2, \quad d_\Lambda^2 \succcurlyeq \theta_\Lambda^{(1)} \succcurlyeq m^- |\Lambda|, \qquad (37)$$

by using (34) with $\Delta = \{z\}$, with $\Delta = \Lambda$ and (**H1**). Then, from (36) and by using the well-known inequality $|\prod_j z_j - \prod_j w_j| \leq \sum_j |z_j - w_j|$ for z_j, w_j complex numbers of modulus at most equal to one, we get for $|\xi| \leq c \ln |\Lambda|$,

$$\left| \widetilde{v}_\Lambda(\xi) - \prod_{z \in \Lambda} \left(1 - \frac{\xi M_z \xi^T}{2 d_\Lambda^2} \right) \right| \preccurlyeq \sum_{z \in \Lambda} \frac{(\ln |\Lambda|)^3 \theta_z^{(2)}}{d_\Lambda^3} \preccurlyeq \frac{(\ln |\Lambda|)^4}{|\Lambda|^{\tau_1/2}},$$

where we used (37) in the last estimation. Besides, we have by Taylor's formula

$$\exp \left(-\frac{\xi M_z \xi^T}{2 d_\Lambda^2} \right) = 1 - \frac{\xi M_z \xi^T}{2 d_\Lambda^2} + O \left(\frac{|\xi|^4 (\theta_z^{(2)})^2}{d_\Lambda^4} \right),$$

and we get similarly for $|\xi| \leq c \ln |\Lambda|$,

$$\left| \exp \left(-\frac{\xi M_\Lambda \xi^T}{2 d_\Lambda^2} \right) - \prod_{z \in \Lambda} \left(1 - \frac{\xi M_z \xi^T}{2 d_\Lambda^2} \right) \right| \preccurlyeq \sum_{z \in \Lambda} \frac{(\ln |\Lambda|)^4 (\theta_z^{(2)})^2}{d_\Lambda^4} \preccurlyeq \frac{(\ln |\Lambda|)^6}{|\Lambda|}.$$

Thus, we have proved that

$$|\xi| \leq c \ln |\Lambda| \implies \widetilde{v}_\Lambda(\xi) = \exp \left(-\frac{\xi M_\Lambda \xi^T}{2 d_\Lambda^2} \right) + O \left(\frac{(\ln |\Lambda|)^4}{|\Lambda|^{\tau_1/2}} \right). \qquad (38)$$

Reduction of the integration domain of I

Note that $I(1)$, $I(g_x^\epsilon)$, $I(g_y^{\epsilon'})$ and $I(g_x^\epsilon g_y^{\epsilon'})$ are all of the form

$$\iint_{[-\pi d_\Lambda, \pi d_\Lambda]^2} f(\xi) \widetilde{v}_\Delta(\xi) d\xi$$

with $|\Delta| \asymp |\Lambda|$ and $|f| \leq 1$. For such an f and Δ, and for $c \leq \frac{\pi d_\Lambda}{\ln |\Lambda|}$, one can find some $\tau = \tau(c) > 0$ such that

$$\iint_{[-\pi d_\Lambda, \pi d_\Lambda]^2} f(\xi) \widetilde{v}_\Delta(\xi) d\xi = \iint_{[-c \ln |\Lambda|, c \ln |\Lambda|]^2} f(\xi) \widetilde{v}_\Delta(\xi) d\xi + O \left(|\Lambda|^{-\tau} \right).$$
$$(39)$$

Proof of (39). Indeed, by using (35) and (**H1**), we have for every $\tau < c' c^2$,

$$\left| \iint_{c \ln |\Lambda| < |\xi| \leq \pi d_\Lambda} f(\xi) \widetilde{v}_\Delta(\xi) \mathrm{d}\xi \right| \leq \iint_{c \ln |\Lambda| < |\xi| \leq \pi d_\Lambda} |\widetilde{v}_\Delta(\xi)| \mathrm{d}\xi$$

$$\leq \iint_{|\xi| > c \ln |\Lambda|} \exp\left(-\frac{2\theta_\Delta^{(1)} |\xi|^2}{\pi^2 d_\Lambda^2} \right) \mathrm{d}\xi$$

$$\leq \iint_{|\xi| > c \ln |\Lambda|} \exp\left(-c' \frac{|\xi|^2}{\ln |\Lambda|} \right) \mathrm{d}\xi$$

$$= 2\pi \int_{c \ln |\Lambda|}^{\infty} \exp\left(-c' \frac{r^2}{\ln |\Lambda|} \right) r \mathrm{d}r$$

$$\preccurlyeq \ln |\Lambda| \exp\left(-c' c^2 \ln |\Lambda| \right)$$

$$\preccurlyeq |\Lambda|^{-\tau}. \quad \square$$

Thus, we can write for $f = 1$, g_x^ϵ, $g_y^{\epsilon'}$ or $g_x^\epsilon g_y^{\epsilon'}$,

$$\mathrm{I}(f) = \mathrm{I}'(f) + O\left(|\Lambda|^{-\tau} \right) \quad \text{with} \quad \mathrm{I}'(f) = \iint_{[-c \ln |\Lambda|, c \ln |\Lambda|]^2} f(\xi) \widetilde{v}_\Lambda(\xi) \, \mathrm{d}\xi, \tag{40}$$

and, since $|\widetilde{v}_\Delta| \leq 1$, we have $|\mathrm{I}'(f)| \leq (2c \ln |\Lambda|)^2$. Moreover, if we prove $|\mathrm{I}'(1)| \succcurlyeq 1$, it will follow from (33) and (40) that for some $\tau' > 0$

$$C_{xy}^{\epsilon\epsilon'} = \mu(\eta_x^\epsilon)\mu(\eta_y^{\epsilon'})\mathrm{I}'(g_x^\epsilon; g_y^{\epsilon'}) + O\left(|\Lambda|^{-\tau'} \right), \tag{41}$$

with

$$\mathrm{I}'(f; g) = \frac{\mathrm{I}'(1)\mathrm{I}'(fg) - \mathrm{I}'(f)\mathrm{I}'(g)}{\mathrm{I}'(1)^2}.$$

The estimate

$$\mathrm{I}'(1) = 2\pi + O\left(|\Lambda|^{-\tau''} \right)$$

for some $\tau'' \in (0,1)$ is a consequence of (38) and a classical Gaussian computation.

Taylor expansion of g_x^ϵ

Recall the definition (32) of g_x^ϵ. Set

$$q_x^\epsilon = \left(\delta_\epsilon^+ - \mu(\eta_x^+), \delta_\epsilon^- - \mu(\eta_x^-) \right),$$

and note that $|q_x^\epsilon| \leq \sqrt{2}$. Taylor's formula $e^{it} = 1 + it - t^2 \int_0^1 (1 - u)e^{iut}\mathrm{d}u$ gives

$$\exp\left\{ i\frac{\xi \cdot q_x^\epsilon}{d_\Lambda} \right\} = 1 + i\frac{\xi \cdot q_x^\epsilon}{d_\Lambda} + u_x^\epsilon(\xi), \quad \text{with} \quad |u_x^\epsilon(\xi)| \preccurlyeq \frac{|\xi|^2}{d_\Lambda^2}. \tag{42}$$

14   A. Dermoune and P. Heinrich

Together with (36) and (37), we get for $|\xi| \leq c \ln |\Lambda|$,

$$g_x^\epsilon(\xi) = \left[1 + i\frac{\xi \cdot q_x^\epsilon}{d_\Lambda} + u_x^\epsilon(\xi)\right]\left[1 + \frac{\xi M_x \xi^T}{2d_\Lambda^2} + O\left(\frac{(\ln|\Lambda|)^4}{|\Lambda|^{1+\tau_1/2}}\right)\right]$$

$$= 1 + i\frac{\xi \cdot q_x^\epsilon}{d_\Lambda} + u_x^\epsilon(\xi) + \frac{\xi M_x \xi^T}{2d_\Lambda^2} + O\left(\frac{(\ln|\Lambda|)^4}{|\Lambda|^{1+\tau_1/2}}\right), \quad (43)$$

with, using (**H1**) and (34),

$$\left|\frac{\xi \cdot q_x^\epsilon}{d_\Lambda}\right| \preccurlyeq \frac{\ln|\Lambda|}{|\Lambda|^{\tau_1/2}}, \quad |u_x^\epsilon(\xi)| \preccurlyeq \frac{(\ln|\Lambda|)^2}{|\Lambda|^{\tau_1}}, \quad \left|\frac{\xi M_x \xi^T}{2d_\Lambda^2}\right| \preccurlyeq \frac{(\ln|\Lambda|)^3}{|\Lambda|}. \quad (44)$$

Expansion of $\mathrm{I}'(g_x^\epsilon; g_y^{\epsilon'})$

From (43), by inspecting the order of each term involved, we get using (44)

$$\mathrm{I}'(g_x^\epsilon; g_y^{\epsilon'}) = -\frac{\mathrm{I}'\left(\xi \cdot q_x^\epsilon; \xi \cdot q_y^{\epsilon'}\right)}{d_\Lambda^2} + O\left(\frac{(\ln|\Lambda|)^8}{|\Lambda|^{3\tau_1/2}}\right).$$

To continue, we have to estimate some integrals. As for $\mathrm{I}'(1)$, we find thanks to (38)

$$\mathrm{I}'(\xi^\epsilon) = \iint_{[-c\ln|\Lambda|, c\ln|\Lambda|]^2} \xi^\epsilon \exp\left(-\frac{\xi M_\Lambda \xi^T}{2d_\Lambda^2}\right) d\xi + O\left(\frac{(\ln|\Lambda|)^7}{|\Lambda|^{\tau_1/2}}\right)$$

$$\preccurlyeq |\Lambda|^{-\tau} + \frac{(\ln|\Lambda|)^7}{|\Lambda|^{\tau_1/2}}$$

$$\preccurlyeq |\Lambda|^{-\tau''} \quad (45)$$

so that $\mathrm{I}'(\xi \cdot q_x^\epsilon)$ is at most $O(|\Lambda|^{-\tau''})$. We get then

$$\mathrm{I}'(g_x^\epsilon; g_y^{\epsilon'}) = -\frac{\mathrm{I}'\left(\xi \cdot q_x^\epsilon \xi \cdot q_y^{\epsilon'}\right)}{\mathrm{I}'(1)d_\Lambda^2} + O\left(\frac{|\Lambda|^{-\tau''}}{d_\Lambda^2}\right) \quad (46)$$

so that we obtain

$$R_{xy}^{\epsilon\epsilon'} = -\frac{\mu(\eta_x^\epsilon)\mu(\eta_y^{\epsilon'})\mathrm{I}'\left(\xi \cdot q_x^\epsilon \xi \cdot q_y^{\epsilon'}\right)}{\mathrm{I}'(1)d_\Lambda^2\sqrt{C_{xx}^{\epsilon\epsilon}}\sqrt{C_{yy}^{\epsilon'\epsilon'}}} + O\left(\frac{|\Lambda|^{-\tau''}}{d_\Lambda^2}\frac{\mu(\eta_x^\epsilon)\mu(\eta_y^{\epsilon'})}{\sqrt{C_{xx}^{\epsilon\epsilon}}\sqrt{C_{yy}^{\epsilon'\epsilon'}}}\right). \quad (47)$$

Arguing as for (45), we find whatever ϵ, ϵ'

$$\mathrm{I}'\left(\xi^\epsilon \xi^{\epsilon'}\right) = \frac{2\pi\epsilon\epsilon'}{d_\Lambda^2}M_\Lambda^{\epsilon\epsilon'} + O\left(|\Lambda|^{-\tau''}\right),$$

and consequently

$$\mathrm{I}'\left(\xi \cdot q_x^\epsilon \xi \cdot q_y^{\epsilon'}\right) = q_x^\epsilon \mathrm{I}'\left(\xi^T \xi\right)(q_y^{\epsilon'})^T = 2\pi d_\Lambda^2 q_x^\epsilon M_\Lambda^{-1}(q_y^{\epsilon'})^T + O\left(|\Lambda|^{-\tau''}\right).$$

Now set

$$b_x^\epsilon = \mu(\eta_x^\epsilon)\sqrt{\frac{2\pi|\Lambda|}{\mathrm{I}'(1)C_{xx}^{\epsilon\epsilon}}}q_x^\epsilon,$$

and note that $|b_x^\epsilon|^2 \asymp m^\epsilon |\Lambda|$. By using $C_{xx}^{\epsilon\epsilon} \asymp \mu(\eta_x^\epsilon) \asymp m^\epsilon$ and $d_\Lambda^2 \succeq m^-|\Lambda|$ together with $(\mathbf{H1})$, we get from (47)

$$R_{xy}^{\epsilon\epsilon'} = -\frac{b_x^\epsilon M_\Lambda^{-1}(b_y^{\epsilon'})^T}{|\Lambda|} + O\left(\frac{\ln|\Lambda|}{|\Lambda|^{1+\tau''}}\right).$$

If we set moreover

$$\beta_x^+ = \frac{b_x^+ - b_x^- R_{xx}^{+-}}{\sqrt{1 - |R_{xx}^{+-}|^2}}, \quad \beta_x^- = b_x^-, \quad \beta_{xy}^{\epsilon\epsilon'} = \beta_x^\epsilon \frac{M_\Lambda^{-1}}{\ln|\Lambda|}\beta_y^{\epsilon'},$$

we deduce from (24), (25) and (26) that

$$\Gamma_{xy}^{\epsilon\epsilon'} = \frac{\beta_{xy}^{\epsilon\epsilon'}\ln|\Lambda|}{|\Lambda|} + O\left(\frac{\ln|\Lambda|}{|\Lambda|^{1+\tau''}}\right).$$

Note that $(\beta_{xy}^{\epsilon\epsilon'})_{(x,y)\in\Lambda^2,(\epsilon,\epsilon')\in\{+,-\}^2}$ is a non negative matrix since M_Λ is, and

$$|\beta_{xx}^{\epsilon\epsilon}| \leq \frac{|\beta_x^\epsilon|^2}{\theta_\Lambda^{(1)}\ln|\Lambda|} \preccurlyeq \frac{|b_x^+|^2 + |b_x^-|^2}{m^-|\Lambda|\ln|\Lambda|} \preccurlyeq \frac{m^+ + m^-}{m^-\ln|\Lambda|} \preccurlyeq 1.$$

Lemma 1 is thus established in this case under $(\mathbf{H1})$.

3.5 Proof under assumption (H2)

The method is the same as in [1]. Keep in mind $C_{xx}^{\epsilon\epsilon} \asymp \nu(\eta_x^\epsilon) \asymp m^\epsilon$ and observe that

$$R_{xy}^{\epsilon\epsilon'} = \frac{\nu(\eta_x^\epsilon)\left[\nu\left(\eta_y^{\epsilon'} \mid \eta_x^\epsilon = 1\right) - \nu\left(\eta_y^{\epsilon'}\right)\right]}{\sqrt{C_{xx}^{\epsilon\epsilon}}\sqrt{C_{yy}^{\epsilon'\epsilon'}}}. \tag{48}$$

Assume that the following holds for $x \in \Lambda$ and $\epsilon \in \{-,+\}$:

$$\nu(\eta_x^\epsilon) = \frac{N^\epsilon r_x^\epsilon}{\displaystyle\sum_{z\in\Lambda} r_z^\epsilon} + O(mm^\epsilon) \quad \text{with} \quad r_x^\epsilon = \frac{\mu(\eta_x^\epsilon)}{1 - \mu(|\eta_x|)} \asymp m^\epsilon. \tag{49}$$

We deduce for $x \neq y$

$$\nu\left(\eta_y^{\epsilon'} \mid \eta_x^{\epsilon} = 1\right) - \nu\left(\eta_y^{\epsilon'}\right) = \frac{(N^{\epsilon'} - \delta_\epsilon^{\epsilon'})r_y^{\epsilon'}}{\sum\limits_{z \neq x} r_z^{\epsilon'}} - \frac{N^{\epsilon'} r_y^{\epsilon'}}{\sum\limits_z r_z^{\epsilon'}} + O(mm^{\epsilon'})$$

$$= -\frac{r_y^{\epsilon'} \delta_\epsilon^{\epsilon'}}{\sum\limits_z r_z^{\epsilon'}} + O(mm^{\epsilon'}). \tag{50}$$

Now set

$$b_x^\epsilon = \frac{\sqrt{N^\epsilon} r_x^\epsilon \sqrt{|\Lambda|}}{\sqrt{C_{xx}^{\epsilon\epsilon}} \sum\limits_z r_z^\epsilon}.$$

Note that $b_x^\epsilon \asymp 1$ and that (48), (49) and (50) yields for $x \neq y$

$$R_{xy}^{\epsilon\epsilon'} = -\frac{b_x^\epsilon b_y^{\epsilon'} \delta_\epsilon^{\epsilon'}}{|\Lambda|} + O\left(\frac{mm^{\epsilon'} m^\epsilon}{\sqrt{C_{xx}^{\epsilon\epsilon}} \sqrt{C_{yy}^{\epsilon'\epsilon'}}}\right) = -\frac{b_x^\epsilon b_y^{\epsilon'} \delta_\epsilon^{\epsilon'}}{|\Lambda|} + O\left(m^2\right). \tag{51}$$

In particular, we have $|R_{xy}^{\epsilon\epsilon'}| \preccurlyeq |\Lambda|^{-1}$. Besides, we note by (48) that $R_{xx}^{\epsilon\epsilon'} \asymp -\sqrt{m^+ m^-}$ if $\epsilon \neq \epsilon'$. It follows from (24), (25) and (26) that for $x \neq y$

$$\Gamma_{xy}^{++} = \frac{-R_{xy}^{++}}{\sqrt{1 - |R_{xx}^{+-}|^2}\sqrt{1 - |R_{yy}^{+-}|^2}} + O\left(m^2\right),$$

$$\Gamma_{xy}^{+-} = O\left(m^2\right),$$

$$\Gamma_{xy}^{--} = -R_{xy}^{--}.$$

Now we define a non negative matrix $(\beta_{xy}^{\epsilon\epsilon'})$ by setting

$$\beta_{xy}^{--} = \frac{b_x^- b_y^-}{\ln|\Lambda|}, \quad \beta_{xy}^{+-} = \beta_{xy}^{-+} = 0, \quad \beta_{xy}^{++} = \frac{b_x^+ b_y^+}{\sqrt{1 - |R_{xx}^{+-}|^2}\sqrt{1 - |R_{yy}^{+-}|^2} \ln|\Lambda|}.$$

It is then easy to see that $\beta_{xy}^{\epsilon\epsilon'} \preccurlyeq (\ln|\Lambda|)^{-1}$ and to deduce from (51) that

$$\Gamma_{xy}^{\epsilon\epsilon'} = \frac{\beta_{xy}^{\epsilon\epsilon'} \ln|\Lambda|}{|\Lambda|} + O\left(\frac{1}{|\Lambda|^{1+\tau_2}}\right).$$

It remains to prove (49). By symmetry, we assume that $\epsilon = +$. By factorization, rewrite (12) as

$$\frac{\nu(\eta_x^+)}{\nu(\eta_y^+)} = \frac{\mu(\eta_x^+)(1 - \mu(|\eta_y|)) \left[r_y^+ t_{11} + r_y^- t_{20} + 1\right]}{\mu(\eta_y^+)(1 - \mu(|\eta_x|)) \left[r_x^+ t_{20} + r_x^- t_{11} + 1\right]} = \frac{r_x^+ \left[r_y^+ t_{11} + r_y^- t_{20} + 1\right]}{r_y^+ \left[r_x^+ t_{20} + r_x^- t_{11} + 1\right]} \tag{52}$$

with

$$t_{jk} = \frac{\mu(E_{jk})}{\mu(E_{10})}.$$

The term t_{20} is $O(1)$. Indeed, set

$$V_{20}(z) = \left\{ \sum_{u \neq x,y,z} \eta_u^+ = N^+ - 2, \ \sum_{u \neq x,y,z} \eta_u^- = N^- \right\}.$$

Partitioning E_{10} with respect to the support of η^+ and using $\mu(\eta_z^+) \asymp m^+$ yields

$$\sum_{z \neq x,y} \mu\left(E_{10}, \eta_z^+ = 1\right) = (N^+ - 1)\mu\left(E_{10}\right)$$

$$= \sum_{z \neq x,y} \mu(\eta_z^+)\mu\left(V_{20}(z)\right) \asymp m^+ \sum_{z \neq x,y} \mu\left(V_{20}(z)\right),$$

and by a similar argument on E_{20} using $1 - \mu(|\eta_z|) \asymp 1 - m$,

$$\sum_{z \neq x,y} \mu\left(E_{20}, \eta_z = 0\right) = (|\Lambda| - N + 2)\mu\left(E_{20}\right) = \sum_{z \neq x,y} (1 - \mu(|\eta_z|))\mu\left(V_{20}(z)\right)$$

$$\asymp (1 - m) \sum_{z \neq x,y} \mu\left(V_{20}(z)\right).$$

It follows that

$$t_{20} \asymp \frac{1-m}{m^+} \frac{N^+ - 1}{|\Lambda| - N + 2} = O(1).$$

Arguments are similar to prove that t_{11} is $O(1)$. Because of $r_x^\epsilon \asymp r_y^\epsilon \asymp m^\epsilon$, we derive from (52) that

$$\frac{\nu(\eta_y^+)}{\nu(\eta_x^+)} = \frac{r_y^+}{r_x^+}\left[1 + O(m)\right], \tag{53}$$

and by summing (53) over y, we get thanks to $\sum_y \nu(\eta_y^+) = N^+$ and $r_x^+ \asymp r_y^+ \asymp m^+$,

$$\nu(\eta_x^+) = \frac{N^+ r_x^+}{\sum_{y \in \Lambda} r_y^+} + O(mm^+)$$

which is (49) for $\epsilon = +$.

Acknowledgments. We thank the referee for his carefully reading of our first version and his comments.

References

1. P. Caputo, *Spectral gap inequalities in product spaces with conservation laws*, in T. Funaki and H. Osada (eds.) Adv. Studies in Pure Math. Japan 2004 .
2. A. Dermoune, P. Heinrich, *A small step towards the hydrodynamic limit of a colored disordered lattice gas*, C. R. Acad. Sci. Paris, Ser. I 339, 507-511 (2004).
3. A. Dermoune, P. Heinrich, *Equivalence of ensembles for colored particles in a disordered lattice gas*, to appear in Markov Process. Related Fields (2005).
4. A. Dermoune, S. Martinez, *Around Multicolour disordered lattice gas*, to appear in Journal of Statistical Physics.
5. A. Faggionato, *Hydrodynamic limit of a disordered system*, Ph. D. Thesis (2002).
6. A. Faggionato, F. Martinelli, *Hydrodynamic limit of a disordered lattice gas*, Probab. Theory Relat. Fields 127, 535-608 (2003).

On large deviations for the spectral measure of discrete Coulomb gas

D. Féral

Université Paul Sabatier, 118 route de Narbonne
31062 Toulouse Cedex 4, France
e-mail: dferal@math.ups-tlse.fr

Summary. We establish a large deviation principle for the spectral measure of a large class of discrete Coulomb gas. The setting includes invariant ensembles from the classical orthogonal polynomials which are the discrete analogues of the continuous random matrix models. The proof requires a refinement of the arguments used in the continuous framework due to the constraint that may appear in the description of the rate functional. Our analysis closely follows the investigations of K. Johansson at the level of the largest eigenvalue, that is recovered here by a change of variables.

Key words: large deviations, discrete Coulomb gas, spectral measure, largest eigenvalue, random matrix models, continuous Coulomb gas.

1 Introduction

During the recent years, several authors (eg. [B-G], [H-P]) have established and handled large deviation principles (LDP) for the spectral measure of the classical random matrix models such as the Gaussian and Laguerre Unitary or Orthogonal Ensembles (known as the GU(O)E and LU(O)E). The analysis is based on the description of the joint law of the eigenvalues as a properly rescaled continuous Coulomb gas. For the special case of the GOE, it has been noticed in [BDG] that those LDP may be used to describe the corresponding large deviations for the associated largest eigenvalue.

More recently, K. Johansson deeply investigated the asymptotic properties of discrete Coulomb gas. In particular, he described in [Jo2] both the fluctuations and large deviations properties of the rightmost charge (or "largest eigenvalue") of discrete Coulomb gas associated to families of classical orthogonal polynomials, thus called orthogonal polynomial ensembles. He carefully examined the Meixner ensemble in [Jo2]. The Charlier ensemble allowed him to study in [Jo3] the asymptotics of weakly increasing subsequences in random words. Using Krawtchouk polynomials, he further obtained a new approach of the Seppäläinen's first passage percolation problem (cf. [Jo3]). With the Hahn polynomials, he was able to investigate the problem of random rhombus tilings (cf. [Jo4]).

Following Johansson's investigations, we consider in this paper LDP for the spectral measure of a general class of discrete Coulomb gas (see below). The results may be seen as an extension of both the LDP for the spectral measure of classical random matrix models and the results of [Jo2] which are concerned with LDP for the largest eigenvalue. In order to point out the characteristics of this discrete setting, we briefly describe in Section 2 the continuous case whose results may be directly adapted from the known random matrix examples. Then, Section 3 develops in full details the discrete case which requires a finer investigation due to the constraint that may appear in the description of the rate functional. To this task, we adapt the arguments of [Jo2] at the level of the spectral measure. In Section 4, we then apply the general results on the spectral measure to recover a LDP for the largest eigenvalue of these ensembles following [BDG]. This approach thus provides another view, although based on the same tools, to the results of [Jo2]. In the last section, we briefly discuss a few concrete examples from the classical orthogonal polynomial ensembles.

2 The continuous case

To introduce the main conclusion of this work, we first briefly present in this section a general continuous framework which includes the classical orthogonal polynomial ensembles from the random matrix models (Section 5 describes more carefully the example of the LOE).

Given an integer $N \geq 1$, consider N real particles $x = (x_1, \ldots, x_N)$ having a law absolutely continuous with respect to Lebesgue measure on $\mathbb{R}^N$ with the Coulomb gas representation

$$dP_{N,V_N,\beta}(x) = \frac{1}{Z_{N,V_N,\beta}} |\Delta_N(x)|^\beta \prod_{i=1}^{N} \exp(-\frac{\beta N}{2} V_N(x_i)) \prod_{i=1}^{N} dx_i. \quad (1)$$

Here $\beta > 0$ is a parameter,

$$\Delta_N(x) = \prod_{1 \leq i < j \leq N} (x_i - x_j)$$

is the Vandermonde determinant, $Z_{N,V_N,\beta}$ is the normalisation constant

$$Z_{N,V_N,\beta} = \int_{\mathbb{R}^N} |\Delta_N(x)|^\beta \prod_{i=1}^{N} \exp(-\frac{\beta N}{2} V_N(x_i)) \prod_{i=1}^{N} dx_i < +\infty$$

and $(V_N)_{N \geq 1}$ is a sequence of potentials satisfying:

(i)   for all $N \geq 1$, $V_N : \mathbb{R} \to \mathbb{R}$ is continuous,

(ii) there are two constants $\xi > 0$ and $T > 0$ such that

$$V_N(t) \geq (1 + \xi) \log(1 + t^2), \quad \forall |t| \geq T$$

(iii) there exists a potential V such that $(V_N)_N \to V$ uniformly on the compact subsets of $\mathbb{R}$.

It should be emphasized that $\beta = 2$ (resp. $\beta = 1$, $\beta = 4$) corresponds to the classical joint law of the eigenvalues of random Hermitian (resp. symmetric, symplectic) matrices which is invariant under the action of the unitary (resp. orthogonal, sympletic) group. In particular, by taking $\beta = 2$ and $V_N(t) = 2t^2$ in (1), we recover the joint density of the N eigenvalues of a random matrix X_N element of the (rescaled) GUE that is X_N is a $N \times N$ random Hermitian matrix with, on and off the diagonal, centered independent Gaussian entries with variances $\mathbb{E}[|X_{i,j}|^2] = 1/4N$.

Furthermore, all the results of this section (as well as those of Section 4.2 below) are also true when replacing $\mathbb{R}^+$ by $\mathbb{R}$. In this way, the LOE and the LUE may also be included in this framework (cf. Section 5.2). Although only the values $\beta = 1, 2$ and 4 correspond to some random matrix models, we interpret the probability distribution $P_{N,V_N,\beta}$ as the joint law of the "eigenvalues" $(x_1, \cdots, x_N)$ of the associated Coulomb gas and we define

$$\hat{\mu}_N = \frac{1}{N} \sum_{i=1}^{N} \delta_{x_i}$$

to be the corresponding spectral measure. As announced, our main purpose will be to establish a LDP for the law of $(\hat{\mu}_N)_N$. We consider $\hat{\mu}_N$ as a random variable taking values in $\mathcal{M}(\mathbb{R})$, the space of probability measures on $\mathbb{R}$ equipped with the weak topology. This topology is compatible with the Lévy metric d defined for two measures μ and ν of $\mathcal{M}(\mathbb{R})$ by

$$d(\mu, \nu) = \inf\{\delta > 0 \ \ / \ \ \mu(F) \leq \nu(F^\delta) + \delta \text{ and } \nu(F) \leq \mu(F^\delta) + \delta, \forall F \text{ closed in } \mathbb{R}\} \tag{2}$$

(where $F^\delta = \{t \in \mathbb{R} : d(t, F) < \delta\}$) and makes $(\mathcal{M}(\mathbb{R}), d)$ a Polish space (cf. Section 3.2 of [De-St]). We denote by $\mathcal{B}(\mu, \epsilon)$ the open ball for d with center μ and radius ϵ.

In order to present the large deviations theorem in this context, set

$$k_V(s, t) = \log |s - t|^{-1} + \frac{1}{2}V(s) + \frac{1}{2}V(t), \quad (s, t) \in \mathbb{R}^2.$$

Set furthermore for all $\mu \in \mathcal{M}(\mathbb{R})$,

$$E_V(\mu) = \int \int_{\mathbb{R}^2} k_V(s, t) \, d\mu(s) d\mu(t),$$

and let $F_V = \inf_{\nu \in \mathcal{M}(\mathbb{R})} E_V(\nu)$.

Theorem 2.1 *1. a) F_V is finite and there is a unique probability measure μ_V compactly supported on $\mathbb{R}$ such that*

$$E_V(\mu_V) = F_V.$$

b) The function I_V^β defined by

$$I_V^\beta(\mu) = \frac{\beta}{2}(E_V(\mu) - F_V), \quad \forall \mu \in \mathcal{M}(\mathbb{R})$$

is a good rate function on $\mathcal{M}(\mathbb{R})$.
2. Under $P_{N,V_N,\beta}$, the law of the spectral measure $(\hat{\mu}_N)_N$ satisfies on $\mathcal{M}(\mathbb{R})$ a LDP with speed N^2 and GRF I_V^β.

This theorem has first been proven in [B-G] for the particular GUE matrix model. It has then been extended in [H-P] (in Theorem 5.4.3) to large families of continuous Coulomb gas of type (1) defined by a single potential rather than a general sequence $(V_N)_N$. Nevertheless, the proof of [H-P] can readily be extended to the current setting thanks to the uniform minoration (ii) of the sequence $(V_N)_N$ as well as its uniform convergence (iii) (the discrete setting detailed below will make clear how to use such assumptions).

A consequence of this theorem (and of the general theory of large deviations [De-St]) is the convergence in probability of the spectral measure $(\hat{\mu}_N)$ to the deterministic extremal measure μ_V. In the particular case of the GUE, we recover the classical Wigner Theorem that is the almost sure convergence of $(\hat{\mu}_N)$ to the so-called semicircle law. Here, in the general setting of Theorem 2.1, one can only assert the convergence in probability. The reason is that in the GUE setting, we can consider X_N for all N defined on the same space of realizations of the infinite matrix $X = (X_{i,j})_{i,j=1}^\infty$ equipped with the infinite product Gaussian measure (a similar remark holds true for the LOE model). Such an embedding does not exist for general random matrix models corresponding to parameters $\beta = 1, 2$ or 4 and clearly neither for all the other parameters $\beta > 0$.

To conclude this section, we would like to mention a few facts about the extremal (or equilibrium) measure μ_V that will be helpful in the comparison with the discrete case studied in the next section. By definition, μ_V solves the following optimisation problem:

$$\mu_V \in \mathcal{M}(\mathbb{R}) \quad \text{and} \quad E_V(\mu_V) = \inf_{\mu \in \mathcal{M}(\mathbb{R})} E_V(\mu) := F_V \in \mathbb{R}. \tag{3}$$

The finiteness of F_V as well as the existence and the uniqueness of a compactly supported solution are well-known and follow from the general theory of "the energy problem" developed by E.B. Saff and V. Totik in their book [S-T]. Actually, this theory gives a variational characterisation of μ_V which, under some regularity assumptions on the potential V (such as V convex, differentiable with a derivative V' Hölder continuous), implies the existence of a

continuous extremal density and allows us to make it explicit as well as the
two endpoints of its compact support (which is then a single interval). Remark
that all the classical continuous orthogonal polynomial ensembles actually fall
in this context.

As announced, the next section presents and establishes the analogue of
Theorem 2.1 in the discrete setting.

3 The discrete case

3.1 The setting and the main statement

Classically, a discrete Coulomb gas corresponds to the joint distribution of N
particles $x = (x_1, \cdots, x_N) \in \mathbb{N}^N$ represented by

$$dP_{N,\beta}(x) = \frac{1}{Z_{N,\beta}} |\Delta_N(x)|^\beta \prod_{i=1}^N d\mu(x_i)$$

where $\beta > 0$, μ is a discrete measure on $\mathbb{N}$ and $Z_{N,\beta}$ is the normalisation
factor.

The classical examples, as in the continuous setting, are built from orthogonal polynomials associated to some discrete measure with $\beta = 2$. In this way,
one obtains the orthogonal discrete polynomial ensembles. For example,

- the Charlier ensemble corresponds to $\mu = \mu_\theta$ the Poisson distribution of
  parameter $\theta > 0$,
- the Meixner ensemble is associated to the negative binomial distribution
  with parameters $\theta > 0$ and $0 \le q \le 1$ given by

$$\mu(\{k\}) = \frac{(\theta)_k}{k!} q^k (1-q)^\theta \quad (\forall k \in \mathbb{N}),$$

where $(\theta)_k = \theta(\theta - 1) \cdots (\theta - k + 1)$.

These two models have been investigated in Johansson's work [Jo2], [Jo3].
In particular, the asymptotic analysis as $N \to \infty$ is made relevant by considering parameters θ which depend on N and by rescaling the particles by
N. Notice that, for the GUE model (with Gaussian entries with a variance of
order $1/N$) the corresponding scaling is $\sqrt{N}$.

These few remarks justify that we consider some rescaled discrete particles
lying in the set

$$\mathbb{A}_N = \left\{ \frac{k}{N}, \, k \in \mathbb{N} \right\}.$$

Following [Jo2], we thus adopt the following general setting (which is in fact
the discrete analogue of the one of Section 2). Given $\beta > 0$, let $(V_N)_{N \ge 1}$ be
potentials $V_N : \mathbb{R}^+ \mapsto \mathbb{R}$ satisfying the following conditions:

(i) for all $N \geq 1$, V_N is continuous,
(ii) there exist $\xi > 0$ and $T > 0$ such that

$$V_N(t) \geq (1 + \xi) \log(1 + t^2), \quad \forall t \geq T$$

(iii) there exists $V : \mathbb{R}^+ \mapsto \mathbb{R}$ such that $(V_N)_N \to V$ uniformly on the compact subsets of $\mathbb{R}^+$.

Consider then N particles $x = (x_1, \cdots, x_N)$ having on $\mathbb{A}_N^N$ the joint distribution

$$P_{N,V_N,\beta}(x) = \frac{1}{Z_{N,V_N,\beta}} \, |\Delta_N(x)|^\beta \prod_{1 \leq i < j \leq N} \exp\left(-\frac{\beta N}{2} V_N(x_i)\right) \, d\delta_x \quad (4)$$

meaning that, for all $B \subset \mathbb{A}_N^N$,

$$P_{N,V_N,\beta}(B) = \frac{1}{Z_{N,V_N,\beta}} \sum_{x=(x_1,\cdots,x_N)\in B} |\Delta_N(x)|^\beta \exp\left(-\frac{\beta N}{2} \sum_{i=1}^{N} V_N(x_i)\right)$$

where $\Delta_N(x) = \prod_{1 \leq i < j \leq N}(x_i - x_j)$ is the Vandermonde determinant and

$$Z_{N,V_N,\beta} = \sum_{x=(x_1,\cdots,x_N)\in \mathbb{A}_N^N} |\Delta_N(x)|^\beta \exp\left(-\frac{\beta N}{2} \sum_{i=1}^{N} V_N(x_i)\right) < +\infty$$

is the normalisation constant. By analogy with the continuous case, we will continue to speak of $(x_1, \cdots, x_N)$ as the eigenvalues of the Coulomb gas $P_{N,V_N,\beta}$ with arbitrary $\beta > 0$, although there might not be any underlying random matrix model. Our purpose is again to investigate large deviations properties of the associated spectral measure $\hat{\mu}_N = \frac{1}{N}\sum_{i=1}^{N} \delta_{x_i}$. Here, under the probability $P_{N,V_N,\beta}$, $\hat{\mu}_N$ is a random measure on the space $\mathcal{M}(\mathbb{R}^+)$ of the probability measures on $\mathbb{R}^+$. Endow $\mathcal{M}(\mathbb{R}^+)$ by the metric topology induced by the Lévy metric d defined on $\mathcal{M}(\mathbb{R})$ by (2). As in the continuous case, the LDP speed is N^2 but the good rate function (GRF) leading the LDP will only be finite inside a certain subclass $\mathcal{M}^\lambda$ of $\mathcal{M}(\mathbb{R}^+)$. Inside $\mathcal{M}^\lambda$, the associated GRF has an expression very close to the one of the continuous case. More precisely, the class $\mathcal{M}^\lambda$ is defined as

$$\mathcal{M}^\lambda = \{\mu \in \mathcal{M}(\mathbb{R}^+) : \mu \leq \lambda\}$$

where λ is Lebesgue measure on $\mathbb{R}^+$ and "$\mu \leq \lambda$" means that for all Borel subsets B of $\mathbb{R}^+$, $\mu(B) \leq \lambda(B)$. In other words, $\mu \in \mathcal{M}^\lambda$ if and only if $\lambda - \mu$ is a positive measure on $\mathbb{R}^+$. The upper-bound λ is called the constraint and $\mathcal{M}^\lambda$ is the class of the measures submitted to the constraint λ.

To state precisely the LDP for the spectral measure, let

$$E_V(\mu) = \int \int_{\mathbb{R}_+^2} k_V(s,t)\, d\mu(s) d\mu(t), \quad \forall \mu \in \mathcal{M}(\mathbb{R}^+),$$

where $k_V(s,t) = \log |s-t|^{-1} + \frac{1}{2}V(s) + \frac{1}{2}V(t)$ and set

$$F_V^\lambda = \inf_{\nu \in \mathcal{M}^\lambda} E_V(\nu).$$

Let us first make some comments about this infimum F_V^λ. As in the preceding unconstrained setting of Section 2, F_V^λ is finite and is reached at a unique measure denoted by μ_V^λ. This fact follows from the general theory of "the constrained variational problem" developed by P.D. Dragnev and E.B. Saff in [Dr-Sa1]. To summarize, this theory is a sort of extension of the classical optimisation problem (3) associated to the continuous and unconstrained case. Here are the main properties of μ_V^λ (see Sections 1 and 2 of [Dr-Sa1]) which will turn to be useful in the next sections.

Proposition 3.1 *(cf. [Dr-Sa1])*

1. *$F_V^\lambda := \inf_{\mu \in \mathcal{M}^\lambda} E_V(\mu)$ is finite and there is a unique mesure μ_V^λ defined by the following optimisation problem:*

$$\mu_V^\lambda \in \mathcal{M}^\lambda \quad and \quad E_V(\mu_V^\lambda) = \inf_{\mu \in \mathcal{M}^\lambda} E_V(\mu) := F_V^\lambda. \qquad (5)$$

   μ_V^λ is compactly supported in $\mathbb{R}^+$ and is called the constrained extremal (or equilibrium) measure.
2. *If $\mu \in \mathcal{M}^\lambda$ has compact support and if there exists a constant ξ_V^λ such that*
   a) $\int \log |s-t|^{-1}\, d\mu(t) + \frac{1}{2}V(s) \geq \xi_V^\lambda, \quad \forall s \in \mathrm{Supp}(\lambda - \mu),$
   b) $\int \log |s-t|^{-1}\, d\mu(t) + \frac{1}{2}V(s) \leq \xi_V^\lambda, \quad \forall s \in \mathrm{Supp}(\mu),$
   then $\mu = \mu_V^\lambda$ and $\xi_V^\lambda = F_V^\lambda - \frac{1}{2}\int V(t)\, d\mu_V^\lambda(t)$.
3. *Let b_V be the upper bound of the compact support of μ_V^λ. The function g_V defined on $\mathbb{R}^+$ by*

$$g_V(t) = \int \log |s-t|^{-1}\, d\mu_V^\lambda(s) + \frac{1}{2}V(t) - \xi_V^\lambda$$

   is continuous at b_V (with $g_V(b_V) = 0$) and in fact on the whole set $\mathbb{R}^+$.

The announced LDP for the spectral measure can now be stated.

Theorem 3.2 *1. a) The function $I_V^{\beta,\lambda}$ given by*

$$I_V^{\beta,\lambda}(\mu) = \begin{cases} \frac{\beta}{2}(E_V(\mu) - F_V^\lambda), & \text{if } \mu \in \mathcal{M}^\lambda, \\ +\infty, & \text{otherwise.} \end{cases}$$

is a good rate function (GRF) on $\mathcal{M}(\mathbb{R}^+)$.

b) $I_V^{\beta,\lambda}$ *reaches its minimum at a unique probability* μ_V^λ *of* $\mathcal{M}^\lambda$ *compactly supported in* $\mathbb{R}^+$ *and having a density* Φ_V^λ *with respect to Lebesgue measure* λ *such that almost everywhere* $0 \leq \Phi_V^\lambda \leq 1$.

2. Under $P_{N,V_N,\beta}$, *the law of the spectral measure* $(\hat{\mu}_N)_N$ *satisfies on* $\mathcal{M}(\mathbb{R}^+)$ *a LDP with speed* N^2 *and GRF* $I_V^{\beta,\lambda}$.

From the preceding, it is straightforward to state point *1*. Indeed, one can observe that $\mathcal{M}^\lambda$ is a closed subset of $\mathcal{M}(\mathbb{R}^+)$ so that point *1.* of Theorem 2.1 (cf. also [H-P] p. 214) combined with the finiteness of F_V^λ imply that $I_V^{\beta,\lambda}$ is a GRF on $\mathcal{M}(\mathbb{R}^+)$. For point *1. b)*, it remains to justify the existence of the density Φ_V^λ of the measure μ_V^λ. This is due to the fact that μ_V^λ belongs to $\mathcal{M}^\lambda$ and follows from Lemma 3.5 stated below.

As in the continuous case, Theorem 3.2 implies the convergence in probability of $(\hat{\mu}_N)_N$ to the deterministic extremal measure μ_V^λ. Note that, as in the unconstrained case and under some additional assumptions on the potential V, the compact support of μ_V^λ (which is then a single interval) as well as its density (which is then continuous) can be computed explicitly. Nevertheless, due to the constraint, the reasoning is more subtle and the computations rather tedious and involved. In Section 5.1 below, we will try to point out the main arguments (by referring to [BKMM] and [Dr-Sal]). We will also give (however in a rough form only) a practical method and illustrate it on the Charlier ensemble.

In the rest of this section, we first explain and justify the appearence of the constraint λ and prove that the GRF is infinite outside the class $\mathcal{M}^\lambda$. The last subsection is devoted to the proof of point *2.* of the main Theorem 3.2.

Remark 3.3 *In all our proofs, we will assume that the potentials* V_N *are non-negative and that* (ii) *is satisfied on the whole set* $\mathbb{R}^+$ *that is: there is* $\xi > 0$ *such that for all* $N \geq 1$,

$$V_N(t) \geq (1+\xi)\log(1+t^2), \quad \forall t \geq 0. \tag{6}$$

One can observe that this does alter the problem and amounts to add a constant to each potential V_N *since, from assumptions* (ii) *and* (iii), *one can find* $A \geq 0$ *such that for all* N,

$$V_N(t) + A \geq (1+\xi)\log(1+t^2), \quad \forall t \geq 0.$$

3.2 The class $\mathcal{M}^\lambda$

To start with, we state a characterisation of the class $\mathcal{M}^\lambda$ and then, we will justify that the GRF is infinite outside $\mathcal{M}^\lambda$.

The class $\mathcal{M}^\lambda$ is a feature of the discrete framework and follows from the definition of the set $\mathbb{A}_N$. In fact, this constraint can be justified thanks to very simple measure theoretic arguments developed in Lemmas 3.4 and 3.5 below. It will then follow (Lemma 3.6) that

$$\forall \mu \notin \mathcal{M}^\lambda, \quad \overline{\lim_{\epsilon \to 0}} \, \overline{\lim_{N}} \, \frac{1}{N^2} \log P_{N,V_N,\beta}\big(\hat{\mu}_N \in \mathcal{B}(\mu, \epsilon)\big) = -\infty \qquad (7)$$

and thus that $I_V^{\beta,\lambda}(\mu) = +\infty$ for all $\mu \notin \mathcal{M}^\lambda$.

Lemma 3.4 *If μ has no atoms, then $\mu \notin \mathcal{M}^\lambda$ if and only if there exist a, b, $0 \leq a < b$, such that $\mu([a, b]) > b - a$.*

Proof: The condition is clearly necessary according to the definition of $\mathcal{M}^\lambda$. Conversely, assume that: $\forall 0 \leq a < b$, $\mu([a, b]) \leq b - a$. Then, denoting by $\mathcal{B}(\mathbb{R}^+)$ the Borel σ-field on $\mathbb{R}^+$, the class $\mathcal{C} = \{B \in \mathcal{B}(\mathbb{R}^+) \ / \ \mu(B) \leq \lambda(B)\}$ contains all the compact intervals of $\mathbb{R}^+$. Since the measures μ and λ are both σ-finite and have no atoms, $\mathcal{C}$ actually contains all the intervals of $\mathbb{R}^+$. As $\mathcal{C}$ is trivially a monotone class, the Monotone Class Theorem implies that $\mathcal{C}$ coincides with the Borel σ-field and thus $\mu \in \mathcal{M}^\lambda$. $\qquad \square$

Observing that all the elements of $\mathcal{M}^\lambda$ are absolutely continuous with respect to λ, the following characterisation can readily be deduced.

Lemma 3.5 *$\mu \in \mathcal{M}^\lambda$ if and only if μ is absolutely continuous with respect to λ with density f such that, almost everywhere, $0 \leq f \leq 1$.*

We are ready to state the announced claim (7) on the class $\mathcal{M}^\lambda$.

Lemma 3.6 *For all $\mu \notin \mathcal{M}^\lambda$, $I_V^{\beta,\lambda}(\mu) = +\infty$.*

Proof: On the one hand, if $\mu \notin \mathcal{M}^\lambda$, then there exist $0 \leq a_0 < b_0$ such that $\mu([a_0, b_0]) > b_0 - a_0$. This is indeed Lemma 1 if μ has no atoms, and is obvious otherwise. On the other hand, by the very definition of $\mathbb{A}_N$,

$$\begin{aligned} \hat{\mu}_N([a, b]) &= \frac{1}{N} \#\{1 \leq i \leq N \ / \ x_i \in \mathbb{A}_N \cap [a, b]\} \\ &\leq \frac{1}{N} \left(\lfloor N(b-a) \rfloor + 1\right) \\ &\leq (b - a) + \frac{1}{N}. \end{aligned}$$

It may then be shown by the definition of the Lévy metric (taking $F = [a_0, b_0]$), that for all $\epsilon > 0$ small enough, there exists $N_\epsilon \geq 1$ such $d(\mu, \hat{\mu}_N) > \epsilon$ for all $N \geq N_\epsilon$. As a consequence, one obtains the limit (7). $\qquad \square$

Remark 3.7 *Our previous analysis and all the results given in this paper can be extended to some general discrete Coulomb gas (4) involving particles lying in a discrete finite set as*

$$\left\{ \frac{k}{N}, 0 \leq k \leq \gamma(N) \right\} \qquad (8)$$

with $\gamma(N) \in \mathbb{N}$ such that $\gamma(N) \geq N - 1$ and $\dfrac{\gamma(N)}{N} \to \gamma \geq 1$.

In this case, the associated constraint is Lebesgue measure on $[0, \gamma]$. Note that the condition "$\gamma(N) \geq N - 1$" follows from the fact that, due to the Vandermonde determinant, the particles x_i are almost surely distinct.

Let us denote by $\mathcal{M}^{\lambda,\gamma}$ the set of the probability measures of $\mathcal{M}(\mathbb{R}^+)$ submitted to the constraint $\lambda_{/[0,\gamma]}$. All our results remain true with $\mathcal{M}^\lambda$ replaced by the class $\mathcal{M}^{\lambda,\gamma}$ noticing that $\mathcal{M}^{\lambda,\gamma}$ is made up of the elements of $\mathcal{M}^\lambda$ compactly supported in $[0, \gamma]$.

3.3 Proof of point 2. of Theorem 3.2

As announced, we describe carefully in this subsection the various steps of the proof of point *2.* of the main Theorem 3.2. We want to emphasize that some arguments have yet been used in [Jo2] for the investigations of the large deviations properties of the corresponding largest eigenvalue.

We are going to proceed classically as in the continuous case (cf. [B-G] or the proof of Theorem 5.4.3 in [H-P]), in four steps, proving first a weak LDP on the open balls centered on the elements of $\mathcal{M}^\lambda$ and then the exponential tightness (note that by Lemma 3.6, we may restrict ourselves to the class $\mathcal{M}^\lambda$). Define $\bar{\mathbb{P}}_N = P_{N,V_N,\beta} \times Z_{N,V_N,\beta}$ and $\mathbb{P}_N = P_{N,V_N,\beta}$.

1st *step:* Let us first prove that, for every $\mu \in \mathcal{M}^\lambda$,

$$\overline{\lim_{\epsilon \to 0}} \, \overline{\lim_{N}} \, \frac{1}{N^2} \log \bar{\mathbb{P}}_N[\hat{\mu}_N \in B(\mu, \epsilon)] \leq -\frac{\beta}{2} E_V(\mu). \tag{9}$$

To this aim, define for all $x = (x_1, \cdots, x_N) \in \mathbb{A}_N^N$,

$$K_{N,V_N}(x) = \sum_{1 \leq i \neq j \leq N} k_{V_N}(x_i, x_j).$$

One easily verifies that

$$K_{N,V_N}(x) = \sum_{1 \leq i < j \leq N} \log |x_i - x_j|^{-2} + (N-1) \sum_i V_N(x_i) \tag{10}$$

and for all $B \subset \mathbb{A}_N^N$,

$$\bar{\mathbb{P}}_N(B) = \sum_{x \in B} \exp[-\frac{\beta}{2} K_{N,V_N}(x)] \times \exp[-\frac{\beta}{2} \sum_{i=1}^N V_N(x_i)]. \tag{11}$$

Write then for any $0 < \rho < 1$,

$$\bar{\mathbb{P}}_N(B) = \sum_{x/\hat{\mu}_N \in \mathcal{B}(\mu,\epsilon)} \exp[-\frac{\beta}{2}(1-\rho)K_{N,V_N}(x)]$$

$$\times \exp[-\frac{\beta}{2}\{\rho K_{N,V_N}(x) + \sum_{i=1}^{N} V_N(x_i)\}]$$

$$\leq \exp[-\frac{\beta}{2}(1-\rho)\inf_{x \in B} K_{N,V_N}(x)]$$

$$\times \sum_{x \in \mathbb{A}_N^N} \exp[-\frac{\beta}{2}\{\rho K_{N,V_N}(x) + \sum_{i=1}^{N} V_N(x_i)\}] \qquad (12)$$

We first show that the second factor in (12) is exponentially negligible. To this task, it is enough to show that there exists a constant $C > 0$ (independent of ρ) and an integer N_ρ such that

$$\forall N \geq N_\rho, \quad \sum_{x \in \mathbb{A}_N^N} \exp\left[-\frac{\beta}{2}(\rho K_{N,V_N}(x) + \sum_{i=1}^{N} V_N(x_i))\right] \leq e^{CN \log N}. \qquad (13)$$

Indeed, combining (10) with (6) and the elementary inequality

$$|t - s|^2 \leq (1 + s^2)(1 + t^2) \qquad (14)$$

it follows that for all $x \in \mathbb{A}_N^N$,

$$K_{N,V_N}(x) \geq \xi(N-1) \sum_{i=1}^{N} \log(1 + x_i^2) \qquad (15)$$

and

$$-\frac{\beta}{2}\left(\rho K_{N,V_N}(x) + \sum_{i=1}^{N} V_N(x_i)\right) \leq e^{-\rho \frac{\beta}{2} K_{N,V_N}(x)}$$

$$\leq e^{-\frac{\beta}{2}(\rho\xi(N-1)) \sum_{i=1}^{N} \log(1+x_i^2)}$$

$$\leq \left\{\sum_{t \in \mathbb{A}_N} (1 + t^2)^{-\beta\rho\xi(N-1)/2}\right\}^N. \qquad (16)$$

This is what we wanted since as soon as $\beta\rho\xi(N-1)/2 > 1$,

$$\sum_{t \in \mathbb{A}_N} (1 + t^2)^{-\beta\rho\xi(N-1)/2} \leq 1 + 2N.$$

We now investigate the first term of the product in (12) for $B = \mathcal{B}(\mu, \epsilon)$ with $\epsilon > 0$. Clearly (due to the Vandermonde determinant), the particles x_i are $\bar{\mathbb{P}}_N$-almost everywhere distinct. In particular, if $\mathbb{D}$ denotes the diagonal of $(\mathbb{R}^+)^2$ then $\hat{\mu}_N \otimes \hat{\mu}_N(\mathbb{D}) = \frac{1}{N}$. Therefore, for all $M > 0$,

$$\frac{1}{N^2} K_{N,V_N}(x) = \int\int_{s\neq t} k_{V_N}(s,t)d\hat{\mu}_N(s)\,d\hat{\mu}_N(t) + M\hat{\mu}_N\otimes\hat{\mu}_N(\mathbb{D}) - \frac{M}{N}$$

$$\geq \int\int_{s,t} k_{V_N}^M(s,t)d\hat{\mu}_N(s)\,d\hat{\mu}_N(t) - \frac{M}{N}$$

from which it follows that on $\{x \ / \ \hat{\mu}_N \in \mathcal{B}(\mu,\epsilon)\}$,

$$\frac{1}{N^2} K_{N,V_N}(x) \geq \inf_{\nu\in\mathcal{B}(\mu,\epsilon)} E_{V_N}^M(\nu) - \frac{M}{N}$$

where we set

$$k_{V_N}^M(s,t) = \min(k_{V_N}(s,t), M) \ \text{ and } \ E_{V_N}^M(\nu) = \int\int_{s,t} k_{V_N}^M(s,t)d\nu(s)\,d\nu(t).$$

Thus, for all N large enough, we get

$$\bar{\mathbb{P}}_N[\hat{\mu}_N \in \mathcal{B}(\mu,\epsilon)]$$
$$\leq \exp\left\{-\frac{\beta}{2}(1-\rho)N^2 \inf_{\nu\in B(\mu,\epsilon)} E_{V_N}^M(\nu) - \frac{\beta(1-\rho)M}{2N}\right\} \times e^{CN\log N}.$$

Since (V_N) converges uniformly to V on any compact subset of $\mathbb{R}^+$, and since the functions V_N and V are continuous and tend to ∞ at $+\infty$, the sequence $(k_{V_N}^M)$ also converges uniformly to k_V^M. Hence,

$$\lim_N \inf_{\nu\in\mathcal{B}(\mu,\epsilon)} E_{V_N}^M(\nu) = \inf_{\nu\in B(\mu,\epsilon)} E_V^M(\nu),$$

and

$$\overline{\lim_N} \frac{1}{N^2}\log\bar{\mathbb{P}}_N[\hat{\mu}_N\in\mathcal{B}(\mu,\epsilon)] \leq -\frac{\beta}{2}(1-\rho)\inf_{\nu\in B(\mu,\epsilon)} E_V^M(\nu).$$

Noticing the continuity of E_V^M on $\mathcal{M}^\lambda$ yields

$$\lim_{\epsilon\to 0^+}\inf_{\nu\in\mathcal{B}(\mu,\epsilon)} E_V^M(\nu) = E_V^M(\mu).$$

As a consequence and for all $M > 0$,

$$\overline{\lim_{\epsilon\to 0}}\ \overline{\lim_N}\ \frac{1}{N^2}\log\bar{\mathbb{P}}_N[\hat{\mu}_N\in\mathcal{B}(\mu,\epsilon)] \leq -\frac{\beta}{2}E_V^M(\mu).$$

Letting $M \to \infty$ completes the proof of (9) since one can easily notice that $\sup_{M>0} E_V^M = E_V$.

2nd *step:* We show here the lower bound:

$$\forall\mu\in\mathcal{M}^\lambda, \forall\epsilon > 0, \quad \underline{\lim_N}\ \frac{1}{N^2}\log\bar{\mathbb{P}}_N[\hat{\mu}_N\in\mathcal{B}(\mu,\epsilon)] \geq -\frac{\beta}{2}E_V(\mu). \qquad (17)$$

Let us first explain that we only need to prove the lower bound for any measure $\mu \in \mathcal{M}^\lambda$ having a compact support. Indeed, given $\mu \in \mathcal{M}^\lambda$, we can consider for all $k \geq 1$, the compactly supported probability measure $\mu_k = (\mu([0,k]))^{-1} \times \mu \mathbb{1}_{[0,k]}$. Obviously, (μ_k) converges weakly towards μ. Furthermore, as k_V is continuous on $\mathbb{R}\backslash\mathbb{D}$ and takes positive values (recall (6)), the Monotone Convergence Theorem implies that as $k \to \infty$,

$$E_V(\mu_k) = \int\int k_V(s,t)\,d\mu_k(s)d\mu_k(t) \to \int\int k_V(s,t)\,d\mu(s)d\mu(t) = E_V(\mu).$$

Thus, we start with a measure $\mu \in \mathcal{M}^\lambda$ having a compact support $[a,b]$ and a density $\varPhi_\mu$ such that $0 \leq \varPhi_\mu \leq 1$. Then, for each $N \geq 1$, we can define $y = y^N = (y_{i,N})_i \in \mathbb{A}_N^N$ by taking for all $1 \leq i \leq N$,

$$y_{i,N} = \max\left\{\frac{j}{N} \ / \ j \in \mathbb{N} \ \text{and} \ \int_a^{\frac{j}{N}} \varPhi_\mu(t)\,dt < \frac{i}{N}\right\}. \tag{18}$$

Clearly, $a \leq y_{1,N} < y_{2,N} < \cdots < y_{N,N} \leq b$ and the sequence $(\hat{\nu}_N^y = \frac{1}{N}\sum_{i=1}^N \delta_{y_{i,N}})_N$ converges weakly to μ (use the Portmanteau Lemma). So as $N \to \infty$,

$$\frac{1}{N}\sum_{i=1}^N V(y_{i,N}) = \int V(t)\,d\hat{\nu}_N^y(t) \to \int V(t)\varPhi_\mu(t)dt$$

and $\hat{\nu}_N^y \in \mathcal{B}(\mu,\epsilon)$ (at least for N large). Thus, writing y_i for $y_{i,N}$,

$$\bar{\mathbb{P}}_N(\hat{\mu}_N \in \mathcal{B}(\mu,\epsilon)) \geq \bar{\mathbb{P}}_N(y) = \exp(-\frac{N}{2}\beta\sum_{i=1}^N V_N(y_i)) \times \prod_{i<j}|y_i - y_j|^\beta$$

$$\geq \prod_{i<j}|y_i - y_j|^\beta \times \exp(-\frac{N\beta}{2}\sum_i V(y_i))$$

$$\times \exp\left(-\frac{N^2\beta}{2}\sup_{t\in[a,b]}|V_N(t) - V(t)|\right).$$

It follows that

$$\frac{1}{N^2}\log\bar{\mathbb{P}}_N[\hat{\mu}_N \in \mathcal{B}(\mu,\epsilon)] \geq -\frac{\beta}{2N^2}\sum_{i\neq j}\log|y_j - y_i|^{-1} - \frac{\beta}{2N}\sum_{i=1}^N V(y_i)$$

$$-\frac{\beta}{2}\sup_{t\in[a,b]}|V_N(t) - V(t)|.$$

The last term vanishes as $N \to \infty$ thanks to the assumption (iii). It remains to prove that

$$\frac{1}{N^2} \sum_{i \neq j} \log |y_j - y_i|^{-1} \to \int \int \log |s - t|^{-1} \, \Phi_\mu(t) \Phi_\mu(s) \, ds \, dt. \qquad (19)$$

To this aim, given $M \geq 1$, we consider the function $f_M(t) = \log(|t|^{-1} \wedge M)$ and then write

$$\frac{1}{N^2} \sum_{i \neq j} \log |y_j - y_i|^{-1} = \frac{1}{N^2} \sum_{i \neq j} f_M(|y_j - y_i|)$$
$$+ \frac{1}{N^2} \sum_{i \neq j} \left\{ \log |y_j - y_i|^{-1} \right.$$
$$\left. - f_M(|y_j - y_i|) \right\} 1_{|y_j - y_i| < 1/M}. \qquad (20)$$

By definition of f_M, one can observe that the last sum in the r.h.s of (20) is bounded for all N by

$$\frac{1}{N^2} \sum_{\substack{0 \leq i,j \leq bN \\ 1 \leq |i-j| \leq N/M}} \log \frac{N}{|i-j|} \leq 2b \frac{\log M}{M}.$$

Moreover as $f_M(0) = \log M$,

$$\frac{1}{N^2} \sum_{i \neq j} f_M(|y_j - y_i|) = \int \int f_M(|s - t|) \, d\hat{\nu}_N^y(s) d\hat{\nu}_N^y(t) - \frac{\log M}{N}$$

which, using the weak convergence of $(\hat{\nu}_N^y)$, implies

$$\frac{1}{N^2} \sum_{i \neq j} f_M(|y_j - y_i|) \to \int \int \log(|s - t|^{-1} \wedge M) \, \Phi_\mu(t) \Phi_\mu(s) \, ds \, dt, \quad \text{as } N \to \infty.$$

Hence, letting $M \to \infty$ in the r.h.s of (20) gives the limit (19).

3rd step: From the preceding, to get that $(\hat{\mu}_N)_N$ satisfies a weak LDP on $\mathcal{M}(\mathbb{R}^+)$, it remains to establish that

$$\lim_N \frac{1}{N^2} \log(Z_{N,V_N,\beta}) = -\frac{\beta}{2} F_V^\lambda \qquad (21)$$

where we recall that $F_V^\lambda = E_V(\mu_V^\lambda)$.

Since, for all $\epsilon > 0$,

$$Z_{N,V_N,\beta} = \bar{\mathbb{P}}_N(\mathbb{A}_N^N) \geq \bar{\mathbb{P}}_N(\mathcal{B}(\mu_V^\lambda, \epsilon))$$

the weak lower bound (17) readily implies that

$$\varliminf_N \frac{1}{N^2} \log(Z_{N,V_N,\beta}) \geq -\frac{\beta}{2} \lim_{\epsilon \to 0} \inf_{\nu \in \mathcal{B}(\mu_V^\lambda, \epsilon)} E_V(\mu_V^\lambda) = -\frac{\beta}{2} F_V^\lambda. \qquad (22)$$

The reverse inequality has yet been proven in [Jo2]: we refer the reader to Corollary 4.3 of [Jo2].

4th *step:* Now, in order to obtain a strong LDP, the last point is to show that $(\hat{\mu}_N)_N$ is exponentially tight under $\mathbb{P}_N$, that is

$$\forall \alpha > 0 \, , \, \exists \mathcal{K}_\alpha \text{ compact of } \mathcal{M}(\mathbb{R}^+) : \overline{\lim_N} \frac{1}{N^2} \log \mathbb{P}_N[\hat{\mu}_N \in \mathcal{K}_\alpha^c] \leq -\alpha.$$

The argument is close to the continuous case and relies on the uniform lower bound (6) on the potentials V_N and V. We first observe that, for every $c_\alpha > 0$,

$$\mathcal{K}_\alpha = \left\{ \mu : \int_{\mathbb{R}^+} \log(1 + t^2) \, d\mu(t) \leq c_\alpha \right\}$$

is a compact set of $\mathcal{M}(\mathbb{R}^+)$. The proof is classical and left to the reader.

We turn to the exponential tightness itself. Write, for $\alpha > 0$,

$$\mathbb{P}_N[\hat{\mu}_N \in \mathcal{K}_\alpha^c] = \frac{1}{Z_{N,V_N,\beta}} \sum_{\hat{\mu}_N \in \mathcal{K}_\alpha^c} \prod_{i<j} |x_i - x_j|^\beta \times e^{-\frac{N\beta}{4} \sum_i V_N(x_i)}$$

$$\times e^{-\frac{\beta N^2}{4} \int V_N(x) \, d\hat{\mu}_N(x)}.$$

By the lower bound (6),

$$\mathcal{K}_\alpha^c \subset \bigcap_N \left\{ \mu : \int_{\mathbb{R}^+} V_N(t) \, d\mu(t) > C \times c_\alpha \right\}.$$

In particular,

$$\{\hat{\mu}_N \in \mathcal{K}_\alpha^c\} \subset \left\{ \int_{\mathbb{R}^+} V_N(t) \, d\hat{\mu}_N(t) > C \times c_\alpha \right\}.$$

which leads to

$$\mathbb{P}_N[\hat{\mu}_N \in \mathcal{K}_\alpha^c] \leq \frac{1}{Z_{N,V_N,\beta}} \sum_{\hat{\mu}_N \in \mathcal{K}_\alpha^c} \prod_{i<j} |x_i - x_j|^\beta \, e^{-\frac{N\beta}{4} \sum_{i=1}^N V_N(x_i)} \, e^{-\frac{C \times c_\alpha}{4} N^2 \beta}$$

$$\leq \frac{Z_{N,\frac{V_N}{2},\beta}}{Z_{N,V_N,\beta}} \, e^{-\frac{C \times c_\alpha}{4} N^2 \beta}.$$

Now, from (21), it remains to properly choose c_α. This yields the exponential tightness. Thus the proof of Theorem 3.2 is complete. □

4 Large deviations for the largest eigenvalue

In this section, we will derive large deviations properties for the largest eigenvalue of both discrete and continuous Coulomb gas from the corresponding LDP for the spectral measure established above. Our approach follows the one developed in [BDG] (in Theorem 6.2) for the particular case of the GO(U)E model. We start with the discrete setting and recover in this way results established first by K. Johansson (in Theorem 2.2 of [Jo2]). Then, we will briefly present the corresponding conclusions in the continuous case and outline how to adapt the main arguments.

4.1 The discrete setting

We consider the general framework described in Section 3 dealing with particles $x = (x_1, \cdots, x_N) \in \mathbb{A}_N^N$ with joint probability $\mathbb{P}_N := P_{N,V_N,\beta}$ represented by (4). Our purpose is to establish, from the LDP of the spectral measure (Theorem 3.2), some large deviations properties for the largest eigenvalue $x_N^* = \max_{1 \le i \le N} x_i$. Some further notations will turn useful. Let μ_V^λ be the extremal measure associated to the potential V and the constraint λ, Φ_V^λ its density and let b_V be the right endpoint of its compact support. Define furthermore, for all $t \in [1, +\infty]$,

$$\mathcal{M}^{\lambda,t} = \mathcal{M}^\lambda \cap \mathcal{M}([0,t]) \text{ and } F_V^{\lambda,t} = \inf_{\mu \in \mathcal{M}^{\lambda,t}} E_V(\mu)$$

where $\mathcal{M}([0,t])$ is the set of probability measures compactly supported in $[0,t]$. With the notations of Section 3, we have: $F_V^{\lambda,\infty} = F_V^\lambda$ and
$$\inf_{\mu \in \mathcal{M}^{\lambda,t}} I_V^{\beta,\lambda}(\mu) = \frac{\beta}{2}(F_V^{\lambda,t} - F_V^\lambda).$$

Let finally F_N be the distribution function of x_N^*. The rate functions L and J which will govern the large deviations of F_N with respective speed N^2 and N will be given on $\mathbb{R}^+$ by

$$L(t) = \begin{cases} \inf_{\mu \in \mathcal{M}^{\lambda,t}} I_V^{\beta,\lambda}(\mu), & \text{if } t \ge 1 \\ +\infty, & \text{if } t < 1. \end{cases}$$

and

$$J(t) = \begin{cases} \inf_{\tau \ge t} \int_{\mathbb{R}^+} k_V(\tau,s)\Phi_V^\lambda(s)\,ds - F_V^\lambda, & \text{if } t > b_V \\ 0, & \text{if } t \le b_V. \end{cases}$$

Remark 4.1 *By definition, L is clearly non-negative and satisfies $L(b_V) = 0$. In the following, we will see that the function J is also non-negative and that*

$$\inf_{\tau \ge b_V} \int_{\mathbb{R}^+} k_V(\tau,s)\Phi_V^\lambda(s)\,ds - F_V^\lambda = \int_{\mathbb{R}^+} k_V(b_V,s)\Phi_V^\lambda(s)\,ds - F_V^\lambda = 0.$$

We now state the precise large deviations statement for the distribution F_N already established in [Jo2].

Theorem 4.2 *1. a) For all $t \geq 0$,*

$$\lim_{N \to +\infty} \frac{1}{N^2} \log F_N(t) = -L(t).$$

b) $\forall \, 1 \leq t < b_V$,

$$L(t) > 0 \quad and \quad \lim_{N \to +\infty} \frac{1}{N} \log(1 - F_N(t)) = 0.$$

2. Assume that $J(t) > 0$ for all $t > b_V$. Then, for all $t > b_V$,

$$\lim_{N \to +\infty} \frac{1}{N} \log(1 - F_N(t)) = -\beta J(t).$$

3. Moreover, if the function J is continuous at b_V, then the limit of part 2. holds also for $t = b_V$.

As we will see below, the rate function L appears rather naturally from the LDP of $(\hat{\mu}_N)$ governed by the GRF $I_V^{\beta,\lambda}$ since for any $t \in \mathbb{R}^+$,

$$\{x_N^* \leq t\} = \{\hat{\mu}_N \in \mathcal{M}([0,t])\}.$$

In particular, it is easy to state the results of Theorem 4.2 whenever $t \in [0,1[$. Indeed, by Lemma 3.5, if $\mu \in \mathcal{M}^\lambda$, the diameter of its support is greater or equal to 1, so that if $t \in [0,1[$ then

$$\mathcal{M}^\lambda \cap \mathcal{M}([0,t]) = \emptyset$$

and $\inf_{\mu \in \mathcal{M}^\lambda \cap \mathcal{M}([0,t])} I_V^{\beta,\lambda}(\mu) = +\infty$. So, Theorem 3.2 gives

$$\forall t \in [0,1[, \quad \lim_{N \to +\infty} \frac{1}{N^2} \log F_N(t) = -\infty \quad and \quad \lim_{N \to +\infty} \frac{1}{N} \log(1 - F_N(t)) = 0.$$

The connection of the function J with $I_V^{\beta,\lambda}$ is more subtle and is based on the variational properties of the extremal measure μ_V^λ (given by point 2. of Proposition 3.1). Before we detail this and complete the proof of Theorem 4.2, let us notice that parts 1. and 2. of Theorem 4.2 imply that x_N^* converges in probability to the upper bound b_V of the support of μ_V^λ. Note that in the classical random matrix theory, it is well-known that the largest eigenvalue x_N^* converges almost everywhere to the upper bound of the compact support of the extremal measure (see for instance [Bai]).

Under some additional assumptions on J, we obtain a complete LDP for x_N^* with speed N (this is the analogue of Theorem 6.2 in [BDG] obtained for the GOE model).

Theorem 4.3 *Assume that J is continuous and strictly increasing on $[b_V; +\infty[$. Then x_N^* satisfies on $\mathbb{R}^+$ a LDP with speed N and GRF $I_{\beta,V}^*$ defined by*

$$I_{\beta,V}^*(t) = \begin{cases} \beta J(t), & if \ t \geq b_V, \\ +\infty, & if \ t < b_V. \end{cases}$$

The continuity of J combined with the fact that J is going to infinity with t ensure that $I^*_{\beta,V}$ is a GRF. The proof is an easy consequence of our previous Theorem 4.2. Indeed, under the assumptions made on J, one readily derives that

$$\lim_N \frac{1}{N} \log \mathbb{P}_N[x^*_N \leq t] = -\infty, \quad \forall t < b_V$$

whereas

$$\lim_N \frac{1}{N} \log \mathbb{P}_N[x^*_N \geq t] = -\beta J(t), \quad \forall t > b_V.$$

Furthermore, one can remark that these two assertions ensure the strong LDP of Theorem 4.3.

Next, we turn to the proof of the main result of this section.

Proof of Theorem 4.2: It remains to treat the case where $t \geq 1$.
Proof of 1.: Let us first show that

$$\forall t > b_V, \quad L(t) > 0 \qquad (23)$$

with $L(t)$ given by $L(t) = F_V^{\lambda,t} - F_V^\lambda$.
In fact, it suffices to observe the following identity

$$\mathcal{M}^{\lambda,t} := \mathcal{M}^\lambda \cap \mathcal{M}([0,t]) = \{\mu \in \mathcal{M}(\mathbb{R}^+) \ / \ 0 \leq \mu \leq \lambda_{/[0,t]}\}$$

which means that $\mathcal{M}^{\lambda,t}$ is the class of probability measures submitted to the constraint $\lambda_{/[0,t]}$. Then according to the general theory of [Dr-Sa1], we know the existence of a unique extremal measure $\mu_V^{\lambda,t}$ in $\mathcal{M}^{\lambda,t}$ such that

$$F_V^{\lambda,t} = E_V(\mu_V^{\lambda,t}) = \inf_{\mu \in \mathcal{M}^{\lambda,t}} E_V(\mu) \in \mathbb{R}.$$

As $\mathcal{M}^{\lambda,t} \subset \mathcal{M}^\lambda$, $F_V^{\lambda,t} \geq F_V$. If one assumes that $F_V^{\lambda,t} = F_V$, then the uniqueness of the extremal measure μ_V^λ yields $\mu_V^{\lambda,t} = \mu_V^\lambda$ and then $\mu_V^\lambda \in \mathcal{M}([0,t])$. Under $t < b_V$, this is impossible since b_V is the upper bound of the support of μ_V^λ. In this way, (23) is justified and, assuming that *a)* holds true, this readily gives

$$\lim_N \frac{1}{N} \log(1 - F_N(t)) = 0, \quad \forall t < b_V.$$

Let us now establish *a)*. First, as $\{x^*_N \leq t\} = \{\hat{\mu}_N \in \mathcal{M}([0,t])\}$, the upper bound of Theorem 3.2 states that

$$\overline{\lim_N} \frac{1}{N^2} \log F_N(t) \leq - \inf_{\mu \in \mathcal{M}^{\lambda,t}} I_V^\beta(\mu) = -L(t).$$

With our notations, the reverse inequality amounts to show

$$\overline{\lim_N} \frac{1}{N^2} \log \bar{\mathbb{P}}_N[\hat{\mu}_N \in \mathcal{M}([0,t])] \geq -\frac{\beta}{2} E_V(\mu_V^{\lambda,t}).$$

But, the measure $\mu_V^{\lambda,t}$ belongs to the class $\mathcal{M}^{\lambda,t}$. So it is compactly supported in $[0,t]$ and admits a density $\Phi_{\lambda,t}$ satisfying $0 \leq \Phi_{\lambda,t} \leq 1$. Then, the reasoning is very close to that done in the 2nd step of the proof of Theorem 3.2 and we do not give the details.

Proof of 3.: Here, we assume that part *2.* is true and that J is continuous at b_V. Obviously, $\overline{\lim}_N \frac{1}{N} \log(1 - F_N(b_V)) \leq 0$. Conversely, using the trivial fact that for all $\epsilon > 0$, $\mathbb{P}_N(x_N^* > b_V) \geq \mathbb{P}_N(x_N^* \geq b_V + \epsilon)$, we have

$$\overline{\lim}_N \frac{1}{N} \log(1 - F_N(b_V)) \geq \sup_{\epsilon > 0} \overline{\lim}_N \frac{1}{N} \log(1 - F_N(b_V + \epsilon))$$
$$= -\beta \inf_{\epsilon > 0} J(b_V + \epsilon) = -\beta J(b_V) = 0.$$

Hence as announced, $\lim_N \frac{1}{N} \log(1 - F_N(b_V)) = 0$.

Proof of 2.: The argument relies on Theorem 3.2 and a change of variables that is taken from [BDG] in the case of the GOE. It also involves the variational properties (Proposition 3.1) of the extremal measure μ_V^λ and some results of [Jo2].

We start by rewriting the function J. Define the map Ψ_V on $\mathbb{R}^+ \times \mathcal{M}(\mathbb{R}^+)$ as

$$\Psi_V(\tau, \mu) = \int_{\mathbb{R}^+} \log|\tau - s|^{-1} \, d\mu(s) + \frac{1}{2} V(\tau)$$

and let

$$\xi_V^\lambda = F_V^\lambda - \frac{1}{2} \int_{\mathbb{R}^+} V(s) \Phi_V^\lambda(s) \, ds.$$

Thus,

$$\int_{\mathbb{R}^+} k_V(\tau, s) d\mu_V^\lambda(s) = \Psi_V(\tau, \mu_V^\lambda) + \frac{1}{2} \int_{\mathbb{R}^+} V(s) \Phi_V^\lambda(s) \, ds$$

and the variational relations on μ_V^λ given in Proposition 3.1 read as the following identity

$$\int_{\mathbb{R}^+} k_V(b_V, s) d\mu_V^\lambda(s) = F_V^\lambda = \inf_{\tau \geq t} \int_{\mathbb{R}^+} k_V(\tau, s) d\mu_V^\lambda(s), \quad \forall t \geq b_V$$

and also imply that the function J is non-negative. As a consequence, we may rewrite

$$\xi_V^\lambda = \inf_\tau \Psi_V(\tau, \mu_V^\lambda)$$

and, for all $t \geq b_V$,

$$J(t) = \inf_{\tau \geq t} \Psi_V(\tau, \mu_V^\lambda) - \xi_V^\lambda. \tag{24}$$

These last expressions will turn useful in our reasoning.

38   D. Féral

Now, let us state two technical lemmas needed for the proof. First, our following change of variables will involve the probability measure $\mathbb{Q}_{N-1}$ defined for all $x' = (x'_1, \cdots, x'_{N-1}) \in \mathbb{A}_N^{N-1}$ by

$$\mathbb{Q}_{N-1}(x') = \frac{1}{Z_{N-1,V_N,\beta}} \prod_{1 \le i < j \le N-1} |x'_i - x'_j|^\beta \exp\{-\frac{N\beta}{2} \sum_{i=1}^{N-1} V_N(x'_i)\} \quad (25)$$

where $Z_{N-1,V_N,\beta}$ is the normalisation factor. We will need to call on the following result.

Lemma 4.4 *(cf. Lemmas 4.5 and 4.6 of [Jo2]) Under the preceding notations,*

$$\lim_{N \to +\infty} \frac{1}{N} \log(\frac{Z_{N-1,V_N,\beta}}{Z_{N,V_N,\beta}}) = \beta \xi_V^\lambda$$

We would like to emphasize that the knowledge of this limit is crucial for the proof below (note that for the GOE, this limit follows from Selberg's formula [BDG]). In the sequel, we let

$$C_{N,\beta} = \frac{Z_{N-1,V_N,\beta}}{Z_{N,V_N,\beta}}.$$

Our analysis will also require the following bound.

Lemma 4.5 *There exists a function $f : \mathbb{R}^+ \mapsto \mathbb{R}^+$ going to infinity at infinity such that for every M large enough, for all N large enough,*

$$\mathbb{P}_N(x_N^* \ge M) \le e^{-Nf(M)}.$$

We postone the proof of this result to the end of this section. Before, let us proceed to the proof of Property 2. We first show that for all $t > b_V$,

$$\overline{\lim_N} \frac{1}{N} \log(1 - F_N(t)) \le -\beta J(t) \quad (26)$$

with $J(t)$ given by (24).
We have, for all M such that $b_V < t \le M$,

$$\mathbb{P}_N[x_N^* > t] = \mathbb{P}_N(x_N^* > M) + \mathbb{P}_N(x_N^* > t; x_N^* \le M).$$

According to Lemma 4.5 and as soon as M is large enough, the left term of this sum is exponentially negligible. Hence,

$$\overline{\lim_N} \frac{1}{N} \log(1 - F_N(t)) = \overline{\lim_N} \frac{1}{N} \log \mathbb{P}_N(x_N^* > t; x_N^* \le M)$$

$$\le \overline{\lim_N} \frac{1}{N} \log \mathbb{P}_N(x_1 \in]t, M]; \max_{2 \le k \le N} x_k \le M)$$

where the last inequality is due to the invariance by permutation of (4).
We perform a change of variables in order to use the results of Theorem 3.2.
Write

$$\mathbb{P}_N(x_1 \in]t, M]; \max_{2 \leq k \leq N} x_k \leq M)$$

$$= \frac{1}{Z_{N,V_N,\beta}} \sum_{(*)} \prod_{k=2}^{N} |x_1 - x_k|^\beta \exp\left(-\frac{N\beta}{2} V_N(x_1)\right)$$

$$\times \prod_{2 \leq i < j \leq N} |x_i - x_j|^\beta \exp\left\{-\frac{N\beta}{2} \sum_{k=2}^{N} V_N(x_k)\right\}$$

where $(*) = \{t \leq x_1 \leq M; x' = (x_2, \cdots, x_N) \in (\mathbb{A}_N \cap [0, M])^{N-1}\}$. Then,
given x_1 and $x' = (x_2, \cdots, x_N)$, we set $\hat{\nu}_{N-1} = \frac{1}{N-1} \sum_{i=1}^{N-1} \delta_{x'_i}$ and

$$\Psi_{V_N}(x_1, \hat{\nu}_{N-1}) = \int \log |x_1 - z|^{-1} d\hat{\nu}_{N-1}(z) + \frac{1}{2} V_N(x_1).$$

Note that on $(*)$,
$$-\Psi_{V_N}(x_1, \hat{\nu}_{N-1}) \leq \log(2M).$$

So, we have

$$\mathbb{P}_N(x_1 \in]t, M]; \max_{2 \leq k \leq N} x_k \leq M) \leq \frac{Z_{N-1,V_N,\beta}}{Z_{N,V_N,\beta}} \sum_{(*)} e^{-\beta N \Psi_{V_N}(x_1, \hat{\nu}_{N-1})} \mathbb{Q}_{N-1}(x').$$

Let now $\delta > 0$ and set for all $r > 0$, $\mathcal{B}_r(\mu_V^\lambda, \delta) = \mathcal{B}(\mu_V^\lambda, \delta) \cap \mathcal{M}([0, r])$. Write
then

$$\mathbb{P}_N(x_1 \in]t, M]; \max_{2 \leq k \leq N} x_k \leq M)$$

$$\leq C_{N,\beta} \sum_{(*)} e^{-\beta(N-1)\Psi_{V_N}(x_1, \hat{\nu}_{N-1})} \{1_{\hat{\nu}_{N-1} \in \mathcal{B}_{2M}(\mu_V^\lambda, \delta)} + 1_{\hat{\nu}_{N-1} \notin \mathcal{B}(\mu_V^\lambda, \delta)}\} \mathbb{Q}_{N-1}(x')$$

$$\leq C_{N,\beta} \left\{ e^{-\beta(N-1)\inf_{(x_1, \mu) \in [t, M] \times \mathcal{B}_{2M}(\mu_V^\lambda, \delta)} \Psi_V(x_1, \mu)} \right.$$

$$\left. + (2M)^N \mathbb{Q}_{N-1}[\hat{\nu}_{N-1} \notin \mathcal{B}(\mu_V^\lambda, \delta)] \right\} \tag{27}$$

It is not hard to see that the measures $\hat{\mu}_{N-1}$(under the law $\mathbb{P}_{N-1} = P_{N-1,V_{N-1},\beta}$) and $\hat{\nu}_{N-1}$ (under $\mathbb{Q}_{N-1}$) are exponentially equivalent. So, by
Theorem 4.2.13 in [De-Ze], the conclusion of our Theorem 3.2.2. remains valid
for $\mathbb{Q}_{N-1}$ and, as μ_V^λ is the unique minimiser of the GRF I_V^λ, the second term
in (27) is exponentially negligible as $N \to \infty$ (and for any $\delta > 0$ and $M < \infty$).
Therefore,

$$\overline{\lim_{N}} \frac{1}{N} \log \mathbb{P}_N(x_1 \geq t; \max_{2 \leq k \leq N} x_k \leq M)$$

$$\leq \overline{\lim_{N}} \frac{1}{N} \log C_{N,\beta} - \beta \lim_{\delta \to 0} \overline{\lim_{N}} \inf_{(x_1,\mu) \in [t,M] \times \mathcal{B}_{2M}(\mu_V^\lambda, \delta)} \Psi_{V_N}(x_1, \mu)$$

$$= \beta \xi_V^\lambda - \beta \inf_{(x_1,\mu) \in [t,M] \times \mathcal{B}_{2M}(\mu_V^\lambda, \delta)} \Psi_V(x_1, \mu).$$

In the last equality, the first term is due to Lemma 4.4 and for the second term, we have used the uniform convergence of (V_N) to V on $[0, 2M]$.

Then, note that the map Ψ_V is lower semicontinuous on $[0, M] \times \mathcal{M}([0, 2M])$. Indeed, one may write Ψ_V as $\Psi_V = \sup_{K>0} \Psi_V^K$ where for all $K > 0$,

$$\Psi_V^K(x, \mu) := \int \log(|x - y|^{-1} \wedge K) \, d\mu(y) + \frac{1}{2} V(x)$$

is clearly continuous Ψ_V^K on $[0, M] \times \mathcal{M}([0, 2M])$. Hence,

$$\lim_{\delta \to 0} \inf_{(x,\mu) \in [t,M] \times \mathcal{B}_{2M}(\mu_V^\lambda, \delta)} \Psi_V(x, \mu) = \inf_{x \in [t,M]} \Psi_V(x, \mu_V^\lambda).$$

We get the announced claim (26) by letting $M \to +\infty$.

To complete the proof of Property 2., it is enough to show (recall the definition (24) of J) that for all $t > b_V$,

$$\lim_{N} \frac{1}{N} \log(1 - F_N(t)) \geq -\beta(\Psi_V(t, \mu_V^\lambda) - \xi_V^\lambda). \tag{28}$$

Fix $t > b_V$. Consider then two real numbers (s, r) such that $s > t > r > b_V$. Then, as

$$\mathbb{P}_N(x_N^* \geq t) \geq \mathbb{P}_N(x_1 \in \mathbb{A}_N \cap [t, s]; \max_{2 \leq k \leq N} x_k \leq r),$$

we deduce from the previous change of variables that, $\forall \delta > 0$,

$$\mathbb{P}_N(x_N^* \geq t) \geq C_{N,\beta} \times \alpha_N(t, s) \times \exp\left(-\beta N \sup_{(x,\nu) \in [t,s] \times \mathcal{B}_r(\mu_V^\lambda, \delta)} \Psi_{V_N}(x, \nu)\right)$$

$$\times \mathbb{Q}_{N-1}[\hat{\nu}_{N-1} \in \mathcal{B}_r(\mu_V^\lambda, \delta)] \tag{29}$$

where $\alpha_N(t, s) = \sum_{x_1 \in \mathbb{A}_N \cap [t,s]} e^{-\frac{\beta}{2} V_N(x_1)}$. From the uniform convergence of (the continuous potentials) (V_N) to V on $[t, s]$, the sequence $\left(\frac{\alpha_N(t,s)}{N}\right)$ is bounded and strictly positive for all (t, s). So, $\lim_N \frac{1}{N} \log \alpha_N(t, s) = 0$ and it remains to study the two last terms in the product in (29). To this task, observe first that

$$\lim_{N} \sup_{(x,\nu) \in [t,s] \times \mathcal{B}_r(\mu_V^\lambda, \delta)} \Psi_{V_N}(x, \nu) = \sup_{(x,\nu) \in [t,s] \times \mathcal{B}_r(\mu_V^\lambda, \delta)} \Psi_V(x, \nu).$$

As we have chosen $t > r > b_V$, the function Ψ_V is continuous on $[t,s] \times \mathcal{M}([0,r])$ and thus

$$\lim_{s \to t} \lim_{\delta \to 0} \sup_{(x,\nu) \in [t,s] \times \mathcal{B}_r(\mu_V^\lambda, \delta)} \Psi_V(x,\nu) = \Psi_V(t, \mu_V^\lambda).$$

At last, we have to state that (at least for every r large enough)

$$\lim_N \frac{1}{N} \log \mathbb{Q}_{N-1}[\hat{\nu}_{N-1} \in \mathcal{B}_r(\mu_V^\lambda, \delta)] = 0.$$

For this, we show that $\lim_N \mathbb{Q}_{N-1}[\hat{\nu}_{N-1} \in \mathcal{B}_r(\mu_V^\lambda, \delta)] = 1$. Recall that $\mathcal{B}_r(\mu_V^\lambda, \delta) = \mathcal{B}(\mu_V^\lambda, \delta) \cap \mathcal{M}([0,r])$. On the one hand,

$$\lim_N \mathbb{Q}_{N-1}[\hat{\nu}_{N-1} \notin \mathcal{B}(\mu_V^\lambda, \delta)] = 0, \quad \text{for every } \delta > 0$$

since we have yet observed that this probability is exponentially negligible. On the other hand, $\lim_N \mathbb{Q}_{N-1}[\hat{\nu}_{N-1} \notin \mathcal{M}([0,r])] = 0$ since

$$\lim_N \frac{1}{N} \log \mathbb{Q}_{N-1}[\hat{\nu}_{N-1} \notin \mathcal{M}([0,r])] = \lim_N \frac{1}{N} \log \mathbb{Q}_{N-1}(x_{N-1}^* > r)$$
$$= -\beta J(r) < 0$$

where the last inequality follows from both $r > b_V$ and (26).

In this way, (28) is established and to complete the proof of Theorem 4.2, it remains to prove Lemma 4.5.

Proof of Lemma 4.5: Note that a similar result was yet used in [Jo2] (p. 465) but not proved.

It is enough to prove that

$$\mathbb{P}_N(x_1 \geq M) \leq e^{-N f(M)} \tag{30}$$

since by invariance by permutation, $\mathbb{P}_N(x_N^* \geq M) \leq N\mathbb{P}_N(x_1 \geq M)$. One has

$$\mathbb{P}_N(x_1 \geq M) = \frac{1}{Z_{N,V_N,\beta}} \sum_{x \,:\, x_1 \geq M} \prod_{j=2}^N |x_1 - x_j|^\beta \, e^{-\frac{N\beta}{2} V_N(x_1)}$$
$$\times \prod_{2 \leq i < j \leq N} |x_i - x_j|^\beta \, e^{-\frac{N\beta}{2} \sum_{i=2}^N V_N(x_i)}.$$

Then, using (14) and the minoration (6), we easily have

$$\prod_{j=2}^N |x_1 - x_j|^\beta \, e^{-\frac{N\beta}{2} V_N(x_1)} \leq \prod_{2 \leq j \leq N} (1 + x_j^2)^{\beta/2} \times (1 + x_1^2)^{-\xi\beta(N-1)/2}$$

and

$$\mathbb{P}_N(x_1 \geq M) \leq C_{N,\beta} \sum_{x_1 \geq M} (1 + x_1^2)^{-\xi\beta(N-1)/2} \times \mathbb{E}_{N-1}\left[\prod_{j=1}^{N-1}(1 + x_j'^2)^{\beta/2}\right]$$

where $\mathbb{E}_{N-1}$ denotes the expectation with respect to the law $\mathbb{Q}_{N-1}$. First, it is an easy fact that there is $A > 0$ such that for all M large,

$$\sum_{x_1 \geq M} (1 + x_1^2)^{-\xi\beta(N-1)/2} \leq e^{-AN \log M}$$

for all N (at least such that $\xi\beta(N-1)4$). Secondly, a computation very similar to that done in the fourth step of the proof of Theorem 3.2 shows that there is $C > 0$ such that for all $A' > 0$ large enough (for all N),

$$\mathbb{Q}_{N-1}\left[\sum_{j=1}^{N-1} \log(1 + x_j'^2) \geq A'N\right] \leq e^{-CA'N^2}.$$

With this, we trivially deduce there exists another constant $A'' > 0$ such that

$$\mathbb{E}_{N-1}\left[\prod_{j=1}^{N-1}(1 + x_j'^2)^{\beta/2}\right] \leq e^{A''N}$$

which together with Lemma 4.4 gives Lemma 4.5. This ends the proof of Theorem 4.2.   □

4.2 The continuous case

As already mentioned, Theorem 4.3 has been established in [BDG] for the GO(U)E model. But, to our knowledge, large deviations properties of the associated largest eigenvalue of general random matrix models do not appear elsewhere in the extensive literature on random matrix theory.

We claim here that the preceding Theorems 4.2 and 4.3 extend (with $\mathcal{M}(\mathbb{R})$ instead of $\mathcal{M}^\lambda$) to general continuous Coulomb gas (1) satisfying assumptions (i) to (iii) of Section 2 and such that the limiting potential $V : \mathbb{R} \mapsto \mathbb{R}$ is as in classical random matrix models that is:

(iv) V is convex and has a derivative V' Hölder continuous.

Note that if the particles are assumed to be positive, the assumption (iv) may be replaced by:

(iv') $V : \mathbb{R}^+ \mapsto \mathbb{R}$ has a derivative V' Hölder continuous and $tV''(t)$ increases.

Here, we want to precise that the condition "V' Hölder continuous" must be understood as "V' Hölder and continuous on the compact support of the extremal measure".

For the proof, the approach is similar to the discrete case. So, a complete rewriting of the closely related arguments would be lengthy and not informative. We shall nevertheless explain why we have add some regularity assumptions on the potential V. The main difficulty is to state the analogue of Lemma 4.4 which has been established in the discrete case by [Jo2] (in its Section 4) and follows in the particular models of the G(O)UE or L(O)UE from Selberg's formula. The analysis made in [Jo2] can be adapted here but this requires to add some assumptions on the potential V (see also Section 4 in [Jo1]). Indeed, the arguments of [Jo2] use points *2.* and *3.* of Proposition 3.1 and the fact that μ_V^λ has a density. It is worth noticing that in the continuous setting, these properties are not necessarily verified for any potential V. But they hold in particular if V is regular enough as above (see [Dr-Sa1] and Chapters I and IV of [S-T]).

5 Applications to the classical discrete and continuous orthogonal polynomial ensembles

In this last section, we illustrate the preceding results through several examples from the classical orthogonal polynomial ensembles. We actually mainly focus on the discrete context. As a complement (at least at the level of the largest eigenvalue), the last short subsection presents the results of the LOE model.

In all our examples, the knowledge of the extremal measure is essential. As already announced, both with or without constraint and thanks to the regularity of the limiting potential V, there are some explicit formulas for the computations of the extremal compact support (with is then a single interval) and the (continuous) density. Remark that all the classical orthogonal polynomial ensembles actually fall in this context (cf. Sections 2 and 3 in [BKMM]). Thus, in the following, we will assume that condition (iv) or (iv') of Section 4.2 is satisfied.

5.1 Discrete examples

Here, we will investigate the classical discrete orthogonal polynomial ensembles and mainly concentrate on the description of the extremal measure μ_V^λ. The main reason is that, due to the constraint λ, the computations of μ_V^λ are rather tedious. Nevertheless, [Jo2] deeply investigated the Meixner ensemble and computed μ_V^λ as well as the GRF associated to the largest eigenvalue. However, he did not detail the derivations of the extremal support. Our purpose here is to explain and illustrate on the Charlier ensemble a practical and general method to obtain the constrained extremal measure (this discussion may be adapted similarly to the Meixner ensemble).

To this aim, two approaches can actually be adopted. The first one uses the method of [K-V] based on the recursion formula for the underlying orthogonal polynomials. Below, we will rather solve the variational problem (5). This approach is very general and completely independent of the underlying orthogonal polynomial ensemble. It requires to solve a Riemann-Hilbert scalar problem for the Cauchy transform of the equilibrium measure and uses the technique of "ansatz". For details, we refer the reader to [Dr-Sa1] and [BKMM] (the latter gives in Section 2 a complete and rigorous explanation of these various theories). In the following, we just outline the general method and illustrate it in the particular case of the Poisson-Charlier orthogonal polynomial ensemble. Next, we will briefly mention some other classical discrete ensembles.

• Let us consider first a general potential V on $\mathbb{R}^+$ regular enough as above. We look for the extremal constrained density Φ_V^λ solution of (5) where λ denotes Lebesgue measure on $\mathbb{R}^+$. For this, we use the technique of "ansatz" which we now briefly explain.

Generally speaking (see [Dr-Sa1]), from the definition of any constraint, two situations can occur: either the constraint is active or it is not. Saying the constraint is active on an interval I means that the extremal measure coincides with λ on I. In both cases, the arguments consist first in finding the endpoints of the compact support and then describing the density. In pratice, this is expressed by the next two cases:

  – if the constraint is not active, we recover the situation without any constraint of the continuous setting. On the one hand, the support is $\mathrm{Supp}(\mu_V^\lambda) = [a, b]$ with $a < b$ solutions of

$$\begin{cases} \dfrac{1}{2\pi} \displaystyle\int_a^b \dfrac{V'(t)}{\sqrt{(t-a)(b-t)}} \, dt = 0, \\[4mm] \dfrac{1}{2\pi} \displaystyle\int_a^b \dfrac{t V'(t)}{\sqrt{(t-a)(b-t)}} \, dt = 1. \end{cases}$$

On the other hand (see Th. IV.1.11 and IV.3.1 of [S-T]), the density Φ_V^λ solves the integral equation

$$PV \left(\int_a^b \frac{\Phi_V^\lambda(t)}{t - x} \, dt \right) = -\frac{1}{2} V'(x), \quad \forall x \in [a, b] \tag{31}$$

(where PV denotes the principal value) and is given by

$$\Phi_V^\lambda(x) = \sqrt{(x-a)(b-x)} \, PV \left(\frac{1}{2\pi^2} \int_a^b \frac{V'(t)}{\sqrt{(t-a)(b-t)}} \frac{dt}{t-x} \right).$$

  – otherwise, if the constraint is active, we will make an ansatz. That is we will assume that the potential V is such that the constraint may only be active on a compact interval of the form $[0, a]$. The reason for such assumption is twofold. First, this is the only case where the extremal measure can be determined explicitly. Secondly, this situation occurs in classical orthogonal discrete ensembles (this fact can be seen using the recursion formula of the underlying polynomials). Nevertheless, and as observed in [BKMM] (cf. their Section 2), it is impossible to make explicit the conditions required on V to lead to such a situation. So as in [BKMM] (see the proof of their Theorem 2.17 in Appendix B), in practice, you begin with an ansatz that the constraint is active only on $[0, a]$ and so that the extremal support is $[a_V, b_V] = [0, b]$. If, under the ansatz you can compute a solution which is consistent with the variational problem, then this is the extremal measure by uniqueness. Now, under the ansatz, if $0 < a < b$ exist, they are solutions of

$$\begin{cases} \dfrac{1}{2\pi} \displaystyle\int_a^b \dfrac{V'(t)}{\sqrt{(t-a)(b-t)}} \, dt = \log\left(\dfrac{\sqrt{b} + \sqrt{a}}{\sqrt{b} - \sqrt{a}}\right), \\[4mm] \dfrac{1}{2\pi} \displaystyle\int_a^b \dfrac{tV'(t)}{\sqrt{(t-a)(b-t)}} \, dt = 1 - \sqrt{ab} - \dfrac{a+b}{2}\log\left(\dfrac{b-a}{(\sqrt{a}+\sqrt{b})^2}\right). \end{cases}$$

Then, the description of Φ_V^λ on $[a, b]$ boils down again to the resolution of an integral equation. Precisely, by Theorem 2.13 of [Dr-Sa1], Φ_V^λ is given by

$$\Phi_V^\lambda(t) = \mathbf{1}_{[0,b]}(t) + (1-b)\Psi_V(t)$$

where Ψ_V is a probability density supported on $[a, b]$ (hence $\Phi_V^\lambda \equiv 1$ on $[0, a]$) and which solves the integral equation (31) associated to the potential W given by

$$W(t) = \frac{1}{1-b}\left[(t-b)\log(b-t) - t\log t - b + \frac{1}{2}V(t)\right], \quad \forall a < t < b.$$

• We now illustrate this procedure on the Charlier ensemble. It seems that such computations do not appear in the literature ([Jo3] investigated the fluctuations of the corresponding largest eigenvalue and [K-V] derived the extremal measure using the recursion formula for Charlier polynomials). Then, we briefly mention the other classical orthogonal polynomial ensembles.

The Charlier ensemble corresponds in the representation (4) to the Poisson measure μ of parameter $\theta = hN0$ and to the potentials

$$V_N(t) = -\frac{1}{N}\log(\theta^{[Nt]}\frac{e^{-\theta}}{[Nt]!}), \quad \forall t > 0.$$

By Stirling's formula, the limiting potential is $V(t) = t\log t - t\log h - t + h$. From the discussion above and the following useful formulas (cf. [Dr-Sa2] p. 129-130):

$$\begin{cases} \dfrac{1}{\pi} \displaystyle\int_a^b \dfrac{\log t}{\sqrt{(t-a)(b-t)}} \, dt = 2\log\left(\dfrac{\sqrt{a} + \sqrt{b}}{2}\right), \\[4mm] \dfrac{1}{\pi} \displaystyle\int_a^b \dfrac{t\log t}{\sqrt{(t-a)(b-t)}} \, dt = (a+b)\log\left(\dfrac{\sqrt{a} + \sqrt{b}}{2}\right) + \dfrac{a+b}{2} - \sqrt{ab} \end{cases}$$

we deduce that the constraint λ is active if and only if $0 < h \leq 1$. Moreover, whatever the parameter $h > 0$ is, the couple (a, b) is given by $a = (1 - \sqrt{h})^2$ and $b = (1 + \sqrt{h})^2$. Together with the fact that for every $a < x < b$ (cf. [Dr-Sa2] p. 138),

$$\begin{cases} PV\left(\dfrac{1}{\pi}\displaystyle\int_a^b \dfrac{1}{t-x} \dfrac{dt}{\sqrt{(t-a)(b-t)}}\right) = 0, \\[3mm] PV\left(\dfrac{1}{\pi}\displaystyle\int_a^b \dfrac{\log t}{t-x} \dfrac{dt}{\sqrt{(t-a)(b-t)}}\right) = \dfrac{2}{\sqrt{(x-a)(b-x)}} \\[3mm] \hspace{3cm}\cdot\left[\arctan\sqrt{\dfrac{b-x}{x-a}} - \arctan\sqrt{\dfrac{a(b-x)}{b(x-a)}}\right], \end{cases}$$

we derive Φ_V^λ:

$$\forall h > 1,\quad \Phi_V^\lambda(x) = \frac{1}{\pi}\left(\arctan\sqrt{\frac{b-x}{x-a}} - \arctan\sqrt{\frac{a(b-x)}{b(x-a)}}\right)\mathbf{1}_{[a,b]}(x).$$

$$\forall 0 < h \leq 1,\quad \Phi_V^\lambda(x) = \mathbf{1}_{[0,a]} + \frac{1}{\pi}\left(\pi - \arctan\sqrt{\frac{x-a}{b-x}} - \arctan\sqrt{\frac{b(x-a)}{a(b-x)}}\right)\mathbf{1}_{[a,b]}.$$

The computations of the GRF driving the large deviations results for both the spectral measure and the largest particle x_N^* of the Charlier ensemble are rather tedious, so we do not give the details. Nevertheless, using only the expression of V, one can easily show that, for all $\epsilon > 0$, there are two real numbers $L(b - \epsilon) > 0$ and $J(b + \epsilon) > 0$ such that

$$\lim_{N \to +\infty} \frac{1}{N^2} \log \mathbb{P}_N[x_N^* \leq b - \epsilon] = -L(b - \epsilon) \quad \text{and}$$

$$\lim_{N \to +\infty} \frac{1}{N} \log \mathbb{P}_N[x_N^* \geq b + \epsilon] = -2J(b + \epsilon).$$

According to Theorem 4.2 and formula (24), we shall only justify that $J(x) > 0$ for all $x > b$ where

$$J(x) = \inf_{t \geq x} g(t) \quad \text{with} \quad g(t) = \int_0^b \log|t - x|^{-1}\, \Phi_V^\lambda(x)\, dx + \frac{1}{2}V(t) - \xi_V^\lambda.$$

One has for all $t > b$,

$$g''(t) = \int_0^b \frac{\Phi_V^\lambda(x)}{(x-t)^2}\, dx + \frac{1}{2}V''(t)$$

with $V''(t) = \frac{1}{t} > 0$. Thus, g is strictly convex on $]b, +\infty[$. Using the (general) fact that g is continuous at b with $g(b) = 0$ (cf. Proposition 3.1), we deduce that g and so J are strictly positive on $]b, +\infty[$.

In the case of the Krawtchouk and Hahn ensembles, our previous method must be somewhat modified (but the main arguments are similar, see Remark 3.7 above). Indeed, the constraint λ is compactly supported on $[0, 1]$ instead of $\mathbb{R}^+$. Consequently, according to [BKMM], the constraint can also be active in a neighborhood of the bound 1. Thus, four situations have to be investigated: the constraint can be active on subintervals of $[0, 1]$ such as $[0, a]$ or/and $[b, 1]$. The various computations are thus more tedious but not more difficult. For the Krawtchouk ensemble, the reader is referred to [Dr-Sa2] and [Jo4] and for the Hahn ensemble, to [BKMM] and [Jo4].

5.2 The LOE

As a complement, and for a matter of comparison, we briefly present the corresponding analysis for the LOE model. As we already mentioned, the GO(U)E has been deeply investigated in [B-G] and [BDG]. For the LOE, the LDP result for the spectral measure has yet been obtained in [H-P] but the large deviations properties for the largest eigenvalue do not appear elsewhere in the litterature.

The LOE is defined as follows. Given $N \geq 1$, let $p(N)$ be an integer such that $p(N) \geq N$ and $\lim_N \frac{p(N)}{N} = \gamma \geq 1$. A real random matrix S_N is said to be element of the LOE (or called Wishart matrix) if it is defined by $S_N = \frac{1}{N} Y_N^t Y_N$ where Y_N is a $p(N) \times N$ real random matrix with standard Gaussian independent entries. The density of the joint distribution of the N eigenvalues $(x_1, \cdots, x_N)$ of S_N is given on $(\mathbb{R}^+)^N$ by (1) with $\beta = 1$ and $V_N(t) = t - (\frac{p(N)}{N} - 1) \log t$. Remark that the limiting potential $V(t) = t - (\gamma - 1) \log t$ fulfills the assumption (iv) of Section 4.2. According to Theorem 5.5.7 of [H-P], the spectral measure μ_{S_N} satisfies on $\mathcal{M}(\mathbb{R}^+)$ a LDP in the speed N^2 with GRF

$$I(\mu) = \frac{1}{2}(E_V(\mu) - F_V)$$

with $F_V = \frac{1}{2}(3\gamma - \gamma^2 \log \gamma + (\gamma - 1)^2 \log(\gamma - 1))$ and the extremal measure is the well-known Marchenko-Pastur law μ_γ with density

$$\Phi_\gamma(x) = \frac{\sqrt{4\gamma - (x - 1 - \gamma)^2}}{2\pi x} \mathbf{1}_{[(1-\sqrt{\gamma})^2, (1+\sqrt{\gamma})^2]}.$$

Moreover, the largest eigenvalue x_N^* of S_N satisfies a LDP of speed N and GRF

$$I^*(x) = \begin{cases} \int_{(1+\sqrt{\gamma})^2}^{x} \dfrac{\sqrt{(t - 1 - \gamma)^2 - 4\gamma}}{2t} \, dt, & \text{if } x \geq (1 + \sqrt{\gamma})^2 \\ +\infty, & \text{otherwise.} \end{cases}$$

To see this, recall that by Theorem 4.3 and (24), the GRF I^* is given for all $x \geq (1 + \sqrt{\gamma})^2$, by $I^*(x) = \inf_{t \geq x} g(t)$ where the function

$$g(t) = \int \log|t - x|^{-1} \, \Phi_\gamma(x) \, dx + \frac{1}{2}(t - (\gamma - 1)\log t) - \xi_V$$

is continuous on $[(1 + \sqrt{\gamma})^2, +\infty[$ and vanishes at $b = (1 + \sqrt{\gamma})^2$. The idea is to prove that g is increasing on $[(1 + \sqrt{\gamma})^2, +\infty[$ and that its derivative is given by

$$g'(t) = \frac{\sqrt{(t - 1 - \gamma)^2 - 4\gamma}}{2t}, \quad \forall t > (1 + \sqrt{\gamma})^2.$$

This formula follows observing that for all $t > (1 + \sqrt{\gamma})^2$, $g'(t) = m_{\mu_\gamma}(t) + \frac{1}{2}V'(t)$ where the Cauchy transform m_{μ_γ} of μ_γ is known (cf. Proposition 5.3.7 in [H-P]) as

$$m_{\mu_\gamma}(t) := \int \frac{1}{z - t} \, \Phi_\gamma(z) \, dz = \frac{\sqrt{(t - 1 - \gamma)^2 - 4\gamma}}{2t} - \frac{1}{2}V'(t),$$

$$\forall t \in \mathbb{R} \backslash [(1 - \sqrt{\gamma})^2, (1 + \sqrt{\gamma})^2].$$

Hence, one may write $I^*(x) = \int_{(1+\sqrt{\gamma})^2}^{x} g'(t) \, dt$ and the result follows.

Acknowledgments. I would like to thank Michel Ledoux for his encouragements and many helpful discussions. I am also grateful to Catherine Donati and Alain Rouault for constructive criticisms that led to an improved presentation of this paper.

References

[Bai]   Z. Bai, *Methodologies in spectral analysis of large-dimensional random matrices, a review,* Statist. Sinica **9**, 611–677 (1999).

[BKMM]   J. Baik, T. Kriecherbauer, K.T-R. McLaughlin and P.D. Miller, *Uniform asymptotics for polynomials orthogonal with respect to a general class of discrete weights and universality results for associated ensembles,* Arxiv math.CA/0310278 (2003).

[BDG]   G. Ben Arous, A. Dembo and A. Guionnet, *Aging of Spherical Spin Glasses,* Probab. Theory Relat. Fields **120**, 1–67 (2001).

[B-G]   G. Ben Arous, and A. Guionnet, *Large deviations for Wigner's law and Voiculescu's Non-Commutative Entropy,* Probab. Theory Relat. Fields **108**, 517–542 (1997).

[De-Ze]   A. Dembo and 0. Zeitouni, *Large deviations techniques and applications,* Springer-Verlag, (1998).

[De-St]   J.D. Deuschel and D.W. Stroock, *Large deviations,* Academic Press-Boston, (1989).

[Dr-Sa1] P.D. Dragnev and E.B.Saff, *Constrained energy problems with applications to orthogonal polynomials of a discrete variable,* J. Anal. Math. **72**, 223–259 (1997).

[Dr-Sa2] P.D. Dragnev and E.B. Saff, *A problem in potential theory and zero asymptotics of Krawtchouk polynomials,* Journal of Approximation Theory **102**, 120–140 (2000).

[H-P]   F. Hiai and D. Petz, *The semicircle law, free random variables and entropy*, Mathematical Surveys and monographs 77, AMS, (2000).

[Jo1]   K. Johansson, *On fluctuations of eigenvalues of random hermitian matrices*, Duke Mathematical Journal **91**, 151–204 (1998).

[Jo2]   K. Johansson, *Shape fluctuations and random matrices*, Comm. Math. Phys. **209**, 437–476 (2000).

[Jo3]   K. Johansson, *Discrete orthogonal polynomial ensembles and the Plancherel measure*, Annals Comm. Math. **153**, 259–296 (2001).

[Jo4]   K. Johansson, *Non-intersecting paths, random tilings and random matrices*, Probab. Theory Relat. Fields **123**, 225–280 (2003).

[K-V]   A.B. Kuijlaars and W. Van Assche, *The asymptotic zero distribution of orthogonal polynomials with varying reccurrence coefficients*, Journal of Approx. Theory **99**, 167–197 (1999).

[S-T]   E.B. Saff and V. Totik, *Logarithmic potentials with external fields*, Grundlehren Mathematischen Wissenschaften 316, Springer, (1997).

Estimates for moments of random matrices with Gaussian elements

Oleksiy Khorunzhiy

LMV, Université Versailles-Saint-Quentin
45 Av. des Etats-Unis, 78035-Versailles Cedex, France
e-mail: Oleksiy.Khorunzhiy@math.uvsq.fr

Summary. We describe an elementary method to get non-asymptotic estimates for the moments of Hermitian random matrices whose elements are Gaussian independent random variables. We derive a system of recurrence relations for the moments and the covariance terms and develop a triangular scheme to prove the recurrence estimates. The estimates we obtain are asymptotically exact in the sense that they give exact expressions for the first terms of $1/N$-expansions of the moments and covariance terms.

As the basic example, we consider the Gaussian Unitary Ensemble of random matrices (GUE). Immediate applications include the Gaussian Orthogonal Ensemble and the ensemble of Gaussian anti-symmetric Hermitian matrices. Finally we apply our method to the ensemble of $N \times N$ Gaussian Hermitian random matrices $H^{(N,b)}$ whose elements are zero outside the band of width b. The other elements are taken from GUE; the matrix obtained is renormalized by $b^{-1/2}$. We derive estimates for the moments of $H^{(N,b)}$ and prove that the spectral norm $\|H^{(N,b)}\|$ remains bounded in the limit $N, b \to \infty$ when $(\log N)^{3/2}/b \to 0$.

1 Introduction

The moments of $N \times N$ Hermitian random matrices H_N are given by the expression

$$M_k^{(N)} = \mathbf{E}\left\{\frac{1}{N} \operatorname{Tr}(H_N)^k\right\},$$

where $\mathbf{E}\{\cdot\}$ denotes the corresponding mathematical expectation. The asymptotic behavior of $M_k^{(N)}$ in the limit $N \to \infty$ is the source of numerous studies and many publications. One can observe three main directions of research; we list and mark them with the references that are earliest in the field up to our knowledge.

The first group of results is related with the limiting transition $N \to \infty$ when the numbers k are fixed. In this case the limiting values of $M_k^{(N)}$, if they exist, determine the moments m_k of the limiting spectral measure σ of the ensemble $\{H_N\}$. This problem was first considered by E. Wigner [20].

Another asymptotic regime, when k goes to infinity at the same time as N does, is more informative and can be considered in two particular cases. In the first one k grows slowly and $1 \ll k \ll N^\gamma$ for any $\gamma > 0$. In particular, if k is of the order $\log N$ or greater, the maximal eigenvalue of H_N dominates in the asymptotic behavior of $M_{2k}^{(N)}$. Then the exponential estimates of $M_{2k}^{(N)}$ provide the asymptotic bounds for the probability of deviations of the spectral norm $\|H_N\|$. This observation due to U. Grenander has originated a series of deep results started by S. Geman [1,7,9].

The second asymptotic regime is related to the limit when $k = O(N^\gamma)$ with $\gamma > 0$. The main subject here is to determine the critical exponent $\tilde{\gamma}$ such that the same estimates for $M_{2k}^{(N)}$ as in the previous case remain valid for all $\gamma \leq \tilde{\gamma}$ and fail otherwise [18]. This gives results on the order of the mean distance between eigenvalues at the border of the support of the limiting spectral density $d\sigma$ [4,19].

In present article we describe a method to obtain estimates for $M_{2k}^{(N)}$ valid for all values of N and k such that $k \leq CN^{\tilde{\gamma}}$ for some constant C. Estimates of this type are called non-asymptotic. However, they remain valid in the limit $N \to \infty$ and in this case they belong to the second asymptotic regime.

As the basic example, we consider the Gaussian Unitary (Invariant) Ensemble of random matrices that is usually abbreviated as GUE. In Section 2 we describe our method and prove the main results for GUE. Immediate applications of our method include the Gaussian Orthogonal (Invariant) Ensemble of random matrices (GOE) and the Gaussian anti-symmetric (or skew-symmetric) Hermitian random matrices with independent elements. A detailed description of these ensembles is given in the monograph [16]. In Section 3 we present the non-asymptotic estimates for the corresponding moments.

Our approach is elementary. We only use the integration by parts formula and generating functions techniques. We do not employ such a powerful method as the orthogonal polynomials technique commonly applied to unitary and orthogonally invariant random matrix ensembles. This allows us to consider more general ensembles of random matrices than GUE and GOE. One of the possible developments is given by the study of the ensemble of Hermitian band random matrices $H^{(N,b)}$. The matrix elements of $H^{(N,b)}$ within the band of the width b along the principal diagonal coincide with those of GUE. Outside this band they are equal to zero; the matrix obtained is normalized by $b^{-1/2}$. In Section 4 we prove non-asymptotic estimates for the moments of $H^{(N,b)}$. These estimates allow us to conclude about the asymptotic behavior of the spectral norm $\|H^{(N,b)}\|$ in the limit $b, N \to \infty$.

In Section 5 we collect auxiliary computations and formulas.

1.1 GUE, recurrence relations and semi-circle law

GUE is determined by the probability distribution over the set of Hermitian matrices $\{H_N\}$ with density proportional to

$$\exp\{-2N \operatorname{Tr} H_N^2\}. \tag{1.1}$$

The odd moments of H_N are zero and the even ones $M_{2k}^{(N)}$ verify the following remarkable recurrence relation discovered by Harer and Don Zagier [11]

$$M_{2k}^{(N)} = \frac{2k-1}{2k+2} M_{2k-2}^{(N)} + \frac{2k-1}{2k+2} \cdot \frac{2k-3}{2k} \cdot \frac{k(k-1)}{4N^2} M_{2k-4}^{(N)}, \tag{1.2}$$

where $M_0^{(N)} = 1$ and $M_2^{(N)} = 1/4$. It follows from (1.2) that the moments $M_{2k}^{(N)}, k = 0, 1, \ldots$ converge as $N \to \infty$ to the limit m_k determined by the relations

$$m_k = \frac{2k-1}{2k+2} m_{k-1}, \quad m_0 = 1. \tag{1.3}$$

The limiting moments $\{m_k, k \geq 0\}$ are proportional to the Catalan numbers C_k:

$$m_k = \frac{1}{4^k} \frac{1}{(k+1)} \binom{2k}{k} = \frac{1}{4^k} C_k \tag{1.4}$$

and therefore verify the following recurrence relation

$$m_k = \frac{1}{4} \sum_{j=0}^{k-1} m_{k-1-j} \, m_j, \quad k = 1, 2, \ldots \tag{1.5}$$

with the obvious initial condition $m_0 = 1$.

In random matrix theory, equality (1.5) was observed for the first time by E. Wigner [20]. Relation (1.5) implies that the generating function of the moments m_k

$$f(\tau) = \sum_{k=0}^{\infty} m_k \cdot \tau^k$$

verifies the quadratic equation $\tau f^2(\tau) - 4f(\tau) + 4 = 0$ and is given by

$$f(\tau) = \frac{1 - \sqrt{1-\tau}}{\tau/2}. \tag{1.6}$$

Using (1.6), Wigner has shown that the measure σ_w determined by the moments $m_k = \int \lambda^{2k} \, d\sigma_w(\lambda)$ has the density of the semicircle form

$$\sigma_w'(\lambda) = \frac{2}{\pi} \begin{cases} \sqrt{1-\lambda^2}, & \text{if } |\lambda| \leq 1, \\ 0, & \text{if } |\lambda| > 1. \end{cases} \tag{1.7}$$

The statement that the moments $M_l^{(N)}$ converge to m_k for $l = 2k$ and to 0 for $l = 2k + 1$ is known as the Wigner semicircle law.

In the present paper we show that the generating function $f(\tau)$ together with its derivatives represents a very convenient tool to estimate the moments $M_{2k}^{(N)}$. Everywhere below, we use the notation $[\cdot]_k$ for the k-th coefficient of the corresponding development, so $[f(\tau)]_k = m_k$.

1.2 Estimates for the moments of GUE

Using relations (1.2) and (1.3), one can easily prove by induction the estimate

$$M_{2k}^{(N)} \leq \left(1 + \frac{k^2}{8N^2}\right)^{2k} m_k. \tag{1.8}$$

Indeed, let us assume inequalities $M_{2l}^{(N)} \leq (1 + l^2/(gN^2))^{2l} m_l$ with some $g > 0$ to hold for all values of l such that $1 \leq l \leq k - 1$. Let us show that this is also true for $l = k$ provided $g \leq 8$.

Considering the right-hand side of (1.2) and replacing $M_{2k-2}^{(N)}$ and $M_{2k-4}^{(N)}$ by the corresponding estimates with $l = k - 1$ and $l = k - 2$, respectively, we bound the right-hand side of (1.2) by the sum of

$$\frac{2k-1}{2k+2}\left(1 + \frac{(k-1)^2}{gN^2}\right)^{2k-2} m_{k-1} = \left(1 + \frac{(k-1)^2}{gN^2}\right)^{2k-2} m_k$$

and

$$\frac{k(k-1)}{4N^2}\left(1 + \frac{(k-2)^2}{gN^2}\right)^{2k-4} m_k.$$

Here we have used identity (1.3). Comparing the expression obtained with the right-hand side of (1.8), we see that the following inequality

$$\left(1 + \frac{(k-1)^2}{gN^2}\right)^2 + \frac{k(k-1)}{4N^2} \leq \left(1 + \frac{k^2}{gN^2}\right)^4$$

is sufficient for (1.8) to be true. Expanding the powers, we see that the condition $g \leq 8$ is sufficient to have (1.8) valid for all values of k and N.

Estimates (1.8) are valid for all values of k and N without any restriction. They allow one to estimate the probability of deviations of the largest eigenvalue of H_N (see, for example [14, 15] and references therein). Then one can study the asymptotic behavior of the maximal eigenvalues and also conclude about spectral scales at the borders of the support of σ'_w (see [18]).

It should be noted that relations (1.2) are obtained in [11] with the help of the orthogonal polynomials technique (see [10] and [15] for a simpler derivation). There are several more random matrix ensembles (see [15] for references) whose moments verify recurrence relations of the type (1.2). But relations of the type (1.2) are rather exceptional than typical. Even in the case of GOE, it is not known whether relations of the type (1.2) exist. As a result, no simple derivation of the estimates of the form (1.8) for GOE has been reported.

We develop one more approach to prove non-asymptotic estimates of the type (1.8). Instead of relations (1.2), we use the system of recurrence relations (1.5) that is of more general character than (1.2). Considering various random matrix ensembles, one can observe that the limiting moments verify either (1.5) itself or one or another system of recurrence relations generalizing (1.5) (see for instance, Section 5 of [3], where the first elements of the present approach were presented).

We derive a system of recurrence relations for the moments $M_{2k}^{(N)}$ that have (1.5) as their limiting form. These relations for $M_{2k}^{(N)}$ involve corresponding covariance terms. Using the generating functions technique, we find the form of the estimates and use the triangle scheme of recurrence estimates to prove the bounds for moments and covariance terms. The final result can be written as

$$M_{2k}^{(N)} \le \left(1 + \alpha \frac{k^3}{N^2}\right) m_k \qquad (1.9)$$

with some $\alpha > 1/12$. The estimates obtained are valid in the domain $k^3 \le \chi N^2$ for some constant χ, i.e. not for all values of k and N, contrary to (1.8). But in this region our estimates are more precise than those of (1.8). If $k^3 \ll N^2$, our estimates provide exact expressions for $1/N$-corrections for the moments $M_{2k}^{(N)}$.

1.3 Band random matrices and the semi-circle law

The Hermitian band random matrices $H^{(N,b)}$ can be obtained from GUE matrices by erasing all elements outside the band of width b along the principal diagonal and by renormalizing the matrix obtained by the factor $b^{-1/2}$. It appears that the limiting values of the moments

$$M_{2k}^{(N,b)} = \mathbf{E}\left\{ \frac{1}{N} \operatorname{Tr}\left(H^{(N,b)}\right)^{2k} \right\}$$

crucially depend of the ratio between b and N when $N \to \infty$ (see [5, 13, 17]).

If $b/N \to 1$ as $N \to \infty$, then $M_{2k}^{(N,b)} \to m_k$ and the semicircle law is valid in this case. If $b/N \to c$ and $0 < c < 1$, then the limiting values of $M_{2k}^{(N,b)}$ differ from m_k. Finally, if $1 \ll b \ll N$, then the semicircle law is valid again.

The last asymptotic regime of (relatively) narrow band width attracts a special interest from researchers. In this case the spectral properties of band random matrices exhibit a transition from one type to another. The first one is characterized by GUE matrices and the second is given by spectral properties of Jacobi random matrices, i.e., the discrete analog of the random Schrödinger operator with $b = 3$ (see [6,8] for the results and references). It is shown that the value $b' = \sqrt{N}$ is critical with respect to this transition [6,8,12].

In the present paper we derive estimates for $M_{2k}^{(N,b)}$ that have the same form as the estimates for GUE with N replaced by b. This can be viewed as an evidence to the fact that the asymptotic behavior of the eigenvalues of $H^{(N,b)}$ at the border of the semi-circle density is similar to that of matrices of the size $b \times b$. The estimates we obtain show that the value $b' = \sqrt{N}$ does not play any particular role with respect to the asymptotic behavior of the spectral norm $\|H^{(N,b)}\|$. We show that if $b \gg (\log N)^{3/2}$, then the spectral norm converges with probability 1 when $N \to \infty$ to the edge of the corresponding semicircle density. To our knowledge, this is the first result on the upper bound of the spectral norm of band random matrices.

2 Gaussian Hermitian Ensembles

Consider the family of complex random variables

$$
h_{xy} = \begin{cases} V_{xy} + iW_{xy}, & \text{if } x \leq y, \\ V_{yx} - iW_{yx}, & \text{if } x > y, \end{cases} \tag{2.1}
$$

where $\{V_{xy}, W_{xy}, 1 \leq x \leq y \leq N\}$ are real jointly independent random variables that have normal (Gaussian) distribution with the properties

$$
\mathbf{E}V_{xy} = \mathbf{E}W_{xy} = 0, \tag{2.2a}
$$

and

$$
\mathbf{E}V_{xy}^2 = (1 + \delta_{xy})\frac{1+\eta}{8}, \quad \mathbf{E}W_{xy}^2 = (1 - \delta_{xy})\frac{1-\eta}{8}, \tag{2.2b}
$$

where δ_{xy} is the Kronecker δ-symbol and $\eta \in [-1, 1]$. Then we obtain the family of Gaussian ensembles of $N \times N$ Hermitian random matrices of the form

$$
(H_N^{(\eta)})_{xy} = \frac{1}{\sqrt{N}} h_{xy}, \quad x, y = 1, \ldots, N \tag{2.3}
$$

that generalizes the Gaussian Unitary Ensemble (1.1). Indeed, it is easy to see that $\{H_N^{(0)}\}$ coincides with the GUE, while $\{H^{(1)}\}$ and $\{H^{(-1)}\}$ reproduce the GOE and Hermitian skew-symmetric Gaussian matrices. In [16], the last ensemble is referred to as the Hermitian anti-symmetric one; below we follow this terminology. The present Section is devoted to the results for GUE and to their proofs. Two other ensembles will be considered in Section 3.

2.1 Main results for GUE and the scheme of the proof

Let us consider the moments $M_{2k}^{(N)}$ of GUE matrices. We prove a slightly more precise estimate than (1.9).

Theorem 2.1
Given any constant $\alpha > 1/12$, there exists $\chi > 0$ such that the estimate

$$M_{2k}^{(N)} \leq \left(1 + \alpha \frac{k(k^2 - 1)}{N^2}\right) m_k \qquad (2.4)$$

holds for all values of k, N under condition that $k^3/N^2 \leq \chi$.

Remark. Using relation (1.2), one can prove (2.4) under condition that

$$\alpha > \frac{1}{12 - \chi}. \qquad (2.5)$$

This relation shows that Theorem 2.1 gives the correct lower bound for α. In our proof we get relations between χ and α more complicated than (2.5), but they are of the same character as (2.5). It follows from (2.5) that the closer α is to $1/12$, the smaller χ has to be chosen and vice versa. Indeed, the following proposition shows that the estimate (2.4) is asymptotically exact.

Theorem 2.2
Given k fixed, the following asymptotic expansion holds:

$$M_{2k}^{(N)} = m_k + \frac{1}{N^2} m_k^{(2)} + O(N^{-4}), \qquad \text{as } N \to \infty, \qquad (2.6a)$$

where

$$m_k^{(2)} = \frac{k(k-1)(k+1)}{12} m_k, \qquad k \geq 1. \qquad (2.6b)$$

If $k \to \infty$ and $\tilde{\chi} = k^3/N^2 \to 0$, then relation (2.6a) remains true with $O(N^{-4})$ replaced by $o(\tilde{\chi})$.

Remark. It follows from (1.2) that the sequence $\{m_k^{(2)}, k \geq 1\}$ is determined by the recurrence relation

$$m_k^{(2)} = \frac{2k-1}{2k+2} \cdot m_{k-1}^{(2)} + \frac{k(k-1)}{4} \cdot m_k, \qquad k = 1, 2, \ldots$$

with the obvious initial condition $m_0^{(2)} = 0$. It is easy to check that (2.6b) is in complete agreement with this recurrence relation for $m_k^{(2)}$.

Let us explain the role of recurrence relations (1.5) in the proof of Theorem 2.1. To do this, let us consider the normalized trace $L_a = \frac{1}{N} \operatorname{Tr} H^a$

$$\mathbf{E}\{L_a\} = \frac{1}{N} \sum_{x,s=1}^{N} \mathbf{E}\{H_{xs} H_{sx}^{a-1}\}$$

and compute the latter mathematical expectation. Here and below we omit subscripts and superscripts N when no confusion can arise. Applying the integration by parts formula (see Section 5 for details), we obtain the equality

$$\mathbf{E}\{L_a\} = \frac{1}{4}\sum_{j=0}^{a-2}\mathbf{E}\{L_{a-2-j}L_j\}. \tag{2.7}$$

Introducing the centered random variables $L_j^o = L_j - \mathbf{E}L_j$, we can write that

$$\mathbf{E}\{L_{a_1}L_{a_2}\} = \mathbf{E}\{L_{a_1}\}\mathbf{E}\{L_{a_2}\} + \mathbf{E}\{L_{a_1}^o L_{a_2}^o\}.$$

Taking into account that $\mathbf{E}L_{2k+1} = 0$, we deduce from (2.7) the relation

$$M_{2k}^{(N)} = \frac{1}{4}\sum_{j=0}^{k-1}M_{2k-2-2j}^{(N)}M_{2j}^{(N)} + \frac{1}{4}\,D_{2k-2}^{(2;N)}, \tag{2.8}$$

where we denoted

$$D_{2k-2}^{(2;N)} = \sum_{a_1+a_2=2k-2}\mathbf{E}\{L_{a_1}^o L_{a_2}^o\}.$$

Obviously, the last summation runs over $a_i > 0$. Comparing (2.8) with (1.5), we see that the problem is to estimate the covariance terms $D^{(2)}$. Here and below we omit superscripts N when no confusion can arise.

In what follows, we prove that under the conditions of Theorem 2.1,

$$|D_{2k}^{(2;N)}| \le \frac{ck}{N^2}, \tag{2.9}$$

with some constant c. Inequality (2.9) represents the main technical result of this paper. It is proved in the next subsection. With (2.9) at hand, we can use relation (2.8) to show that (2.4) holds.

Now let us explain the use of the generating function $f(\tau)$ (1.6). Regarding the right-hand side of (2.4), one can observe that the third derivative of $f(\tau)$ could be useful in computations because of the equality

$$[f'''(\tau)]_k = (k+3)(k+2)(k+1)m_{k+3}\,.$$

Indeed, more accurate computations (see identity (5.12) of Section 5) show that the function

$$f(\tau) + \frac{A}{N^2}\frac{\tau^2}{(1-\tau)^{5/2}} = \varPhi_N(\tau)\quad\text{with}\quad A = \frac{3a}{4} \tag{2.10a}$$

is a very good candidate to generate the estimating expressions. This is not by a mere coincidence or an artificial choice. Later we will see that the form of $\varPhi_N(\tau)$ is in certain sense optimal. It is dictated by the iteration scheme we use to get $1/N$-corrections for the moments and covariance terms (see subsection 2.5, proof of Theorem 2.2).

Let us now show how (2.9) implies the estimate

$$M_{2k}^{(N)} \le [\Phi_N(\tau)]_k. \tag{2.10b}$$

Assuming that this estimate and (2.9) are valid for all the terms of the right-hand side of (2.8), we can estimate it with the help of the inequalities

$$\frac{1}{4} \sum_{j=0}^{k-1} M_{2k-2-2j} M_{2j} + \frac{1}{4}|D_{2k-2}^{(2)}| \le \frac{1}{4} \left[\Phi_N^2(\tau)\right]_{k-1} + \frac{c}{4N^2} \left[\frac{1}{(1-\tau)^2}\right]_{k-2}.$$

Denoting by $\Theta(k; N)$ the terms of order $O(N^{-4})$, we can write

$$\left[\frac{\tau}{4}\Phi_N^2(\tau)\right]_k = \left[\frac{\tau f^2(\tau)}{4} + \frac{\tau^3 f(\tau)}{2}\frac{A}{N^2(1-\tau)^{5/2}}\right]_k + \Theta(k; N).$$

Rewriting (1.6) and the quadratic equation for $f(\tau)$ in convenient forms

$$\frac{\tau f^2(\tau)}{4} = f(\tau) - 1 \quad \text{and} \quad \frac{\tau f(\tau)}{2} = 1 - \sqrt{1-\tau}, \tag{2.11}$$

we transform the expression in brackets:

$$\left[f(\tau) + \frac{A}{N^2}\frac{\tau^2}{(1-\tau)^{5/2}} - \frac{A}{N^2}\frac{\tau^2}{(1-\tau)^2}\right]_k = [\Phi_N(\tau)]_k - \frac{A}{N^2}\left[\frac{\tau^2}{(1-\tau)^2}\right]_k.$$

Remembering that $[\Phi_N(\tau)]_k$ reproduces the expression to estimate $M_{2k}^{(N)}$, we conclude that (2.10) is valid provided

$$\frac{A}{N^2}\left[\frac{\tau^2}{(1-\tau)^2}\right]_k \ge \frac{c}{4N^2}\left[\frac{\tau^2}{(1-\tau)^2}\right]_k = \frac{c(k-1)}{4N^2}. \tag{2.12}$$

This requires the inequality $A \ge c/4$.

Our final comment is related to the role of the terms $\Theta(k; N)$. They are of the form

$$\Theta(k; N) = \frac{A^2}{4N^4}\left[\frac{\tau^5}{(1-\tau)^5}\right]_k \le \frac{A^2 k^4}{N^4}.$$

If one wants these terms not to violate inequality (2.12) involving terms of the form k/N^2, one has to set the ratio $k^3/N^2 = \tilde{\chi}$ sufficiently small. This explains the last condition of Theorem 2.1.

It should be noted that the same comments concern the proof of the estimate of covariance terms (2.9), where the recurrence relations, generating functions and terms of the type $\tilde{\chi}$ appear. In the proofs, we constantly use relations (2.11).

2.2 Main technical result

In this subsection we prove estimates of the covariance terms of the type $D_{2k}^{(2)} = \sum \mathbf{E}\{L_{a_1}^o L_{a_2}^o\}$. The main idea is that these terms are determined by a system of recurrence relations similar to (2.8). These relations involve terms of more complicated structure than $D^{(2)}$. The variables we study are defined as

$$D_{2k}^{(q)} = \sum_{a_1+\cdots+a_q=2k} D_{a_1,\ldots,a_q}^{(q)} = \sum_{a_1+\cdots+a_q=2k} \mathbf{E}\left\{L_{a_1}^o L_{a_2}^o \cdots L_{a_q}^o\right\}, \quad q \geq 2.$$

Here and everywhere below, we assume that the summation runs over all positive integers $a_i > 0$.

Our main technical result is given by the following statement.

Proposition 2.1.
Given $A > 1/16$, there exists $\chi > 0$ such that estimate (2.10) holds for all values of $1 \leq k \leq k_0$, where k_0 verifies the condition

$$\frac{k_0^3}{N^2} \leq \chi . \tag{2.13}$$

Also there exists C

$$\frac{1}{4!} < C < \max\{\frac{2A}{3}, 4!\} \tag{2.14}$$

such that the inequalities

$$|D_{2k}^{(2s)}| \leq C \frac{(3s)!}{N^{2s}} \left[\frac{\tau}{(1-\tau)^{2s}}\right]_k, \tag{2.15a}$$

and

$$|D_{2k}^{(2s+1)}| \leq C \frac{(3s+3)!}{N^{2s+2}} \left[\frac{\tau}{(1-\tau)^{2s+5/2}}\right]_k, \tag{2.15b}$$

are true for all k, s such that

$$2k + q \leq 2k_0 \tag{2.16}$$

with $q = 2s$ and $q = 2s + 1$, respectively.

Remark. The form of estimates (2.15) is dictated by the structure of the recurrence relations we derive below. The bounds for the constants A and C and the form of the factorial terms of (2.15) are explained in subsection 2.4.

We prove Proposition 2.1 in the next subsection using recurrence relations for $D^{(q)}$ that we derive now. Let us use the identity $\mathbf{E}\{X^o Y^o\} = \mathbf{E}\{XY^o\}$ for centered random variables and consider the equality

$$\mathbf{E}\left\{L_{a_1}^o L_{a_2}^o \cdots L_{a_q}^o\right\} = \mathbf{E}\left\{L_{a_1}[L_{a_2}^o \cdots L_{a_q}^o]^o\right\}$$

$$= \frac{1}{N}\sum_{x,s=1}^{N}\mathbf{E}\left\{H_{xs}(H_{sx}^{a_1-1}[L_{a_2}^o \cdots L_{a_q}^o]^o\right\} \quad (2.17)$$

We apply to the last expression the integration by parts formula (5.1) and obtain the equality

$$D_{a_1,\dots,a_q}^{(q)} = \frac{1}{4}\sum_{j=0}^{a_1-2}\mathbf{E}\left\{L_{a_1-2-j}L_j[L_{a_2}^o \cdots L_{a_q}^o]^o\right\}$$

$$+ \frac{1}{4N^2}\sum_{i=2}^{q}\mathbf{E}\left\{L_{a_2}^o \cdots L_{a_{i-1}}^o \, a_i \, L_{a_i+a_1-2}L_{a_{i+1}}^o \cdots L_{a_q}^o\right\},\ (2.18)$$

with the help of formulas (5.7a) and (5.7b), respectively. The detailed derivation of (2.18) is presented in subsection 5.2.

Consider the first term from the right-hand side of (2.18). We can rewrite it in terms of variables D with the help of the following identity

$$\mathbf{E}\{L_1 L_2 Q^o\} = \mathbf{E}\{L_1\}\mathbf{E}\{L_2^o Q\}+\mathbf{E}\{L_2\}\mathbf{E}\{L_1^o Q\}+\mathbf{E}\{L_1^o L_2^o Q\}-\mathbf{E}\{L_1^o L_2^o\}\mathbf{E}\{Q\},$$

where $Q = L_{a_2}^o \cdots L_{a_q}^o$. For the last term of (2.18), we use (2.17) and obtain the relation

$$D_{a_1,\dots,a_q}^{(q)} = \frac{1}{4}\sum_{j=0}^{a_1-2}M_j D_{a_1-2-j,a_2,\dots,a_q}^{(q)} + \frac{1}{4}\sum_{j=0}^{a_1-2}M_{a_1-2-j}D_{j,a_2,\dots,a_q}^{(q)}$$

$$+ \frac{1}{4}\sum_{j=0}^{a_1-2}D_{j,a_1-2-j,a_2,\dots,a_q}^{(q+1)} - \frac{1}{4}\sum_{j=0}^{a_1-2}D_{j,a_1-2-j}^{(2)}D_{a_2,\dots,a_q}^{(q-1)}$$

$$+ \frac{1}{4N^2}\sum_{i=2}^{q}a_i M_{a_1+a_i-2}D_{a_2,\dots,a_{i-1},a_{i+1},\dots,a_q}^{(q-2)}$$

$$+ \frac{1}{4N^2}\sum_{i=2}^{q}a_i D_{a_2,\dots,a_{i-1},a_i+a_1-2,a_{i+1},\dots,a_q}^{(q-1)}. \quad (2.19)$$

Taking into account that $M_{2k+1}^{(N)} = 0$, it is easy to deduce from (2.19) by induction on k that

$$D_{a_1,\dots,a_q}^{(q)} = 0 \quad \text{whenever} \quad a_1 + \cdots + a_q = 2k + 1.$$

Introduce the variables

$$\bar{D}_{2k}^{(q)} = \sum_{a_1+\cdots+a_q=2k}\left|D_{a_1,\dots,a_q}^{(q)}\right|.$$

Using the positivity of M_{2j}, we derive from (2.19) the second main relation

$$\bar{D}_{2k}^{(q)} \leq \frac{1}{2} \sum_{j=0}^{k-1} \bar{D}_{2k-2-2j}^{(q)} M_{2j} + \frac{q-1}{4N^2} \sum_{j=0}^{k-1} \bar{D}_{2k-2-2j}^{(q-2)} \cdot \frac{(2j+2)(2j+1)}{2} \cdot M_{2j}$$

$$+\frac{1}{4}\bar{D}_{2k-2}^{(q+1)} + \frac{1}{4}\sum_{j=0}^{k-1} \bar{D}_{2k-2-2j}^{(q-1)} \bar{D}_{2j}^{(2)} + \frac{2k(2k-1)}{2} \cdot \frac{(q-1)}{4N^2} \cdot \bar{D}_{2k-2}^{(q-1)}, \quad (2.20)$$

where $1 \leq k, 2 \leq q \leq 2k$. When dealing with the last two terms of (2.19), we have used the obvious equality

$$\sum_{a_1+a_2=a'} a_2 \, F_{a_1+a_2-2} = \left(\sum_{a_2=1}^{a'-1} a_2 \right) F_{a'-2} = \frac{a'(a'-1)}{2} F_{a'-2}.$$

Using this relation with F replaced by M and $a' = 2j+2$, we obtain that

$$\sum_{a_1+\cdots+a_q=2k} a_i M_{a_1+a_i-2} |D_{a_1,\dots,a_{i-1},a_{i+1},\dots,a_q}^{(q)}|$$

$$= \sum_{j=0}^{k-1} \bar{D}_{2k-2-2j}^{(q-2)} \frac{(2j+2)(2j+1)}{2} \cdot M_{2j}.$$

Also we can write that

$$\sum_{a_1+\cdots+a_q=2k} a_i |D_{a_2,\dots,a_{i-1},a_i+a_1-2,a_{i+1},\dots,a_q}^{(q-1)}|$$

$$= \sum_{b_2+\cdots+b_q=2k-2} |D_{b_2,\dots,b_q}^{(q-1)}| \times \sum_{1\leq a_1\leq b_i+1} (b_i - a_1 + 2)$$

$$\leq \sum_{a_1+\cdots+a_{q-1}=2k-2} |D_{a_1,\dots,a_{q-1}}^{(q-1)}| \times \frac{2k(2k-1)}{2}$$

and get the last term of (2.20).

The upper bounds of sums in (2.20) are written under the convention that $\bar{D}_{2k}^{(q)} = 0$ whenever $q > 2k$. Also we note that the form of inequalities (2.20) is slightly different when we consider particular values of q and k. Indeed, some terms are missing when the left-hand side is $\bar{D}_{2k}^{(2)}$, $\bar{D}_{2k}^{(3)}$, $\bar{D}_{2k}^{(2k)}$, $\bar{D}_{2k}^{(2k-1)}$, and $\bar{D}_{2k}^{(2k-2)}$. However, the convention that $\bar{D}_{2k}^{(q)} = 0$ whenever $q > 2k$ and that $\bar{D}_{2k}^{(1)} = 0$ and $\bar{D}_{2k}^{(0)} = \delta_{k,0}$ makes (2.20) valid in these cases.

Obviously, we have that

$$M_{2k} \leq \frac{1}{4} \sum_{j=0}^{k-1} M_{2k-2-2j} M_{2j} + \frac{1}{4}\bar{D}_{2k-2}^{(2)}. \quad (2.21)$$

2.3 Recurrent relations and estimates

To estimate M and $\bar{D}^{(q)}$, we introduce auxiliary numbers $B_k^{(N)} \geq 0$ and $R_{2k}^{(q;N)} \geq 0$ determined by a system of two recurrence relations induced by (2.20) and (2.21). This system is given by the following equalities (we omit superscripts N)

$$B_k = \frac{1}{4}(B * B)_{k-1} + \frac{1}{4}R_{k-1}^{(2)}, \qquad (2.22)$$

and

$$R_k^{(q)} = \frac{1}{2}\left(R^{(q)} * B\right)_{k-1} + \frac{q-1}{4N^2}\left(R^{(q-2)} * B''\right)_{k-1} + \frac{1}{4}R_{k-1}^{(q+1)}$$

$$+ \frac{1}{4}\left(R^{(q-1)} * R^{(2)}\right)_{k-1} + \frac{k^2 q}{2N^2}R_{k-1}^{(q-1)}, \qquad (2.23)$$

considered in the domain

$$\Delta = \{(k, q): \ k \geq 1, \ 2 \leq q \leq 2k\}$$

with denotation

$$B_k'' = \frac{(2k+2)(2k+1)}{2}B_k$$

and the convolutions as follows

$$(B * B)_{k-1} = \sum_{j=0}^{k-1} B_{k-1-j}B_j.$$

The initial values for (2.22)-(2.23) coincide with those of M and D:

$$B_0^{(N)} = 1, \qquad R_1^{(2;N)} = \frac{1}{4N^2}.$$

Let us note that one can consider relations (2.22) and (2.23) for all integers k and q with the obvious convention that outside Δ the values of R are zero except the origin $R_0^{(0;N)} = 1$. The system (2.22)-(2.23) plays a fundamental role in our method for proving Proposition 2.1. This proof is composed of the following three statements.

Lemma 2.1.
Given a fixed N, the family of numbers $\{B_k, R_k^{(q)}, \ (k, q) \in \Delta\}$ exist; it is uniquely determined by the system of relations (2.22)-(2.23).

Lemma 2.2.
The inequalities

$$M_{2k}^{(N)} \leq B_k^{(N)} \quad and \quad \bar{D}_{2k}^{(q;N)} \leq R_k^{(q;N)} \qquad (2.24)$$

hold for all N and $(k, q) \in \Delta$.

Lemma 2.3.

Under the conditions of Proposition 2.1, the numbers B_k and $R_k^{(q)}$ are estimated by the right-hand sides of inequalities (2.10) and (2.15), respectively; that is:

$$B_k^{(N)} \leq \left[f(\tau) + AN^{-2}\tau^2(1-\tau)^{-5/2}\right]_k \equiv [\Phi_N(\tau)]_k \qquad (2.25)$$

and

$$R_k^{(q;N)} \leq \begin{cases} C(3s)! N^{-2s}\left[\tau(1-\tau)^{-2s}\right]_k, & \text{if } q = 2s; \\ C(3s+3)! N^{-2s-2}\left[\tau(1-\tau)^{-(4s+5)/2}\right]_k, & \text{if } q = 2s+1. \end{cases} \qquad (2.26)$$

Lemma 2.3 represents the main technical result concerning the system (2.22)-(2.23). Lemmas 2.1 and 2.2 look like a simple consequence of the recurrence procedure applied to relations (2.22)-(2.23) and (2.20)-(2.21), respectively. However, the form of the recurrence relations (2.22)-(2.23) is not usual because the relations for B involve the values of R and vice-versa. The ordinary scheme of recurrence has to be modified. This modification is described in the next subsection. Lemma 2.3 is also proved on the basis of this modified scheme of recurrence.

The triangular scheme of recurrence estimates

Let us show on the example of Lemma 3 that the ordinary scheme of recurrence estimates can be applied to the system (2.20)-(2.23). By the ordinary scheme we mean the following reasoning. Assume that the estimates we need are valid for the terms entering the right-hand side of the inequalities derived. Apply these estimates to all terms there and show that the sum of the expressions obtained is smaller than what we assume for the terms of the left-hand side; check the estimates of the initial terms. Then all estimates we need are true. Considering the plane of integers (k, q), assume that estimates (2.26) are valid for all variables R with (k, q) lying inside of the triangle domain $\Delta(m), m \geq 3$

$$\Delta(m) = \{(k, q) : 1 \leq k, 2 \leq q \leq 2k, k + q \leq m\}$$

and that estimates (2.25) are valid for all variables B_l with $1 \leq l \leq m - 2$.

Then we proceed to complete the next line $k + q = m + 1$ step by step starting from the top point $T(m + 1)$ of $\Delta(m + 1)$ and ending at the bottom point $(m - 1, 2)$ of this side line. This means that on each step, we assume estimates (2.25) and (2.26) to be valid for all terms entering the right-hand sides of relations (2.23) and show that the same estimate is valid for the term standing on the left hand side of (2.23).

Once the bottom point $(m - 1, 2)$ achieved, we turn to relation (2.22) and prove that estimate (2.25) is valid for B_m. Again, this is done by assuming that all terms entering the right-hand side of (2.22) verify estimates (2.25) and (2.26) with $q = 2$, and showing that the expression obtained is bounded by the

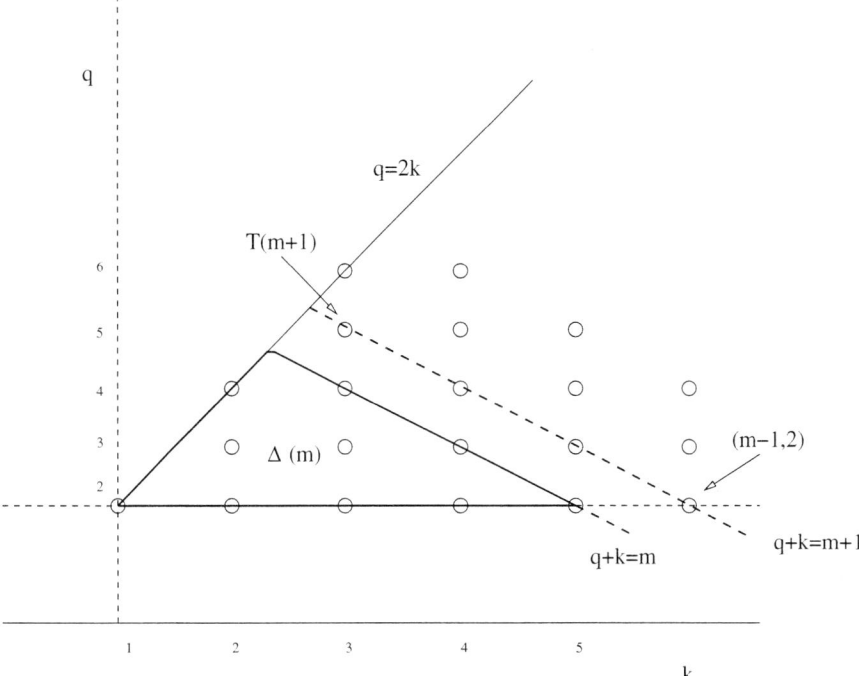

Fig. 1. The triangle domain $\Delta(m)$ with $m = 7$ and the long dotted line $k+q = m+1$

right-hand side of (2.25). This completes the triangular scheme of recurrence estimates.

It is easy to see that the reasoning described above proves, with obvious changes, Lemmas 2.1 and 2.2.

Estimates for B

Assuming that the terms standing in the right-hand side of (2.22) are estimates (2.25) and (2.26) with $s = 1$, we can write the inequality

$$\frac{1}{4}(B * B)_{k-1} + \frac{1}{4}R_{k-1}^{(2)}$$

$$\leq \left[\frac{\tau f^2(\tau)}{4} + \frac{A}{N^2}\frac{\tau^3 f(\tau)}{2(1-\tau)^{5/2}} + \frac{A^2}{4N^4}\frac{\tau^5}{(1-\tau)^5}\right]_k + \frac{3C}{2N^2}\left[\frac{\tau^2}{(1-\tau)^2}\right]_k. \quad (2.27)$$

Taking into account relations (2.11), we transform the first bracket of (2.27) into the expression

$$\left[f(\tau) + \frac{A}{N^2}\frac{\tau^2}{(1-\tau)^{5/2}}\right]_k - \frac{A}{N^2}\left[\frac{\tau^2}{(1-\tau)^2}\right]_k + \frac{A^2}{4N^4}\left[\frac{\tau^5}{(1-\tau)^5}\right]_k.$$

Here, the first term reproduces the expression $[\Phi_N(\tau)]_k$; the second term is negative and this allows us to show that the estimate wanted is true. Then we see that the estimate $B_k \leq [\Phi_N(\tau)]_k$ is true whenever the inequality

$$A\left[\frac{\tau^2}{(1-\tau)^2}\right]_k \geq \frac{3C}{2}\left[\frac{\tau^2}{(1-\tau)^2}\right]_k + \frac{A^2}{4N^2}\left[\frac{\tau^5}{(1-\tau)^5}\right]_k \qquad (2.28)$$

holds. This is equivalent to the condition

$$A \geq \frac{3C}{2} + \frac{A^2}{4N^2}\frac{(k-4)(k-3)(k-2)}{4!}.$$

Remembering that $k^3 \leq \chi N^2$, we see that the estimate (2.25) of B_k is true provided

$$A \geq \frac{3C}{2} + \frac{A^2\chi}{96}. \qquad (2.29)$$

Estimates for $R^{(2s)}$

Let us rewrite (2.8) with $q = 2s, s \geq 2,\ k \geq 1$ in the form

$$R_k^{(2s)} = \frac{1}{2}\left(R^{(2s)} * B\right)_{k-1} + \frac{2s-1}{4N^2}\left(R^{(2s-2)} * B''\right)_{k-1} + X + Y + Z, \quad (2.30)$$

where we denote

$$X = \frac{1}{4}R_{k-1}^{(2s+1)}, \quad Y = \frac{1}{4}\left(R^{(2s-1)} * R^{(2)}\right)_{k-1}, \quad Z = \frac{k^2 s}{N^2}R_{k-1}^{(2s-1)}. \quad (2.31)$$

The first term in the right-hand side of (2.30) admits the following estimate

$$\frac{1}{2}\left(R^{(2s)} * B\right)_{k-1} \leq \frac{C(3s)!}{N^{2s}}\left[\frac{\tau^2 f(\tau)}{2(1-\tau)^{2s}} + \frac{A}{2N^2}\frac{\tau^4}{(1-\tau)^{2s+5/2}}\right]_k.$$

Using (2.11), we transform the last expression to the form

$$\frac{C(3s)!}{N^{2s}}\left[\frac{\tau}{(1-\tau)^{2s}} - \frac{\tau}{(1-\tau)^{2s-1/2}} + \frac{A}{2N^2}\frac{\tau^4}{(1-\tau)^{2s+5/2}}\right]_k. \qquad (2.32)$$

The first term reproduces the expression needed in order to estimate $R_k^{(2s)}$.

Let us consider the second terms of the right-hand side of (2.30). Assuming (2.25) and using the identities of subsection 5.1, it is not hard to show that

$$B_k'' \leq \left[\frac{1}{(1-\tau)^{3/2}} + \frac{18A}{N^2}\frac{1}{(1-\tau)^{9/2}}\right]_k. \qquad (2.33)$$

Indeed, it follows from (5.11) that

$$\frac{(2k+2)(2k+1)}{2}[f(\tau)]_k = \frac{(2k+2)(2k+1)}{2}m_k = \left[\frac{1}{(1-\tau)^{3/2}}\right]_k .$$

Next, identity (5.12) implies the relation

$$\frac{(2k+2)(2k+1)}{2}\left[\frac{\tau^2}{(1-\tau)^{5/2}}\right]_k = \frac{(2k+2)(2k+1)}{2} \cdot \frac{2k(2k-1)(2k+1)}{3!}m_k.$$

Now, regarding (5.9) with $r = 4$, it is easy to see that

$$\left[\frac{1}{(1-\tau)^{9/2}}\right]_k = \frac{(2k+1)(2k+2)(2k+3)(2k+5)(2k+7)}{5\cdot 6\cdot 7}m_k.$$

Then (2.33) follows.

Returning to the right-hand side of (2.30), thanks to (2.33) we can write the inequality

$$\left(R^{(2s-2)} * B''\right)_{k-1} \leq \frac{C(3s-3)!}{4N^{2s-2}}\left[\frac{\tau}{(1-\tau)^{2s-1/2}} + \frac{18A}{N^2}\frac{\tau}{(1-\tau)^{2s+5/2}}\right]_k .$$
(2.34)

Here and below we use the relation $[\tau^j g(\tau)]_k \leq [g(\tau)]_k$, which is valid for the generating functions under consideration. Let us stress that (2.34) remains valid in the case when $s = 1$ with C replaced by 1.

Let us turn to (2.31). We estimate the sum of X and Y by

$$X + Y \leq \frac{C(1+C)(3s+3)!}{4N^{2s+2}}\left[\frac{\tau}{(1-\tau)^{2s+5/2}}\right]_k . \qquad (2.35)$$

For the last term of (2.31) we can write the inequality

$$Z \leq \frac{Ck^2(3s+1)!}{N^{2s+2}}\left[\frac{\tau}{(1-\tau)^{2s+1/2}}\right]_k . \qquad (2.36)$$

Comparing the second term of (2.32) with the sum of the last term of (2.32) and the right-hand sides of (2.34), (2.35) and (2.36), we arrive at the following inequality to hold

$$C \geq \frac{(2s-1)(3s-3)!}{(3s)!} \cdot \frac{\delta_{s,1} + C(1-\delta_{s,1})}{4} + \frac{k^2(3s+1)}{3N^2} \cdot \frac{\left[\tau(1-\tau)^{-2s-1/2}\right]_k}{\left[\tau(1-\tau)^{-2s+1/2}\right]_k}$$

$$+ C\frac{(1+C)(3s+3)!+18A(3s-2)!+2A(3s)!}{4N^2(3s)!} \cdot \frac{\left[\tau(1-\tau)^{-2s-5/2}\right]_k}{\left[\tau(1-\tau)^{-2s+1/2}\right]_k} . \qquad (2.37)$$

Using identity (5.10), we see that

$$\frac{\left[\tau(1-\tau)^{-2s-1/2}\right]_k}{\left[\tau(1-\tau)^{-2s+1/2}\right]_k} = \frac{2k+4s-2}{4s-1} \leq \frac{4k_0}{4s-1}.$$

Similarly

$$\frac{\left[\tau(1-\tau)^{-2s-5/2}\right]_k}{\left[\tau(1-\tau)^{-2s+1/2}\right]_k} \leq \frac{(4k_0)^3}{(4s-1)(4s+1)(4s+3)}.$$

Inserting these inequalities into (2.37), maximizing the expressions obtained with respect to s, and using (2.13), we get the following sufficient condition

$$C \geq \frac{\delta_{s,1}+C(1-\delta_{s,1})}{24} + 2\chi\left(1+10C(1+C)+2AC\right). \tag{2.38}$$

Estimates for $R^{(2s+1)}$

Let us turn to the case $q=2s+1$ and rewrite (2.8) in the form

$$R_k^{(2s+1)} = \frac{1}{2}\left(R^{(2s+1)} * B\right)_{k-1} + \frac{s}{2N^2}\left(R^{(2s-1)} * B''\right)_{k-1} + X_1 + Y_1 + Z_1, \tag{2.39}$$

where

$$X_1 = \frac{1}{4}R_{k-1}^{(2s+2)}, \quad Y_1 = \frac{1}{4}\left(R^{(2s-1)} * R^{(2)}\right)_{k-1}, \quad Z_1 = \frac{k^2 s}{N^2}R_{k-1}^{(2s)}. \tag{2.40}$$

Regarding the first term of (2.39), we can write the inequality

$$\frac{1}{2}\left(R^{(2s+1)} * B\right)_k \leq \frac{C(3s+3)!}{N^{2s+2}}\left[\frac{\tau^2 f(\tau)}{2(1-\tau)^{2s+5/2}} + \frac{A\tau}{2N^2(1-\tau)^{2s+5}}\right]_k$$

$$= \frac{C(3s+3)!}{N^{2s+2}}\left[\frac{\tau}{(1-\tau)^{2s+5/2}} - \frac{\tau}{(1-\tau)^{2s+2}} + \frac{A\tau}{2N^2(1-\tau)^{2s+5}}\right]_k. \tag{2.41}$$

The first term in the right-hand side of (2.41) reproduces the expression needed to estimate $R_k^{(2s+1)}$.

Let us consider the second term of (2.39). It is estimated as follows:

$$\frac{s}{2N^2}\left(R^{(2s-1)} * B''\right)_{k-1} \leq \frac{Cs(3s)!}{2N^{2s+2}}\left[\frac{\tau}{(1-\tau)^{2s+2}}\right]_k + \frac{9ACs(3s)!}{N^{2s+4}}\left[\frac{\tau}{(1-\tau)^{2s+5}}\right]_k.$$

Regarding two first terms of (2.40), we can write that

$$X_1 + Y_1 \leq \frac{C(3s+3)! + 6C^2(3s)!}{4N^{2s+2}}\left[\frac{\tau}{(1-\tau)^{2s+2}}\right]_k,$$

and

$$Z_1 \le \frac{Ck^2 s(3s)!}{N^{2s+2}} \left[\frac{\tau}{(1-\tau)^{2s}} \right]_k .$$

Comparing the negative term of (2.41) with the sum of the last term of (2.41) and the estimates for the terms of (2.40), we obtain the inequality

$$C \left(\frac{3}{4} - \frac{s(3s)!}{2(3s+3)!} \right) \ge \frac{3C^2(3s)!}{2(3s+3)!} + \frac{k^2 s(3s)!}{(3s+3)!} \cdot \frac{[\tau(1-\tau)^{-2s}]_k}{[\tau(1-\tau)^{-2s-2}]_k}$$
$$+ \frac{AC}{2N^2} \cdot \left(1 + \frac{18s(3s)!}{(3s+3)!} \right) \cdot \frac{[\tau(1-\tau)^{-2s-5}]_k}{[\tau(1-\tau)^{-2s-2}]_k}. \quad (2.42)$$

Equality (5.13) implies that

$$\frac{[\tau(1-\tau)^{-2s}]_k}{[\tau(1-\tau)^{-2s-2}]_k} = \frac{2s(2s+1)}{(k-1+2s)(k+2s)}$$

and that

$$\frac{[\tau(1-\tau)^{-2s-5}]_k}{[\tau(1-\tau)^{-2s-2}]_k} \le \frac{8k_0^3}{(2s+2)(2s+3)(2s+4)}.$$

Inserting these two relations into (2.42) and maximizing with respect to s, we obtain, after elementary transformations, the following sufficient condition

$$C \le \frac{4!}{1+4A\chi}. \quad (2.43)$$

2.4 Proof of Theorem 2.1

Let us repeat that inequalities (2.29), (2.38), and (2.43) represent sufficient conditions for the recurrence estimates (2.25) and (2.26) to be true. Let $A > 1/16$. Then for any constant $C < 4!$ verifying condition

$$\frac{1}{24} < C < \frac{2A}{3},$$

there exists some $\chi > 0$ such that (2.38) is true. Indeed, it suffices to take $\chi \le \chi'$, where χ' is such that

$$2\chi' K < \min\{C - \frac{1}{24}, \frac{23}{24}C\},$$

with $K = 1 + 10C(1+C) + 2AC$. Also there exists χ'' such that (cf. (2.29))

$$A \ge 3C/2 + A^2\chi''.$$

The choice of $\chi \le \min\{\chi', \chi''\}$ makes (2.29) and (2.38) true. Condition (2.43) is obviously verified. Thus, conditions (2.13), (2.14), and $A > 1/16$ of Proposition 2.1 are sufficient for (2.29), (2.38), and (2.43) to hold. This completes the proof of Lemma 2.3.

Lemma 2.2 together with Lemma 2.3 implies estimates (2.10) and (2.15). Then Proposition 2.1 follows. The statement of Theorem 1.1 is a simple consequence of the estimate (2.10) and Proposition 2.1.

We complete this subsection with the discussion of the form of the estimates (2.26) and the constants A and C. First let us note that the upper bound 4! for C imposed by (2.14) represents a technical restriction; it can be avoided, for example, by modifying estimates (2.26) for $R^{(2s)}$ and $R^{(2s+1)}$, where C is replaced by C^s and C^{s+1}, respectively. However, in this case the lower bounds $1/16$ for A and $1/24$ for C are to be replaced by $1/6$ and $1/9$, respectively.

The closer A and C are to their optimal values $1/16$ and $1/24$, the smaller χ is to be chosen. The inverse is also correct. Namely, in the next subsection we prove that estimates (2.9) and (2.10) become asymptotically exact in the limit $\chi \to 0$. In this case the factorials $(3s)!$ and $(3s+3)!$ in the right-hand sides of (2.26) can be replaced by other expressions $g(s)$ and $h(s)$ that provide more precise estimates for $R^{(q)}$. Indeed, repeating the computations of subsections 2.3.3 and 2.3.4, one can see that in the limit $\chi \to 0$ the function $g(s)$ can be chosen close to $(2s-1)!!/4^s$. This makes an evidence for the central limit theorem to hold for the centered random variables

$$NL_a^o = \operatorname{Tr} H^a - \mathbf{E}\{\operatorname{Tr} H^a\}.$$

This observation explains also the fact that the odd "moments" of the variable L_a^o decrease faster than the even ones as $N \to \infty$. That is why the estimates for $R^{(2s)}$ have a different form than those of $R^{(2s+1)}$ and are proved separately.

For finite values of χ, the use of some expression proportional to $(3s)!$ is unavoidable.

2.5   Proof of Theorem 2.2

We present the proof of Theorem 2.2 for the case when k is fixed and $N \to \infty$. Regarding relation (2.19) with $q = 2$, we obtain the relation

$$D_{2k}^{(2)} = \frac{1}{2} \left(D^{(2)} * M \right)_{2k-2} + \frac{1}{4N^2} \cdot \frac{2k(2k-1)}{2} M_{2k-2} + \frac{1}{4} D_{2k-2}^{(3)}. \qquad (2.44)$$

Proposition 2.1 implies that $D_{2k}^{(3)} = O(N^{-4})$ and that $M_{2k}^{(N)} - m_k = O(N^{-2})$. Then we easily arrive at the conclusion that

$$D_{2k}^{(2)} = \frac{r_k}{N^2} + O\left(\frac{1}{N^4}\right), \qquad (2.45)$$

where the r_k are determined by the relations $r_0 = 0$ and

$$r_k = \frac{1}{2} (r * m)_{k-1} + \frac{1}{4} \cdot \frac{2k(2k-1)}{2} m_{k-1}, \quad k \geq 1. \qquad (2.46)$$

Passing to the generating functions and using relations (2.11) and (5.11), we obtain the equality

$$r_k = \frac{1}{4} \left[\frac{\tau}{(1-\tau)^2} \right]_k = \frac{k}{4}.$$

Returning to relation (2.8), we conclude that

$$M_{2k}^{(N)} = m_k + \frac{1}{N^2} m_k^{(2)} + O\left(\frac{1}{N^4}\right).$$

Indeed, the difference between $M_{2k}^{(N)}$ and m_k is of order N^{-2} and the next correction is of order N^{-4}. Regarding $m_k^{(2)}$ and using (2.45), we obtain the equality

$$m_k^{(2)} = \frac{1}{2} \left[m^{(2)} * m \right]_{k-1} + \frac{1}{16} \left[\frac{\tau^2}{(1-\tau)^2} \right]_k, \quad k \geq 1, \tag{2.47}$$

and $m_1^{(2)} = 0$. Solving (2.47) with the help of (2.11), we get

$$m_k^{(2)} = \frac{1}{16} \left[\frac{\tau^2}{(1-\tau)^{5/2}} \right]_k. \tag{2.48}$$

It is easy to see that (2.48) implies the relation

$$\frac{1}{16} \left[\frac{\tau^2}{(1-\tau)^{5/2}} \right]_k = \frac{1}{16} \frac{(2k-3)(2k-2)(2k-1)}{3!} m_{k-2}$$

and hence (2.6b). Theorem 2.2 is proved.

2.6 More about asymptotic expansions

The system (2.22)–(2.23) of recurrence relations is the main technical tool in the proof of the Proposition 2.1, where the estimates for B and R are given. However, the crucial question is to find the correct form of these estimates. The first terms of the asymptotic expansions described in previous subsection give a solution of this problem. Indeed, repeating the proof of Theorem 2.2, we see that formulas (2.46) and (2.48) suggest the form of the estimates to be proved. Then the proof of Proposition 2.1 is reduced to elementary computations, where the most important part is related with the correct choice of the factorial terms in inequalities (2.15).

The next observation is that relation (2.23) resembles inequality (2.20) obtained from (2.19) by considering the absolute values of the variables $D_{a_1,\ldots,a_q}^{(q)}$ and by replacing in the right-hand side of (2.19) the minus sign by a plus sign. So, relation (2.23) determines the estimating terms $R^{(q)}$ with a certain error. However, it is not difficult to deduce from estimates (2.25) and (2.26) that if $q = 2s$, then this error is of the order smaller than the order of $R^{(2s)}$. This means that relations (2.23) determine correctly the first terms of the

$1/N$-expansions of all $R^{(2s)}, s \geq 1$ and not only of $R^{(2)}$ as mentioned by Theorem 2.2. The same is true for the $1/N$ expansions of $D_{2k}^{(2s)}$. It is easy to show by using (2.23) and results of Proposition 2.1 that these corrections are given by the formulas

$$D_{2k}^{(2s)} = r_k^{(2s)} + o(k^{2s-1}/N^{2s}),$$

where $r_k^{(2s)}$ are such that the corresponding generating function $\tilde{r}^{(2s)}(\tau) = \sum_{k \geq 0} r_k^{(2s)} \tau^k$ verifies the equation

$$\tilde{r}^{(2s)}(\tau) = \frac{\tau f(\tau)}{2} \tilde{r}^{(2s)}(\tau) + (2s-1)\tilde{r}^{(2s-2)}(\tau) \frac{d^2}{2N^2 d\tau^2}(\tau f(\tau)). \qquad (2.49)$$

Using equalities (2.11) and resolving (2.49), we obtain the expression

$$r_k^{(2s)} = \frac{(2s-1)!!}{(4N^2)^s} \left[\frac{\tau^s}{(1-\tau)^{2s}} \right]_k.$$

The left-hand side of relation (2.23) for $R_k^{(q)}$ involves variables $R_j^{(q)}$, $R_j^{(q-1)}$, and $R^{(q+1)}$. This can lead to the idea of using generating functions with two variables $G(\tau, \mu)$ to describe the family of numbers R. In this connection, the following comment on the structure of the variables $D^{(q)}$ could be useful. Introducing a generating function $F(\tau) = \sum_{j \geq 0} \tau^j L_j$, we see that

$$\sum_{k \geq 1} D_{2k}^{(q)} \tau^{2k} = \mathbf{E}\{[F^o(\tau)]^q\},$$

where $F^o(\tau) = F(\tau) - \mathbf{E}F(\tau)$. Then the function mentioned above can have the form

$$G_D(\tau, \mu) = \sum_{k \geq 1, q \geq 2} D_{2k}^{(q)} \tau^{2k} \frac{\mu^q}{q!} = \mathbf{E}\left\{ e^{\mu F^o(\tau)} \right\} - 1.$$

In particular, concerning such a generating function of $r_k^{(2s)}$, one arrives at the expression

$$G_r(\tau, \mu) = \sum_{k \geq 1, s \geq 1} r_k^{(2s)} \tau^{2k} \frac{\mu^{2s}}{(2s)!} = \exp\left\{ \frac{\mu^2}{4N^2} \frac{\tau}{(1-\tau)^2} \right\}.$$

This expression shows that the central limit theorem can be proved for the random variable $NF^o(\tau)$ in the asymptotic regime $k^3/N^2 \ll 1$ mentioned in Theorem 2.2. This asymptotic regime can be compared with the mesoscopic regime for the resolvent of H_N and the central limit theorem valid there [2].

3 Orthogonal and anti-symmetric ensembles

In this Section we return to Hermitian random matrix ensembles $H^{(\eta)}$ with $\eta = 1$ and $\eta = -1$ introduced in Section 2. Let us consider the moments of $H^{(1)}$. Using the method developed in Section 2, we prove the following statements.

Theorem 3.1 (GOE).
Given $A > 1/2$, there exists χ such that

$$M_{2k}^{(N)} \leq m_k + A\frac{1}{N} \tag{3.1}$$

for all k, N such that $k \leq k_0$ and (2.13) hold. If k is fixed and $N \to \infty$, then

$$M_{2k}^{(N)} = m_k + \frac{1 - (k+1)m_k}{2N} + o(N^{-1}) \tag{3.2}$$

and

$$D_{2k}^{(2;N)} = \sum_{a+b=2k} \mathbf{E}\{L_a^o L_b^o\} = \frac{k}{2N^2} + O(N^{-3}). \tag{3.3}$$

The proof of Theorem 3.1 is obtained by the method described in Section 2. Briefly speaking, we derive recurrence inequalities for $M_{2k}^{(N)}$ and $D_{2k}^{(q)}$, then we introduce related auxiliary numbers B and R determined by a system of recurrence relations. Using the triangular scheme of recurrence estimates we prove the estimates we need. Corresponding computations are somehow different from those of Section 2. We describe this difference below (see Subsection 3.1).

Let us turn to the ensemble $H^{(-1)}$. Regarding the recurrence relations for the moments of these matrices, we will see that $M_{2k}^{(\eta=-1)}$ are bounded by $M_{2k}^{(\eta=1)}$. A slight modification of the computations performed in the proof of Theorem 3.1 yields the following result.

Theorem 3.2 (Gaussian anti-symmetric Hermitian matrices).
Given $A > 1/2$, there exists $\chi > 0$ such that the moments of Gaussian skew-symmetric Hermitian ensemble $H_N^{(-1)}$ admit the estimate

$$M_{2k}^{(N)} \leq m_k + A\frac{1}{N} \tag{3.4}$$

for all values of k, N such that (2.13) holds. Also

$$|D_{2k}^{(2)}| = O\left(\frac{1}{N^2}\right) \quad and \quad |D_{2k}^{(3)}| = O\left(\frac{1}{N^3}\right). \tag{3.5}$$

Given k fixed, the following asymptotic expansions are true for the moments of $H^{(-1)}$

$$M_{2k}^{(N)} = m_k + \frac{\delta_{k,0} - (k+1)m_k}{2N} + o(N^{-1}), \tag{3.6}$$

and for the covariance terms

$$D_{2k}^{(2;N)} = \sum_{a_1+a_2=2k} \mathbf{E}\{L_{a_1}^o L_{a_2}^o\} = \frac{k+1}{4N^2} + O(N^{-3}). \tag{3.7}$$

3.1 Proof of Theorem 3.1

Using the integration by parts formula (5.7) with $\eta = 1$ and repeating computations of the previous section, we obtain recurrence relation for $M_{2k} = \mathbf{E}L_{2k}$;

$$M_{2k} = \frac{1}{4}\sum_{j=0}^{k-1} M_{2k-2-2j}M_{2j} + \frac{2k-1}{4N}M_{2k-2} + \frac{1}{4}\sum_{a_1+a_2=2k-2}\mathbf{E}\left\{L_{a_1}^o L_{a_2}^o\right\}. \tag{3.8}$$

Regarding the variables

$$D_{2k}^{(q)} = \sum_{a_1,\ldots,a_q}^{2k} D_{a_1,\ldots,a_q}^{(q)} = \sum_{a_1,\ldots,a_q}^{2k} \mathbf{E}\left\{L_{a_1}^o L_{a_2}^o \cdots L_{a_q}^o\right\}$$

and using formulas (5.6) and (5.8) with $\eta = 1$, we obtain relation

$$D_{a_1,\ldots,a_q}^{(q)} = \frac{1}{4}\sum_{j=0}^{a_1-2} M_j D_{a_1-2-j,a_2,\ldots,a_q}^{(q)} + \frac{1}{4}\sum_{j=0}^{a_1-2} M_{a_1-2-j}D_{j,a_2,\ldots,a_q}^{(q)}$$

$$+\frac{1}{4}\sum_{j=0}^{a_1-2} D_{j,a_1-2-j,a_2,\ldots,a_q}^{(q+1)} - \frac{1}{4}\sum_{j=0}^{a_1-2} D_{j,a_1-2-j}^{(2)}D_{a_2,\ldots,a_q}^{(q-1)}$$

$$+\frac{1}{4N}(a_1-1)\mathbf{E}\left\{L_{a_1-2}^o L_{a_2}^o \cdots L_{a_q}^o\right\}$$

$$+\frac{1}{2N^2}\sum_{i=2}^{q} a_i M_{a_1+a_i-2}D_{a_2,\ldots,a_{i-1},a_{i+1},\ldots,a_q}^{(q-2)}$$

$$+\frac{1}{2N^2}\sum_{i=2}^{q} a_i D_{a_2,\ldots,a_{i-1},a_i+a_1-2,a_{i+1},\ldots,a_q}^{(q-1)}. \tag{3.9}$$

Introducing the variables

$$\bar{D}_{2k}^{(q)} = \sum_{a_1,\ldots,a_q}^{2k} \left|\mathbf{E}\left\{L_{a_1}^o \cdots L_{a_q}^o\right\}\right|,$$

we derive from (3.9) the inequality

$$\bar{D}_{2k}^{(q)} \leq \frac{1}{2}\sum_{j=0}^{k-1} \bar{D}_{2k-2-2j}^{(q)}M_{2j} + \frac{1}{4}\bar{D}_{2k-2}^{(q+1)} + \frac{1}{4}\sum_{j=0}^{k-1} \bar{D}_{2k-2-2j}^{(q-1)}\bar{D}_{2j}^{(2)} + \frac{k}{2N}\bar{D}_{2k-2}^{(q)}$$

$$+\frac{q-1}{2N^2}\sum_{j=0}^{k-1} \bar{D}_{2k-2-2j}^{(q-2)}\frac{(2j+2)(2j+1)}{2}M_{2j} + \frac{(q-1)k^2}{N^2}\cdot\bar{D}_{2k-2}^{(q-1)}. \tag{3.10}$$

We have used here the same transformations as when passing from equality (2.19) to inequality (2.20).

Now we proceed as in Section 2 and introduce the auxiliary numbers B and R that verify the relations

$$B_k = \frac{1}{4}(B * B)_{k-1} + \frac{k}{2N} B_{k-1} + \frac{1}{4} R_{k-1}^{(2)}, \quad k \geq 1, \tag{3.11}$$

and

$$R_k^{(q)} = \frac{1}{2}\left(B * R^{(q)}\right)_{k-1} + \frac{q-1}{2N^2}\left(B'' * R^{(q-2)}\right)_{k-1-j}$$
$$+ \frac{k}{2N} R_{k-1}^{(q)} + \frac{1}{4} R_{k-1}^{(q+1)} + \frac{1}{4}\left(R^{(2)} * R^{(q-1)}\right)_{k-1} + \frac{qk^2}{N^2} R_{k-1}^{(q-1)}. \tag{3.12}$$

The initial conditions are: $B_0 = 1$, $R_1^{(2)} = 1/(2N^2)$. The triangular scheme of recurrence estimates implies the inequalities

$$M_{2k}^{(N)} \leq B_k^{(N)}, \quad \text{and} \quad |D_{2k}^{(q)}| \leq \bar{D}_{2k}^{(q)} \leq R_k^{(q)}. \tag{3.13}$$

The main technical result for GOE is given by the following proposition.

Proposition 3.1
Let us consider B and R for the case of GOE ($\eta = 1$). Given $A > 1/2$ and $1/4 < C < 2 \cdot 6!$, there exists χ such that the following estimates

$$B_k^{(N)} \leq m_k + \frac{A}{N}, \quad k \geq 2,$$

or equivalently

$$B_k^{(q)} \leq [f(\tau)]_k + \frac{A}{N}\left[\frac{\tau}{1-\tau}\right]_k, \quad k \geq 2, \tag{3.14}$$

and

$$R_k^{(2s)} \leq \frac{C(3s)!}{N^{2s}}\left[\frac{\tau}{(1-\tau)^{2s}}\right]_k, \tag{3.15a}$$

and

$$R_k^{(2s+1)} \leq \frac{C(3s+3)!}{N^{2s+2}}\left[\frac{\tau}{(1-\tau)^{2s+5/2}}\right]_k, \tag{3.15b}$$

hold for all values of k, q and N such that $k \leq k_0$ and (2.13) and (2.16) hold.

The proof of this proposition resembles very much that of Proposition 2.1. However, there is a difference in the formulas that leads to a somewhat different condition on A. To show this, let us consider the estimate for B_k. Substituting (3.14) and (3.15) into the right-hand side of (3.11) and using (2.11), we arrive at the following inequality (cf. (2.28))

$$\frac{A}{N}\left[\frac{\tau}{\sqrt{1-\tau}}\right]_k \geq \frac{k}{2N} m_{k-1} + \frac{Ak}{2N^2}\left[\frac{\tau}{1-\tau}\right]_k + \frac{A^2+6C}{4N^2}\left[\frac{\tau^2}{(1-\tau)^2}\right]_k$$

which is sufficient for the estimate (3.14) to be true. Taking into account that

$$\left[\frac{\tau}{\sqrt{1-\tau}}\right]_k = k m_{k-1}, \tag{3.16}$$

we obtain the inequality

$$A \geq \frac{1}{2} + \frac{2A + A^2 + 6C}{4N m_{k-1}}.$$

It is easy to show that $m_{k-1}\sqrt{k} \geq (2k)^{-1}$. Then the last inequality is reduced to the condition

$$A \geq \frac{1}{2} + (A + A^2 + 3C)\sqrt{\chi}. \tag{3.17}$$

The estimates for $R^{(q)}$ also include the values $\sqrt{\chi}$ and χ. We do not present these computations.

Let us prove the second part of Theorem 3.1. Regarding relation (3.8) and taking into account estimate (3.15a) with $q = 2$, we conclude that

$$M_{2k}^{(N)} = m_k + \frac{1}{N} m_k^{(1)} + o(N^{-1}), \quad \text{as} \ \ N \to \infty.$$

It is easy to see that the numbers $m_k^{(1)}$ are determined by the relations

$$m_k^{(1)} = \frac{1}{2}\left(m^{(1)} * m\right)_{k-1} + \frac{2k-1}{4} m_{k-1} \tag{3.19}$$

and $m_0^{(1)} = 0$. Passing to the generating functions, we deduce from (3.19) the equality

$$m_k^{(1)} = \frac{1}{2}\left[\frac{\tau}{1-\tau}\right]_k - \frac{1}{2}\left[\frac{1-\sqrt{1-\tau}}{\sqrt{1-\tau}}\right]_k = \frac{1}{2} - \frac{(k+1)m_k}{2}.$$

Relation (3.2) is proved.

Let us consider the covariance term $D^{(2)}$. It follows from the results of Proposition 3.1 that

$$D_{2k}^{(2)} = \frac{r_k}{N^2} + o(N^{-2}).$$

Then we deduce from (3.12) with $q = 2$ that r_k is determined by the following recurrence relations

$$r_k = \frac{1}{2}\left(r * m\right)_{k-1} + \frac{1}{2}\frac{2k(2k-1)}{2} m_{k-1}.$$

Solving this equation, we get

$$r_k = \left[\frac{\tau}{2(1-\tau)^2}\right]_k.$$

This completes the proof of Theorem 3.1.

3.2 Proof of Theorem 3.2

In present Section we consider the ensemble $H^{(\eta)}$ with $\eta = -1$. In this case the elements of H (2.3) are given by imaginary numbers

$$(H)_{xy} = \frac{i}{\sqrt{N}} W_{xy}, \quad x < y$$

and the skew-symmetric condition holds:

$$(H)_{xy} = -(H)_{yx} = (-H)_{yx}$$

Regarding the last term of the formula (5.7) and using the previous identity, we can write that

$$\frac{\eta}{4N} \sum_{j=1}^{2k-1} \sum_{x,y=1}^{N} \mathbf{E}\left\{ (H^{j-1})_{yx} (H^{2k-1-j})_{yx} \right\} = -\frac{1}{4} \sum_{j=1}^{2k-1} (-1)^{j-1} \mathbf{E}\left\{ \frac{1}{N} \operatorname{Tr} H^{2k-2} \right\}.$$

Then we derive from (5.7) the equality

$$\mathbf{E} L_{2k} = \frac{1}{4} \sum_{j=0}^{2k-2} \mathbf{E}\left\{ L_j L_{2k-2-j} \right\} - \frac{1}{4N} \mathbf{E} L_{2k-2}$$

that gives the recurrence relation for the moments of the matrices $H_N^{(-1)}$

$$M_{2k} = \frac{1}{4} \sum_{j=0}^{k-1} M_{2j} M_{2k-2-2j} - \frac{1}{4N} M_{2k-2} + \frac{1}{4} D_{2k-2}^{(2)}, \qquad (3.20)$$

where the term $D^{(2)}$ is determined as usual. Regarding the general case of $D^{(q)}, q \geq 2$, and using (5.8), we obtain the following relation

$$
\begin{aligned}
D_{a_1,\dots,a_q}^{(q)} &= \frac{1}{4} \sum_{j=0}^{a_1-2} \mathbf{E}\left\{ L_{a_1-2-j} L_j [L_{a_2}^o \cdots L_{a_q}^o]^o \right\} \\
&+ \frac{(-1)^{a_1+1}}{4N} \mathbf{E}\left\{ L_{a_1-2}^o L_{a_2}^o \cdots L_{a_q}^o \right\} \\
&+ \frac{1 + (-1)^{a_i+1}}{4N^2} \sum_{i=2}^{q} \mathbf{E}\left\{ L_{a_2}^o \cdots L_{a_{i-1}}^o \, a_i \, L_{a_i+a_1-2}^o L_{a_{i+1}}^o \cdots L_{a_q}^o \right\}.
\end{aligned}
$$

$$(3.21)$$

Comparing this equality with (3.9) and then with (3.10), we see that $M_{2k}^{(\eta=-1)}$ and $D_{2k}^{(q,\eta=-1)}$ are bounded by the elements B and $R^{(q)}$ of the recurrence relations determined by equalities (3.11) and (3.12), respectively. Indeed, taking into account the positivity of M_{2j} and the fact that $1/4N \leq k/4N$ and $1 + (-1)^{a_i+1} \leq 2$, we obtain the inequalities

$$M_{2k} \leq \frac{1}{4} \sum_{j=0}^{k-1} M_{2j} M_{2k-2-2j} + \frac{1}{4N} M_{2k-2} + \frac{1}{4} \bar{D}_{2k-2}^{(2;\eta=-1)},$$

where

$$\bar{D}_{2k}^{(q,\eta=-1)} \leq \bar{D}_{2k}^{(q,\eta=1)}.$$

Then we conclude that

$$M_{2k}^{(\eta=-1)} \leq M_{2k}^{(\eta=1)} \leq B_k \quad \text{and} \quad \bar{D}_{2k}^{(q,\eta=-1)} \leq R_k^{(q)}. \tag{3.22}$$

This completes the proof of the first part of Theorem 3.2.

Now let us turn to the asymptotic expansion of M_{2k} and $D_{2k}^{(2)}$ for fixed k and $N \to \infty$. Regarding (3.21) with $q = 2$, we obtain the following relation

$$D_{2k}^{(2)} = \frac{1}{2} \left(D^{(2)} * M \right)_{2k-2} + \frac{1}{4} D_{2k-2}^{(3)} + \frac{(-1)^{a_1+1}}{4N} \sum_{a_1+a_2=2k} \mathbf{E} \left\{ L_{a_1-1}^o L_{a_2}^o \right\}$$

$$+ \frac{1+(-1)^{a_2+1}}{4N^2} \sum_{a_1+a_2=2k} a_2 \mathbf{E} L_{a_1+a_2-2}.$$

Now, introducing the variable r_k

$$D_{2k}^{(2)} = \frac{r_k}{N^2} + O(N^{-3})$$

and taking into account estimates (3.4) and (3.5), we conclude after simple computations that r is determined by the recurrence relations

$$r_k = \frac{1}{2} (r * m)_{k-1} + \frac{k^2}{2} m_{k-1}.$$

Using relation (5.10), we can write that

$$\frac{k^2}{2} m_{k-1} = \frac{2k(2k-1)}{2 \cdot 4} m_{k-1} + \frac{k}{4} m_{k-1} = \frac{1}{4} \left[\frac{\tau}{(1-\tau)^{3/2}} \right]_k + \frac{1}{4} \left[\frac{\tau}{(1-\tau)^{1/2}} \right]_k.$$

Then

$$r_k = \frac{1}{4} \left[\frac{\tau}{(1-\tau)^2} \right]_k + \frac{1}{4} \left[\frac{\tau}{1-\tau} \right]_k = \frac{k+1}{4}. \tag{3.23}$$

Now let us consider the $1/N$-expansion for M_{2k}

$$M_{2k} = m_k + \frac{1}{N} m_k^{(1)}.$$

It is easy to see that equality (3.20) together with estimates (3.5) implies the following recurrence relation for $m_k^{(1)}$

$$m_k^{(1)} = \frac{1}{2}\left(m^{(1)} * m\right)_{k-1} - \frac{m_{k-1}}{4}.$$

Then

$$m_k^{(1)} = -\left[\frac{\tau f(\tau)}{4\sqrt{1-\tau}}\right]_k = \frac{\delta_{k,0} - (k+1)m_k}{2}. \tag{3.24}$$

These computations prove the second part of Theorem 3.2.

Theorems 3.1 and 3.2 show that there exists essential difference between GOE and Gaussian anti-symmetric ensemble. Let us illustrate this by the direct computation of $M_2^{(N)}$ for these two ensembles.

In the case of GOE, we have

$$\frac{1}{N}\sum_{x,y=1}^N \mathbf{E}H_{xy}^2 = \frac{2}{N^2}\sum_{x<y}\mathbf{E}V_{xy}^2 + \frac{1}{N^2}\sum_{x=1}^N \mathbf{E}V_{xx}^2 = \frac{N(N-1)}{4N^2} + \frac{1}{2N} = \frac{1}{4} + \frac{1}{4N}.$$

This relation reproduces (3.2) with $k = 1$.

The first nontrivial moment of anti-symmetric matrices reads as

$$\frac{1}{N}\sum_{x,y=1}^N \mathbf{E}H_{xy}^2 = \frac{2}{N^2}\sum_{x<y}\mathbf{E}W_{xy}^2 = \frac{N(N-1)}{4N^2} = \frac{1}{4} - \frac{1}{4N};$$

this agrees with (3.24).

Finally, let us point out the difference between GOE and anti-symmetric ensemble with respect to the first term of the expansion of $D^{(2)}$ given by (3.3) and (3.23), respectively. This indicates that Gaussian Hermitian anti-symmetric ensemble represents a universality class of spectral properties of random matrices different from that of GOE (see for example, the monograph [16]).

4 Gaussian Band random matrices

Now let us consider the ensemble of Hermitian random matrices given by the formula

$$\left[H^{(N,b)}\right]_{xy} = h_{xy}\sqrt{U_{xy}}, \quad x,y = 1,\ldots,N, \tag{4.1}$$

where $\{h_{xy}, \ x \leq y\}$ are determined by (2.1) and (2.2) with $\eta = 0$. The elements of non-random matrix $U = U^{(N,b)}$ are determined by the relation

$$U_{xy} = \frac{1}{b}\,u\left(\frac{x-y}{b}\right), \quad x,y = 1,\ldots,N,$$

where $u(t)$, $t \in \mathbf{R}$ is a positive even piece-wise continuous function such that

$$\sup_{t \in \mathbf{R}} u(t) = u_0 < \infty \quad \text{and} \quad \int_{-\infty}^{\infty} u(t) \, dt = u_1.$$

Without loss of generality, we can consider $u_0 = 1$. We assume also that $u(t), t \geq 0$ is monotone. If $u(t)$ is given by the indicator function of the interval $(-1/2, 1/2)$, then matrices (4.1) are of the band form. We keep the term of band random matrices when regarding the ensemble (4.1) in the general case.

It is known (see for instance [5, 17]) that the moments of $H^{(N,b)}$ converge in the limit of $1 \ll b \ll N$ to the moments of the semicircle law;

$$M_{2k+1}^{(N,b)} = 0, \quad M_{2k}^{(N,b)} = \mathbf{E} \left\{ \frac{1}{N} \operatorname{Tr} \left[H^{(N,b)} \right]^{2k} \right\} \to m_k(u_1), \qquad (4.3)$$

where the numbers $\{m_k(u_1), \; k \geq 0\}$ are given by recurrence relations

$$m_k(u_1) = \frac{u_1}{4} \sum_{j=0}^{k-1} m_{k-1-j}(u_1) \, m_j(u_1), \quad m_0(u_1) = 1.$$

The generating function $f_1(\tau) = \sum \tau^k m_k(u_1)$ is related with $f(\tau)$ (1.7) by equality $f_1(\tau) = f(\tau u_1)$ and therefore

$$m_k(u_1) = u_1^k \, m_k.$$

4.1 Main results

In this Section we present a non-asymptotic estimate for the moments of $M_{2k}^{(N,b)}$. This improves proposition (4.3). Set

$$\hat{u}_1 = \hat{u}_1^{(b)} = \frac{1}{b} + \frac{1}{b} \sum_{l=-\infty}^{+\infty} u \left(\frac{l}{b} \right).$$

Clearly $\hat{u}_1 \geq u_1$ and $\hat{u}_1^{(b)} \to u_1$ as $b \to \infty$.

Theorem 4.1.
Given $\alpha > 1/12$, there exists $\theta > 0$ such that the estimate

$$M_{2k}^{(N,b)} \leq \left(1 + \alpha \hat{u} \frac{(k+1)^3}{b^2} \right) m_k(\hat{u}_1), \qquad (4.4)$$

where $\hat{u} = \max\{\hat{u}_1, 1/8\}$, holds for all values of k, b such that $\dfrac{(k+1)^3}{b^2} \leq \theta$ and $b \leq N$.

The proof of this theorem is obtained by the method described in Section 2. We consider the mathematical expectations of the variables $L_{2k}(x) = (H^{2k})_{xx}$ and derive recurrence relations for them and related covariance variables. Certainly, these relations are of more complicated structure than those derived for GUE in Section 2. However, regarding the estimates for $M_{2k} = \mathbf{E}L_{2k}(x)$ by auxiliary numbers $\bar{B}_k$, one can observe that equalities for $\bar{B}_k$ and related numbers $\bar{R}_k^{(q)}$ are almost the same as the system (2.22)–(2.23) derived for GUE. This allows us to say that the system (2.22)–(2.23) plays an important role in random matrix theory and is of somewhat canonical character. The estimates for the moments $M_{2k}^{(N,b)}$ follow immediately.

4.2 Moment relations and estimates

In what follows, we omit superscripts (N, b) when no confusion can arise. It follows from the integration by parts formula (5.8) that

$$\mathbf{E}\left\{H_{xy}\left(H^l\right)_{yx}\right\} = \frac{1}{4} \cdot U_{xy} \sum_{j=0}^{l-1} \mathbf{E}\left\{(H^j)_{yy}\left(H^{l-j}\right)_{xx}\right\}. \qquad (4.5)$$

Then, regarding $L_k(x) = (H^k)_{xx}$, we obtain the equality

$$\mathbf{E}L_{2k}(x) = \frac{1}{4}\sum_{j=0}^{2k-2} \mathbf{E}\left\{L_{2k-2-j}(x)\, L_j[x]\right\},$$

where we denoted

$$L_j[x] = \frac{1}{b}\sum_{y=1}^{N} u\left(\frac{x-y}{b}\right)(H^j)_{yy}.$$

Introducing the variables $M_k(x) = \mathbf{E}L_k(x)$ and $M_k[x] = \mathbf{E}L_k[x]$, we obtain the equality

$$M_{2k}(x) = \frac{1}{4}\sum_{j=0}^{k-1} M_{2k-2-2j}(x)\, M_{2j}[x] + \frac{1}{4}D_{2k-2}^{(2)}(x,[x]), \qquad (4.6)$$

where we denoted

$$D_{2k-2}^{(2)}(x,[x]) = \sum_{a_1+a_2=2k-2} \mathbf{E}\left\{L_{a_1}^o(x)\, L_{a_2}^o[x]\right\}. \qquad (4.7)$$

In (4.6) we have used obvious equality $M_{2k+1}(x) = 0$.

To get the estimates on the terms on the right-hand sides of (4.6) and (4.7), we need to consider more general expressions than M and D introduced above. Let us consider the following variables

$$M_{2k}^{(\pi_r, \bar{y}_r)}(x) = \mathbf{E}\left\{ \left(H^{p_1} \Psi_{y_1} H^{p_2} \cdots \Psi_{y_r} H^{p_{r+1}}\right)_{xx} \right\}, \qquad (4.8)$$

where we denoted $\pi_r = (p_1, p_2, \ldots, p_{r+1})$ with $\sum_{i=1}^{r+1} p_i = 2k$, the vector $\bar{y}_r = (y_1, \ldots, y_r)$ and Ψ_y denotes the diagonal matrix

$$(\Psi_y)_{st} = \delta_{st}\, U\left(\frac{t - y}{b}\right), \quad s, t = 1, \ldots, N.$$

One can associate the right-hand side of (4.8) with $2k$ white balls separated into $r + 1$ groups by r black balls.

The second variable we need is

$$D_{a_1, a_2, \ldots, a_q}^{(q, \pi_r(\alpha_q), \bar{y}_r)}(\bar{x}_q) = \mathbf{E}\{ \underbrace{L_{a_1}^o(x_1) L_{a_2}^o[x_2] \cdots L_{a_q}^o[x_q]}_{\pi_r(\bar{y}_r)} \}, \qquad (4.9)$$

where $\alpha_q = (a_1, \ldots, a_q)$ and $\bar{x}_q = (x_1, \ldots, x_q)$. We also denote $|\alpha_q| = \sum_{i=1}^q a_i$. So, we have a set of $|\alpha_q|$ white balls separated into q boxes by $q - 1$ walls.

The brace under the last product means that the set $\{a_1|a_2|\cdots|a_q\}$ of walls and white balls is separated into $r + 1$ groups by r black balls. The places where the black balls are inserted depend on the vector α_q.

Let use derive recurrence relations for (4.8) and (4.9). These relations resemble very much those obtained in Section 2. First, we write the identity

$$M_{2k}^{(\pi_r, \bar{y}_r)}(x) = \sum_{s=1}^{N} \mathbf{E}\left\{ H_{xs}\left(H^{p_1 - 1} \Psi_{y_1} H^{p_2} \cdots \Psi_{y_r} H^{p_{r+1}}\right)_{sx} \right\},$$

and apply the integration by parts formula (4.5). We obtain the equality

$$M_{2k}^{(\pi_r, \bar{y}_r)}(x) = \sum_{a_1 + a_2 = 2k - 2} \mathbf{E}\{ \underbrace{L_{a_1}(x) L_{a_2}[x]}_{\pi_r'(\bar{y}_r, \alpha_2)} \}.$$

In this relation the partition π' is different from the original π from the left-hand side. It is not difficult to see that π' depends on particular values of a_1 and a_2, i.e., on the vector (a_1, a_2). Returning to the denotation $M = \mathbf{E}\{L\}$, we obtain the first main relation

$$M_{2k}^{(\pi_r, \bar{y}_r)}(x) = \sum_{a_1 + a_2 = 2k - 2} \underbrace{M_{a_1}(x)\, M_{a_2}[x]}_{\pi'(y_r, \alpha_2)} + \sum_{a_1 + a_2 = 2k - 2} D_{a_1, a_2}^{(2, \pi'(y_r, \alpha_2))}(x, [x]).$$

$$(4.10)$$

Let us consider

$$D_{a_1, a_2, \ldots, a_q}^{(q, \pi_r(\alpha_q), \bar{y}_r)}(\bar{x}_q) = \sum_{s=1}^{N} \mathbf{E}\{ \underbrace{H_{x_1 s}(H^{a_1 - 1})_{sx_1} \left[L_{a_2}^o[x_2] \cdots L_{a_q}^o[x_q]\right]^o}_{\pi_r(\bar{y}_r, \alpha_q)} \}$$

and apply (4.5) to the latter mathematical expectation. We get

$$D_{a_1,a_2,\ldots,a_q}^{(q,\pi_r(\alpha_q)),\bar{y}_r)}(\bar{x}_q) = \frac{1}{4}\sum_{a'=0}^{a_1-2}\mathbf{E}\{\underbrace{L_{a_1-2-a'}(x_1)L_{a'}[x_1]\left[L_{a_2}^o[x_2]\cdots L_{a_q}^o[x_q]\right]^o}_{\pi_r'(\bar{y}_r,\alpha_{q+1}')}\}$$

$$+\frac{1}{4b^2}\sum_{i=2}^{q}\sum_{j=0}^{a_i-1}\mathbf{E}\{\underbrace{\left(H^j\Psi_{x_i}H^{a_i-1-j}\Psi_{x_1}H^{a_1-1}\right)_{x_1x_1}L_{a_2}^o[x_2]\cdots\times_i\cdots L_{a_q}^o[x_q]}_{\pi_{r+2}''(\bar{y}_{r+2}',\alpha_{q+1}''(i))}\}.$$

$$(4.11)$$

In these expressions, π' and π'' designate partitions different from π; they depend on the vectors $\alpha'_{q+1} = (a_1 - 2 - a', a', a_2, \ldots, a_q)$ and

$$\alpha''_{q+1}(i) = (j, a_i - 1 - j, a_1 - 1, a_2, \ldots, a_{i-1}, a_{i-1}, \ldots, a_q),$$

respectively; also $\bar{y}_{r+2}' = (x_i, x_1, y_1, y_2, \ldots, y_r)$. The notation $\times_i$ in the last product of (4.11) means that the factor L_{a_i} is absent there. Repeating the computations of Section 2, we arrive at the second main relation

$$D_{a_1,a_2,\ldots,a_q}^{(q,\pi_r(\alpha_q)),\bar{y}_r)}(\bar{x}_q) = \sum_{l=1}^{6}T_l,\qquad (4.12)$$

where

$$T_1 = \frac{1}{4}\sum_{a'=0}^{a_1-2}\underbrace{M_{a_1-2-a'}(x_1)\,D_{a',a_2,\ldots,a_q}^{(q)}}_{\pi_r'(\bar{y}_r,\alpha_{q+1}')}([x_1],[x_2],\ldots,[x_q]);$$

$$T_2 = \frac{1}{4}\sum_{a'=0}^{a_1-2}\underbrace{M_{a_1-2-a'}[x_1]\,D_{a',a_2,\ldots,a_q}^{(q)}}_{\pi_r'(\bar{y}_r,\alpha_{q+1}')}(x_1,[x_2],\ldots,[x_q]);$$

$$T_3 = \frac{1}{4}\sum_{a'=0}^{a_1-2}D_{a_1-2-a',a',a_2,\ldots,a_q}^{(q+1,\pi_r'(\bar{y}_r,\alpha_{q+1}'))}(x_1,[x_1],[x_2],\ldots,[x_q]);$$

$$T_4 = -\frac{1}{4}\sum_{a'=0}^{a_1-2}\underbrace{D_{a_1-2-a',a'}^{(2)}(x_1,[x_1])\,D_{a_2,\ldots,a_q}^{(q-1)}}_{\pi_r'(\bar{y}_r,\alpha_{q+1}')}([x_2],\ldots,[x_q]);$$

$$T_5 = \frac{1}{4b^2}\sum_{i=2}^{q}\sum_{j=0}^{a_i-1}\underbrace{M_{a_1+a_i-2}(x_1)D_{a_2,\ldots,a_{i-1},a_{i+1},\ldots,a_q}^{(q-2)}}_{\pi_{r+2}''(\bar{y}_{r+2}',\alpha_{q+2}''(i))}([x_2],\ldots,[x_{i-1}],[x_{i+1}],\ldots,[x_q]);$$

and finally

$$T_6 = \frac{1}{4b^2}\sum_{i=2}^{q}\sum_{j=0}^{a_i-1}D_{a_1+a_i-2,a_2,\ldots,a_{i-1},a_{i+1},\ldots,a_q}^{(q-1,\pi_{r+2}''(\bar{y}_{r+2}',\alpha_q''(i)))}(x_1,[x_2],\ldots,[x_{i-1}],[x_{i+1}],\ldots,[x_q]).$$

Now let us introduce some auxiliary numbers $\{\hat{B}_k^{(N,b)}, \ k \geq 0\}$ and

$$\hat{R}_{\alpha_q}^{(q;N,b)} = \hat{R}_{a_1,\ldots,a_q}^{(q;N,b)}, \quad \text{for} \quad q \geq 0 \text{ and } a_i \geq 0,$$

determined for all integer k, q and a_i by the following recurrence relations (in $\hat{B}$ and $\hat{R}$, we omit superscripts N and b). Regarding $\{\hat{B}\}$, we set $\hat{B}_0 = 1$ and determine $\hat{B}_k$ by the relation

$$\hat{B}_k = \frac{\hat{u}_1}{4} \sum_{j=0}^{k-1} \hat{B}_{k-1-j} \hat{B}_j + \frac{1}{4} \sum_{a_1+a_2=2k-2} \hat{R}_{a_1,a_2}^{(2)}, \quad k \geq 1. \quad (4.13)$$

Regarding $\{\hat{R}\}$, we set $\hat{R}^{(0)} = 1$ and $\hat{R}_a^{(1)} = 0$. We also assume that $\hat{R}_{\alpha_q}^{(q)} = 0$ when either $q > |\alpha_q|$ or one of the variables a_i is equal to zero. The recurrence relation for $\hat{R}$ is

$$\hat{R}_{a_1,\ldots,a_q}^{(q)} = \frac{\hat{u}_1}{2} \sum_{j=0}^{a_1-2-j} \hat{B}_{a_1-2-j} \hat{R}_{j,a_2,\ldots,a_q}^{(q)}$$

$$+ \frac{1}{4} \sum_{j=0}^{a_1-2-j} \hat{R}_{j,a_1-2-j,a_2,\ldots,a_q}^{(q+1)} + \frac{1}{4} \sum_{j=0}^{a_1-2-j} \hat{R}_{a_1-2-j,j}^{(2)} \hat{R}_{a_2,\ldots,a_q}^{(q-1)}$$

$$+ \frac{\hat{u}_1}{4b^2} \sum_{i=2}^{q} a_i \hat{B}_{a_1+a_I-2} \hat{R}_{a_2,\ldots,a_{i-1},a_{i+1},\ldots,a_q}^{(q-2)}$$

$$+ \frac{1}{4b^2} \sum_{i=2}^{q} a_i \hat{R}_{a_2,\ldots,a_{i-1},a_{i+1},\ldots,a_q}^{(q-1)}. \quad (4.14)$$

Existence and uniqueness of the numbers $\hat{B}$ and $\hat{R}$ follow from the triangular scheme described above in Section 2.

Using the triangular scheme of Section 2, it is easy to deduce from relations (4.10) and (4.11) that

$$\sup_{x,\bar{y}_r} M_{2k}^{(\pi_r,\bar{y}_r)}(x) \leq \hat{B}_k \quad (4.15)$$

and

$$\sup_{\bar{x}_q,\bar{y}_r} |D_{a_1,a_2,\ldots,a_q}^{(q,\pi_r(\alpha_q),\bar{y}_r)}(x_1, [x_2], \ldots, [x_q])| \leq \hat{R}_{a_1,a_2,\ldots,a_q}^{(q)}. \quad (4.16)$$

Let us note that when regarding (4.15) with $k = 0$, we have used the property of u (4.2)

$$M_0^{(\pi_r,\bar{y}_r)}(x) = \prod_{i=1}^{r} u\left(\frac{x-y_i}{b}\right) \leq u_0^r \leq 1.$$

Now, let us introduce two more auxiliary sets of numbers $\bar{B}_k$ and $\bar{R}(q)_k$. We determine them by the relations

$$\bar{B}_k = \frac{\hat{u}_1}{4} \sum_{j=0}^{k-1} \bar{B}_{k-1-j} \bar{B}_j + \frac{1}{4} \bar{R}_{k-1}^{(2)}, \quad \bar{B}_0 = 1, \tag{4.17}$$

and

$$\bar{R}_k^{(q)} = \frac{\hat{u}_1}{2} \sum_{j=0}^{k-1} \bar{R}_{k-1-j}^{(q)} \bar{B}_j + \frac{\hat{u}_1(q-1)}{4b^2} \sum_{j=0}^{k-1} \bar{R}_{k-1-j}^{(q-2)} \frac{(2j+2)(2j+1)}{2} \bar{B}_j$$

$$+ \frac{1}{4} \bar{R}_{k-1}^{(q+1)} + \frac{1}{4} \sum_{j=0}^{k-1} \bar{R}_{k-1-j}^{(2)} \bar{R}_j^{(q-1)} + \frac{2k^2(q-1)}{4b^2} \bar{R}_{k-1}^{(q-1)}. \tag{4.18}$$

It is clear that

$$\hat{B}_k \le \bar{B}_k \qquad \text{and} \qquad \sum_{a_1 + \dots a_q = 2k} \hat{R}_{a_1,\dots,a_q}^{(q)} \le \bar{R}_k^{(q)}. \tag{4.19}$$

The main technical result of this Section is as follows.

Proposition 4.1.
Let $\hat{u} = \max\{\hat{u}_1, 1/8\}$. Given $A > 1/16$, there exists $\theta > 0$ such that the estimate

$$\bar{B}_k \le \left[f_1(\tau) + \frac{A\hat{u}}{b^2} \frac{\tau^2}{(1 - \tau\hat{u}_1)^{5/2}} \right]_k \tag{4.20}$$

holds for all values of $k \le k_0$, where k_0 verifies condition $k_0^3 \le \theta b^2$. Also there exists C

$$\frac{1}{24} < C < \max\{\frac{3A}{2}, 4!\}$$

such that the inequalities

$$\bar{R}_k^{(2s)} \le C \frac{\hat{u}^s(3s)!}{b^{2s}} \left[\frac{\tau}{(1 - \tau\hat{u}_1)^{2s}} \right]_k \tag{4.21a}$$

and

$$R_k^{(2s+1)} \le C \frac{\hat{u}^{s+1}(3s+3)!}{b^{2s+2}} \left[\frac{\tau}{(1 - \tau\hat{u}_1)^{2s+1}} \right]_k, \tag{4.21b}$$

hold for all values of k and s such that

$$2k + q \le 2k_0$$

with $q = 2s$ and $q = 2s + 1$, respectively.

The proof of this proposition can be obtained by repeating the proof of Proposition 2.1 with obvious changes. The only difference is due to the presence of the factors $\hat{u}_1$ in (4.17) and (4.18). This implies corresponding changes in the generating functions used in estimates (4.20) and (4.21). Also, the conditions for A (2.29) and C (2.38), (2.43) are replaced by the conditions

$$A > \frac{3C}{2} + \frac{A^2 \hat{u}_1}{16} \theta,$$

$$C > \frac{\delta_{s,1} + C(1 - \delta_{s,1})}{24} + 2\theta \left(1 + 10\hat{u}C(1 + C) + 2\hat{u}_1 AC\right)$$

and

$$179\hat{u} > 20 + 3\hat{u}C + 18\theta\hat{u}\hat{u}_1 A.$$

The latter inequality forces us to use $\hat{u}$ instead of $\hat{u}_1$ in the proof. Otherwise, we should assume that $\hat{u}_1 > 1/8$. We believe this condition is technical and can be avoided.

4.3 Spectral norm of band random matrices

Using this result, we can estimate the lower bound for b to have the spectral norm of $\|H^{(N,b)}\| = \lambda_{\max}^{(N,b)}$ bounded.

Theorem 4.2 If $1 \ll (\log N)^{3/2} \ll b$, then $\lambda_{\max}^{(N,b)} \to \sqrt{u_1}$ with probability 1.

Proof. Using the standard inequality

$$P\left\{\lambda_{\max}^{(N,b)} > \sqrt{u_1}(1 + \varepsilon)\right\} \leq N \frac{M_{2k}^{(N,b)}}{u_1^k (1 + \varepsilon)^{2k}},$$

we deduce from (4.4) the estimate

$$P\left\{\lambda_{\max}^{(N,b)} > \sqrt{u_1}(1 + \varepsilon)\right\} \leq N \frac{\left(1 + \alpha\hat{u}\dfrac{(k+1)^2}{b^2}\right)^k}{u_1^k (1 + \varepsilon)^{2k}} \hat{u}_1^k \qquad (4.22)$$

which holds for all $k + 1 \leq \theta^{1/3} b^{2/3}$, where θ is as in Theorem 4.1. In (4.22), we have used the inequalities $m_k(\hat{u}_1) \leq \hat{u}_1^k m_{2k}$ and $m_{2k} \leq 1$.

Assuming that $b = \phi_N (\log N)^{3/2}$, where $\phi_N \to \infty$ as $N \to \infty$, and taking $k + 1 = t\theta^{1/3} b^{2/3}$, $0 < t \leq 1$, we obtain the estimate

$$P\left\{\lambda_{\max}^{(N,b)} > \sqrt{u_1}(1 + \varepsilon)\right\} \leq N \exp\left\{-2t\theta^{1/3} b^{2/3} \log(1 + \varepsilon) + 2\alpha\hat{u}t^3\right\} \cdot \left(\frac{\hat{u}_1}{u_1}\right)^k. \tag{4.23}$$

Using the relation $\hat{u}_1 = u_1(1 + 1/b)$, we easily deduce from (4.23) that

$$P\left\{\lambda_{\max}^{(N,b)} > \sqrt{u_1}(1 + \varepsilon)\right\} \leq N^{1 - C \log(1+\varepsilon)\phi_N^{2/3}}$$

with some positive C. Then the corresponding series of probability converges and the Borel-Cantelli lemma implies convergence of $\lambda_{\max}^{(N,b)}$ to $\sqrt{u_1}$. Theorem 4.2 is proved.

Let us complete this subsection with the following remark. If one optimizes the right-hand side of (4.23), one can see that the choice of $t = t_0 = b^{1/3}\sqrt{\log(1+\varepsilon)}(\alpha\hat{u})^{-1/2}\theta^{-1/3}$ gives the best possible estimate in the form

$$N \exp\{-b\frac{1}{\sqrt{2\alpha\hat{u}}}(\log(1+\varepsilon))^{3/2}\}.$$

Once this estimate shown, convergence $\lambda_{\max}^{(N,b)} \to \sqrt{u_1}$ would be true provided $b = O(\log N)$. However, one cannot use the optimal value of t_0 mentioned above because this choice makes k to be $k = O(b)$. This asymptotic regime is out of reach for the method of this paper.

5 Auxiliary relations

5.1 Integration by parts for complex random variables

Let us consider matrices H with elements $H_{xy} = v_{xy} + iw_{xy}$, where the family $\{v_{xy}, w_{xy}, 1 \le x \le y \le N\}$ is given by jointly independent Gaussian random variables with zero mean value. We denote

$$\mathbf{E}v_{xy}^2 = \xi_{xy}, \quad \mathbf{E}w_{xy}^2 = \zeta_{xy}.$$

Let us assume that $x < y$. Then integration by parts formula says that

$$\mathbf{E}H_{xy}(H^l)_{st} = \xi_{xy}\mathbf{E}\left\{\frac{\partial(H^l)_{st}}{\partial v_{xy}}\right\} + i\zeta_{xy}\mathbf{E}\left\{\frac{\partial(H^l)_{st}}{\partial w_{xy}}\right\} \quad (5.1)$$

It is easy to see that

$$\frac{\partial(H^l)_{st}}{\partial v_{xy}} = \sum_{j=1}^{l}\sum_{s',t'=1}^{N} H_{ss'}^{j-1} \cdot \frac{\partial H_{s't'}}{\partial v_{xy}} \cdot H_{vt}^{l-j} = \sum_{j=1}^{l}\left[H_{sx}^{j-1}H_{yt}^{l-j} + H_{sy}^{j-1}H_{xt}^{l-j}\right].$$
$$(5.2)$$

Similarly

$$\frac{\partial(H^l)_{st}}{\partial w_{xy}} = i\sum_{j=1}^{l}\left[H_{sx}^{j-1}H_{yt}^{l-j} - H_{sy}^{j-1}H_{xt}^{l-j}\right]. \quad (5.3)$$

Substituting (5.2) and (5.3) into (5.1), we get the equality

$$\mathbf{E}H_{xy}(H^l)_{st} = (\xi_{xy} - \zeta_{xy}) \sum_{j=1}^{l} \mathbf{E}\{H_{sx}^{j-1} H_{yt}^{l-j}\}$$

$$+ (\xi_{xy} + \zeta_{xy}) \sum_{j=1}^{l} \mathbf{E}\{H_{sy}^{j-1} H_{xt}^{l-j}\}, \quad x < y. \qquad (5.4)$$

It is not hard to check that the same relation is true when $x > y$. Also

$$\mathbf{E}H_{xx}(H^l)_{st} = \xi_{xx} \sum_{j=1}^{l} \mathbf{E}\{H_{sx}^{j-1} H_{xt}^{l-j}\}. \qquad (5.5)$$

Gaussian Ensembles $\{H^{(\eta)}\}$

Regarding formulas (2.1)-(2.3), we see that

$$v_{xy} = \frac{V_{xy}}{\sqrt{N}}, \quad w_{xy} = \frac{W_{xy}}{\sqrt{N}}$$

and

$$\xi_{xy} + \zeta_{xy} = \frac{1 + \delta_{xy}\eta}{4N}, \quad \xi_{xy} - \zeta_{xy} = \frac{\eta + \delta_{xy}}{4N}.$$

Regarding the sum of (5.5) with doubled (5.4), we obtain a relation valid for all values of x and y:

$$\mathbf{E}H_{xy}(H^l)_{st} = \frac{1}{4N} \sum_{j=1}^{l} \mathbf{E}\{H_{sy}^{j-1} H_{xt}^{l-j}\} + \frac{\eta}{4N} \sum_{j=1}^{l} \mathbf{E}\{H_{sx}^{j-1} H_{yt}^{l-j}\}. \qquad (5.6)$$

Let us mention two useful formulas that follow from (5.6); these are

$$\mathbf{E}\,\mathrm{Tr}(H^{l+1}) = \frac{1}{4N} \sum_{j=1}^{l} \mathbf{E}\left\{\mathrm{Tr}\,H^{j-1}\,\mathrm{Tr}\,H^{l-j}\right\}$$

$$+ \frac{\eta}{4N} \sum_{j=1}^{l} \sum_{x,y=1}^{N} \mathbf{E}\left\{(H^{j-1})_{yx}(H^{l-j})_{yx}\right\} \qquad (5.7a)$$

and

$$\mathbf{E}H_{xy}\,\mathrm{Tr}\,H^l = \mathbf{E}\left\{H_{xy}\sum_{s=1}^{N}(H^l)_{ss}\right\} = \frac{l}{4N}\mathbf{E}H_{xy}^{l-1} + \frac{\eta l}{4N}\mathbf{E}H_{yx}^{l-1}. \qquad (5.7b)$$

Band Random Matrices

Using (5.4) and (5.5) in the case of matrices (4.1), we see that

$$\xi_{xy} = \frac{1 + \delta_{xy}}{8} U_{xy}, \quad \zeta_{xy} = \frac{1 - \delta_{xy}}{8} U_{xy}.$$

Then (5.4) and (5.5) imply the equality

$$\mathbf{E} H_{xy}(H^l)_{st} = \frac{U_{xy}}{4} \sum_{j=1}^{l} \mathbf{E}\{H_{sy}^{j-1} H_{xt}^{l-j}\}. \tag{5.8}$$

Regarding this relation, one can easily obtain analogs of formulas (5.7a) and (5.7b).

5.2 Derivation of Equality (2.18)

We consider the case of Hermitian matrices $\eta = 0$ only. Regarding (2.17), we can write that

$$\mathbf{E}\{L_{a_1}^o \cdots L_{a_q}^o\} = \mathbf{E}\{L_{a_1} Q^o\},$$

where $Q = L_{a_2}^o \cdots L_{a_q}^o$. Using integration by parts formula, we obtain as in (5.1) that

$$\mathbf{E}\{H_{xy}(H^{a_1-1})_{yx} Q^o\} = \xi_{xy} \mathbf{E}\left\{\frac{\partial H_{yx}^{a_1-1} Q^o}{\partial v_{xy}}\right\} + i\zeta_{xy} \mathbf{E}\left\{\frac{\partial H_{yx}^{a_1-1} Q^o}{\partial w_{xy}}\right\}.$$

Obviously,

$$\frac{\partial H_{yx}^{a_1-1} Q^o}{\partial v_{xy}} = \sum_{j=1}^{a_1-1} \mathbf{E}\left\{\left[H_{yx}^j H_{yx}^{a_1-1-j} + H_{yy}^j H_{xx}^{a_1-1-j}\right] Q^o\right\} + H_{yx}^{a_1-1} \frac{\partial Q^o}{\partial v_{xy}}.$$

It is clear that

$$\frac{\partial Q^o}{\partial v_{xy}} = \frac{\partial Q}{\partial v_{xy}} = \sum_{i=2}^{q} L_{a_2}^o \cdots L_{a_{i-1}}^o \frac{\partial L_{a_i}^o}{\partial v_{xy}} L_{a_{i+1}}^o \cdots L_{a_q}^o$$

and

$$\frac{\partial L_{a_i}}{\partial v_{xy}} = \frac{1}{N} \sum_{t=1}^{N} \sum_{j=1}^{a_i} \left[H_{tx}^{j-1} H_{yt}^{a_i-j} + H_{ty}^{j-1} H_{xt}^{a_i-j}\right] = \frac{a_i}{N}\left[H_{xy}^{a_i-1} + H_{yx}^{a_i-1}\right].$$

Also we have

$$\frac{\partial H_{yx}^{a_1-1} Q^o}{\partial w_{xy}} = i \sum_{j=1}^{a_1-1} \mathbf{E}\left\{\left[H_{yx}^{j-1} H_{yx}^{a_1-1-j} - H_{yy}^{j-1} H_{xx}^{a_1-1-j}\right] Q^o\right\} + H_{yx}^{a_1-1} \frac{\partial Q^o}{\partial w_{xy}}.$$

It is clear that

$$\frac{\partial Q^o}{\partial w_{xy}} = \frac{\partial Q}{\partial w_{xy}} = \sum_{i=2}^{q} L^o_{a_2} L^o_{a_{i-1}} \cdots \frac{\partial L^o_{a_i}}{\partial w_{xy}} L^o_{a_{i+1}} \cdots L^o_{a_q}$$

and

$$\frac{\partial L_{a_i}}{\partial w_{xy}} = \frac{i}{N} \sum_{t=1}^{N} \sum_{j=1}^{a_i} \left[H^{j-1}_{tx} H^{a_i-j}_{yt} - H^{j-1}_{ty} H^{a_i-j}_{xt} \right] = \frac{i a_i}{N} \left[H^{a_i-1}_{yx} - H^{a_i-1}_{xy} \right].$$

Gathering these terms, we finally obtain that

$$\mathbf{E}\{H_{xy}(H^{a_1-1})_{yx} Q^o\} = \frac{1}{4N} \sum_{j=1}^{a_1-1} \mathbf{E}\left\{ H^{j-1}_{yy} H^{a_1-1-j}_{xx} Q^o \right\}$$

$$+ \frac{1}{4N^2} \sum_{i=2}^{q} a_i \mathbf{E}\left\{ H^{a_1-1}_{yx} L^o_{a_2} \cdots L^o_{a_{i-1}} H^{a_i-1}_{xy} L^o_{a_{i+1}} \cdots L^o_{a_q} \right\}.$$

Now (2.18) easily follows.

5.3 Catalan numbers and related identities

In the proofs, we have used the following identity for any integer $r \geq 1$,

$$\left[\frac{1}{(1-\tau)^{r+1/2}} \right]_k = r \frac{\binom{2k+2r}{2k}}{\binom{k+r}{k+1}} m_k , \qquad (5.9)$$

or in equivalent form,

$$\left[\frac{1}{(1-\tau)^{r+1/2}} \right]_k = \frac{1}{2^{2k} k!} \cdot \frac{(2k+2r)!}{(2r)!} \cdot \frac{r!}{(k+r)!}. \qquad (5.10)$$

Two particular cases are important:

$$\frac{(2k+2)(2k+1)}{2} m_k = \left[\frac{1}{(1-\tau)^{3/2}} \right]_k. \qquad (5.11)$$

and

$$\frac{(2k+1)(2k+2)(2k+3)}{3!} m_k = \left[\frac{1}{(1-\tau)^{5/2}} \right]_k. \qquad (5.12)$$

We also use the equality

$$\left[\frac{1}{(1-\tau)^{l+1}} \right]_k = \frac{(k+1)\cdots(k+l)}{l!} = \frac{(k+l)!}{k!\, l!}. \qquad (5.13)$$

Acknowledgments. The author is grateful to Prof. M. Ledoux for the constant interest to this work and to Prof. A. Rouault for numerous remarks and comments. The author also thanks the anonymous referee for the careful reading of the manuscript and for a number of corrections and useful suggestions that improve the presentation. This work was partially supported by the "Fonds National de la Science (France)" via the ACI program "Nouvelles Interfaces des Mathématiques", project MALCOM n° 205.

References

1. Bai, Z.D. and Yin, Y. Q. Necessary and sufficient conditions for almost sure convergence of the largest eigenvalue of a Wigner matrix. *Ann. Probab.* **16** (1988) 1729-1741

2. Boutet de Monvel, A. and Khorunzhy, A. Asymptotic distribution of smoothed eigenvalue density. I. Gaussian random matrices. *Random Oper. Stochastic Equations*, **7** (1999) 1–22

3. Boutet de Monvel, A. and Khorunzhy, A. On the norm and eigenvalue distribution of large random matrices, *Ann. Probab.*, **27** (1999) 913-944

4. Bronk, B. V. Accuracy of the semicircle approximation for the density of eigenvalues of random matrices, *J. Math. Phys.* **5** (1964) 215-220

5. Casati, G. and Girko, V. Wigner's semicircle law for band random matrices, *Rand. Oper. Stoch. Equations* **1** (1993) 15-21

6. Casati G., Molinari, L., and Izrailev, F. Scaling properties of band random matrices, *Phys. Rev. Lett.* **64** (1990) 1851

7. Furedi, Z. and Komlos, J. The eigenvalues of random symmetric matrices, *Combinatorica* **1** (1981) 233-241

8. Fyodorov, Y. V. and Mirlin, A. D. Scaling properties of localization in random band matrices: a σ-model approach, *Phys. Rev. Lett* **67** (1991) 2405

9. Geman, S. A limit theorem for the norm of random matrices, *Ann. Probab.* **8** (1980) 252-261

10. Haagerup, U. and Thornbjørnsen, S. Random matrices with complex gaussian entries, *Expo. Math.* **21** (2003) 293-337

11. Harer, J. and Zagier, D. The Euler characteristics of the moduli space of curves, *Invent. Math.* **85** (1986) 457-485

12. Khorunzhy, A. and Kirsch, W. On asymptotic expansions and scales of spectral universality in band random matrix ensembles, *Commun. Math. Phys.* **231** (2002) 223-255

13. Kuś, M., Lewenstein, M., and Haake, F. Density of eigenvalues of random band matrices. *Phys. Rev. A* **44** (1991) 2800–2808

14. Ledoux, M. A remark on hypercontractivity and tail inequalities for the largest eigenvalues of random matrices, *Séminaire de Probabilités XXXVII,* Lecture Notes in Mathematics 1832, 360-369. Springer (2003).

15. Ledoux, M. Deviation inequalities on largest eigenvalues. *Summer School on the Connections between Probability and Geometric Functional Analysis*, Jerusalem, 14-19 June 2005.

16. Mehta, M.L. *Random Matrices*, Academic Press, New York (1991)

17. S.A. Molchanov, L.A. Pastur, A.M. Khorunzhy. Eigenvalue distribution for band random matrices in the limit of their infinite rank, *Theoret. and Math. Phys.* **90** (1992) 108–118
18. Soshnikov, A. Universality at the edge of the spectrum in Wigner random matrices, *Comm. Math. Phys.* **207** (1999) 697-733
19. Tracy, C.A. and Widom, H. Level spacing distribution and the Airy kernel. *Commun. Math. Phys.* **161** (1994) 289-309
20. Wigner, E. Characteristic vectors of bordered matrices with infinite dimensions, *Ann. Math.* **62** (1955) 548-564

Geometric interpretation of the cumulants for random matrices previously defined as convolutions on the symmetric group

M. Capitaine[1] and M. Casalis[2]

[1] CNRS, LSP, Université Paul Sabatier
118 route de Narbonne, 31062 Toulouse Cedex, France
e-mail: capitain@cict.fr
[2] LSP, Université Paul Sabatier
118 route de Narbonne, 31062 Toulouse Cedex, France
e-mail: casalis@cict.fr

Summary. We show that, dealing with an appropriate basis, the cumulants for $N \times N$ random matrices $(A_1, \ldots, A_n)$, previously defined in [2] and [3], are the coordinates of $\mathbb{E}\{\Pi(A_1 \otimes \cdots \otimes A_n)\}$, where Π denotes the orthogonal projection of $A_1 \otimes \cdots \otimes A_n$ on the space of invariant vectors of $\mathcal{M}_N^{\otimes n}$ under the natural action of the unitary, respectively orthogonal, group. In this way we make the connection between [5] and [2], [3]. We also give a new proof in that context of the properties satisfied by these matricial cumulants.

Mathematics Subject Classification (2000): 15A52, 46L54.

Key words: Cumulants, Random matrices, free probability, invariant vectors under the action of the unitary, orthogonal or symplectic group

1 Introduction

For any $N \times N$ complex matrix X, we have constructed matricial cumulants $(C_n^U(X))_{n \leq N}$ in [2] (resp. $(C_n^O(X))_{n \leq N}$ in [3]) such that if X, Y are $N \times N$ independent complex matrices and U (resp. O) is a Haar distributed unitary (resp. orthogonal) $N \times N$ matrix independent of X, Y, then for any $n \leq N$,

$$C_n^U(X + UYU^*) = C_n^U(X) + C_n^U(Y),$$

$$C_n^O(X + OYO^t) = C_n^O(X) + C_n^O(Y).$$

We defined the $C_n^U(X)$ (resp. $C_n^O(X)$) as the value on the single cycle $(1 \ldots n)$ of a cumulant function $C^U(X)$ (resp. $C^O(X)$) on the symmetric group $\mathcal{S}_n$

(resp. $\mathcal{S}_{2n}$) of the permutations on $\{1,\dots,n\}$ (resp. $\{1,\dots,n,\bar{1},\dots,\bar{n}\}$). Note that we defined more generally cumulant functions for a n-tuple $(X_1,\dots,X_n)$ of $N\times N$ complex matrices. The aim of this paper is to give a geometrical interpretation of the values of the cumulant function $C^U(X_1,\dots,X_n)$ (resp. $C^O(X_1,\dots,X_n)$). It derives from the necessary confrontation of our results with the work of Collins and Sniady on the "Integration with respect to the Haar measure on unitary, orthogonal and symplectic group", see [5].

Let us roughly explain the key ideas of this interpretation and first introduce briefly some notations. Let π be a permutation in $\mathcal{S}_n$, denote by $\mathcal{C}(\pi)$ the set of all the disjoint cycles of π and by $\gamma_n(\pi)$ the number of these cycles. Let $\varepsilon=(\varepsilon_1,\dots,\varepsilon_n)\in\{-1,1\}^n$. We set for any n-tuple $\mathbf{X}=(X_1,\dots,X_n)$ of $N\times N$ complex matrices

$$r_\pi(\mathbf{X})=r_\pi(X_1,\dots,X_n):=\prod_{C\in\mathcal{C}(\pi)}\mathrm{Tr}\left(\prod_{j\in C}X_j\right). \tag{1}$$

and

$$M^{\pm}_{\mathbf{X}}(g_{(\varepsilon,\pi)}):=r_\pi(X_1^{\varepsilon_1},\dots,X_n^{\varepsilon_n}).$$

In this last expression we set X^{-1} for the transpose X^t of the matrix X and $g_{(\varepsilon,\pi)}$ denotes some particular permutation on the symmetric group $\mathcal{S}_{2n}$ which will be made precise in Section 3.1.

These n-linear forms $r_\pi,\pi\in\mathcal{S}_n$ or $M^{\pm}(g_{(\pi,\varepsilon)}),\pi\in\mathcal{S}_n,\varepsilon\in\{-1,1\}^n$, introduced on $\mathcal{M}^n_N$ for any integer $n\geq 1$, are respectively invariant under the action of the unitary group $\mathbb{U}_N$ for the first ones and the orthogonal group $\mathbb{O}_N$ for the second ones. From the point of view of [5], they canonically define linear forms on the tensor product $\mathcal{M}^{\otimes n}_N$ which also are invariant under the corresponding action of $\mathbb{U}_N$, respectively $\mathbb{O}_N$. As $\mathcal{M}^{\otimes n}_N$ is naturally endowed with a non degenerate quadratic form $(u,v)\mapsto\langle u,v\rangle$, these linear forms correspond in the first case to vectors u_π, $\pi\in\mathcal{S}_n$, of $\mathcal{M}^{\otimes n}_N$ which are $\mathbb{U}_N$-invariant, and in the second one to vectors $u_{\eta(g_{\varepsilon,\pi})}$, $\varepsilon\in\{-1;1\}^n$, $\pi\in\mathcal{S}_n$, which are $\mathbb{O}_N$-invariant (η will be defined in Section 3.3). Thus they satisfy

$$r_\pi(X_1,\dots,X_n)=\langle X_1\otimes\dots\otimes X_n,u_\pi\rangle$$

respectively

$$M^+_{\mathbf{X}}(g_{(\varepsilon,\pi)})=\langle X_1\otimes\dots\otimes X_n,u_{\eta(g_{\varepsilon,\pi})}\rangle.$$

Actually, for $n\leq N$, $\{u_\pi\ ;\ \pi\in\mathcal{S}_n\}$ forms a basis of the space $[\mathcal{M}^{\otimes n}_N]^{\mathbb{U}_N}$ of $\mathbb{U}_N$-invariant vectors, while a basis of the space $[\mathcal{M}^{\otimes n}_N]^{\mathbb{O}_N}$ of $\mathbb{O}_N$-invariant vectors can be extracted from $\{u_{\eta(g_{\varepsilon,\pi})};\varepsilon\in\{-1;1\}^n,\ \pi\in\mathcal{S}_n\}$. Note that this last one needs the double parametrization by $\mathcal{S}_n$ and some ε in $\{-1,1\}^n$. This is the reason why, contrary to the unitary case where the adjoints are not involved, the transposes of matrices naturally occur in the orthogonal case.

We then prove that our matricial cumulants $C^U(X_1, \ldots, X_n)$ (respectively $C^O(X_1, \ldots, X_n)$) are the coordinates in this appropriate basis of $\mathbb{E}\left\{ \int U X_1 U^* \otimes \ldots \otimes U X_n U^* dU \right\}$ (respectively $\mathbb{E}\{ \int O X_1 O^t \otimes \ldots \otimes O X_n O^t dO\}$), where integration is taken with respect to the Haar measure on $\mathbb{U}_N$ (resp. $\mathbb{O}_N$).

The paper is split into two parts. The first one concerns the matricial $\mathbb{U}$-cumulants and the second one is devoted to the $\mathbb{O}$-cumulants. In each part we first recall the definition and fundamental properties satisfied by these cumulants (Sections 2.1, 2.2 and similarly 3.1, 3.2). Then we describe a basis of $[\mathcal{M}_N^{\otimes n}]^G$ in each case ($G = \mathbb{U}_N$ in Section 2.3 and $G = \mathbb{O}_N$ in Section 3.3) before giving the geometrical interpretation of our cumulants and ending with a new proof in that context of the properties they satisfy (Sections 2.4 and 3.4).

Note that the same development as for the orthogonal group can be carried out for the symplectic group $Sp(N)$. We just provide the corresponding basis of Sp-invariant vectors of $\mathcal{M}_N^{\otimes n}$ in the final section without giving more details.

Throughout the paper, we suppose $N \geq n$.

Before starting we would like to underline that the description of the subspace of invariant vectors relies on the following ideas. Note this first simple remark:

Lemma 1.1 *Let G and G' be two groups acting on a vector space V through the actions ρ and ρ' and let $[V]^G$ denote the subspace of G-invariant vectors of V. Then, when ρ and ρ' commute, for any vector $v \neq 0$ in $[V]^G$, $\{\rho'(g') \cdot v \;: g' \in G'\} \subset [V]^G$.*

Hence $[V]^G$ is known as soon as we can find a suitable group G' and some vector v in $[V]^G$ for which we get $\{\rho'(g') \cdot v \;; g' \in G'\} = [V]^G$. For the considered groups, the Schur-Weyl duality leads to the right G'. Thus for $G = GL(N, \mathbb{C})$ and $\mathbb{U}_N$, G' is chosen to be equal to $\mathcal{S}_n$. For $G = \mathbb{O}_N$ or $Sp(N)$, G' is $\mathcal{S}_{2n}$. This is well described in [8], see Theorem 4.3.1 for $GL(N, \mathbb{C})$ and Theorem 4.3.3 or Proposition 10.1.1 for $\mathbb{O}_N$ and $Sp(N)$. As for $\mathbb{U}_N$, note that any analytic function invariant by $\mathbb{U}_N$ is invariant by $GL(N, \mathbb{C})$ too (see Weyl's Theorem about analytic functions on $GL(N, \mathbb{C})$, [9]). For any $u \in [\mathcal{M}_N^{\otimes n}]^{\mathbb{U}_N}$, the analytic function on V, $A \mapsto \langle A, u \rangle$ is $\mathbb{U}_N$-invariant, hence is $GL(N, \mathbb{C})$-invariant. Thus, for any A in $\mathcal{M}_N^{\otimes n}$ and any $G \in GL(N, \mathbb{C})$, $\langle A, u \rangle = \langle A, G^{-1} u G \rangle$ and hence $u \in [\mathcal{M}_N^{\otimes n}]^{GL(N, \mathbb{C})}$. It readily comes that $[\mathcal{M}_N^{\otimes n}]^{GL(N, \mathbb{C})} = [\mathcal{M}_N^{\otimes n}]^{\mathbb{U}_N}$.

2 Matricial $\mathbb{U}$-cumulants

We refer the reader to [2] where the present section is developed and we just recall here the fundamental results.

2.1 Definition and first properties

Denote by $*$ the classical convolution operation on the space of complex functions on $\mathcal{S}_n$,

$$f * g(\pi) = \sum_{\sigma \in \mathcal{S}_n} f(\sigma)g(\sigma^{-1}\pi) = \sum_{\rho \in \mathcal{S}_n} f(\pi\rho^{-1})g(\rho),$$

and by id the identity of $\mathcal{S}_n$. Recall that the $*$-unitary element is

$$\delta_{id} := \pi \rightarrow \begin{cases} 1 \text{ if } \pi = id \\ 0 \text{ else} \end{cases},$$

that is $f * \delta_{id} = \delta_{id} * f = f$ for all f. The inverse function of f for $*$, if it exists, is denoted by $f^{(-1)}$ and satisfies $f * f^{(-1)} = f^{(-1)} * f = \delta_{id}$. In particular the function $\pi \mapsto x^{\gamma_n(\pi)}$ is $*$-invertible for $n - 1 < |x|$ (see [6]). Moreover, since γ_n is central (that is, constant on the conjugacy classes), x^{γ_n} and thus $(x^{\gamma_n})^{(-1)}$ commute with any function f defined on $\mathcal{S}_n$.
Recall the definition of the $\mathbb{U}$-cumulants introduced in [2].

Definition 2.1 *For $n \leq N$, for any n-tuple $\mathbf{X} = (X_1, \ldots, X_n)$ of random $N \times N$ complex matrices, the n-th $\mathbb{U}$-cumulant function $C^U(\mathbf{X}) : \mathcal{S}_n \rightarrow \mathbb{C}$, $\pi \mapsto C_\pi^U(\mathbf{X})$ is defined by the relation*

$$C^U(\mathbf{X}) := \mathbb{E}(r(\mathbf{X})) * (N^{\gamma_n})^{(-1)}.$$

The $\mathbb{U}$-cumulants of $\mathbf{X}$ are the $C_\pi^U(\mathbf{X})$ for single cycles π of $\mathcal{S}_n$.
For a single matrix X, $C^U(\mathbf{X})$ where $\mathbf{X} = (X, \cdots, X)$ will be simply denoted by $C^U(X)$.

For example, if $\mathrm{tr}_N = \frac{1}{N}\mathrm{Tr}$,

$$C_{(1)}^U(X) = \mathbb{E}(\mathrm{tr}_N(X))$$

$$C_{(1)(2)}^U(X_1, X_2) = \frac{N\mathbb{E}\{\mathrm{Tr}(X_1)\mathrm{Tr}(X_2)\} - \mathbb{E}\{\mathrm{Tr}(X_1 X_2)\}}{N(N^2 - 1)}$$

$$C_{(1\,2)}^U(X_1, X_2) = \frac{-\mathbb{E}\{\mathrm{Tr}(X_1)\mathrm{Tr}(X_2)\} + N\mathbb{E}\{\mathrm{Tr}(X_1 X_2)\}}{N(N^2 - 1)}.$$

Here are some basic properties remarked in [2]. First, for each π in $\mathcal{S}_n$, $(X_1, \ldots, X_n) \mapsto C_\pi^U((X_1, \ldots, X_n))$ is obviously n-linear. Moreover it is clear that for any unitary matrix U,

$$C_\pi^U(U^*X_1 U, \ldots, U^*X_n U) = C_\pi^U(X_1, \ldots, X_n).$$

Now,

1. For any π and σ in $\mathcal{S}_n$,

$$C_\pi^U((X_{\sigma(1)},\ldots,X_{\sigma(n)})) = C_{\sigma\pi\sigma^{-1}}^U((X_1,\ldots,X_n)). \tag{2}$$

2. $C_\pi^U(X)$ depends only of the conjugacy class of π.

Thus the cumulants $C_\pi^U(X)$ of a matrix X for single cycles π of $\mathcal{S}_n$ are all equal so that we denote by $C_n^U(X)$ this common value. We call it *cumulant of order n of the matrix X*. In particular, $C_1^U(X) = \mathbb{E}(\mathrm{tr}_N X)$ and $C_2^U(X) = \frac{N}{N^2-1}\left[\mathbb{E}\{\mathrm{tr}_N(X^2)\} - \mathbb{E}\{(\mathrm{tr}_N X)^2\}\right]$. We also proved the following

Proposition 2.1 *For any $k < n \le N$, any π in $\mathcal{S}_n$, then*

$$C_\pi^U(X_1,\ldots,X_k,I_N,\ldots,I_N)$$
$$= \begin{cases} C_\rho^U(X_1,\ldots,X_k) & \text{if } \pi = (n)\ldots(k+1)\rho \text{ for some } \rho \in \mathcal{S}_k, \\ 0 & \text{else.} \end{cases}$$

Now recall the fundamental properties we proved in [2] and which motivated the terminology of cumulants.

2.2 Fundamental properties

2.2.1 Mixed moments of two independent tuples

In [2] we have proved the following theorem with great analogy with the results of [10] about the multiplication of free n-tuples.

Theorem 2.1 *Let $\mathbf{X} = (X_1,\ldots,X_n)$ and $\mathbf{B} = (B_1,\ldots,B_n)$ two independent n-tuple of $N \times N$ random complex matrices such that the distribution of $\mathbf{X}$ is invariant under unitary conjugations, namely $\forall U \in \mathbb{U}_N$, $\mathcal{L}(UX_1U^*,\ldots,UX_nU^*) = \mathcal{L}(X_1,\ldots,X_n)$. Then we have for any π in $\mathcal{S}_n$:*

$$\mathbb{E}\left(r_\pi(B_1X_1,\ldots,B_nX_n)\right) = \{\mathbb{E}(r(\mathbf{B})) * C^U(\mathbf{X})\}(\pi) = \{C^U(\mathbf{B}) * \mathbb{E}(r(\mathbf{X}))\}(\pi)$$

From Theorem 2.1 we readily get the following convolution relation which has to be related to Theorem 1.4 in [10].

Corollary 2.1 *With the hypothesis of Theorem 2.1,*

$$C^U(X_1B_1,\ldots,X_nB_n) = C^U(\mathbf{X}) * C^U(\mathbf{B}).$$

If $\mathbf{X} = (X_1,\ldots,X_n)$ and $\mathbf{B} = (B_1,\ldots,B_n)$ are two independent n-tuple of $N \times N$ random complex matrices such that the distribution of $\mathbf{X}$ is invariant under orthogonally conjugations, namely $\forall O \in \mathbb{O}_N$, $\mathcal{L}(OX_1O^t,\ldots,OX_nO^t) = \mathcal{L}(X_1,\ldots,X_n)$, the mixed moments $\mathbb{E}\left(r_\pi(B_1X_1,\ldots,B_nX_n)\right)$ can still be expressed by a convolution relation but on $\mathcal{S}_{2n}$; consequently we were led to introduce in [3] another cumulant function $C^O : \mathcal{S}_{2n} \to \mathbb{C}$, recalled in Section 3.

2.2.2 Linearizing property

Proposition 2.1 together with Corollary 2.1 imply that the cumulants $C_n^U(X_1, \ldots, X_n)$ vanish as soon as the involved matrices $(X_1, \ldots, X_n)$ are taken in two independent sets, one having distribution invariant under unitary conjugation; therefore they do linearize the convolution, namely if X_1, X_2 are two independent matrices such that $\mathcal{L}(UX_1U^*) = \mathcal{L}(X_1)$, $\forall U \in \mathbb{U}_N$, then

$$C_n^U(X_1 + X_2) = C_n^U(X_1) + C_n^U(X_2).$$

2.2.3 Asymptotic behavior

We refer the reader to [12] for noncommutative probability space and freeness and to [11] and [10] for free cumulants. Let $(\mathcal{A}, \Phi)$ be a noncommutative probability space. For any noncommutative random variables $(a_1, \ldots, a_n)$ in $(\mathcal{A}, \Phi)$ and for any $\pi = \prod_{i=1}^{r} \pi_i$ in $\mathcal{S}_n$ with $\pi_i = (l_{i,1}, l_{i,2}, \ldots, l_{i,n_i})$, we write

$$\phi_\pi(a_1, \ldots, a_n) := \prod_{i=1}^{r} \phi(a_{l_{i,1}} a_{l_{i,2}} \cdots a_{l_{i,n_i}}),$$

$$k_\pi(a_1, \ldots, a_n) := \prod_{i=1}^{r} k_{n_i}(a_{l_{i,1}}, a_{l_{i,2}}, \ldots, a_{l_{i,n_i}}),$$

where $(k_n)_{n \in \mathbb{N}}$ stand for the free cumulants. For any n-tuple $(X_1, \ldots, X_n)$ of $N \times N$ matrices, we define the *normalized generalized moments* $\mathbb{E}(r_\pi^{(N)}(X_1, \ldots, X_n))$ where π is in $\mathcal{S}_n$ by setting

$$\mathbb{E}(r_\pi^{(N)}(X_1, \ldots, X_n)) = \frac{1}{N^{\gamma_n(\pi)}} \mathbb{E}(r_\pi(X_1, \ldots, X_n)) = \mathbb{E}\left(\prod_{C \in \mathcal{C}(\pi)} \frac{1}{N} \mathrm{Tr}\left(\prod_{j \in C} X_j \right) \right).$$

We also define the normalized cumulants by

$$(C_\pi^U)^{(N)}(X_1, \ldots, X_n) := N^{n - \gamma_n(\pi)} C_\pi^U(X_1, \ldots, X_n).$$

In [2] we prove the following equivalence.

Proposition 2.2 *Let $(X_1, \ldots, X_n)$ be a n-tuple of $N \times N$ matrices. Let $(x_1, \ldots, x_n)$ be non commutative variables in $(\mathcal{A}, \phi)$. The following equivalence holds,*

$$\mathbb{E}(r_\pi^{(N)}(X_1, \ldots, X_n)) \underset{N \to +\infty}{\longrightarrow} \phi_\pi(x_1, \ldots, x_n), \ \forall \ \pi \in \mathcal{S}_n$$

$$\Leftrightarrow (C_\pi^U)^{(N)}(X_1, \ldots, X_n) \underset{N \to +\infty}{\longrightarrow} k_\pi(x_1, \ldots, x_n), \ \forall \ \pi \in \mathcal{S}_n.$$

2.3 Action of the unitary group on the space of complex matrices

We first need to precisely state some basic generalities and notations. Let $(e_1, \ldots, e_N)$ be the canonical basis of $\mathbb{C}^N$. Endow $\mathbb{C}^N$ with the usual Hermitian product $\langle \sum_i u_i e_i, \sum_i v_i e_i \rangle_{\mathbb{C}^N} = \sum_i u_i \overline{v_i}$. Thus the dual space $(\mathbb{C}^N)^*$ is composed by the linear forms $v^* : \mathbb{C}^N \to \mathbb{C}, u \mapsto \langle u, v \rangle_{\mathbb{C}^N}$ with $v \in \mathbb{C}^N$. Let $(e_1^*, \ldots, e_N^*)$ be the dual basis. First consider the tensor product $\mathbb{C}^N \otimes (\mathbb{C}^N)^*$ with orthonormal basis $e_i \otimes e_j^*$, $i, j = 1, \ldots, N$ with respect to the Hermitian product

$$\langle u_1 \otimes v_1^*, u_2 \otimes v_2^* \rangle_{\mathbb{C}^N \otimes (\mathbb{C}^N)^*} = \langle u_1, u_2 \rangle_{\mathbb{C}^N} \langle v_2, v_1 \rangle_{\mathbb{C}^N}.$$

The unitary group $\mathbb{U}_N$ acts on $\mathbb{C}^N \otimes (\mathbb{C}^N)^*$ as follows:

$$\widetilde{\rho}(U)(e_i \otimes e_j^*) = U e_i \otimes (U e_j)^*.$$

Now consider $\mathcal{M}_N$ with canonical basis $(E_{a,b})_{a,b=1,\ldots,N}$ defined by $(E_{a,b})_{ij} = \delta_{a,i}\delta_{b,j}$, and with Hermitian product $\langle A, B \rangle_{\mathcal{M}_N} = \mathrm{Tr}(AB^*)$. It is well-known that $\mathcal{M}_N$ and $\mathbb{C}^N \otimes (\mathbb{C}^N)^*$ are isomorphic Hermitian vector spaces when we identify any $M = (M_{ij})_{1 \leq i,j \leq N} \in \mathcal{M}_N$ with $\tilde{M} = \sum_{1 \leq i,j \leq N} M_{ij} e_i \otimes e_j^*$ (and hence $\tilde{E}_{a,b} = e_a \otimes e_b^*$). Besides the action $\widetilde{\rho}$ corresponds on $\mathcal{M}_N$ to

$$\rho(U)(M) = U M U^*.$$

Note also that the inner product AB in $\mathcal{M}_N$ corresponds to the product defined by

$$(u_1 \otimes v_1^*).(u_2 \otimes v_2^*) = \langle u_2, v_1 \rangle_{\mathbb{C}^N} \; u_1 \otimes v_2^*,$$

and the adjoint A^* to the following rule: $(u \otimes v^*)^* = v \otimes u^*$.

More generally, for any n, the tensor products $\mathcal{M}_N^{\otimes n}$ and $(\mathbb{C}^N \otimes (\mathbb{C}^N)^*)^{\otimes n}$ are isomorphic through the map: $A = A_1 \otimes \cdots \otimes A_n \mapsto \tilde{A} = \tilde{A}_1 \otimes \cdots \otimes \tilde{A}_n$ and with Hermitian product

$$\langle A_1 \otimes \cdots \otimes A_n, B_1 \otimes \cdots \otimes B_n \rangle_{\mathcal{M}_N^{\otimes n}}$$
$$= \prod_{i=1}^{n} \mathrm{Tr}(A_i B_i^*) = \prod_{i=1}^{n} \langle \tilde{A}_i, \tilde{B}_i \rangle_{\mathbb{C}^N \otimes (\mathbb{C}^N)^*}$$
$$= \langle \tilde{A}_1 \otimes \cdots \otimes \tilde{A}_n, \tilde{B}_1 \otimes \cdots \otimes \tilde{B}_n \rangle_{(\mathbb{C}^N \otimes (\mathbb{C}^N)^*)^{\otimes n}}.$$

Here again the following actions of $\mathbb{U}_N$ are equivalent:

$$\text{on } (\mathbb{C}^N \otimes (\mathbb{C}^N)^*)^{\otimes n} \quad \widetilde{\rho}_n(U)(e_{i_1} \otimes e_{i_{\bar{1}}}^* \otimes \cdots \otimes e_{i_n} \otimes e_{i_{\bar{n}}}^*)$$
$$= U e_{i_1} \otimes (U e_{i_{\bar{1}}})^* \otimes \cdots \otimes U e_{i_n} \otimes (U e_{i_{\bar{n}}})^*,$$
$$\text{on } \mathcal{M}_N^{\otimes n} \quad \rho_n(U)(A_1 \otimes \cdots \otimes A_n) = U A_1 U^* \otimes \cdots \otimes U A_n U^*.$$

Denote by $[V]^{\mathbb{U}_N}$ the subspace of $\mathbb{U}_N$-invariant vectors of V with $V = \mathcal{M}_N^{\otimes n}$ or $(\mathbb{C}^N \otimes (\mathbb{C}^N)^*)^{\otimes n}$. Clearly $[\mathcal{M}_N^{\otimes n}]^{\mathbb{U}_N}$ and $[(\mathbb{C}^N \otimes (\mathbb{C}^N)^*)^{\otimes n}]^{\mathbb{U}_N}$ are isomorphic too. Consequently from now on we identify $\mathcal{M}_N^{\otimes n}$ and $(\mathbb{C}^N \otimes (\mathbb{C}^N)^*)^{\otimes n}$. We also simply denote the Hermitian product by $\langle .,. \rangle$ from now on throughout Section 2. Note lastly that the inner product in $\mathcal{M}_N^{\otimes n}$ is defined by

$$(A_1 \otimes \cdots \otimes A_n).(B_1 \otimes \cdots \otimes B_n) = A_1 B_1 \otimes \cdots \otimes A_n B_n,$$

and the adjunction by $(A_1 \otimes \cdots \otimes A_n)^* = A_1^* \otimes \cdots \otimes A_n^*$. They satisfy for any $u, v, w \in \mathcal{M}_N^{\otimes n}$:

$$\langle u.v, w \rangle = \langle v, u^*.w \rangle = \langle u, w.v^* \rangle. \tag{3}$$

In the following proposition we determine a basis of $[\mathcal{M}_N^{\otimes n}]^{\mathbb{U}_N}$. We use the previous identification in the proof.

Proposition 2.3 *For any permutation σ in $\mathcal{S}_n$, define*

$$u_\sigma := \sum_{i_1, \ldots, i_n} E_{i_{\sigma^{-1}(1)} i_1} \otimes \cdots \otimes E_{i_{\sigma^{-1}(n)} i_n}.$$

Then $\{u_\sigma \; ; \; \sigma \in \mathcal{S}_n\}$ generates $[\mathcal{M}_N^{\otimes n}]^{\mathbb{U}_N}$. Moreover when $N \geq n$, it is a basis of $[\mathcal{M}_N^{\otimes n}]^{\mathbb{U}_N}$.

Proof: The first part of Proposition 2.3 derives from Theorem 4.3.1 in [8]. We briefly recall how this set is introduced before showing that it forms a basis of $[\mathcal{M}_N^{\otimes n}]^{\mathbb{U}_N}$. We work on $(\mathbb{C}^N \otimes (\mathbb{C}^N)^*)^{\otimes n}$ where we consider another group action and a specific invariant vector in order to apply lemma 1.1. Define

$$\Theta_n := \underbrace{I_N \otimes \ldots \otimes I_N}_{n \text{ times}} = \sum_{i_1, \ldots, i_n} e_{i_1} \otimes e_{i_1}^* \otimes \cdots \otimes e_{i_n} \otimes e_{i_n}^*.$$

It is clear that $\Theta_n \in [\mathcal{M}_N^{\otimes n}]^{\mathbb{U}_N}$. Consider now the natural action ρ' of $\mathcal{S}_n \times \mathcal{S}_n$ on $(\mathbb{C}^N \otimes (\mathbb{C}^N)^*)^{\otimes n}$ defined for any permutations σ and τ in $\mathcal{S}_n$ acting respectively on $\{1, \ldots, n\}$ and $\{\bar{1}, \ldots, \bar{n}\}$ by

$$\rho'((\sigma, \tau))(e_{i_1} \otimes e_{i_{\bar{1}}}^* \otimes \cdots \otimes e_{i_n} \otimes e_{i_{\bar{n}}}^*)$$
$$= e_{i_{\sigma^{-1}(1)}} \otimes e_{i_{\tau^{-1}(\bar{1})}}^* \otimes \cdots \otimes e_{i_{\sigma^{-1}(n)}} \otimes e_{i_{\tau^{-1}(\bar{n})}}^*.$$

The actions $\tilde{\rho}_n$ and ρ' obviously commute. Hence, according to Lemma 1.1, for all (σ, τ) in $\mathcal{S}_n \times \mathcal{S}_n$, $\rho'((\sigma, \tau)) \cdot \Theta_n$ belongs to $[\mathcal{M}_N^{\otimes n}]^{\mathbb{U}_N}$. Note that, since $(\sigma, \tau) = (\sigma\tau^{-1}, id)(\tau, \tau)$ and $\rho'((\tau, \tau)) \cdot \Theta_n = \Theta_n$, then

$$\{\rho'((\sigma, \tau)) \cdot \Theta_n \; ; \; (\sigma, \tau) \in \mathcal{S}_n \times \mathcal{S}_n\} = \{\rho'((\sigma, id)) \cdot \Theta_n \; ; \; \sigma \in \mathcal{S}_n\}.$$

Thus we simply denote $\rho'((\sigma, id))$ by $\rho'(\sigma)$ and we set

$$u_\sigma = \rho'(\sigma) \cdot \Theta_n.$$

Remark that $u_{id} = \Theta_n$. Note also that u_σ corresponds to $\rho_{\mathcal{S}_n}^N(\sigma)$ in [5].

From Theorem 4.3.1 in [8], the set $\{\rho'(\sigma) \cdot \Theta_n \; ; \; \sigma \in \mathcal{S}_n\}$ generates $[\mathcal{M}_N^{\otimes n}]^{GL(N,\mathbb{C})} = [\mathcal{M}_N^{\otimes n}]^{\mathbb{U}_N}$ (see [9]). We now prove that it is a basis when $N \geq n$.

One can easily see that the adjoint of $\rho'((\sigma, \tau))$ satisfies $\rho'((\sigma, \tau))^* = \rho'((\sigma^{-1}, \tau^{-1}))$ so that

$$\langle u_\sigma, u_{\sigma'} \rangle = \langle \Theta_n, u_{\sigma^{-1}\sigma'} \rangle = \langle u_{\sigma'^{-1}\sigma}, \Theta_n \rangle .$$

Now from (1) we get:

$$\langle \Theta_n, u_\sigma \rangle = \sum_{i_1,\dots,i_n} \prod_{l=1}^n \delta_{i_l, i_{\sigma(l)}} = \sum_{i_1,\dots,i_n} r_\sigma(E_{i_1,i_1}, \dots, E_{i_n,i_n})$$
$$= r_\sigma(I_N, \dots, I_N) = N^{\gamma(\sigma)},$$

so that

$$\langle u_\sigma, u_{\sigma'} \rangle = N^{\gamma(\sigma^{-1}\sigma')}.$$

Let $G = (\langle u_\sigma, u_{\sigma'} \rangle)_{\sigma, \sigma' \in \mathcal{S}_n \times \mathcal{S}_n}$ be the Gramm matrix of $\{u_\sigma \; ; \; \sigma \in \mathcal{S}_n\}$. Let $a = (a_\sigma)_{\sigma \in \mathcal{S}_n}$ and $b = (b_\sigma)_{\sigma \in \mathcal{S}_n}$ be in $\mathbb{C}^{n!}$. We have:

$$Ga = b \Leftrightarrow \sum_{\sigma' \in \mathcal{S}_n} \langle u_\sigma, u_{\sigma'} \rangle a_{\sigma'} = b_\sigma \; \forall \sigma \in \mathcal{S}_n$$
$$\Leftrightarrow \sum_{\sigma' \in \mathcal{S}_n} N^{\gamma(\sigma'^{-1}\sigma)} a_{\sigma'} = b_\sigma \; \forall \sigma \in \mathcal{S}_n$$
$$\Leftrightarrow b = a * N^\gamma$$
$$\Leftrightarrow a = b * (N^\gamma)^{(-1)}$$

when $N \geq n$ since in that case N^γ is $*$-invertible. Therefore G is invertible when $N \geq n$ and $\{u_\sigma \; ; \; \sigma \in \mathcal{S}_n\}$ is a free system of vectors of $[(\mathcal{M}_N)^{\otimes n}]^{\mathbb{U}_N}$. $\square$

Here are some basic properties satisfied by the $u_\sigma, \sigma \in \mathcal{S}_n$, which can be easily proved. For any σ and τ in $\mathcal{S}_n$ and $A_1, \dots, A_n \in \mathcal{M}_N$,

$$u_\sigma^* = u_{\sigma^{-1}}, \tag{4}$$

$$u_\sigma . u_\tau = u_{\sigma\tau}, \tag{5}$$

$$\langle u_\sigma . (A_1 \otimes \cdots \otimes A_n), u_\tau \rangle = \langle A_1 \otimes \cdots \otimes A_n, u_{\sigma^{-1}\tau} \rangle, \tag{6}$$
$$\langle (A_1 \otimes \cdots \otimes A_n) . u_\sigma, u_\tau \rangle = \langle A_1 \otimes \cdots \otimes A_n, u_{\tau\sigma^{-1}} \rangle,$$

the two last ones coming from (3), (4) and (5).

Moreover, for any $k < n$, if π in $\mathcal{S}_n$ is such that $\pi = (n) \cdots (k+1)\rho$ for some ρ in $\mathcal{S}_k$, then

$$u_\pi = u_\rho \otimes \underbrace{I_N \otimes \dots \otimes I_N}_{n-k \text{ times}} \tag{7}$$

and more generally, if $\pi = \rho_1 \rho_2$ with $\rho_1 \in \mathcal{S}\{1, \ldots, k\}$ and $\rho_2 \in \mathcal{S}\{k + 1, \ldots, n\}$, then

$$u_\pi = u_{\rho_1} \otimes u_{\rho_2}. \tag{8}$$

Lastly note the following straightforward equality:

$$\rho'((\sigma, \sigma)) \cdot u_\pi = u_{\sigma \pi \sigma^{-1}}. \tag{9}$$

Here is an immediate interpretation of the generalized moments in terms of Hermitian products with the u_π.

Lemma 2.1 *For any* $A_1 \otimes \cdots \otimes A_n$ *in* $\mathcal{M}_N^{\otimes n}$ *and any* $\pi \in \mathcal{S}_n$

$$r_\pi(A_1, \ldots, A_n) = \langle A_1 \otimes \ldots \otimes A_n, u_\pi \rangle. \tag{10}$$

Proof: We have:

$$
\begin{aligned}
\langle A_1 \otimes \ldots \otimes A_n, u_\pi \rangle &= \sum_{i_1, \ldots, i_n} \mathrm{Tr}(A_1 E_{i_1 i_{\pi^{-1}(1)}}) \cdots \mathrm{Tr}(A_n E_{i_n i_{\pi^{-1}(n)}}) \\
&= \sum_{i_1, \ldots, i_n} (A_1)_{i_{\pi^{-1}(1)} i_1} \cdots (A_n)_{i_{\pi^{-1}(n)} i_n} \\
&= \sum_{j_1, \ldots, j_n} (A_1)_{j_1 j_{\pi(1)}} \cdots (A_n)_{j_n j_{\pi(n)}} \\
&= r_\pi(A_1, \ldots, A_n). \qquad \square
\end{aligned}
$$

2.4 Geometrical interpretation of the $\mathbb{U}$-cumulants

In [5] the authors introduce the linear map Π of $\mathcal{M}_N^{\otimes n}$ on $[\mathcal{M}_N^{\otimes n}]^{\mathbb{U}_N}$ defined for any $A_1 \otimes \cdots \otimes A_n$ by:

$$\Pi(A_1 \otimes \ldots \otimes A_n) := \int_{\mathbb{U}_N} U A_1 U^* \otimes \ldots \otimes U A_n U^* dU = \int_{\mathbb{U}_N} \rho_n(U)(A_1 \otimes \cdots \otimes A_n) dU$$

where integration is performed with respect to the Haar measure on $\mathbb{U}_N$. Note that they call it the conditional expectation onto $[\mathcal{M}_N^{\otimes n}]^{\mathbb{U}_N}$ and denote it by $\mathbb{E}(A_1 \otimes \ldots \otimes A_n)$ but we prefer to adopt the previous notation $\Pi(A_1 \otimes \ldots \otimes A_n)$ in order to stay faithful to our notations of the expectation in [1] and [3] and also to underline the property of orthogonal projection mentioned in [5] instead of conditional expectation. Indeed it is easy to verify that for any $\mathbf{B} \in [\mathcal{M}_N^{\otimes n}]^{\mathbb{U}_N}$ and any $\mathbf{A} \in \mathcal{M}_N^{\otimes n}$,

$$\langle \Pi(\mathbf{A}), \mathbf{B} \rangle = \int_{\mathbb{U}_N} \langle \rho_n(U)(\mathbf{A}), \mathbf{B} \rangle dU = \int_{\mathbb{U}_N} \langle \mathbf{A}, \rho_n(U^*)(\mathbf{B}) \rangle dU = \langle \mathbf{A}, \mathbf{B} \rangle.$$

We first get the following proposition in the same spirit as formula (10) in [5]. It will be one of the key tools when recovering of the properties of Section 2.2.

Proposition 2.4 *Let* $\mathbf{A} = (A_1, \ldots, A_n)$ *and* $\mathbf{B} = (B_1, \ldots, B_n)$ *be two independent sets of* $N \times N$ *matrices such that the distribution of* $\mathbf{A}$ *is invariant under unitary conjugation, i.e., for any deterministic unitary matrix* U, $(UA_1U^*, \ldots, UA_nU^*)$ *and* $(A_1, \ldots, A_n)$ *are identically distributed. Then*

$$\mathbb{E}\left(\Pi(A_1 B_1 \otimes \ldots \otimes A_n B_n)\right)$$
$$= \mathbb{E}\left(\Pi(A_1 \otimes \ldots \otimes A_n)\right) . \mathbb{E}\left(\Pi(B_1 \otimes \ldots \otimes B_n)\right). \tag{11}$$

Proof:

$$\mathbb{E}\left(\Pi(A_1 B_1 \otimes \ldots \otimes A_n B_n)\right)$$
$$= \mathbb{E}\left(\int U_1 A_1 B_1 U_1^* \otimes \ldots \otimes U_1 A_n B_n U_1^* \, dU_1\right)$$
$$\overset{(a)}{=} \int \mathbb{E}\left(\int U_1 U_2 A_1 U_2^* B_1 U_1^* \otimes \ldots \otimes U_1 U_2 A_n U_2^* B_n U_1^* \, dU_1\right) dU_2$$
$$\overset{(b)}{=} \mathbb{E}\left(\int \int U A_1 U^* U_1 B_1 U_1^* \otimes \ldots \otimes U A_n U^* U_1 B_n U_1^* \, dU_1 dU\right)$$
$$\overset{(c)}{=} \mathbb{E}\left(\Pi(A_1 \otimes \ldots \otimes A_n)\right) . \mathbb{E}\left(\Pi(B_1 \otimes \ldots \otimes B_n)\right),$$

where we used the invariance under unitary conjugaison of the distribution of $\mathbf{A}$ in (a), a change of variable U for $U_1 U_2$ in (b) and the independence of $\mathbf{A}$ and $\mathbf{B}$ in (c). $\qquad\square$

Here is the main result of the section:

Theorem 2.2 *Let* $A_1, \cdots, A_n$ *be in* $\mathcal{M}_N$, $N \geq n$. *Then the matricial* $\mathbb{U}$-*cumulants of* $(A_1, \cdots, A_n)$, $C_\sigma^U(A_1, \ldots, A_n)$ *with* $\sigma \in \mathcal{S}_n$, *are the coordinates of* $\mathbb{E}\left(\Pi(A_1 \otimes \ldots \otimes A_n)\right)$ *in the basis* $\{u_\sigma, \sigma \in \mathcal{S}_n\}$:

$$\mathbb{E}\left(\Pi(A_1 \otimes \ldots \otimes A_n)\right) = \sum_{\sigma \in \mathcal{S}_n} C_\sigma^U(A_1, \ldots, A_n) u_\sigma.$$

Proof: According to Proposition 2.3, there exist $\{\tilde{C}_\sigma(A_1, \ldots, A_n), \sigma \in \mathcal{S}_n\}$ in $\mathbb{C}$ such that
$$\Pi(A_1 \otimes \ldots \otimes A_n) = \sum_{\sigma \in \mathcal{S}_n} \tilde{C}_\sigma(A_1, \ldots, A_n) u_\sigma.$$

Then, using (10),

$$r_\pi(A_1, \ldots, A_n) = \langle \Pi(A_1 \otimes \ldots \otimes A_n), u_\pi \rangle = \sum_{\sigma \in \mathcal{S}_n} \tilde{C}_\sigma(A_1, \ldots, A_n) \langle u_\sigma, u_\pi \rangle$$

$$= \sum_{\sigma \in \mathcal{S}_n} \tilde{C}_\sigma(A_1, \ldots, A_n) N^{\gamma(\sigma^{-1}\pi)} = \tilde{C}(A_1, \ldots, A_n) * N^\gamma(\pi).$$

Thus,

$$\mathbb{E}\left(r_\pi(A_1,\ldots,A_n)\right) = \mathbb{E}\left(\tilde{C}(A_1,\ldots,A_n)\right) * N^\gamma(\pi).$$

On the other hand, by definition of the $C^U(A_1,\ldots,A_n)$, we have

$$\mathbb{E}\left(r_\pi(A_1,\ldots,A_n)\right) = C^U(A_1,\ldots,A_n) * N^\gamma(\pi).$$

Since N^γ is invertible for the $*$-convolution, we can deduce that for any $\sigma \in \mathcal{S}_n$,

$$\mathbb{E}\left(\tilde{C}_\sigma(A_1,\ldots,A_n)\right) = C_\sigma^U(A_1,\ldots,A_n). \qquad \square$$

The key properties of these cumulants taken from [2] and recalled in Section 2.1 can be recovered using this geometric interpretation.

- Proof of Formula (2) (or Lemma 3.1 in [2]):
  Note that $A_{\sigma(1)} \otimes \cdots \otimes A_{\sigma(n)} = \rho'(\sigma^{-1},\sigma^{-1})(A_1 \otimes \cdots \otimes A_n)$. Thus since the actions ρ_n and ρ' commute we have $\Pi(A_{\sigma(1)} \otimes \cdots \otimes A_{\sigma(n)}) = \rho'(\sigma^{-1},\sigma^{-1})\Pi(A_1 \otimes \cdots \otimes A_n)$. Using (9) and Theorem 2.2, Formula (2) follows from the linear independence of the u_π, $\pi \in \mathcal{S}_n$. $\qquad \square$

- Proof of Proposition 2.1:
  On the one hand, from Theorem 2.2 we have

$$\mathbb{E}(\Pi(A_1 \otimes \ldots \otimes A_k \otimes I_N \otimes \cdots \otimes I_N)) = \sum_{\sigma \in \mathcal{S}_n} C_\sigma^U(A_1,\ldots,A_k,I_N,\ldots,I_N)u_\sigma.$$

  On the other hand, we also have

$$\begin{aligned}
\mathbb{E}\left(\Pi(A_1 \otimes \ldots \otimes A_k \otimes I_N \otimes \cdots \otimes I_N)\right) \\
= \mathbb{E}\left(\Pi(A_1 \otimes \ldots \otimes A_k)\right) \otimes I_N \otimes \cdots \otimes I_N \\
= \left(\sum_{\rho \in \mathcal{S}_k} C_\rho^U(A_1,\ldots,A_k)\, u_\rho\right) \otimes I_N \otimes \cdots \otimes I_N \\
= \sum_{\substack{\sigma \in \mathcal{S}_n \\ \sigma = (n)\cdots(k+1)\rho \\ \text{for some } \rho \in \mathcal{S}_k}} C_\rho^U(A_1,\ldots,A_k)\, u_\sigma,
\end{aligned}$$

  the last equality coming from (7). The result follows by the linear independence of all the u_σ. $\qquad \square$

- From the two previous points we easily get Corollary 3.1 in [2] that we recall here:
  Let $V = \{i \in \{1,\ldots,n\}, A_i \neq I_N\} = \{i_1 < \cdots < i_k\}$. Then

$$C_\pi^U(A_1,\ldots,A_n) = \begin{cases} C_\rho^U(A_{i_1},\ldots,A_{i_k}) & \text{if } \pi_{|V^c} = id \text{ and } \pi_{|V} = \rho, \\ 0 & \text{else.} \end{cases}$$

- Proof of Theorem 2.1:
  Write:

$$\mathbb{E}\left(r_\pi(A_1B_1,\dots,A_nB_n)\right)$$
$$= \mathbb{E}\left(\langle \Pi(A_1B_1 \otimes \dots \otimes A_nB_n), u_\pi \rangle\right)$$
$$\overset{(a)}{=} \langle \mathbb{E}\left(\Pi(A_1 \otimes \dots \otimes A_n)\right).\mathbb{E}\left(\Pi(B_1 \otimes \dots \otimes B_n)\right), u_\pi \rangle$$
$$\overset{(b)}{=} \sum_{\sigma \in \mathcal{S}_n} C_\sigma^U(\mathbf{A}) \langle u_\sigma.\mathbb{E}\left(\Pi(B_1 \otimes \dots \otimes B_n)\right), u_\pi \rangle$$
$$\overset{(c)}{=} \sum_{\sigma \in \mathcal{S}_n} C_\sigma^U(\mathbf{A}) \mathbb{E}\left(\langle \Pi(B_1 \otimes \dots \otimes B_n), u_{\sigma^{-1}\pi} \rangle\right)$$
$$= \sum_{\sigma \in \mathcal{S}_n} C_\sigma^U(\mathbf{A}) \mathbb{E}(r_{\sigma^{-1}\pi}(\mathbf{B})),$$

where (a) comes from (11), (b) from Theorem 2.2 and (c) from (6). Similarly, developing $\mathbb{E}\left(\Pi(B_1 \otimes \dots \otimes B_n)\right)$, we also get

$$\mathbb{E}\left(r_\pi(A_1B_1,\dots,A_nB_n)\right)$$
$$= \sum_{\sigma \in \mathcal{S}_n} C_\sigma^U(\mathbf{B}) \mathbb{E}\left(\langle \Pi(A_1 \otimes \dots \otimes A_n), u_{\pi\sigma^{-1}} \rangle\right)$$
$$= \sum_{\sigma \in \mathcal{S}_n} C_\sigma^U(\mathbf{B}) \mathbb{E}(r_{\pi\sigma^{-1}}(\mathbf{A})) = \sum_{\tau \in \mathcal{S}_n} \mathbb{E}(r_\tau(\mathbf{A})) C_{\tau^{-1}\pi}^U(\mathbf{B}).$$

$\square$

- Proof of Corollary 2.1:
  Using (11), Theorem 2.2 and then (5), we get

$$\mathbb{E}\left(\Pi(\mathbf{AB})\right) = \sum_{\sigma,\tau} C_\sigma^U(\mathbf{A}) C_\tau^U(\mathbf{B}) u_\sigma.u_\tau = \sum_\pi \Big(\sum_\sigma C_\sigma^U(\mathbf{A}) C_{\sigma^{-1}\pi}^U(\mathbf{B}) \Big) u_\pi.$$

The result follows from the linear independence of the u_π. $\square$
Note that Theorem 2.1 or Corollary 2.1 enable to compute the coordinates of $\mathbb{E}\{\Pi(\mathbf{AB})\}$ in the basis $\{u_\pi, \pi \in \mathcal{S}_n\}$. This also was the aim of formula (10) in [5].

- The linearizing property followed from Proposition 5.1 in [2]. We propose here a slightly modified version of this proposition:

Proposition 2.5 *Let $\mathcal{A}$ and $\mathcal{B}$ be two independent sets of $N \times N$ matrices such that the distribution of $\mathcal{A}$ is invariant under unitary conjugation. Let $X_1,\dots,X_n$ be in $\mathcal{A} \cup \mathcal{B}$ and define $V = \{i \in \{1,\dots,n\}, X_i \in \mathcal{A}\}$. Denote X_i by A_i if $i \in V$ and by B_i else. Denote also by $\mathbf{A}_{|V}$ the tuple composed by the $X_i, i \in V$ and by $\mathbf{B}_{|V^c}$ the complementary tuple. We assume that $V \neq \emptyset$ and $V \neq \{1,\dots,n\}$. Then*

$$C_\pi^U(X_1,\dots,X_n) = \begin{cases} C_{\pi_{|V}}^U(\mathbf{A}_{|V}) C_{\pi_{|V^c}}^U(\mathbf{B}_{|V^c}) & \text{if } \pi(V) = V, \\ 0 & \text{else.} \end{cases}$$

Proof: Without lost of generality, thanks to formula (2), we can assume that $V = \{1, \ldots, k\}$, $1 < k < n$, so that $(X_1, \ldots, X_n) = (A_1, \ldots, A_k, B_{k+1}, \ldots, B_n)$. Then write

$$\mathbb{E}\left(\Pi(X_1 \otimes \ldots \otimes X_n)\right)$$
$$= \mathbb{E}\left(\Pi(A_1 I_N \otimes \ldots \otimes A_k I_N \otimes I_N B_{k+1} \otimes \ldots \otimes I_N B_n)\right)$$
$$\overset{(a)}{=} \mathbb{E}\left(\Pi(A_1 \otimes \ldots \otimes A_k \otimes I_N \otimes \ldots \otimes I_N)\right)$$
$$. \; \mathbb{E}\left(\Pi(I_N \otimes \ldots \otimes I_N \otimes B_{k+1} \otimes \ldots \otimes B_n)\right)$$
$$= \left\{\mathbb{E}\left(\Pi(A_1 \otimes \ldots \otimes A_k)\right) \otimes I_N \otimes \ldots \otimes I_N\right\}$$
$$. \; \left\{I_N \otimes \ldots \otimes I_N \otimes \mathbb{E}\left(\Pi(B_{k+1} \otimes \ldots \otimes B_n)\right)\right\}$$
$$= \mathbb{E}\left(\Pi(A_1 \otimes \ldots \otimes A_k)\right) \otimes \mathbb{E}\left(\Pi(B_{k+1} \otimes \ldots \otimes B_n)\right)$$
$$\overset{(b)}{=} \sum_{\sigma \in \mathcal{S}\{1,\ldots,k\}, \tau \in \mathcal{S}\{k+1,\ldots,n\}} C_\sigma^U(A_1, \ldots, A_k) \, C_\tau^U(B_{k+1}, \ldots, B_n) \, u_{\sigma\tau}$$

where $(a), (b)$ respectively come from (11), (8). Thus the coordinates of $\mathbb{E}\left(\Pi(X_1 \otimes \ldots \otimes X_n)\right)$ in the basis $\{u_\pi, \pi \in \mathcal{S}_n\}$ are null unless $\pi = \sigma\tau$ with $\sigma \in \mathcal{S}\{1, \ldots, k\}$, $\tau \in \mathcal{S}\{k + 1, \ldots, n\}$. In that case they are $C_\sigma^U(A_1, \ldots, A_k) \, C_\tau^U(B_{k+1}, \ldots, B_n)$. □

In particular if π is a single cycle we have $C_\sigma^U(X_1, \ldots, X_n) = 0$ from which the linearisation property follows.

3 Matricial $\mathbb{O}$-cumulants

In order to underline the parallel with the previous section, we first begin with a summary of the definitions and main results of [3]. Note that this work [3] has been greatly inspired by the paper of Graczyk P., Letac G., Massam H. [7].

3.1 Definitions

Let us introduce some objects. Let $\mathcal{S}_{2n}$ be the group of permutations of $\{1, \ldots, n, \bar{1}, \ldots, \bar{n}\}$. Denote by $(i\,j)$ the transposition sending i onto j and j onto i. Define

$$\theta := \prod_{i=1}^{n} (i\,\bar{i}),$$

$$H_n = \{h \in \mathcal{S}_{2n}, \theta h = h\theta\}.$$

H_n is the hyperoctahedral group. For $\varepsilon = (\varepsilon_1, \cdots, \varepsilon_n)$ in $\{-1, 1\}^n$, set

$$\tau_\varepsilon = \prod_{i; \, \varepsilon_i = -1} (i\,\bar{i}).$$

For any $\pi \in \mathcal{S}_n$, define the permutation $s_\pi \in \mathcal{S}_{2n}$ as follows: for all $j = 1, \ldots, n$,

$$s_\pi(j) = \pi(j) \qquad s_\pi(\bar{j}) = \overline{\pi(j)}.$$

Note that $H_n = \{s_\pi \tau_\varepsilon, (\pi, \varepsilon) \in \mathcal{S}_n \times \{-1, 1\}^n\}$. If $\pi \in \mathcal{S}_n$, we still denote by π its extension on $\mathcal{S}_{2n}$ which is equal to the identity on $\{\bar{1}, \cdots, \bar{n}\}$. For ε in $\{-1, 1\}^n$ and $\pi \in \mathcal{S}_n$, we define

$$g_{(\varepsilon, \pi)} := \tau_\varepsilon \pi \tau_\varepsilon.$$

Note that it is easy to deduce $g_{(\varepsilon, \pi)}$ from π, since one just has to put a bar on i if $\varepsilon_i = -1$ in the writing of π.

Example: $\pi = (134)(25)$, $\tau_{(1,-1,-1,1,1)} = (2\bar{2})(3\bar{3})$ then $g_{((1,-1,-1,1,1),(134)(25))} = (1\bar{3}4)(\bar{2}5)$.

Definition 3.1 *A pair $(\varepsilon, \pi) \in \{-1; 1\}^n \times \mathcal{S}_n$ is particular if for any cycle c of π we have $\varepsilon_i = 1$ when i is the smallest element of c. The permutation $g_{(\varepsilon, \pi)}$ is called particular too.*

There are $K = \frac{(2n)!}{n! 2^n}$ particular pairs $(\varepsilon(l), \pi_l)$ which define K particular permutations $g_l = g_{(\varepsilon(l), \pi_l)}$ and it is easy to deduce from Theorem 8 in [7] (see also [3]) that we have the partition

$$\mathcal{S}_{2n} / H_n = \bigcup_{l=1}^{K} g_l H_n.$$

We are going to extend the generalized moments (1) defined on $\mathcal{S}_n$ into two functions defined on $\mathcal{S}_{2n}$, respectively H_n-right and H_n-left invariant:

Definition 3.2 *Let $g_l, l = 1 \ldots, K$ be the particular permutations of $\mathcal{S}_{2n}$. For any n-tuple $\mathbf{X} = (X_1, \ldots, X_n)$ of complex random matrices, set for any $g \in \mathcal{S}_{2n}$*

$$M_{\mathbf{X}}^+(g) := r_{\pi_l}(X_1^{\varepsilon_1(l)}, \ldots, X_n^{\varepsilon_n(l)}) \text{ when } g \in g_l H_n,$$
$$\mathbb{M}_{\mathbf{X}}^+(g) := \mathbb{E}\{M_{\mathbf{X}}^+(g)\},$$
$$M_{\mathbf{X}}^-(g) := r_{\pi_l}(X_1^{\varepsilon_1(l)}, \ldots, X_n^{\varepsilon_n(l)}) \text{ when } g \in H_n g_l,$$
$$\mathbb{M}_{\mathbf{X}}^-(g) := \mathbb{E}\{M_{\mathbf{X}}^-(g)\}.$$

Note that $M_{(I_N, \ldots, I_N)}^+ = M_{(I_N, \ldots, I_N)}^-$ and we will denote this H_n-bi-invariant function by M_{I_N}. Note also that

$$M_{I_N}(g_{(\varepsilon, \pi)}) = N^{\gamma_n(\pi)}. \tag{12}$$

We denote by $\mathcal{A}^-$ the space of H_n-left invariant functions on $\mathcal{S}_{2n}$, by $\mathcal{A}^+$ the space of II_n-right invariant functions and by $\mathcal{A}_0$ the space of H_n-bi-invariant functions. For any ϕ in $\mathcal{A}^+$ and any ψ in $\mathcal{A}^-$, define the convolution $\circledast$ on $\mathcal{A}^+ \times \mathcal{A}^-$ by

$$\phi \circledast \psi(g) := \frac{1}{|H_n|} \phi * \psi(g) = \sum_{l=1}^{K} \phi(g_l)\psi(g_l^{-1}g),$$

where $*$ stands for the classical convolution on $\mathcal{S}_{2n}$.

We showed in [3] that M_{I_N} is $\circledast$-invertible when $n \leq N$ and its $\circledast$-inverse relies on the Weingarten function Wg introduced in [5]. Denoting by $(M_{I_N})^{\circledast(-1)}$ this inverse function, we introduced two cumulant functions $C_{\mathbf{X}}^{O+}$, $C_{\mathbf{X}}^{O-}$: $\mathcal{S}_{2n} \to \mathbb{C}$ by setting

$$C_{\mathbf{X}}^{O+} = \mathbb{M}_{\mathbf{X}}^+ \circledast (M_{I_N})^{\circledast(-1)},$$
$$C_{\mathbf{X}}^{O-} = (M_{I_N})^{\circledast(-1)} \circledast \mathbb{M}_{\mathbf{X}}^-.$$

(We slightly modified the notation $C^{O\pm}(\mathbf{X})$ we adopted in the introduction and for the $\mathbb{U}$-cumulant functions $C^U(\mathbf{X})$ in order to lighten the indices when we consider for instance $C_{\mathbf{X}}^{O\pm}(g_{(\varepsilon,\pi)})$, that seems more readable than $C_{g_{(\varepsilon,\pi)}}^{O\pm}(\mathbf{X}))$.)

Note that

$$C_{\mathbf{X}}^{O+}(g) = C_{\mathbf{X}}^{O-}(\theta g^{-1}\theta).$$

These functions are respectively H_n-right and H_n-left invariant and coincide on the $g_{(\varepsilon,\pi)}$, $(\varepsilon,\pi) \in \{-1,1\}^n \times \mathcal{S}_n$.

Definition 3.3 *The functions $C_{\mathbf{X}}^{O+}$ and $C_{\mathbf{X}}^{O-}$ are respectively called the right and left $\mathbb{O}$-cumulant functions of order n.*

Thus, for example,

$$C_X^{O+}((1)) = \frac{1}{N}\mathbb{E}(\mathrm{Tr}(X)),$$

$$C_{(X_1,X_2)}^{O+}((1)(2)) = \frac{(N+1)\mathbb{E}\{\mathrm{Tr}(X_1)\mathrm{Tr}(X_2)\} - \mathbb{E}\{\mathrm{Tr}(X_1X_2)\} - \mathbb{E}\{\mathrm{Tr}(X_1{}^tX_2)\}}{N(N-1)(N+2)},$$

$$C_{(X_1,X_2)}^{O+}((1\,2)) = \frac{-\mathbb{E}\{\mathrm{Tr}(X_1)\mathrm{Tr}(X_2)\} + (N+1)\mathbb{E}\{\mathrm{Tr}(X_1X_2)\} - \mathbb{E}\{\mathrm{Tr}(X_1{}^tX_2)\}}{N(N-1)(N+2)}.$$

The analogues of formula (2) and Proposition 2.1 are the following:

Lemma 3.1 *If $\mathbf{X}^\varepsilon = (X_1^{\varepsilon_1}, \cdots, X_n^{\varepsilon_n})$ and if $\mathbf{X}_\pi = (X_{\pi(1)}, \cdots, X_{\pi(n)})$, then*

$$M_{\mathbf{X}^\varepsilon}^+(g) = M_{\mathbf{X}}^+(\tau_\varepsilon g) \quad and \quad M_{\mathbf{X}_\pi}^+(g) = M_{\mathbf{X}}^+(s_\pi g).$$
$$C_{\mathbf{X}^\varepsilon}^{O+}(g) = C_{\mathbf{X}}^{O+}(\tau_\varepsilon g) \quad and \quad C_{\mathbf{X}_\pi}^{O+}(g) = C_{\mathbf{X}}^{O+}(s_\pi g). \tag{13}$$

Proposition 3.1 *Let $X_1, \cdots, X_k$ be k $N \times N$ matrices. Then*

$$C_{(X_1,\cdots,X_k,I_N,\cdots,I_N)}^{O+}(g)$$
$$= \begin{cases} C_{(X_1,\cdots,X_k)}^{O+}(g') & \text{if there exists } g' \text{ in } \mathcal{S}_{2k} \text{ such that } g \in g'H_n \\ 0 & \text{else.} \end{cases}$$

3.2 Fundamental properties

3.2.1 Mixed moments of independent tuples

In [3] we established the general convolution formula for mixed moments involving the cumulant functions C^{O+} or C^{O-}.

Theorem 3.1 *Let $\mathcal{X}$ and $\mathcal{B}$ be two independent sets of $N \times N$ random matrices such that $\mathcal{B}$ is deterministic and $\mathcal{X}$ is random whose distribution is invariant under orthogonal conjugation. Then for any $1 \leq n \leq N$, $\mathbf{X} = (X_1, \ldots, X_n)$ a n-tuple in $\mathcal{X}$, $\mathbf{B} = (B_1, \ldots, B_n)$ in $\mathcal{B}$, and for any $(\varepsilon, \varepsilon', \pi) \in \{-1; 1\}^n \times \{-1; 1\}^n \times \mathcal{S}_n$,*

$$\mathbb{E}\{r_\pi(B_1^{\varepsilon_1} X_1^{\varepsilon_1'}, \ldots, B_n^{\varepsilon_n} X_n^{\varepsilon_n'})\} = \left(M_{\mathbf{B}}^+ \circledast C_{\mathbf{X}}^{O-}\right)(\tau_\varepsilon \pi \tau_{\varepsilon'}) = \left(C_{\mathbf{B}}^{O+} \circledast M_{\mathbf{X}}^-\right)(\tau_\varepsilon \pi \tau_{\varepsilon'}).$$

In particular, we have

$$\mathbb{E}\{r_\pi(B_1 X_1, \ldots, B_n X_n)\} = \left(M_{\mathbf{B}}^+ \circledast C_{\mathbf{X}}^{O-}\right)(\pi)$$
$$= \left(C_{\mathbf{B}}^{O+} \circledast M_{\mathbf{X}}^-\right)(\pi).$$

3.2.2 Linearizing property

Note that unlike the $\mathbb{U}$-cumulants the $\mathbb{O}$-cumulants $C_{\mathbf{X}}^{O\pm}(\pi)$ of a matrix X do not depend only on the class of conjugation of π (Nevertheless, when X is symmetric, $M_{\mathbf{X}}^\pm$ and $C_{\mathbf{X}}^\pm$ are bi-invariant). Thus the linearizing property has the following meaning.

Proposition 3.2 *Let A and B be two independent $N \times N$ matrices such that the distribution of A is invariant under orthogonal conjugation. Then for any single cycle π in $\mathcal{S}_n$ and any $\varepsilon \in \{-1, 1\}^n$,*

$$C_{A+B}^{O+}(g_{(\varepsilon,\pi)}) = C_A^{O+}(g_{(\varepsilon,\pi)}) + C_B^{O+}(g_{(\varepsilon,\pi)}).$$

3.2.3 Asymptotic behavior

We now come to the asymptotic behavior of the moment and cumulant functions. We need the following normalization:

Definition 3.4 *Let $\mathbf{X}$ be a n-tuple of $N \times N$ complex random matrices. The functions defined for all $g \in \mathcal{S}_{2n}$ by:*

$$M_{\mathbf{X}}^{\pm(N)}(g) := \frac{1}{N^{\tilde{\gamma}_n(g)}} M_{\mathbf{X}}^\pm(g)$$
$$(C_{\mathbf{X}}^{O\pm})^{(N)}(g) := N^{n-\tilde{\gamma}_n(g)} C_{\mathbf{X}}^{O\pm}(g)$$

where

$$\tilde{\gamma}_n(g) = \gamma_n(\pi) \quad \text{if} \quad g \in g_{(\varepsilon,\pi)} H_n$$

are respectively called the normalized right/left moment and $\mathbb{O}$-cumulant functions of $\mathbf{X}$ on $\mathcal{S}_{2n}$.

Proposition 3.3 *Let* $\mathcal{X} = \{X_i, i \in \mathbb{N}^*\}$ *be a set of* $N \times N$ *complex random matrices and let* $\mathrm{x} = \{x_i, i \in \mathbb{N}^*\}$ *be a set of noncommutative random variables in some noncommutative probability space* $(\mathcal{A}, \phi)$. *Denote by* k *the corresponding free cumulant functions. Then for all* $n, i_1, \ldots, i_n \in \mathbb{N}^*$, *the two following assertions are equivalent:*

$$i)\ \forall\, \varepsilon, \pi\ \mathrm{M}^{\pm(N)}_{X_{i_1}, \ldots, X_{i_n}} (g_{(\varepsilon, \pi)}) \underset{N \to \infty}{\longrightarrow} \phi_\pi(x_{i_1}^{\varepsilon_1}, \ldots, x_{i_n}^{\varepsilon_n}),$$

$$ii)\ \forall\, \varepsilon, \pi\ (C^{O\pm}_{X_{i_1}, \ldots, X_{i_n}})^{(N)}(g_{(\varepsilon, \pi)}) \underset{N \to \infty}{\longrightarrow} k_\pi(x_{i_1}^{\varepsilon_1}, \ldots, x_{i_n}^{\varepsilon_n}).$$

3.3 Action of the orthogonal group on the space of complex matrices

We start again with some basic generalities and notations. Endow now $\mathbb{C}^N$ with the symmetric non degenerate bilinear form $\widetilde{B}(\sum_i u_i e_i, \sum_i v_i e_i) = \sum_i u_i v_i$ so that $(e_1, \ldots, e_N)$ is $\widetilde{B}$-orthonormal. Then the tensor product $\mathbb{C}^N \otimes \mathbb{C}^N$ is endowed with the bilinear form

$$\widetilde{B}_2(u_1 \otimes v_1, u_2 \otimes v_2) = \widetilde{B}(u_1, u_2)\widetilde{B}(v_1, v_2)$$

and $e_i \otimes e_j$, $i, j = 1, \ldots, N$ is a $\widetilde{B}_2$-orthonormal basis of $(\mathbb{C}^N)^{\otimes 2}$. The orthogonal group $\mathbb{O}_N$ acts on $(\mathbb{C}^N)^{\otimes 2}$ as follows:

$$\widetilde{\rho}(O)(e_i \otimes e_j) = Oe_i \otimes Oe_j.$$

On the other hand, endow $\mathcal{M}_N$ now with the symmetric non degenerate bilinear form $B(X, Y) = \mathrm{Tr}(XY^t)$. Here again, $\mathcal{M}_N$ and $\mathbb{C}^N \otimes \mathbb{C}^N$ are isomorphic vector spaces when we identify any $X = (X_{ij})_{1 \leq i, j \leq N} \in \mathcal{M}_N$ with $\tilde{X} = \sum_{1 \leq i, j \leq N} X_{ij} e_i \otimes e_j$ (and hence $\tilde{E}_{a,b} = e_a \otimes e_b$). The action $\tilde{\rho}$ gives on $\mathcal{M}_N$

$$\rho(O)(X) = OXO^t.$$

Note also that the inner product XY in $\mathcal{M}_N$ corresponds to the product defined by

$$(u_1 \otimes v_1).(u_2 \otimes v_2) = \widetilde{B}(v_1, u_2)\, u_1 \otimes v_2, \tag{14}$$

and the transposition X^t to the following rule: $(u \otimes v)^t = v \otimes u$.

Now for any n, the tensor products $\mathcal{M}_N^{\otimes n}$ and $(\mathbb{C}^N \otimes \mathbb{C}^N)^{\otimes n} = (\mathbb{C}^N)^{\otimes 2n}$ are isomorphic through the map: $X = X_1 \otimes \cdots \otimes X_n \mapsto \tilde{X} = \tilde{X}_1 \otimes \cdots \otimes \tilde{X}_n$ and with bilinear forms

$$B_n(X_1 \otimes \cdots \otimes X_n, Y_1 \otimes \cdots \otimes Y_n) = \prod_{i=1}^n \mathrm{Tr}(X_i Y_i^t) = \prod_{i=1}^n \widetilde{B}_2(\tilde{X}_i, \tilde{Y}_i)$$

$$= \widetilde{B}_{2n}(\tilde{X}_1 \otimes \cdots \otimes \tilde{X}_n, \tilde{Y}_1 \otimes \cdots \otimes \tilde{Y}_n).$$

Here again the following actions of $\mathbb{O}_N$ are equivalent:

$$\text{on } (\mathbb{C}^N)^{\otimes 2n} \quad \widetilde{\rho}_n(O)(e_{i(1)} \otimes e_{i(\bar{1})} \cdots \otimes e_{i(n)} \otimes e_{i(\bar{n})})$$
$$= Oe_{i(1)} \otimes Oe_{i(\bar{1})} \otimes \cdots \otimes Oe_{i(n)} \otimes Oe_{i(\bar{n})},$$
$$\text{on } \mathcal{M}_N^{\otimes n} \quad \rho_n(O)(X_1 \otimes \cdots \otimes X_n) = OX_1O^t \otimes \cdots \otimes OX_nO^t.$$

Denote by $[V]^{\mathbb{O}_N}$ the subspace of $\mathbb{O}_N$-invariant vectors of V with $V = \mathcal{M}_N^{\otimes n}$ or $(\mathbb{C}^N)^{\otimes 2n}$. Then $[\mathcal{M}_N^{\otimes n}]^{\mathbb{O}_N}$ and $[(\mathbb{C}^N)^{\otimes 2n}]^{\mathbb{O}_N}$ are still isomorphic and we identify $\mathcal{M}_N^{\otimes n}$ and $(\mathbb{C}^N)^{\otimes 2n}$. We also simply denote the bilinear form B_n or $\widetilde{B}_{2n}$ by $\langle .,. \rangle$ (even if it is not a scalar nor a Hermitian product). Note lastly that the inner product in $\mathcal{M}_N^{\otimes n}$ is defined by

$$(X_1 \otimes \cdots \otimes X_n).(Y_1 \otimes \cdots \otimes Y_n) = X_1Y_1 \otimes \cdots \otimes X_nY_n,$$

and the transposition by $(X_1 \otimes \cdots \otimes X_n)^t = X_1^t \otimes \cdots \otimes X_n^t$. They satisfy for any $u, v, w \in \mathcal{M}_N^{\otimes n}$:

$$\langle u.v, w \rangle = \langle v, u^t.w \rangle = \langle u, w.v^t \rangle. \tag{15}$$

In order to present a basis of $[\mathcal{M}_N^{\otimes n}]^{\mathbb{O}_N}$ in Proposition 3.4 below, we need to introduce the second action of group. We always use the notation

$$\Theta_n := \underbrace{I_N \otimes \ldots \otimes I_N}_{n \text{ times}} = \sum_{i_1, \ldots, i_n} e_{i_1} \otimes e_{i_1} \otimes \cdots \otimes e_{i_n} \otimes e_{i_n}$$

and we now consider the natural action ρ' of $\mathcal{S}_{2n}$ on $(\mathbb{C}^N)^{\otimes 2n}$ defined for any permutation g in $\mathcal{S}_{2n}$ acting on $\{1, \ldots, n, \bar{1}, \ldots, \bar{n}\}$ by

$$\rho'(g)(e_{i(1)} \otimes e_{i(\bar{1})} \otimes \cdots \otimes e_{i(n)} \otimes e_{i(\bar{n})})$$
$$= e_{i(g^{-1}(1))} \otimes e_{i(g^{-1}(\bar{1}))} \otimes \cdots \otimes e_{i(g^{-1}(n))} \otimes e_{i(g^{-1}(\bar{n}))}. \tag{16}$$

Note first that

$$\langle \rho'(g)u, v \rangle = \langle u, \rho'(g^{-1})v \rangle. \tag{17}$$

Now the actions ρ and ρ' commute. Hence, according to Lemma 1.1,

$$\{\rho'(g) \cdot \Theta_n; g \in \mathcal{S}_{2n}\} \subset [\mathcal{M}_N^{\otimes n}]^{\mathbb{O}_N}.$$

But writing

$$\rho'(g) \cdot \Theta_n = \sum_{i(1), \ldots, i(n), i(\bar{1}), \ldots, i(\bar{n})} \left(\prod_{l=1}^n \delta_{i(l)i(\bar{l})} \right)$$
$$\cdot e_{i(g^{-1}(1))} \otimes e_{i(g^{-1}(\bar{1}))} \otimes \cdots e_{i(g^{-1}(n))} \otimes e_{i(g^{-1}(\bar{n}))},$$

it is easy to see that

$$\rho'(g) \cdot \Theta_n = \Theta_n \iff \forall l, \ g^{-1}(l) = \overline{g^{-1}(\bar{l})} = \theta g^{-1} \theta(l) \quad (\text{where } \theta = \prod_{i=1}^{n}(i\,\bar{i}).)$$
$$\iff \theta = g\theta g^{-1}$$
$$\iff g \in H_n,$$

so that $g \mapsto \rho'(g) \cdot \Theta_n$ is H_n-right invariant. Actually Theorem 4.3.4 in [8] makes this first result more precise:

Lemma 3.2 *Let $\Xi_n \subset \mathcal{S}_{2n}$ be a collection of representatives for the cosets $\mathcal{S}_{2n}/H_n$. Then*

$$[\mathcal{M}_N^{\otimes n}]^{\mathbb{O}_N} = Span\{\rho'(g) \cdot \Theta_n; g \in \Xi_n\}.$$

We will use the parametrization of $\mathcal{S}_{2n}/H_n$ by the subset $\mathcal{P}_{2n}$ of $\mathcal{S}_{2n}$ composed with the pairings of $\{1, \ldots, 2n\}$. Let

$$\eta : \mathcal{S}_{2n} \to \mathcal{P}_{2n}$$
$$g \quad \mapsto \eta(g) = g\theta g^{-1} = \prod_{i=1}^{n} (g(i)\, g(\bar{i})).$$

Clearly $\eta(g) = \eta(g') \iff g' \in gH_n$. We thus get a bijection from $\mathcal{S}_{2n}/H_n$ onto $\mathcal{P}_{2n}$ (see Proposition 17, [7] or Lemma 4.1, [2] for more details). Therefore we set for any $p \in \mathcal{P}_{2n}$:

$$u_p = \rho'(g) \cdot \Theta_n \quad \text{if} \quad \eta(g) = p. \tag{18}$$

The vector u_p corresponds to $\rho_B(p)$ in [5]. Note that $\eta(id) = \theta$ and $u_\theta = \Theta_N$. The $u_p, p \in \mathcal{P}_{2n}$, satisfy the following properties:

Lemma 3.3 *1. For all $A_1, \cdots, A_n$ in $\mathcal{M}_N$, for any $\pi \in \mathcal{S}_n$ and $\varepsilon \in \{-1, 1\}^n$, we have:*

$$r_\pi(A_1^{\varepsilon_1}, \cdots, A_n^{\varepsilon_n}) = \langle A_1 \otimes \cdots \otimes A_n, u_{\eta(g_{(\varepsilon,\pi)})}\rangle,$$

and more generally:

$$M_{\mathbf{A}}^+(g) = \langle A_1 \otimes \cdots \otimes A_n, u_{\eta(g)}\rangle \tag{19}$$

2.

$$\langle \Theta_n, u_{\eta(g_{(\varepsilon,\pi)})}\rangle = N^{\gamma(\pi)} = M_{I_N}(g_{(\varepsilon,\pi)})$$

and hence

$$\langle u_{\eta(g)}, u_{\eta(g')}\rangle = \langle \Theta_n, u_{\eta(g^{-1}g')}\rangle = M_{I_N}(g^{-1}g'). \tag{20}$$

Proof: 1.) Write $\mathbf{j} = (j(1), \cdots, j(n), j(\bar{1}), \cdots, j(\bar{n}))$ a $2n$-tuple of integers in $\{1, \ldots, N\}$ and

$$u_{\eta(g)} = \sum_{\mathbf{j}} \left(\prod_{l=1}^{n} \delta_{j(l)j(\bar{l})}\right) \bigotimes_{l=1}^{n} \left(e_{j(g^{-1}(l))} \otimes e_{j(g^{-1}(\bar{l}))}\right).$$

Thus

$$\langle A_1 \otimes \cdots \otimes A_n, u_{\eta(g)} \rangle$$

$$= \sum_{i,j} \left(\prod_{k=1}^{n} (A_k)_{i(k)i(\bar{k})} \right) \left(\prod_{l=1}^{n} \delta_{j(l)j(\bar{l})} \right)$$

$$\cdot \langle \bigotimes_{l=1}^{n} \left(e_{i(l)} \otimes e_{i(\bar{l})} \right), \bigotimes_{l=1}^{n} \left(e_{j(g^{-1}(l))} \otimes e_{j(g^{-1}(\bar{l}))} \right) \rangle$$

$$= \sum_{i,j} \left(\prod_{k=1}^{n} (A_k)_{i(k)i(\bar{k})} \right) \left(\prod_{l=1}^{n} \delta_{j(l)j(\bar{l})} \right) \left(\prod_{l=1}^{n} \delta_{i(l)j(g^{-1}(l))} \right) \left(\prod_{l=1}^{n} \delta_{i(\bar{l})j(g^{-1}(\bar{l}))} \right).$$

Thus for any s in $\{1, \ldots, n, \bar{1}, \ldots, \bar{n}\}$, $i(s) = j(g^{-1}(s)) = j(\theta g^{-1}(s))$, and setting $s = g(t)$ we get $i(g(t)) = i(g\theta(t)) = j(t)$ for all t in $\{1, \ldots, n, \bar{1}, \ldots, \bar{n}\}$. Hence

$$\langle A_1 \otimes \cdots \otimes A_n, u_{\eta(g)} \rangle = \sum_{i} \left(\prod_{k=1}^{n} (A_k)_{i(k)i(\bar{k})} \right) \left(\prod_{l=1}^{n} \delta_{i(g(l))i(g(\bar{l}))} \right)$$

In particular for $g = g_{(\varepsilon, \pi)}$, this is formula (18) in [3] (or formula (2.10) in [7]) which gives $r_\pi(A_1^{\varepsilon 1}, \cdots, A_n^{\varepsilon n})$. Now (19) comes from definition 3.2.

2.) The first line follows by taking the A_i equal to I_N and from the definition of M_{I_N} (see (12)). The second one comes from (17). □

The following proposition is essential for our purpose. It relies on a result in [5] that we found in a different way in [3] from mixed moments.

Proposition 3.4 *The set $\{u_p; p \in \mathcal{P}_{2n}\}$ is a basis of $[\mathcal{M}_N^{\otimes n}]^{\mathbb{O}_N}$ (when $N \geq n$).*

Proof: let $p_l = \eta(g_l), l = 1, \ldots, K$ and $G = (\langle u_{p_k}, u_{p_l} \rangle)_{k,l=1}^{K}$ be the Gramm-matrix of $\{u_p, p \in \mathcal{P}_{2n}\}$. It exactly corresponds to the matrix of the operator $\tilde{\Phi}$ in [5] which is shown to be invertible with inverse operator the Weingarten function Wg (see Proposition 3.10 in [5]). □

Here are some differences with the unitary case which can explain the intricate development we did for the $\mathbb{O}$-cumulants. We give the proof below.

1. We have $u^t_{\eta(g_{(\varepsilon,\pi)})} = u_{\eta(g_{(\varepsilon,\pi)}^{-1})}$. In particular

$$u^t_{\eta(\pi)} = u_{\eta(\pi^{-1})}. \tag{21}$$

But in general $u^t_{\eta(g)} \neq u_{\eta(g^{-1})}$. Instead we have

$$u^t_{\eta(g)} = u_{\eta(\theta g)}. \tag{22}$$

In fact define the transposition in $\mathcal{P}_{2n}$ by setting $p^t = \eta(\theta g)$ for $p = \eta(g)$. Then $u_p^t = u_{p^t}$. Note that this corresponds to the parametrization of $\mathcal{P}_{2n}$ by $H_n \backslash \mathcal{S}_{2n} = \bigcup_{l=1}^K H_n g_l$. Indeed consider $\eta^- : g \mapsto \eta^-(g) := \eta(\theta g^{-1})$ from $\mathcal{S}_{2n}$ on $\mathcal{P}_{2n}$. It induces a one-to-one mapping from $H_n \backslash \mathcal{S}_{2n}$ onto $\mathcal{P}_{2n}$ such that $\eta^-(g_{(\varepsilon,\pi)}) = \eta(g_{(\varepsilon,\pi)})$. Then $u_{\eta(g)}^t = u_{\eta^-(g^{-1})}$. Consequently

$$M_{\mathbf{A}}^-(g) = \langle A_1 \otimes \cdots \otimes A_n, u_{\eta^-(g)} \rangle = \langle A_1 \otimes \cdots \otimes A_n, u_{\eta(\theta g^{-1})} \rangle. \quad (23)$$

2. In general $u_{\eta(g_1)}.u_{\eta(g_2)} \neq u_{\eta(g_1 g_2)}$, but

$$u_{\eta(\pi)}.u_{\eta(g)} = u_{\eta(\pi g)} \quad \text{and} \quad u_{\eta(g)}.u_{\eta(\pi)} = u_{\eta(\theta \pi^{-1} \theta g)}. \quad (24)$$

Here again this relation could be understood by introducing the inner product in $\mathbb{C}(\mathcal{P}_{2n}) = \{ \sum_{p \in \mathcal{P}_{2n}} a_p p; a_p \in \mathbb{C} \}$ described in [8] Section 10.1.2, for which $\mathbb{C}(\mathcal{P}_{2n})$ is called the Brauer algebra. This product is of the form $p.q = N^{\alpha(p,q)} r(p,q)$ with $\alpha(p,q) \in \mathbb{N}$ and $r(p,q) \in \mathcal{P}_{2n}$ and we get

$$u_p.u_q = N^{\alpha(p,q)} u_{r(p,q)}. \quad (25)$$

As we do not use it in the following, we choose not to detail it here.

3.

$$\langle u_{\eta(\pi)}.(A_1 \otimes \cdots \otimes A_n), u_{\eta(g)} \rangle = \langle A_1 \otimes \cdots \otimes A_n, u_{\eta(\pi^{-1}g)} \rangle = M_{\mathbf{A}}^+(\pi^{-1}g),$$
$$\langle u_{\eta(g)}.(A_1 \otimes \cdots \otimes A_n), u_{\eta(\pi)} \rangle = \langle A_1 \otimes \cdots \otimes A_n, u_{\eta(\theta \pi^{-1}g)} \rangle \quad (26)$$
$$= \langle A_1 \otimes \cdots \otimes A_n, u_{\eta^-(g^{-1}\pi)} \rangle = M_{\mathbf{A}}^-(g^{-1}\pi).$$

4. If $p \in \mathcal{P}(\{1, \ldots, k, \bar{1}, \ldots, \bar{k}\})$, if $\theta_k = \prod_{l=n-k}^n (l \, \bar{l})$, then

$$u_{p\theta_k} = u_p \otimes I_N \cdots \otimes I_N \quad (27)$$

and more generally if $p \in \mathcal{P}(\{1, \ldots, k, \bar{1}, \ldots, \bar{k}\})$ and $q \in \mathcal{P}(\{k+1, \ldots, n, \overline{k+1}, \ldots, \bar{n}\})$, then the juxtaposition pq is in $\mathcal{P}_{2n}$ and

$$u_{pq} = u_p \otimes u_q. \quad (28)$$

Proof: 1.) We simply write:

$$u_{\eta(g)}^t = \sum_{\mathbf{j}} \left(\prod_{l=1}^n \delta_{j(l)j(\bar{l})} \right) \bigotimes_{l=1}^n \left(e_{j(g^{-1}(\bar{l}))} \otimes e_{j(g^{-1}(l))} \right)$$
$$= \sum_{\mathbf{j}} \left(\prod_{l=1}^n \delta_{j(l)j(\bar{l})} \right) \bigotimes_{l=1}^n \left(e_{j(g^{-1}\theta(l))} \otimes e_{j(g^{-1}\theta(\bar{l}))} \right)$$
$$= u_{\eta(\theta g)}$$

Now $\eta(\theta g) = \eta(\theta g \theta) \neq \eta(g^{-1})$ in general. For instance if $g = (12\bar{2})$, then $\theta g \theta = (\bar{1}\bar{2}2)$ and $\eta(\theta g \theta) = (1\bar{2})(\bar{1}2)$. On the other hand $g^{-1} = (1\bar{2}2)$ and $\eta(g^{-1}) = (12)(\bar{1}\bar{2}) \neq \eta(\theta g \theta)$.

Nevertheless $\eta(\theta g_{(\varepsilon,\pi)}) = \eta(g_{(\varepsilon,\pi)}^{-1})$ since $\theta g_{(\varepsilon,\pi)}\theta = \tau_\varepsilon\theta\pi\theta\tau_\varepsilon = \tau_\varepsilon\pi^{-1}\tau_\varepsilon$ $(\tau_\varepsilon s_\pi\tau_\varepsilon) \in g_{(\varepsilon,\pi)}^{-1}H_n$.

2.) Take $g = (1\bar{2})$ so that $g^2 = id$, $\eta(g^2) = \theta$ and $u_\theta = I_N \otimes I_N$. Now $u_{\eta(g)} = \sum_{i_1,i_2} e_{i_2} \otimes e_{i_1} \otimes e_{i_2} \otimes e_{i_1}$ and therefore, with (14), $u_{\eta(g)}.u_{\eta(g)} = \sum_{i_1,i_2,j_1,j_2}(\delta_{i_1 j_2})e_{i_2} \otimes e_{j_1} \otimes e_{i_2} \otimes e_{j_1} = Nu_{\eta(g)} \neq I_N \otimes I_N$.
Now write

$$u_\pi.u_{\eta(g)} = \left(\sum_{i(1),\ldots,i(n)} \bigotimes_{l=1}^{n} e_{i(\pi^{-1}(l))} \otimes e_{i(l)} \right)$$

$$\cdot \left(\sum_{\mathbf{j}} \left(\prod_{l=1}^{n} \delta_{j(l)j(\bar{l})} \right) \bigotimes_{l=1}^{n} \left(e_{j(g^{-1}(l))} \otimes e_{j(g^{-1}(\bar{l}))} \right) \right)$$

$$= \sum_{i(1),\ldots,i(n),\mathbf{j}} \left(\prod_{l=1}^{n} \delta_{j(l)j(\bar{l})} \right) \left(\prod_{l=1}^{n} \delta_{i(l)j(g^{-1}(l))} \right) \bigotimes_{l=1}^{n} \left(e_{i(\pi^{-1}(l))} \otimes e_{j(g^{-1}(\bar{l}))} \right)$$

$$= \sum_{\mathbf{j}} \left(\prod_{l=1}^{n} \delta_{j(l)j(\bar{l})} \right) \bigotimes_{l=1}^{n} \left(e_{j(g^{-1}\pi^{-1}(l))} \otimes e_{j(g^{-1}(\bar{l}))} \right)$$

$$= u_{\eta(\pi g)}.$$

For the second one we have

$$u_{\eta(g)}.u_\pi = \left(\sum_{\mathbf{j}} \left(\prod_{l=1}^{n} \delta_{j(l)j(\bar{l})} \right) \bigotimes_{l=1}^{n} \left(e_{j(g^{-1}(l))} \otimes e_{j(g^{-1}(\bar{l}))} \right) \right)$$

$$\cdot \left(\sum_{i(1),\ldots,i(n)} \bigotimes_{l=1}^{n} e_{i(\pi^{-1}(l))} \otimes e_{i(l)} \right)$$

$$= \sum_{i(1),\ldots,i(n),\mathbf{j}} \left(\prod_{l=1}^{n} \delta_{j(l)j(\bar{l})} \right) \left(\prod_{l=1}^{n} \delta_{j(g^{-1}(\bar{l}))i(\pi^{-1}(l))} \right) \bigotimes_{l=1}^{n} \left(e_{j(g^{-1}(l))} \otimes e_{i(l)} \right)$$

$$= \sum_{\mathbf{j}} \left(\prod_{l=1}^{n} \delta_{j(l)j(\bar{l})} \right) \bigotimes_{l=1}^{n} \left(e_{j(g^{-1}(l))} \otimes e_{j(g^{-1}(\overline{\pi(l)}))} \right)$$

$$\sum_{\mathbf{j}} \left(\prod_{l=1}^{n} \delta_{j(l)j(\bar{l})} \right) \bigotimes_{l=1}^{n} \left(e_{j(g^{-1}(l))} \otimes e_{j(g^{-1}(\theta\pi\theta(\bar{l})))} \right)$$

$$= u_{\eta(\theta\pi^{-1}\theta g)}.$$

3.) comes from (15), (21) or (22), and (24). Finally Property 4.) is clear from the definition of the u_p. $\qquad\square$

3.4 Geometrical interpretation of the $\mathbb{O}$-cumulants

Consider now, as in [5], the orthogonal projection Π of $\mathcal{M}_N^{\otimes n}$ onto $[\mathcal{M}_N^{\otimes n}]^{\mathbb{O}_N}$ defined by

$$\Pi(A_1 \otimes \ldots \otimes A_n) := \int_{\mathbb{O}_N} O A_1 O^t \otimes \ldots \otimes O A_n O^t \, dO = \int_{\mathbb{O}_N} \rho_n(O)(A_1 \otimes \cdots \otimes A_n) dO$$

where integration is performed with respect to the Haar measure on $\mathbb{O}_N$. As was the case for the unitary case, it corresponds to the conditional expectation on $[\mathcal{M}_N^{\otimes n}]^{\mathbb{O}_N}$ (which is still denoted by $\mathbb{E}(A)$ in [5]).
Note first that Π commutes with the action of ρ': for any A in $\mathcal{M}_N^{\otimes n}$ and g in $\mathcal{S}_{2n}$,

$$\rho'(g)\Pi(A) = \Pi\rho'(g)(A). \tag{29}$$

Here is Proposition 2.4 which we have completely translated for models invariant under orthogonal conjugation. Its proof can be carried on in a very similar way.

Proposition 3.5 *Let* $\mathbf{A} = (A_1, \ldots, A_n)$ *and* $\mathbf{B} = (B_1, \ldots, B_n)$ *be two independent sets of* $N \times N$ *matrices such that the distribution of* $\mathbf{A}$ *is invariant under orthogonal conjugation. Then*

$$\mathbb{E}\left(\Pi(A_1 B_1 \otimes \ldots \otimes A_n B_n)\right) = \mathbb{E}\left(\Pi(A_1 \otimes \ldots \otimes A_n)\right)$$
$$\cdot \, \mathbb{E}\left(\Pi(B_1 \otimes \ldots \otimes B_n)\right). \tag{30}$$

Now we get:

Theorem 3.2 *Let* $g_l, l = 1, \ldots, K$ *be all the particular permutations of* $\mathcal{S}_{2n}$*; denote by* p_l *the pairing* $\eta(g_l)$*. For any* $A_1, \cdots, A_n$ *in* $\mathcal{M}_N$*, denote by* $C_{\mathbf{A}}^O(g_l)$ *the matricial* $\mathbb{O}$*-cumulants* $C_{\mathbf{A}}^{O\pm}(g_l)$ *of* $\mathbf{A} = (A_1, \cdots, A_n)$*. Then* $\{C_{\mathbf{A}}^O(g_l), l = 1, \ldots, K\}$ *is the set of coordinates of* $\mathbb{E}\left(\Pi(A_1 \otimes \ldots \otimes A_n)\right)$ *in the basis* $\{u_{p_l}, l = 1, \ldots, K\}$*:*

$$\mathbb{E}\left(\Pi(A_1 \otimes \ldots \otimes A_n)\right) = \sum_{l=1}^{K} C_{\mathbf{A}}^O(g_l) u_{p_l}. \tag{31}$$

Proof: As $\{u_l, l = 1, \ldots, K\}$ is a basis of $[\mathcal{M}_N^{\otimes n}]^{\mathbb{O}_N}$, we can write $\mathbb{E}\big(\Pi(A_1 \otimes \ldots \otimes A_n)\big) = \sum_{l=1}^{K} \alpha_l(\mathbf{A}) u_{p_l}$, and hence, using (19),

$$\mathbb{M}_{\mathbf{A}}^+(g_k) = \mathbb{E}(\langle \Pi(A_1 \otimes \ldots \otimes A_n), u_{p_k} \rangle)$$
$$= \sum_{l=1}^{K} \alpha_l(\mathbf{A}) \langle u_{p_l}, u_{p_k} \rangle = \sum_{l=1}^{K} \alpha_l(\mathbf{A}) M_{I_N}(g_l^{-1} g_k)$$

from (20). Define $\tilde{C}_{\mathbf{A}}$ on $\mathcal{S}_{2n}$ by $\tilde{C}_{\mathbf{A}}(g) = \alpha_l(\mathbf{A})$ if $g \in g_l H_n$ so that the previous equality gives $\mathbb{M}_{\mathbf{A}}^+(g_k) = \tilde{C}_{\mathbf{A}} \circledast M_{I_N}(g_k)$. Since M_{I_N} is $\circledast$-invertible, it follows that $\tilde{C}_{\mathbf{A}} = C_{\mathbf{A}}^{O+}$ and hence $\alpha_l(\mathbf{A}) = C_{\mathbf{A}}^O(g_l)$. $\qquad\square$

We now review the properties of the $\mathbb{O}$-cumulants expressed in Sections 3.1 and 3.2.

- Proof of lemma 3.1:
  Note that $\mathbf{X}^\varepsilon = \rho'(\tau_\varepsilon)\mathbf{X}$ and $\mathbf{X}_\pi = \rho'((s_\pi)^{-1})\mathbf{X}$. Then use (19), (17) and the definition (18). We get the expression of $M_{\mathbf{X}^\varepsilon}^+(g)$ and $M_{\mathbf{X}_\pi}^+(g)$. Then use (29) in writing

$$
\mathbb{E}\left\{\Pi(\mathbf{X}^\varepsilon)\right\} = \rho'(\tau_\varepsilon)\mathbb{E}\left\{\Pi(\mathbf{X})\right\}
$$
$$
= \sum_{l=1}^{K} C_{\mathbf{X}}^{O+}(g_l)\rho'(\tau_\varepsilon)u_{p_l}
$$
$$
= \sum_{l=1}^{K} C_{\mathbf{X}}^{O+}(g_l)u_{\eta(\tau_\varepsilon g_l)}
$$
$$
= \sum_{k=1}^{K} C_{\mathbf{X}}^{O+}(\tau_\varepsilon g_k)u_{\eta(g_k)}, \tag{32}
$$

what gives $C_{\mathbf{X}^\varepsilon}^{O+}$. And a similar development can be led with $\mathbf{X}_\pi$.
- Proof of Proposition 3.1: It is the same to the proof of Proposition 2.1 in using (27).
- Proof of Theorem 3.1:

$$
\mathbb{E}\{r_\pi(B_1^{\varepsilon_1}X_1^{\varepsilon'_1},\ldots,B_n^{\varepsilon_n}X_n^{\varepsilon'_n})\} = \mathbb{E}\left(\langle\Pi(B_1^{\varepsilon_1}X_1^{\varepsilon'_1}\otimes\ldots\otimes B_n^{\varepsilon_n}X_n^{\varepsilon'_n}), u_{\eta(\pi)}\rangle\right)
$$
$$
\overset{(a)}{=} \langle\mathbb{E}\left(\Pi(\mathbf{B}^\varepsilon)\right).\mathbb{E}\left(\Pi(\mathbf{X}^{\varepsilon'})\right), u_{\eta(\pi)}\rangle
$$
$$
\overset{(b)}{=} \sum_{l=1}^{K} C_{\mathbf{B}}^{O+}(g_l)\langle u_{\eta(\tau_\varepsilon g_l)}.\mathbb{E}\left(\Pi(\mathbf{X}^{\varepsilon'})\right), u_{\eta(\pi)}\rangle
$$
$$
\overset{(c)}{=} \sum_{l=1}^{K} C_{\mathbf{B}}^{O+}(g_l)\langle\mathbb{E}\left(\Pi(\mathbf{X}^{\varepsilon'})\right), u_{\eta(\theta\pi^{-1}\tau_\varepsilon g_l)}\rangle
$$
$$
\overset{(d)}{=} \sum_{l=1}^{K} C_{\mathbf{B}}^{O+}(g_l)\langle\mathbb{E}\left(\Pi(\mathbf{X})\right), u_{\eta(\theta\tau_{\varepsilon'}\pi^{-1}\tau_\varepsilon g_l)}\rangle
$$
$$
\overset{(e)}{=} \sum_{l=1}^{K} C_{\mathbf{B}}^{O+}(g_l)\langle\mathbb{E}\left(\Pi(\mathbf{X})\right), u_{\eta^-(g_l^{-1}\tau_\varepsilon\pi\tau_{\varepsilon'})}\rangle
$$
$$
= \sum_{l=1}^{K} C_{\mathbf{B}}^{O+}(g_l)\mathrm{M}_{\mathbf{X}}^-(g_l^{-1}\tau_\varepsilon\pi\tau_{\varepsilon'})
$$
$$
= C_{\mathbf{B}}^{O+} \circledast \mathrm{M}_{\mathbf{X}}^-(\tau_\varepsilon\pi\tau_{\varepsilon'}),
$$

(a) comes from (30), (b) from (32), (c) from (26), (d) uses $\theta\tau_\varepsilon = \tau_\varepsilon\theta$ and finally (e) comes from (23).
We conduct the second equality in an identical way.

- Lastly the linearizing property can be led in a very similar manner as for the $\mathbb{U}$-cumulants. Just translate Proposition 2.5 in using (28) and Proposition (3.5).

Note that Theorem 3.1 here again gives $\langle \mathbb{E}(\mathbf{AB}), u_p \rangle$ but only for the particular $p = \eta(\pi)$, $\pi \in \mathcal{S}_n$. Similarly formula (19) in [5] only gave $\langle \mathbb{E}(\mathbf{AB}), \Theta_n \rangle$. Actually it is impossible to get $\langle \mathbb{E}(\mathbf{AB}), u_p \rangle$ for all p as a convolution formula, although we did it for $\mathbb{U}$-invariant models. This is due to the structure of $\mathcal{P}_{2n}$ as Brauer algebra that we briefly mentioned in (25). In fact we have:

$$\mathbb{E}(\Pi(\mathbf{AB})) = \sum_{k,l} C_{\mathbf{A}}^O(g_k) C_{\mathbf{B}}^O(g_l) u_{p_k} . u_{p_l}$$

$$= \sum_{k,l} C_{\mathbf{A}}^O(g_k) C_{\mathbf{B}}^O(g_l) N^{\alpha(p_k, p_l)} u_{r(p_k, p_l)}.$$

3.5 About matricial Sp-cumulants

Let us end this section with some words about the symplectic case. Here N is even. Recall that if $J = \begin{pmatrix} 0 & I_{\frac{N}{2}} \\ -I_{\frac{N}{2}} & 0 \end{pmatrix}$, then $Sp(N) = \{T \in GL(N, \mathbb{C});$ $T^t J T = J\}$. Now identify $\mathcal{M}_N$ and $\mathbb{C}^N \otimes \mathbb{C}^N$ through

$$X = (X_{ij})_{1 \leq i,j \leq N} \in \mathcal{M}_N \mapsto \tilde{X} = \sum_{1 \leq i,j \leq N} X_{ij} \, e_i \otimes J^{-1} e_j. \qquad (33)$$

Endow $\mathcal{M}_N^{\otimes n}$ with the non degenerate skew-symmetric bilinear form

$$\Omega_n(X_1 \otimes \cdots \otimes X_n, Y_1 \otimes \cdots \otimes Y_n) = \prod_{i=1}^n \mathrm{Tr}(X_i Y_i^*)$$

where $Y_i^* = J Y_i^t J^{-1}$ and consider both following group actions: first the action of $Sp(N)$ defined by $\rho(T)(X_1 \otimes \cdots \otimes X_n) = T X_1 T^* \otimes \cdots \otimes T X_n T^*$, second the action of $\mathcal{S}_{2n}$ corresponding to (16) on $(\mathbb{C}^N \otimes \mathbb{C}^N)^{\otimes n}$ via the previous identification (33) and which we still denote by ρ'.
Then the fit basis of $[\mathcal{M}_N^{\otimes n}]^{Sp(N)}$ is composed by the vectors $u_p, p \in \mathcal{P}_{2n}$ now defined by

$$u_p = \mathsf{sgn}(g) \rho'(g) \cdot \Theta_n \quad \text{if} \quad \eta(g) = p$$

where $\mathsf{sgn}(g)$ denotes the signature of the permutation g in $\mathcal{S}_{2n}$ and where $\Theta_n = I_N \otimes \cdots \otimes I_N$. It can be proved that, denoting A_i^* by A_i^{-1},

$$\Omega_n(A_1 \otimes \cdots \otimes A_n, u_{\eta(g_{(\varepsilon, \pi)})}) = \mathsf{sgn}(\pi) r_\pi(A_1^{\varepsilon_1}, \cdots, A_n^{\varepsilon_n}).$$

We thus are led to introduce:

$$M_{\mathbf{X}}^{Sp+}(g) := \Omega_n(\mathbf{X}, u_{\eta(g)})$$
$$M_{\mathbf{X}}^{Sp-}(g) := \Omega_n(\mathbf{X}, u_{\eta^-(g)}) = \Omega_n(\mathbf{X}, u_{\eta(g^{-1})}^*)$$
$$\mathbb{M}_{\mathbf{X}}^{Sp+}(g) := \mathbb{E}\{M_{\mathbf{X}}^{Sp+}(g)\}$$
$$\mathbb{M}_{\mathbf{X}}^{Sp-}(g) := \mathbb{E}\{M_{\mathbf{X}}^{Sp-}(g)\},$$
$$C_{\mathbf{X}}^{Sp+}(g) := \{\mathbb{M}_{\mathbf{X}}^{Sp+} \circledast (M_{I_N}^{Sp-})^{\circledast(-1)}\}(g)$$
$$C_{\mathbf{X}}^{Sp-}(g) := \{(M_{I_N}^{Sp+})^{\circledast(-1)} \circledast \mathbb{M}_{\mathbf{X}}^{Sp-}\}(g).$$

With these definitions the geometrical interpretation of the Sp-cumulants as in (31) holds true and similar properties as those exposed in Section 3.2 can be proved like in Section 3.4.

References

1. Capitaine M., Casalis M. (2004). Asymptotic freeness by generalized moments for Gaussian and Wishart matrices. Application to Beta random matrices. *Indiana Univ. Math. J.*, **53**, N 2, 397-431.
2. Capitaine M., Casalis M. (2006). Cumulants for random matrices as convolutions on the symmetric group. *Probab. Theory Relat. Fields*, **136**, 19-36.
3. Capitaine M., Casalis M. (2006). Cumulants for random matrices as convolutions on the symmetric group II. to appear in *J. Theoret. Probab.*
4. Collins B. (2003). Moments and cumulants of polynomial random variables on unitary groups, the Itzykson-Zuber integral and free probability. *Int. Math. Res. Not.*, **17**, 953-982.
5. Collins B., Sniady P. (2006). Integration with respect to the Haar measure on unitary, orthogonal and symplectic group, *Commun. Math. Phys.*, **204**, 773-795.
6. Graczyk P., Letac G., Massam H. (2003). The complex Wishart distribution and the symmetric group, *Annals of Statistics*, **31**, 287-309.
7. Graczyk P., Letac G., Massam H. (2005). The Hyperoctahedral group, symmetric group representations and the moments of the real Wishart distribution, *J. Theoret. Probab*, **18**, 1-42.
8. Goodman R., Wallach N.R. *Representations and Invariants of the Classical Groups*, Cambridge, 1998.
9. Mneimé R., Testard F. *Introduction à la théorie des groupes de Lie classiques*, Hermann, 1986.
10. Nica A., Speicher R. (1996). On the multiplication of free N-uples of noncommutative random variables, *Am. Journ. of Math.*, **118**, 799-837.
11. Speicher R. (1994). Multiplicative functions on the lattice of non-crossing partitions and free convolution, *Math Ann.*, **298**, 611-628.
12. Voiculescu D.V., Dykema K.J. and Nica A. *Free random variables*, CRM Monographs Series, Vol. 1, Amer. Math. Soc., Providence, 1992.

Fluctuations of spectrally negative Markov additive processes

Andreas E. Kyprianou[1] and Zbigniew Palmowski[2,3]

[1] Department of Mathematical Sciences, The University of Bath
   Claverton Down, Bath, UK
[2] Mathematical Institute, University of Wrocław
   pl. Grunwaldzki 2/4, 50-384 Wrocław, Poland
[3] Mathematical Institute, Utrecht University
   P.O. Box 80.010, 3508 TA Utrecht, The Netherlands
e-mails: a.kyprianou@bath.ac.uk, zpalma@math.uni.wroc.pl

Summary. For spectrally negative Markov Additive Processes (MAPs) we generalize classical fluctuation identities developed in Zolotarev (1964), Takács (1967), Bingham (1975), Suprun (1976), Emery (1973), Rogers (1990) and Bertoin (1997) which concern one and two sided exit problems for spectrally negative Lévy processes.

1 Spectrally Negative Markov Additive Processes

This paper presents some fluctuation identities for a special, but none the less quite general, class of Markov Additive Processes (MAP). Before entering our discussion on the subject we shall simply begin by defining the class of processes we intend to work with and its properties.

Following Asmussen and Kella (2000) we consider a process $X(t)$, where $X(t) = X^{(1)}(t) + X^{(2)}(t)$, and the independent processes $X^{(1)}(t)$ and $X^{(2)}(t)$ are specified by the characteristics: $q_{ij}, G_{ij}, \sigma_i, a_i, \nu_i(dx)$ which we shall now define. Let $J(t)$ be a right-continuous, ergodic, finite state space continuous time Markov chain, with states $\mathcal{I} = \{1, \ldots, N\}$, and with intensity matrix $\mathbf{Q} = (q_{ij})$. We denote the jumps of the process $J(t)$ by $\{T_i\}$ (with $T_0 = 0$). Let $\{U_n^{(ij)}\}$ be i.i.d. random variables, which are also independent of J, with distribution function $G_{ij}(\cdot)$ ($U^{(ii)} \equiv 0$). Define the jump process by

$$X^{(1)}(t) = \sum_{n \geq 1} \sum_{i,j} U_n^{(ij)} \mathbf{1}_{\{J(T_{n-1})=i,\ J(T_n)=j,\ T_n \leq t\}}.$$

For each $i \in \mathcal{I}$, let $X^i(t)$ be a Lévy process, independent of all other stochastic quantities, with Laplace exponent

$$\log E(\exp \alpha X^i(1)) = \psi_i(\alpha) = a_i\alpha + \frac{\sigma_i^2\alpha^2}{2} + \int_{-\infty}^{0} \left(e^{\alpha y} - 1 - \alpha y 1_{(-1,0)}(y)\right)\nu_i(dy),$$

where $1_{(-1,0)}(y)$ is valued 1 if $y \in (-1,0)$ and valued 0 otherwise and $\int_{-\infty}^{0}(1 \wedge |y|^2)\nu_i(dy) < \infty$. By $X^{(2)}(t)$ we denote the process which behaves in law like $X^i(t)$, when $J(t) = i$. Note that each of the measures ν_i are supported on $(-\infty, 0)$ as well as the distributions of each $U^{(ij)}$ and in this respect we say that X is a *spectrally negative* MAP.

Letting $\mathbf{Q} \circ \widehat{\mathbf{G}}(\alpha) = (q_{ij}\widehat{G}_{ij}(\alpha))$, where $\widehat{G}_{ij}(\alpha) = E\left(\exp(\alpha U^{(ij)})\right)$, we define *matrix cumulant generating function* of MAP $X(t)$:

$$\mathbf{F}(\alpha) = \mathbf{Q} \circ \widehat{\mathbf{G}}(\alpha) + \mathrm{diag}(\psi_1(\alpha), \ldots, \psi_N(\alpha)) . \tag{1}$$

Note then that $\mathbf{F}(\alpha)$ is well defined and finite at least for $\alpha \geq 0$. Within this regime of α, Perron-Frobenius theory identifies $\mathbf{F}(\alpha)$ as having a real-valued eigenvalue with maximal absolute value which we shall label $\kappa(\alpha)$. The corresponding left and right $1 \times N$ eigenvectors we label $\mathbf{v}(\alpha)$ and $\mathbf{h}(\alpha)$ respectively. In this text we shall always write vectors in their horizontal form and use the usual ^{T} to mean transpose. Since $\mathbf{v}(\alpha)$ and $\mathbf{h}(\alpha)$ are given up to multiplying constants, we are free to normalize them such that

$$\mathbf{v}(\alpha)\mathbf{h}(\alpha)^{\mathrm{T}} = 1 \text{ and } \pi\mathbf{h}(\alpha)^{\mathrm{T}} = 1 ,$$

where $\pi = \mathbf{v}(0)$ is the stationary distribution of J. Note also that $\mathbf{h}(0) = \mathbf{e}$, the $1 \times N$ vector consisting of a row of ones. We shall write $h_i(\alpha)$ for the i-th element of $\mathbf{h}(\alpha)$. The eigenvalue $\kappa(\alpha)$ is a convex function (this can also be easily verified) such that $\kappa(0) = 0$, $\kappa(\infty) = \infty$ and $\kappa'(0)$ is the asymptotic drift of X in the sense that for each $i \in \mathcal{I}$ we have $\lim_{t\uparrow\infty} E(X(t)|J(0) = i, X(0) = x)/t = \kappa'(0)$. The sign of $\kappa'(0)$ also determines the asymptotic behaviour of X. When $\kappa'(0) > 0$, the process drifts to infinity, $\lim_{t\uparrow\infty} X(t) = \infty$, when $\kappa'(0) < 0$, the process drifts to minus infinity, $\lim_{t\uparrow\infty} X(t) = -\infty$, and when $\kappa'(0) = 0$ the process oscillates, $\limsup_{t\uparrow\infty} X(t) = -\liminf_{t\uparrow\infty} X(t) = \infty$. For the right inverse of κ we shall write Φ on $[0,\infty)$. That is to say, for each $q \geq 0$,

$$\Phi(q) = \sup\{\alpha \geq 0 : \kappa(\alpha) = q\}.$$

Note that the properties of κ imply that $\Phi(q) > 0$ for $q > 0$. Further $\Phi(0) = 0$ if and only if $\kappa'(0) \geq 0$ and otherwise $\Phi(0) > 0$.

We shall assume the afore mentioned class of MAPs are defined on a probability space with probabilities $\{\mathbb{P}_{i,x} : i \in \mathcal{I}, x \in \mathbb{R}\}$ and right-continuous natural filtration $\mathbb{F} = \{\mathcal{F}_t : t \geq 0\}$. It can be checked that under the following Girsanov change of measure

$$\left.\frac{d\mathbb{P}_{i,x}^\gamma}{d\mathbb{P}_{i,x}}\right|_{\mathcal{F}_t} := e^{\gamma(X(t)-x)-\kappa(\gamma)t}\frac{h_{J(t)}(\gamma)}{h_i(\gamma)}, \text{ for } \gamma \text{ such that } \kappa(\gamma) < \infty, \tag{2}$$

the process $(X, \mathbb{P}^\gamma_{i,x})$ is again a spectrally negative MAP whose intensity matrix $\mathbf{F}_\gamma(\alpha)$ is well defined and finite for $\alpha \geq -\gamma$; see for example Palmowski and Rolski (2002). If $\mathbf{F}_\gamma(\alpha)$ has largest eigenvalue $\kappa_\gamma(\alpha)$ and associated right eigenvector $\mathbf{h}_\gamma(\alpha)$, the triple $(\mathbf{F}_\gamma(\alpha), \kappa_\gamma(\alpha), \mathbf{h}_\gamma(\alpha))$ is related to the original triple $(\mathbf{F}(\alpha), \kappa(\alpha), \mathbf{h}(\alpha))$ via

$$\mathbf{F}_\gamma(\alpha) = \mathbf{\Delta_h}(\gamma)^{-1} \mathbf{F}(\alpha + \gamma) \mathbf{\Delta_h}(\gamma) - \kappa(\gamma)\mathbf{I} \text{ and } \kappa_\gamma(\alpha) = \kappa(\alpha + \gamma) - \kappa(\gamma) , \tag{3}$$

where $\mathbf{I}$ is the $N \times N$ identity matrix and

$$\mathbf{\Delta_h}(\gamma) := \mathrm{diag}(h_1(\gamma), ..., h_N(\gamma)).$$

We shall also use a similar definition for the matrix $\mathbf{\Delta_v}(\gamma)$.

As much as possible we shall prefer to work with matrix notation. For a random variable Y and (random) time τ, we shall understand $\mathbf{E}_x(Y; J(\tau))$ to be the matrix with (i,j)-th elements $\mathbb{E}_{i,x}(Y; J(\tau) = j)$. For an event, A, $\mathbf{P}_x(A; J(\tau))$ will be understood in a similar sense. For simplicity we shall follow the tradition that $\mathbf{E}(\cdot) = \mathbf{E}_0(\cdot)$ and $\mathbf{P}(\cdot) = \mathbf{P}_0(\cdot)$. For shorthand we will denote $\mathbf{I}_{ij}(q) = \mathbb{P}_{i,0}(J(\mathbf{e}_q) = j)$, in other words $\mathbf{I}(q) = q(q\mathbf{I} - \mathbf{Q})^{-1}$.

These details and more concerning the basic characterization of MAPs can be found in Chapter XI of Asmussen (2003).

2 Time reversal

Predominant in the forthcoming discussion will be the use of the bivariate process $(\widehat{J}, \widehat{X})$, representing the process (J, X) time reversed from a fixed moment in the future when $J(0)$ has the stationary distribution π. For definitiveness, we mean

$$\widehat{J}(s) = J((t - s)^-) \text{ and } \widehat{X}(s) = X(t) - X((t - s)^-), \ 0 \leq s \leq t$$

under $\mathbb{P}_{\pi,0} = \sum_{i \in \mathcal{I}} \pi_i \mathbb{P}_{i,0}$. The characteristics of $(\widehat{J}, \widehat{X})$ will be indicated by using a hat over the existing notation for the characteristics of (J, X). For example $\widehat{\mathbf{F}}, \widehat{\mathbf{h}}, \widehat{\kappa}$ and so on. To relate these characteristics to the original ones, recall that the intensity matrix of $\widehat{J}$ must satisfy

$$\widehat{\mathbf{Q}} = \mathbf{\Delta}_\pi^{-1} \mathbf{Q}^\mathrm{T} \mathbf{\Delta}_\pi ,$$

where $\mathbf{\Delta}_\pi$ is the diagonal matrix whose entries are given by the vector π. Hence according to (1) we find that when it exists

$$\widehat{\mathbf{F}}(\alpha) = \mathbf{\Delta}_\pi^{-1} \mathbf{F}(\alpha)^\mathrm{T} \mathbf{\Delta}_\pi.$$

Since $(\mathbf{v}(\alpha)\mathbf{F}(\alpha))^\mathrm{T} = \mathbf{F}(\alpha)^\mathrm{T} \mathbf{v}(\alpha)^\mathrm{T} = \kappa(\alpha)\mathbf{v}(\alpha)^\mathrm{T}$ we have that

$$\widehat{\mathbf{F}}(\alpha)\mathbf{\Delta}_\pi^{-1}\mathbf{v}(\alpha)^\mathrm{T} = \mathbf{\Delta}_\pi^{-1}\mathbf{F}(\alpha)^\mathrm{T}\mathbf{\Delta}_\pi\mathbf{\Delta}_\pi^{-1}\mathbf{v}(\alpha)^\mathrm{T} = \kappa(\alpha)\mathbf{\Delta}_\pi^{-1}\mathbf{v}(\alpha)^\mathrm{T}$$

showing that $\widehat{\kappa}(\alpha) \geq \kappa(\alpha)$. On the other hand a similar calculation reveals that

$$\widehat{\kappa}(\alpha)\,\boldsymbol{\Delta}_\pi\widehat{\mathbf{h}}(\alpha)^{\mathrm{T}} = \boldsymbol{\Delta}_\pi\widehat{\mathbf{F}}(\alpha)\,\widehat{\mathbf{h}}(\alpha) = \mathbf{F}(\alpha)^{\mathrm{T}}\,\boldsymbol{\Delta}_\pi\widehat{\mathbf{h}}(\alpha)^{\mathrm{T}}$$

so that $\widehat{\kappa}(\alpha) \leq \kappa(\alpha)$ and hence $\widehat{\kappa} = \kappa$ and $\boldsymbol{\Delta}_\pi\widehat{\mathbf{h}}(\alpha)^{\mathrm{T}} = \mathbf{v}(\alpha)^{\mathrm{T}}$.

Instead of talking about the process $(\widehat{J}, \widehat{X})$ we shall talk about the process (J, X) under probabilities $\{\widehat{\mathbb{P}}_{i,0} : i \in \mathcal{I}\}$ meaning the MAP whose characteristics are given by $\widehat{\mathbf{F}}$. Note also for future use, following classical time reversed path analysis, for $y \geq 0$,

$$\mathbb{P}_{i,0}\left(-I(t) \in dy | J(t) = j\right) = \mathbb{P}_{j,0}\left(\widehat{S}(t) - \widehat{X}(t) \in dy | \widehat{J}(t) = i\right)$$

$$= \widehat{\mathbb{P}}_{j,0}\left(S(t) - X(t) \in dy | J(t) = i\right), \qquad (4)$$

where $I(t) = \inf_{0 \leq s \leq t} X(s)$, $S(t) = \sup_{0 \leq s \leq t} X(s)$ and $\widehat{S}(t) = \sup_{0 \leq s \leq t} \widehat{X}(s)$. (A diagram may help to explain the last identity). Asmussen (1989, 2000) gives a more thorough discussion on time reversal.

3 The intensity matrix $\boldsymbol{\Lambda}(q)$

Also important for the main results of this paper will be a brief summary of the classical analysis of first passage upward with the help of exponential change of measure given in (2).

Define for each $x \geq 0$

$$\tau_x^+ := \inf\{t \geq 0 : X(t) \geq x\}.$$

Note that for each $q \geq 0$ and $x \geq 0$,

$$\boldsymbol{\Delta}_{\mathbf{h}}(\Phi(q))^{-1}\mathbf{E}\left(e^{\Phi(q)x - q\tau_x^+}1_{(\tau_x^+ < \infty)}; J(\tau_x^+)\right)\boldsymbol{\Delta}_{\mathbf{h}}(\Phi(q)) = \mathbf{I}^{\Phi(q)}(\tau_x^+).$$

Here

$$\mathbf{I}_{ij}^{\Phi(q)}(\tau_x^+) = \mathbb{P}_{i,0}^{\Phi(q)}(J(\tau_x^+) = j, \tau_x^+ < \infty) = \mathbb{P}_{i,0}^{\Phi(q)}(J(\tau_x^+) = j),$$

where the last equality follows since $(X, \mathbb{P}^{\Phi(q)})$ drifts to infinity on account of the fact that $\kappa'_{\Phi(q)}(0) = \kappa'(\Phi(q))$ which is strictly positive by convexity of κ. Hence

$$\mathbf{E}\left(e^{-q\tau_x^+}1_{(\tau_x^+ < \infty)}; J(\tau_x^+)\right) = e^{-\Phi(q)x}\boldsymbol{\Delta}_{\mathbf{h}}(\Phi(q))\mathbf{I}^{\Phi(q)}(\tau_x^+)\boldsymbol{\Delta}_{\mathbf{h}}(\Phi(q))^{-1}.$$

A little thought reveals that process $\{J(\tau_x^+) : x \geq 0\}$ is again an ergodic Markov chain whenever X does not drift to $-\infty$. Let us suppose that under $\mathbf{P}^{\Phi(q)}$, for $q \geq 0$, its intensity matrix is given by $\boldsymbol{\Lambda}(q)$ so that $\mathbf{I}^{\Phi(q)}(\tau_x^+) = \exp(\boldsymbol{\Lambda}(q)x)$. We thus obtain the following result.

Theorem 1. *For $q \geq 0$,*

$$\mathbf{E}\left(e^{-q\tau_x^+}1_{(\tau_x^+ < \infty)}; J(\tau_x^+)\right) = \boldsymbol{\Delta}_{\mathbf{h}}(\varPhi(q))e^{-(\varPhi(q)\mathbf{I}-\boldsymbol{\Lambda}(q))x}\boldsymbol{\Delta}_{\mathbf{h}}(\varPhi(q))^{-1}. \quad (5)$$

Unfortunately, it does not seem that an explicit expression for $\boldsymbol{\Lambda}(q)$ can be derived from existing literature. Indeed establishing an expression for $\boldsymbol{\Lambda}(q)$ for spectrally negative MAPs is an open problem. Recent progress has been made however by Miyazawa and Takada (2002) (see also the references therein) who consider a special case of the spectrally negative MAP that we have here. That is, the case when the jump part of each of the processes X^i are of bounded variation and finite mean. For this class, they prove that $\boldsymbol{\Lambda}(0)$ solves $\mathbf{F}(-\boldsymbol{\Lambda}(0)) = \mathbf{0}$ where we should understand the latter to mean that

$$-\boldsymbol{\Delta}_{\mathbf{a}}\boldsymbol{\Lambda}(0) + \boldsymbol{\Delta}_{\sigma^2/2}\boldsymbol{\Lambda}(0)^2 + \int_{-\infty}^{0} \boldsymbol{\Delta}_\nu(dy)(e^{-\boldsymbol{\Lambda}(0)y} - \mathbf{I} + \boldsymbol{\Lambda}(0)y1_{[-1,0]}(y))$$

$$+ \int_{-\infty}^{0} \mathbf{Q} \circ \mathbf{G}(dy)e^{-\boldsymbol{\Lambda}(0)y} = \mathbf{0}, \quad (6)$$

where $\boldsymbol{\Delta}_{\mathbf{a}}$ is the diagonal matrix with entries $a_1, ..., a_N$ along the diagonal, similarly $\boldsymbol{\Delta}_{\sigma^2/2}$ is diagonal with elements $\sigma_1^2/2, ..., \sigma_N^2/2$, matrix $\boldsymbol{\Delta}_\nu(\cdot)$ is diagonal with $\nu_1(\cdot), \ldots, \nu_N(\cdot)$ on the diagonal and the matrix $\mathbf{G}$ has entries $G_{ij}(\cdot)$ corresponding to the distributions of each U_{ij}. It can be seen from the straightforward martingale techniques described by Pistorius (2006) however, that in general, one may characterize $\boldsymbol{\Lambda}(q)$ as a solution to the matrix equation $\mathbf{F}_{\varPhi(q)}(-\boldsymbol{\Lambda}(q)) = \mathbf{0}$. (We are grateful to M. Pistorius for pointing this out). For further discussion concerning uniqueness of the solution see Rogers (1994), Asmussen (2000) and Miyazawa and Takada (2002).

4 Main results

In this article we shall establish fluctuation identities concerning the exit times τ_a^+ for $a \geq 0$ (defined at the beginning of the previous section) and

$$\tau_0^- := \inf\{t \geq 0 : X(t) \leq 0\}.$$

Our results will be expressed in terms of $N \times N$ matrix functions $\mathbf{W}^{(q)}(x)$ and $\mathbf{M}^{(q)}(x)$ both of which are mappings from $\mathbb{R}$ to $[0, \infty)$ with a parameter range $q \geq 0$.

Before moving to the main results, we shall devote a little time to establishing some further notation. Here and throughout we work with the definition that $\mathbf{e}_q$ is random variable which is exponentially distributed with mean $1/q$ and independent of (J, X).

For $q > 0$, let

$$\widehat{\mathbf{D}}(q) = \mathbf{\Delta_v}\left(\Phi\left(q\right)\right)\left(\Phi(q)\mathbf{I} - \widehat{\mathbf{\Lambda}}(q)\right)^{-1}\mathbf{\Delta_v}\left(\Phi\left(q\right)\right)^{-1},$$

where the matrix $\widehat{\mathbf{\Lambda}}(q)$ takes the same definition as $\mathbf{\Lambda}(q)$ but for the time reversed process $\widehat{X}$. We shall also be interested in the following related limiting quantity

$$\mathbf{D} := \lim_{q \downarrow 0} q\widehat{\mathbf{D}}(q)^{\mathrm{T}}\mathbf{I}(q)^{-1}.$$

There are two cases to consider.

When X drifts to $-\infty$ we know that $\Phi(0) > 0$ then simply

$$\mathbf{D} = \mathbf{\Delta_v}\left(\Phi\left(0\right)\right)^{-1}\left(\Phi(0)\mathbf{I} - \widehat{\mathbf{\Lambda}}(0)^{\mathrm{T}}\right)^{-1}\mathbf{\Delta_v}\left(\Phi\left(0\right)\right)\mathbf{Q}.$$

On the other hand, when X does not drift to ∞, denote by π^+ the stationary distribution of $\{J(\tau_x^+), x \geq 0\}$ under $\widehat{\mathbf{P}}$. Let $\mathbf{\Pi}^+ = \mathbf{e}^{\mathrm{T}}\pi^+$. By Coolen-Schrijner and van Doorn (2002) there exists a matrix

$$\mathbf{A} = \int_0^\infty \left(e^{\widehat{\mathbf{\Lambda}}(0)t} - \mathbf{\Pi}^+\right)\, dt$$

called a deviation matrix (or Drazin inverse) of $\widehat{\mathbf{\Lambda}}(0)$. It solves uniquely the equations

$$-\widehat{\mathbf{\Lambda}}(0)\mathbf{X} = \mathbf{X}\widehat{\mathbf{\Lambda}}(0), \quad \mathbf{X}(-\widehat{\mathbf{\Lambda}}(0))\mathbf{X} = \mathbf{X} \quad \text{and} \quad (-\widehat{\mathbf{\Lambda}}(0))\mathbf{X}(-\widehat{\mathbf{\Lambda}}(0)) = -\widehat{\mathbf{\Lambda}}(0).$$

Note that $\Phi(0) = 0$ and hence $\mathbf{\Delta_v}(\Phi(0)) = \mathbf{\Delta}_\pi$. Also

$$q\widehat{\mathbf{D}}(q)^{\mathrm{T}}\mathbf{I}(q)^{-1} \sim \mathbf{\Delta}_\pi^{-1}[(\Phi(q)\mathbf{I} - \widehat{\mathbf{\Lambda}}(q))^{-1}]^{\mathrm{T}}\mathbf{\Delta}_\pi(q\mathbf{I} - \mathbf{Q})\mathbf{\Delta}_\pi^{-1}\mathbf{\Delta}_\pi$$

$$= \mathbf{\Delta}_\pi^{-1}\left[(q\mathbf{I} - \widehat{\mathbf{Q}})(\Phi(q)\mathbf{I} - \widehat{\mathbf{\Lambda}}(q))^{-1}\right]^{\mathrm{T}}\mathbf{\Delta}_\pi,$$

where $\mathbf{B}(q) \sim \mathbf{C}(q)$ means that $\mathbf{B}(q)\mathbf{C}(q)^{-1} \to \mathbf{I}$ as $q \to 0$. Morover,

$$(q\mathbf{I} - \widehat{\mathbf{Q}})(\Phi(q)\mathbf{I} - \widehat{\mathbf{\Lambda}}(q))^{-1} = (q\mathbf{I} - \widehat{\mathbf{Q}})\int_0^\infty e^{\widehat{\mathbf{\Lambda}}(q)t}e^{-\Phi(q)t}\, dt$$

$$\sim q\mathbf{A} - \widehat{\mathbf{Q}}\mathbf{A} + q\int_0^\infty e^{-\Phi(q)t}\mathbf{\Pi}^+\, dt$$

$$- \int_0^\infty e^{-\Phi(q)t}\widehat{\mathbf{Q}}\mathbf{e}^{\mathrm{T}}\pi^+\, dt$$

$$= q\mathbf{A} - \widehat{\mathbf{Q}}\mathbf{A} + \frac{q}{\Phi(q)}\mathbf{\Pi}^+$$

$$\sim \kappa'(0)\mathbf{\Pi}^+ - \widehat{\mathbf{Q}}\mathbf{A}.$$

Hence

$$\mathbf{D} = \mathbf{\Delta}_\pi^{-1}\left[\kappa'(0)\mathbf{\Pi}^+ - \widehat{\mathbf{Q}}\,\mathbf{A}\right]^{\mathrm{T}}\mathbf{\Delta}_\pi.$$

We now give the main results which are uniformly given under the following assumption.

Assumption 2 *None of the processes X^i are downward subordinators.*

The assumption that none of the processes X^i are downward subordinators offers the convenience that matrices such as $\widehat{\mathbf{\Lambda}}(q)$ have no zero columns.

Theorem 3. *For each $q \geq 0$ there exist $N \times N$ matrix functions $\mathbf{W}^{(q)}(\cdot)$ and $\mathbf{M}^{(q)}(\cdot)$ such that the following hold (for convenience we shall write $\mathbf{W}$ for $\mathbf{W}^{(0)}$).*

(i) *The matrix $\mathbf{W}^{(q)}$ almost everywhere differentiable on $(0,\infty)$, equal to zero on $(-\infty,0]$ and satisfies*

$$\mathbf{W}^{(q)}(x) = \mathbf{\Delta}_{\mathbf{h}}(\Phi(q))e^{\Phi(q)x}\mathbf{W}_{\Phi(q)}(x)\mathbf{\Delta}_{\mathbf{h}}(\Phi(q))^{-1}$$

for all $x \in \mathbb{R}$ where $\mathbf{W}_{\Phi(q)}$ plays the role of $\mathbf{W}$ under $\mathbb{P}^{\Phi(q)}$.

(ii) *The matrix $\mathbf{M}^{(q)}$ is characterized by its Laplace transform on $(0,\infty)$,*

$$\int_{[0,\infty)} e^{-\beta x}\mathbf{M}^{(q)}(dx) = (\mathbf{F}(\beta) - q\mathbf{I})^{-1}(\mathbf{I} - \beta\widehat{\mathbf{D}}(q)^T)(q\mathbf{I} - \mathbf{Q}) \qquad (7)$$

for sufficiently large β and it is equal to $\mathbf{I}$ on $(-\infty,0]$.

(iii) *For $x \geq 0$,*

$$\mathbf{E}_x\left(e^{-q\tau_0^-}1_{(\tau_0^- < \infty)}; J(\tau_0^-)\right) = \mathbf{M}^{(q)}(x). \qquad (8)$$

(iv) *For $x \leq a$,*

$$\mathbf{E}_x\left(e^{-q\tau_a^+}1_{(\tau_a^+ < \tau_0^-)}; J(\tau_a^+)\right) = \mathbf{W}^{(q)}(x)\,\mathbf{W}^{(q)}(a)^{-1}.$$

(v) *For $x \leq a$,*

$$\mathbf{E}_x\left(e^{-q\tau_0^-}1_{(\tau_a^+ > \tau_0^-)}; J(\tau_0^-)\right) = \mathbf{M}^{(q)}(x) - \mathbf{W}^{(q)}(x)\mathbf{W}^{(q)}(a)^{-1}\mathbf{M}^{(q)}(a).$$

To some extent the results in Theorem 3 generalize known expressions for spectrally negative Lévy processes established over a number of years by Zolotarev (1964), Takács (1967), Emery (1973) Bingham (1975), Suprun (1976), Rogers (1990) and Bertoin (1997).

Aside from being a natural generalization of the known results for spectrally negative Lévy processes, there are genuine reasons for wanting to know the conclusions presented in Theorems 3. In recent years classical models in the theory of financial mathematics, risk and queues have been replaced by ones involving Markov modulation and general classes of Lévy processes. In the latter case one particular class that has proved to be quite successful in this respect is spectrally negative Lévy processes, on account of the robustness of their fluctuation theory; see for example Schmidli (1999), Asmussen (2000), Asmussen and Kella (2000), Asmussen et al. (2004), Avram et al. (2004), Klüppelberg et al. (2004), Pistorius (2003), Huzak et al. (2004), Chan (2005), Chiu and Yin (2005) and Dube et al. (2004) to name but a few. Our results justify the reasoning that, in general, one should hope to be able to work with spectrally negative MAPs in these models (see e.g. Miyazawa (2004)).

5 Asmussen-Kella martingale

The basis of this section is the new martingale introduced by Asmussen and Kella (2000). This martingale is closely related to other martingales of an exponential type found in Kella and Whitt (1992) and Jacod and Shiryaev (1987) for example. For our purposes we shall simply introduce the Asmussen-Kella martingale in the form that we shall use it; however the reader is urged to return to the original presentation in Asmussen and Kella (2000) in order to appreciate that it can take a more general form. The martingale in question is zero mean, vector valued and takes the form

$$
\int_0^t e^{-\beta Z(u)} \mathbf{1}_{J(u)} du \cdot \mathbf{F}\left(\beta\right) + e^{-\beta Z(0)} \mathbf{1}_{J(0)} - e^{-\beta Z(t)} \mathbf{1}_{J(t)} - \beta \int_0^t \mathbf{1}_{J(u)} dS(u) \, ,
$$

where $\mathbf{1}_i$ is the $1 \times N$ vector whose elements are zero except at the i-th position where it is 1, $\beta \geq 0$ and $Z(t) = S(t) - X(t)$. From this martingale we shall deduce an identity, below, which forms the basis of our proofs.

Theorem 4. *For $\beta \geq 0$*

$$
\mathbf{E}\left(e^{\beta I(\mathbf{e}_q)}; J(\mathbf{e}_q)\right)^T (\mathbf{F}\left(\beta\right) - q\mathbf{I})^T
$$

$$
= q \mathbf{\Delta_v}(\Phi\left(q\right))[\beta(\Phi(q)\mathbf{I} - \widehat{\mathbf{\Lambda}}(q))^{-1} - \mathbf{I}]\mathbf{\Delta_v}(\Phi\left(q\right))^{-1} \, , \qquad (9)
$$

where the $N \times N$ matrix $\widehat{\mathbf{\Lambda}}(q)$ is the intensity matrix of the process $\{J(\tau_x^+) : x \geq 0\}$ under $\widehat{\mathbf{P}}^{\Phi(q)}$.

Proof From (4) we have that

$$
\mathbb{E}_{i,0}\left(e^{\beta I(\mathbf{e}_q)} \mathbf{1}_{(J(\mathbf{e}_q)=j)}\right) = \mathbb{E}_{i,0}\left(e^{\beta I(\mathbf{e}_q)} | J(\mathbf{e}_q) = j\right) \mathbb{P}_{i,0}\left(J(\mathbf{e}_q) = j\right)
$$

$$
= \widehat{\mathbb{E}}_{j,0}\left(e^{-\beta Z(\mathbf{e}_q)} | J(\mathbf{e}_q) = i\right) \widehat{\mathbb{P}}_{j,0}\left(J(\mathbf{e}_q) = i\right) \frac{\pi_j}{\pi_i}
$$

$$
= \pi_i^{-1} \widehat{\mathbb{E}}_{j,0}\left(e^{-\beta Z(\mathbf{e}_q)} \mathbf{1}_{(J(\mathbf{e}_q)=i)}\right) \pi_j.
$$

That is to say, the matrix $\widehat{\mathbf{E}}\left(e^{-\beta Z(\mathbf{e}_q)}; J(\mathbf{e}_q)\right)$ fulfills

$$
\mathbf{\Delta}_\pi^{-1} \mathbf{E}\left(e^{\beta I(\mathbf{e}_q)}; J(\mathbf{e}_q)\right)^{\mathrm{T}} \mathbf{\Delta}_\pi = \widehat{\mathbf{E}}\left(e^{-\beta Z(\mathbf{e}_q)}; J(\mathbf{e}_q)\right) \, . \qquad (10)
$$

Now note that a simple calculation involving Fubini's Theorem shows that

$$
\frac{1}{q} \widehat{\mathbb{E}}_{i,0}\left(e^{-\beta Z(\mathbf{e}_q)} \mathbf{1}_{J(\mathbf{e}_q)}\right) = \widehat{\mathbb{E}}_{i,0}\left(\int_0^{\mathbf{e}_q} e^{-\beta Z(u)} \mathbf{1}_{J(u)} du\right).
$$

Making use of the Asmussen and Kella's martingale (as given above) applied to the time reversed MAP, it follows by taking expectation with respect to

$\widehat{\mathbb{P}}_{i,0}$ that

$$
\widehat{\mathbb{E}}_{i,0}\left(e^{-\beta Z(\mathbf{e}_q)}\mathbf{1}_{J(\mathbf{e}_q)}\right)\left(\widehat{\mathbf{F}}(\beta)-q\mathbf{I}\right)=-q\mathbf{1}_i+\beta q\widehat{\mathbb{E}}_{i,0}\left(\int_0^{\mathbf{e}_q}\mathbf{1}_{J(u)}dS(u)\right).
\tag{11}
$$

Now note, with the help of Fubini's Theorem and the fact that $(X,\widehat{\mathbb{P}}_{i,0})$ can only creep upwards thus allowing τ^+ to be considered the inverse of $S(\cdot)$,

$$
\begin{aligned}
\widehat{\mathbb{E}}_{i,0}\left(\int_0^{\mathbf{e}_q}\mathbf{1}_{J(u)}dS(u)\right)&=\widehat{\mathbb{E}}_{i,0}\left(\int_0^{\infty}e^{-qu}\mathbf{1}_{J(u)}dS(u)\right)\\
&=\widehat{\mathbb{E}}_{i,0}\left(\int_0^{\infty}\mathbf{1}_{(\tau_a^+<\infty)}e^{-q\tau_a^+}\mathbf{1}_{J(\tau_a^+)}da\right).
\end{aligned}
$$

Further, applying a change of measure in the spirit of (2) we have

$$
\begin{aligned}
&\widehat{\mathbb{E}}_{i,0}\left(\int_0^{\mathbf{e}_q}\mathbf{1}_{(J(u)=j)}dS(u)\right)\\
&=\int_0^{\infty}e^{-\Phi(q)a}\widehat{\mathbb{E}}_{i,0}^{\Phi(q)}\left(\mathbf{1}_{(\tau_a^+<\infty)}\mathbf{1}_{(J(\tau_a^+)=j)}\right)\frac{\widehat{h}_i(\Phi(q))}{\widehat{h}_j(\Phi(q))}da.
\end{aligned}
\tag{12}
$$

(Recall that $\kappa=\widehat{\kappa}$ and hence $\Phi=\widehat{\Phi}$). Finally noting from (3) that $\kappa'_{\Phi(q)}(0)=\kappa'(\Phi(q))>0$, it follows that under $\widehat{\mathbb{P}}_{i,0}^{\Phi(q)}$ the process X drifts to infinity thus allowing the removal of the indicator $\mathbf{1}_{(\tau_a^+<\infty)}$ in (12). Replacing (10) and (12) in (11) and considering the latter vector equality as the i-th row of a matrix, we obtain

$$
\begin{aligned}
&\boldsymbol{\Delta}_{\pi}^{-1}\mathbf{E}\left(e^{\beta I(\mathbf{e}_q)};J(\mathbf{e}_q)\right)^{\mathrm{T}}\boldsymbol{\Delta}_{\pi}\left(\widehat{\mathbf{F}}(\beta)-q\mathbf{I}\right)\\
&=-q\mathbf{I}+\beta q\boldsymbol{\Delta}_{\widehat{\mathbf{h}}}(\Phi(q))\widehat{\mathbf{U}}(q)\boldsymbol{\Delta}_{\widehat{\mathbf{h}}}(\Phi(q))^{-1},
\end{aligned}
$$

where

$$
\widehat{\mathbf{U}}_{ij}(q)=\int_0^{\infty}da\cdot e^{-\Phi(q)a}\widehat{\mathbb{P}}_{i,0}^{\Phi(q)}(J(\tau_a^+)=j).
$$

Note that matrix $\widehat{\mathbf{U}}(q)$ has no zero columns by the assumption that none of the Lévy processes is downward subordinator. Recalling that $\widehat{\mathbf{F}}(\alpha)=\boldsymbol{\Delta}_{\pi}^{-1}\mathbf{F}(\alpha)^{\mathrm{T}}\boldsymbol{\Delta}_{\pi}$ we have

$$
\begin{aligned}
&\mathbf{E}\left(e^{\beta I(\mathbf{e}_q)};J(\mathbf{e}_q)\right)^{\mathrm{T}}\left(\mathbf{F}(\beta)-q\mathbf{I}\right)^{\mathrm{T}}\\
&=-q\mathbf{I}+\beta q\boldsymbol{\Delta}_{\pi}\boldsymbol{\Delta}_{\widehat{\mathbf{h}}}(\Phi(q))\widehat{\mathbf{U}}(q)\boldsymbol{\Delta}_{\widehat{\mathbf{h}}}(\Phi(q))^{-1}\boldsymbol{\Delta}_{\pi}^{-1}.
\end{aligned}
$$

Further noting that $\boldsymbol{\Delta}_{\pi}\widehat{\mathbf{h}}(\alpha)^{\mathrm{T}}=\mathbf{v}(\alpha)^{\mathrm{T}}$ implies that $\boldsymbol{\Delta}_{\pi}\boldsymbol{\Delta}_{\widehat{\mathbf{h}}}(\alpha)=\boldsymbol{\Delta}_{\mathbf{v}}(\alpha)$, the diagonal matrix associated with vector $\mathbf{v}(\alpha)$, we have that

$$\mathbf{E}\left(e^{\beta I(\mathbf{e}_q)}; J(\mathbf{e}_q)\right)^{\mathrm{T}} (\mathbf{F}(\beta) - q\mathbf{I})^{\mathrm{T}} = -q\mathbf{I} + \beta q \mathbf{\Delta_v}(\Phi(q))\widehat{\mathbf{U}}(q)\mathbf{\Delta_v}(\Phi(q))^{-1}. \tag{13}$$

The theorem is proved once we note that $\widehat{\mathbf{U}}(q)$ is a resolvent and hence equal to $(\Phi(q)\mathbf{I} - \widehat{\mathbf{\Lambda}}(q))^{-1}$. □

6 Proof of Theorem 3

(ii) and (iii). We first prove that the Laplace transform in (7) is well-defined, that is $\mathbf{F}(\beta) - q\mathbf{I}$ is invertible for sufficiently large β. From Theorem 4 it follows that $\mathbf{F}(\beta) - q\mathbf{I}$ is invertible for β for which RHS of (9) is invertible. This holds β such that $\beta - \Phi(q)$ is greater than spectrum of matrix $\widehat{\mathbf{\Lambda}}(q)$ since:

$$[\beta(\Phi(q)\mathbf{I} - \widehat{\mathbf{\Lambda}}(q))^{-1} - \mathbf{I}]^{-1} = -[\mathbf{I} + \beta((\Phi(q) - \beta)\mathbf{I} - \widehat{\mathbf{\Lambda}}(q))^{-1}]. \tag{14}$$

Note also that $\mathbf{F}(\beta) - q\mathbf{I}$ is always invertible for $\beta < \Phi(q)$.

According to the definition of $\mathbf{M}^{(q)}$ via its Laplace transform simple manipulation of (9) yields

$$\mathbf{P}_x(I(\mathbf{e}_q) < 0; J(\mathbf{e}_q)) = \mathbf{M}^{(q)}(x)\mathbf{I}(q). \tag{15}$$

Using the law of total probability and first conditioning the probability on the left hand side of (15) with respect to $\mathcal{F}_{\tau_0^-}$ we have

$$\mathbf{E}_x(e^{-q\tau_0^-} 1_{(\tau_0^- < \infty)}; J(\tau_0^-))\mathbf{I}(q) = \mathbf{M}^{(q)}(x)\mathbf{I}(q).$$

and hence the first result follows for $q > 0$.

We may now take limits as $q \downarrow 0$ and note that the limit exists on the right hand side of (8) because it exists on the left hand side. □

(i) and (iv). Assume momentarily that X drifts to infinity. With this assumption, it is clear that X hits every level $a > 0$ with probability one. Next note that for any $0 < x \leq a$,

$$\{\mathbf{P}_x(\tau_a^+ < \tau_0^-; J(\tau_a^+))\}_{ij} = \{\mathbb{P}_{i,x}(J(\tau_a^+) = j, \tau_a^+ < \tau_0^-)\}$$

is the transition matrix of the Feller Markov chain $\{J^*(\tau_a^+) : a \geq 0\}$ of the modulating state on the upward ladder-height process of X^*, where X^* is the process X killed on exiting $(0, \infty)$. Note that the transitions of $J^*(\tau_a^+)$ depend on the position of $X^*(\tau_a^+)$ and hence on the level a since on $\tau_a^+ < \tau_0^-$ we have $X^*(\tau_a^+) = a$. Therefore, for each $y > 0$ there exists some $N \times N$ sub-stochastic intensity matrix $\mathbf{\Lambda}^*(y)$ such that

$$\mathbf{P}_x(\tau_a^+ < \tau_0^-; J(\tau_a^+)) = \exp\left\{\int_x^a \mathbf{\Lambda}^*(y)dy\right\}, \quad x > 0, \tag{16}$$

which has inverse $\exp\{-\int_x^a \mathbf{\Lambda}^*(y)dy\}$. It is also now clear that there exists invertible matrix

$$\mathbf{W}(x) = \exp\left\{-\int_x^b \mathbf{\Lambda}^*(y)dy\right\}$$

for some arbitrary $b > a$ (and therefore determined up to a pre- or post-multiplicative constant, invertible matrix) which is obviously almost everywhere differentiable on $(0,\infty)$. We have thus proved part (ii) of Theorem 3 for the case that $q = 0$ and X drifts to infinity.

Using a change of measure (2) we have for any $q > 0$ that

$$\mathbb{E}_{i,x}\left(e^{-q\tau_a^+}1_{(\tau_a^+ < \tau_0^-, J(\tau_a^+)=j)}\right)$$

$$= \mathbb{E}_{i,x}\left(e^{\Phi(q)(a-x)-q\tau_a^+}1_{(\tau_a^+ < \tau_0^-, J(\tau_a^+)=j)}\frac{h_j(\Phi(q))}{h_i(\Phi(q))}\right)\frac{h_i(\Phi(q))}{h_j(\Phi(q))}e^{-\Phi(q)(a-x)}$$

$$= h_i(\Phi(q))\mathbb{P}_{i,x}^{\Phi(q)}\left(\tau_a^+ < \tau_0^-, J(\tau_a^+) = j\right)\frac{1}{h_j(\Phi(q))}e^{-\Phi(q)(a-x)}. \tag{17}$$

Now $\kappa'_{\Phi(q)}(0) = \kappa'(\Phi(q)) > 0$. So that under $\mathbb{P}_{i,x}^{\Phi(q)}$ the process X always drifts to infinity. We may thus use the conclusions of the previous paragraph in (17) to deduce

$$\mathbf{E}_x\left(e^{-q\tau_a^+}1_{(\tau_a^+ < \tau_0^-)}; J(\tau_a^+)\right)$$

$$= [\mathbf{\Delta_h}(\Phi(q))e^{\Phi(q)x}\mathbf{W}_{\Phi(q)}(x)\mathbf{\Delta_h}(\Phi(q))^{-1}]$$

$$\times [\mathbf{\Delta_h}(\Phi(q))e^{\Phi(q)a}\mathbf{W}_{\Phi(q)}(a)\mathbf{\Delta_h}(\Phi(q))^{-1}]^{-1}$$

$$= \mathbf{W}^{(q)}(x)\mathbf{W}^{(q)}(a)^{-1} \tag{18}$$

where in the first equality we understand $\mathbf{W}_{\Phi(q)}(x)$ to be the scale function associated with $(X, P^{\Phi(q)})$ and the equality itself follows from the analysis in the case that X drifts to infinity above. Further, in the last line of (18) uses the definition

$$\mathbf{W}^{(q)}(x) = \mathbf{\Delta_h}(\Phi(q))e^{\Phi(q)x}\mathbf{W}_{\Phi(q)}(x)\mathbf{\Delta_h}(\Phi(q))^{-1}. \tag{19}$$

Note that $\mathbf{W}^{(q)}(x)$ is almost everywhere differentiable since $\mathbf{W}_{\Phi(q)}$ is. □

(v). This result follows from elementary linear algebra. Simply note from the Strong Markov Property that for $q > 0$,

$$\mathbf{E}_x(e^{-q\tau_0^-}1_{(\tau_0^- < \infty)}; J(\tau_0^-))$$

$$= \mathbf{E}_x(e^{-q\tau_0^-}1_{(\tau_0^- < \tau_a^+)}; J(\tau_0^-))$$

$$+ \mathbf{E}_x(e^{-q\tau_0^-}1_{(\tau_0^- > \tau_a^+)}; J(\tau_0^-))$$

$$= \mathbf{E}_x(e^{-q\tau_0^-}1_{(\tau_0^- < \tau_a^+)}; J(\tau_0^-))$$

$$+ \mathbf{E}_x(e^{-q\tau_a^+}1_{(\tau_0^- > \tau_a^+)}; J(\tau_a^+))\mathbf{E}_a(e^{-q\tau_0^-}1_{(\tau_0^- < \infty)}; J(\tau_0^-)).$$

Rearranging and substituting in the conclusions of part (i) and (ii) gives the required result for $q > 0$. To deal with the case that $q = 0$, take the limit above as q tends to zero.   □

7 An example

In general, it is difficult to identify the scale matrix $\mathbf{W}^{(q)}(x)$. Note that using the Strong Markov Property we have for $0 < x \leq a$,

$$\mathbf{P}_x(\tau_a^+ < \tau_0^-; J(\tau_a^+)) = \mathbf{P}_x(J(\tau_a^+)) - \mathbf{P}_x(\tau_0^- < \tau_a^+; J(\tau_a^+))$$

$$= \mathbf{P}_x(J(\tau_a^+)) - \int_{-\infty}^0 \mathbf{P}_x(\tau_0^- < \tau_a^+; X(\tau_0^-) \in dy; J(\tau_0^-))\mathbf{P}_y(J(\tau_a^+))$$

$$= e^{\mathbf{\Lambda}(0)(a-x)} - \int_{-\infty}^0 \mathbf{P}_x(\tau_0^- < \tau_a^+; X(\tau_0^-) \in dy; J(\tau_0^-))e^{\mathbf{\Lambda}(0)(a-y)}$$

$$= \mathbf{W}(x)\mathbf{W}(a)^{-1},$$

where $\mathbf{W}(x) = \mathbf{W}^{(0)}(x)$. Thus

$$\mathbf{W}(x)[e^{\mathbf{\Lambda}(0)a}\mathbf{W}(a)]^{-1}$$

$$= e^{-\mathbf{\Lambda}(0)x} - \int_{-\infty}^0 \mathbf{P}_x(\tau_0^- < \tau_a^+; X(\tau_0^-) \in dy; J(\tau_0^-))e^{-\mathbf{\Lambda}(0)y}. \quad (20)$$

Note the intensity matrix $\mathbf{\Lambda}(0)$ has eigenvalues with non-positive real part and hence the integral on the right hand side of the above identity makes sense. Further, it has a limit as $a \to \infty$. This implies that the left hand side of (20) also has a limit. Thus up to some invertible constant matrix $\mathbf{V}$ we have

$$\mathbf{W}(x) = \mathbf{V}\left[e^{-\mathbf{\Lambda}(0)x} - \int_{-\infty}^0 \mathbf{P}_x(\tau_0^- < \infty; X(\tau_0^-) \in dy; J(\tau_0^-))e^{-\mathbf{\Lambda}(0)y}\right]. \quad (21)$$

Without loss of generality we can assume further that $\mathbf{V} = \mathbf{I}$. Note also that

$$\mathbf{E}_x\left[e^{-qX(\tau_0^-)}\mathbf{1}_{(\tau_0^- < \infty)}; J(\tau_0^-)\right]$$

$$= \mathbf{\Delta}_\mathbf{h}(q)\mathbf{E}_x^q\left[e^{\kappa(q)\tau_0^-}\mathbf{1}_{(\tau_0^- < \infty)}; J(\tau_0^-)\right]\mathbf{\Delta}_\mathbf{h}(q)^{-1}, \quad (22)$$

and the latter can be derived from (8).

Here is an example of $\mathbf{W}^{(q)}(x)$ that can be identified. We consider *Brownian motion* and independent *Markov chain*. In other words $X^{(1)}(t) = 0$ for all $t \geq 0$ and $\psi_i(\beta) = \beta^2/2$ for each $i = 1, ..., N$. For this case, because the trajectories of the Brownian motion are continuous,

$$\mathbf{W}(x) = e^{-\mathbf{\Lambda}(0)x} - \mathbf{P}_x(\tau_0^- < \infty; J(\tau_0^-))$$

$$= e^{-\mathbf{\Lambda}(0)x} - \mathbf{M}^{(0)}(x).$$

Denote by λ_i the eigenvalues of $\mathbf{Q}$ (which have non-positive real part by Theorem 2.5 of Seneta (1973)) and denote by $\mathbf{H}$ the matrix of eigenvectors of $\mathbf{Q}$. Hence $\mathbf{Q} = \mathbf{H}^{-1}\mathrm{diag}(\lambda_i)\mathbf{H}$. Then $\mathbf{F}(\beta) = \mathbf{H}^{-1}\mathrm{diag}(\beta^2/2 + \lambda_i)\mathbf{H}$ and thus $\mathbf{\Lambda}(0) = \mathbf{H}^{-1}\mathrm{diag}((-2\lambda_i)^{1/2})\mathbf{H}$. Straightforward calculations based on the definition of matrix $\mathbf{M}^{(0)}(x)$ gives that $\int_0^\infty e^{-\beta x}\mathbf{W}(x)\,dx = \mathbf{C}\mathbf{F}(\beta)^{-1}$, where $\mathbf{C} = \mathbf{H}^{-1}\mathrm{diag}((-2\lambda_i/2)^{1/2})\mathbf{H}$. From the definition of the matrix $\mathbf{W}^{(q)}(x)$ given in (19) we have

$$\int_0^\infty e^{-\beta x}\mathbf{W}^{(q)}(x)\,dx = \Delta_{\mathbf{h}}\left(\Phi\left(q\right)\right)\int_0^\infty e^{-(\beta-\Phi(q))x}\mathbf{W}_{\Phi(q)}(x)\,dx\Delta_{\mathbf{h}}\left(\Phi\left(q\right)\right)^{-1}$$

$$= \Delta_{\mathbf{h}}\left(\Phi\left(q\right)\right)\mathbf{C}_{\Phi(q)}\Delta_{\mathbf{h}}\left(\Phi\left(q\right)\right)^{-1}.$$

$$\times \Delta_{\mathbf{h}}\left(\Phi\left(q\right)\right)\mathbf{F}_{\Phi(q)}\left(\beta-\Phi\left(q\right)\right)^{-1}\Delta_{\mathbf{h}}\left(\Phi\left(q\right)\right)^{-1},$$

where $\mathbf{C}_{\Phi(q)}$ plays the role of the matrix $\mathbf{C}$ for $(X, \mathbf{P}^{\Phi(q)})$. From (3) we can check that

$$\Delta_{\mathbf{h}}\left(\Phi\left(q\right)\right)\mathbf{F}_{\Phi(q)}\left(\beta-\Phi\left(q\right)\right)^{-1}\Delta_{\mathbf{h}}\left(\Phi\left(q\right)\right)^{-1} = \left(\mathbf{F}\left(\beta\right)-q\mathbf{I}\right)^{-1}.$$

Thus defining $\mathbf{C}(q) = \Delta_{\mathbf{h}}\left(\Phi\left(q\right)\right)\mathbf{C}_{\Phi(q)}\Delta_{\mathbf{h}}\left(\Phi\left(q\right)\right)^{-1}$ we deduce that

$$\int_0^\infty e^{-\beta x}\mathbf{W}^{(q)}(x)\,dx = \mathbf{C}(q)\times\left(\mathbf{F}\left(\beta\right)-q\mathbf{I}\right)^{-1}.$$

From above we derive $\mathbf{W}^{(q)}(x) = \mathbf{H}^{-1}\mathrm{diag}(\sinh(x(2q-2\lambda_i)^{1/2}))\mathbf{H}$ up to multiplicative constant invertible matrix. Thus from Theorem 3 we obtain e.g. the state of the independent Markov chain at the exit time of the Brownian motion from the interval:

$$\mathbf{E}_x\left(e^{-q\tau_a^+}1_{(\tau_a^+<\tau_0^-)};J(\tau_a^+)\right) = \mathbf{H}^{-1}\mathrm{diag}\left(\frac{\sinh(x(2q-2\lambda_i)^{1/2})}{\sinh(a(2q-2\lambda_i)^{1/2})}\right)\mathbf{H}.$$

Remark 5. For $N = 1$, that is for spectrally negative Lévy process with Laplace exponent $\psi(\beta) = F(\beta)$, from (21) we have $W^{(0)}(x) = V(1-M^{(0)}(x))$ for some constant $V > 0$ and hence from Theorem 3 and (19) we derive (up to a multiplicative constant) $\int_0^\infty e^{-\beta x}W^{(q)}(x)\,dx = (\psi(\beta)-q)^{-1}$ (compare with Theorem 1 of Kyprianou and Palmowski (2004)).

Acknowledgments. Part of this collaborative work was carried out when the A.E.K. was visiting Risklab, ETH Zürich and Z.P. was a researcher at EU-RANDOM. Both authors would like to express respective thanks to these institutions for their hospitality and support. In addition, both authors gratefully acknowledge grant nr. 613.000.310 from Nederlandse Organisatie voor Wetenschappelijk Onderzoek. Z.P. also acknowledges support by KBN 1P03A03128. Thanks are also due to anonymous referees for their valuable comments, one of whom uncovered a serious error in a previous version of this manuscript.

References

1. Asmussen, S. (1989) Aspects of matrix Wiener-Hopf factorization in applied probability. *Math. Scientist.* **14**, 101–116.
2. Asmussen, S. (2000) *Ruin Probabilities.* World Scientific, Singapore.
3. Asmussen, S. (2003) *Applied Probability and Queues* Second edition. Springer.
4. Asmussen, A., Avram, F. and Pistorius, M. R. (2004) Russian and American put options under exponential phase-type Lévy models. *Stoch. Proc. Appl.* **109**, 79–111.
5. Asmussen, S. and Kella, O. (2000) Multi-dimensional martingale for Markov additive processes and its applications. *Adv. Appl. Probab.* **32(2)**, 376–380.
6. Avram, F., Kyprianou, A.E. and Pistorius, M.R. (2004) Exit problems for spectrally negative Lévy processes and applications to (Canadized) Russian options. *Ann. Appl. Probab.* **14**, 215–238.
7. Bertoin, J. (1997) Exponential decay and ergodicity of completely asymmetric Lévy processes in a finite interval, *Ann. Appl. Probab.* **7**, 156–169.
8. Bingham, N.H. (1975) Fluctuation theory in continuous time. *Adv. Appl. Probab.* **7**, 705–766.
9. Chan, T. (1998) Some applications of Lévy processes to stochastic investment models for actuarial use. *ASTIN Bulletin* **28**, 77–93.
10. Chiu, S.N. and Yin, C. (2005) Passage times for a spectrally negative Lévy process with applications to risk theory. *Bernoulli* **11(3)**, 511–522.
11. Coolen-Schrijner, P. and van Doorn, E.A. (2002) The deviation matrix of a continuous-time Markov chain. *Probab. Engrg. Inform. Sci.* **16(3)**, 351–366.
12. Dube, P., Guillemin, F. and Mazumdar, R. (2004) Scale functions of Lévy processes and busy periods of finite capacity $M/GI/1$ queues. *J. Appl. Probab.* **41(4)**, 1250-1254.
13. Emery, D.J. (1973) Exit problems for a spectrally positive process. *Adv. Appl. Prob.* **5**, 498–520.
14. Huzak, M., Perman, M., Šikić, H. and Vondraček, Z. (2004), Ruin probabilities and decompositions for general perturbed risk processes. *Ann. Appl. Probab.* **14**, 1378–1397.
15. Jacod, J. and Shiryaev, A.N. (1987) *Limit Theorems for Stochastic Processes.* Springer.
16. Kella, O. and Whitt, W. (1992) Useful martingales for stochastic storage processes with Lévy input. *J. Appl. Probab.* **29**, 396–403.
17. Klüppelberg, C., Kyprianou, A.E. and Maller, R.A. (2003) Ruin probabilties and overshoots for general Lévy insurance risk process. *Ann. Appl. Probab.* **14(4)**, 1766–1801.
18. Kyprianou, A. E. and Palmowski, Z. (2004) A martingale review of some fluctuation theory for spectrally negative Lévy processes. *Séminaire de Probabilités XXXVIII*, 16–29.
19. Miyazawa, M. (2004) Hitting probabilities in a Markov Additive process with linear movements and upward jumps: applications to risk and queueing processes. *Ann. Appl. Probab.* **14(2)**, 1029–1054.
20. Miyazawa, M. and Takada, H. (2002) A matrix exponential form for hitting probabilities and its application to a Markov modulated fluid queue with downward jumps. *J. Appl. Probab.* **39(3)**, 604-618.
21. Palmowski, Z. and Rolski, T. A technique for exponential change of measure for Markov processes. *Bernoulli* **8**, 767-785

22. Pistorius, M.R. (2003) *Exit problems for Lévy processes and applications to mathematical finance.* Ph.D Thesis, University of Utrecht.
23. Pistorius, M.R. (2006) On maxima and ladder processes for a dense class of Lévy processes. To appear in *Journal of Applied Probability* **43**.
24. Rogers, L.C.G. (1990) The two-sided exit problem for spectrally positive Lévy processes. *Adv. Appl. Probab.* **22**, 486–487.
25. Rogers, L.C.G. (1994) Fluid models in queueing theory and Wiener-Hopf factorization of Markov chains. *Ann. Appl. Probab.* **4(2)**, 390–413.
26. Schmidli, H. (1999) Perturbed risk processes: a review. *Theor. Stoch. Proc.* **5**, 145–165.
27. Seneta, E. (1973) *Non-negative Matrices. An Introduction to Theory and Applications.* George Allen & Unwin Ltd, London.
28. Suprun, V.N. (1976), Problem of destruction and resolvent of terminating processes with independent increments.*Ukranian Math. J.* **28**, 39–45.
29. Takács, L. (1967) *Combinatorial methods in the theory of stochastic processes.* John Wiley & Sons, Inc.
30. Zolotarev, V.M. (1964) The first passage time of a level and the behavior at infinity for a class of processes with independent increments. *Theory Prob. Appl.* **9**, 653–661.

On Continuity Properties of the Law
of Integrals of Lévy Processes

Jean Bertoin[1], Alexander Lindner[2], and Ross Maller[3]

[1] Laboratoire de Probabilités, Université Paris VI
175 rue du Chevaleret, 75013 Paris, France
email: jbe@ccr.jussieu.fr
[2] Department of Mathematics and Computer Science, University of Marburg
Hans-Meerwein-Straße, D-35032 Marburg, Germany
email: lindner@mathematik.uni-marburg.de
[3] Centre for Mathematical Analysis, and School of Finance & Applied Statistics,
Australian National University
Canberra, ACT
email: Ross.Maller@anu.edu.au

Summary. Let (ξ, η) be a bivariate Lévy process such that the integral $\int_0^\infty e^{-\xi_{t-}}\, d\eta_t$ converges almost surely. We characterise, in terms of their Lévy measures, those Lévy processes for which (the distribution of) this integral has atoms. We then turn attention to almost surely convergent integrals of the form $I := \int_0^\infty g(\xi_t)\, dt$, where g is a deterministic function. We give sufficient conditions ensuring that I has no atoms, and under further conditions derive that I has a Lebesgue density. The results are also extended to certain integrals of the form $\int_0^\infty g(\xi_t)\, dY_t$, where Y is an almost surely strictly increasing stochastic process, independent of ξ.

1 Introduction

The aim of this paper is to study continuity properties of stationary distributions of generalised Ornstein-Uhlenbeck processes and of distributions of random variables of the form $\int_0^\infty g(\xi_t)\, dt$ for a Lévy process ξ and a general function $g : \mathbb{R} \to \mathbb{R}$.

For a bivariate Lévy process $(\xi, \zeta) = (\xi_t, \zeta_t)_{t \geq 0}$, the *generalised Ornstein-Uhlenbeck (O-U) process* $(V_t)_{t \geq 0}$ is defined as

$$V_t = e^{-\xi_t}\left(\int_0^t e^{\xi_{s-}}\, d\zeta_s + V_0\right), \quad t \geq 0,$$

where V_0 is a finite random variable, independent of (ξ, ζ). This process appears as a natural continuous time generalisation of random recurrence equations, as shown by de Haan and Karandikar [11], and has applications in many areas, such as risk theory (e.g. Paulsen [19]), perpetuities (e.g. Dufresne [6]), financial time series (e.g. Klüppelberg et al. [14]) or option pricing (e.g. Yor [23]), to name just a few. See also Carmona et al. [2,3] for further properties of this process. Lindner and Maller [17] have shown that the existence of a stationary solution to the generalised O-U process is closely related to the almost sure convergence of the stochastic integral $\int_0^t e^{-\xi_{s-}} \, d\eta_s$ as $t \to \infty$, where (ξ, η) is a bivariate Lévy process, and η can be explicitly constructed in terms of (ξ, ζ). The stationary distribution is then given by $\int_0^\infty e^{-\xi_{s-}} \, d\eta_s$. Necessary and sufficient conditions for the convergence of $\int_0^\infty e^{-\xi_{s-}} \, d\eta_s$ were obtained by Erickson and Maller [7]. Distributional properties of the limit variable and hence of the stationary distribution of generalised O-U processes are of particular interest. Gjessing and Paulsen [9] determined the distribution in many cases when ξ and η are independent and the Lévy measure of (ξ, η) is finite. Carmona et al. [2] considered the case when $\eta_t = t$ and the jump part of ξ is of finite variation. Under some additional assumptions, they showed that $\int_0^\infty e^{-\xi_{s-}} \, ds$ is absolutely continuous, and its density satisfies a certain integro-differential equation. In Section 2 we shall be concerned with continuity properties of the limit variable $\int_0^\infty e^{-\xi_{s-}} \, d\eta_s$ without any restrictions on (ξ, η), assuming only convergence of the integral. We shall give a complete characterisation of when this integral has atoms, in terms of the characteristic triplet of (ξ, η). This characterisation relies on a similar result of Grincevičius [10] for "perpetuities" which are a kind of discrete time analogue of Lévy integrals.

Then, in Section 3, we turn our attention to continuity properties of the distribution of the integral $\int_0^\infty g(\xi_t) \, dt$, where $\xi = (\xi_t)_{t \geq 0}$ is a one-dimensional Lévy process with non-zero Lévy measure and g is a general deterministic Borel function. Such integrals appear in a variety of situations, for example concerning shattering phenomena in fragmentation processes, see, e.g., Haas [12].

Fourier analysis and Malliavin calculus are classical tools for establishing the absolute continuity of distributions of functionals of stochastic processes. In a different direction, the book of Davydov et al. [5] treats three different methods for proving absolute continuity of such functionals: the "stratification method", the "superstructure method" and the "method of differential operators". Chapter 4 in [5] pays particular attention to Poisson functionals, which includes integrals of Lévy processes. While it may be hard to check the conditions and apply these methods in general (in particular to find admissible semigroups for the stratification method), it has been carried out in some cases. For example, Davydov [4] gives sufficient conditions for absolute continuity of integrals of the form $\int_0^1 g(X_t) \, dt$ for *strictly stationary* processes $(X_t)_{t \geq 0}$ and quite general g. Concerning integrals of Lévy processes, Lifshits [16], p. 757, has shown that $\int_0^1 g(\xi_t) \, dt$ is absolutely continuous if ξ is a Lévy process with infinite and absolutely continuous Lévy measure, and g is locally Lipschitz-continuous and such that on a set of full Lebesgue measure in $[0,1]$ the derivative g' of g exists and is continuous and non-vanishing; see also Problem 15.1 in [5]. For our study of atoms of the distributions of integrals such as $\int_0^\infty g(\xi_t)\,dt$, we will impose less restrictive assumptions on g in Section 3. Note also that [5] and the references given there are usually concerned with the absolute continuity of functionals such as $\int_0^1 g(\xi_t)\,dt$ on the *compact* interval $[0,1]$, while we are concerned with integrals over $(0,\infty)$. That absolute continuity of the distribution of integrals over compact sets and over $(0,\infty)$ can be rather different topics is straightforward by considering the special case of compound Poisson processes. See also part (iii) of Theorem 2.2 below for situations where the integral over every finite time horizon may be absolutely continuous, while the limit variable can degenerate to a constant.

Section 3 is organised as follows: we start with some motivating examples, in some of which $\int_0^\infty g(\xi_t)\,dt$ has atoms while in others it does not. Then, in Section 3.2 we present some general criteria which ensure the continuity of the distribution of $\int_0^\infty g(\xi_t)\,dt$. The proofs there are based on the sample path behaviour and on excursion theory for Lévy processes. Then, in Section 3.3 we use a simple form of the stratification method to obtain absolute continuity of $\int_0^\infty g(\xi_t)\,dt$ for certain cases of g and ξ (which assume however no differentiability properties of g); the results are also extended to more general integrals of the form $\int_0^\infty g(\xi_t)\,dY_t$, where $Y = (Y_t)_{t\geq 0}$ is a strictly increasing stochastic process, independent of the Lévy process ξ.

Observe that our focus will be on continuity properties of the distribution of the integral $\int_0^\infty g(\xi_t)\,dt$ (or similar integrals), under the assumption that it is finite a.s. A highly relevant question is to ask under which conditions the integral does converge. It is important of course that any conditions we impose to ensure continuity of the integral, or its absence, be compatible with convergence. We only occasionally address this issue, when it is possible to give some simple sufficient (or, sometimes, necessary) conditions for convergence. Our approach is essentially to assume convergence and study the properties of the resulting integral. For a much fuller discussion of conditions for convergence *per se* we refer to Erickson and Maller [7,8], who give an overview of known results as well as new results on the finiteness of Lévy integrals.

We end this section by setting some notation. Recall that a *Lévy process* $X = (X_t)_{t\geq 0}$ in $\mathbb{R}^d$ ($d \in \mathbb{N}$) is a stochastically continuous process having independent and stationary increments, which has almost surely càdlàg paths and satisfies $X_0 = 0$. For each Lévy process, there exists a unique constant $\gamma = \gamma_X = (\gamma_1,\ldots,\gamma_d) \in \mathbb{R}^d$, a symmetric positive semidefinite matrix $\Sigma = \Sigma_X$, and a Lévy measure $\Pi = \Pi_X$ on $\mathbb{R}^d \setminus \{0\}$ satisfying $\int_{\mathbb{R}^d} \min\{1,|x|^2\}\,\Pi_X(dx) < \infty$, such that for all $t > 0$ and $\theta \in \mathbb{R}^d$ we have

$$(1/t)\log E\exp(i\langle\theta,X_t\rangle) = i\langle\gamma,\theta\rangle - \frac{1}{2}\langle\theta,\Sigma\theta\rangle$$
$$+ \int_{\mathbb{R}^d}(e^{i\langle z,\theta\rangle} - 1 - i\langle z,\theta\rangle\mathbf{1}_{|z|\leq 1})\,\Pi_X(dz).$$

Here, $\langle \cdot, \cdot \rangle$ and $|\cdot|$ denote the inner product and Euclidian norm in $\mathbb{R}^d$, and $\mathbf{1}_A$ is the indicator function of a set A. Together, (γ, Σ, Π) form the *characteristic triplet* of X. The Brownian motion part of X is described by the covariance matrix Σ_X. If $d = 1$, then we will also write σ_X^2 for Σ_X, and if $d = 2$ and $X = (\xi, \eta)$, the upper and lower diagonal elements of Σ_X are given by σ_ξ^2 and σ_η^2. We refer to Bertoin [1] and Sato [21] for further definitions and basic properties of Lévy processes. Integrals of the form $\int_a^b e^{-\xi_{t-}} d\eta_t$ for a bivariate Lévy process (ξ, η) are interpreted as the usual stochastic integral with respect to its completed natural filtration as in Protter [20], where $\int_a^b$ denotes integrals over the set $[a, b]$, and $\int_{a+}^b$ denotes integrals over the set $(a, b]$. If η (or a more general stochastic process $Y = (Y_t)_{t \geq 0}$ as an integrator) is of bounded variation on compacts, then the stochastic integral is equal to the pathwise computed Lebesgue-Stieltjes integral, and will also be interpreted in this sense. Integrals such as $\int_0^\infty$ are to be interpreted as limits of integrals of the form $\int_0^t$ as $t \to \infty$, where the convergence will typically be almost sure. The *jump* of a càdlàg process $(Z_t)_{t \geq 0}$ at time t will be denoted by $\Delta Z_t := Z_t - Z_{t-} = Z_t - \lim_{u \uparrow t} Z_u$, with the convention $Z_{0-} := 0$. The symbol "$\overset{D}{=}$" will be used to denote equality in distribution of two random variables, and "$\overset{P}{\to}$" will denote convergence in probability. Almost surely holding statements will be abbreviated by "a.s.", and properties which hold almost everywhere by "a.e.". The Lebesgue measure on $\mathbb{R}$ will be denoted by λ. Throughout the paper, in order to avoid trivialities, we will assume that ξ and η are different from the zero process $t \mapsto 0$.

2 Atoms of exponential Lévy integrals

Let $(\xi, \eta) = (\xi_t, \eta_t)_{t \geq 0}$ be a bivariate Lévy process. Erickson and Maller [7] characterised when the exponential integral $I_t := \int_0^t e^{-\xi_{s-}} d\eta_s$, $t > 0$, converges almost surely to a finite random variable I as $t \to \infty$. They showed that this happens if and only if

$$\lim_{t \to \infty} \xi_t = +\infty \text{ a.s.}, \quad \text{and} \quad \int_{\mathbb{R} \setminus [-e, e]} \left(\frac{\log |y|}{A_\xi(\log |y|)} \right) \Pi_\eta(dy) < \infty. \quad (2.1)$$

Here, the function A_ξ is defined by

$$A_\xi(y) := 1 + \int_1^y \Pi_\xi((z, \infty)) \, dz, \quad y \geq 1.$$

As a byproduct of the proof, they obtained that I_t converges almost surely to a finite random variable I if and only if it converges in distribution to I, as $t \to \infty$. Observe that the convergence condition (2.1) depends on the marginal distributions of ξ and η only, but not on the bivariate dependence structure of ξ and η.

In this section we shall be interested in the question of whether the limit random variable I can have a distribution with atoms. A complete characterisation of this will be given in Theorem 2.2. A similar result for the characterisation of the existence of atoms for discrete time perpetuities was obtained by Grincevičius [10], Theorem 1. We will adapt his proof to show that $\int_0^\infty e^{-\xi_{t-}} d\eta_t$ has atoms if and only if it is constant. This will be a consequence of the following lemma, which is formulated for certain families of random fixed point equations.

Lemma 2.1 *For every $t \geq 0$, let Q_t, M_t and ψ_t be random variables such that $M_t \neq 0$ a.s., and ψ_t is independent of (Q_t, M_t). Suppose ψ is a random variable satisfying*

$$\psi = Q_t + M_t \psi_t \quad \text{for all } t \geq 0,$$

and such that

$$\psi \stackrel{D}{=} \psi_t \quad \text{for all } t \geq 0,$$

and suppose further that

$$Q_t \stackrel{P}{\to} \psi \quad \text{as} \quad t \to \infty.$$

Then ψ has an atom if and only if it is a constant random variable.

Proof. We adapt the proof of Theorem 1 of [10]. Suppose that ψ has an atom at $a \in \mathbb{R}$, so that

$$P(\psi = a) =: \beta > 0.$$

Then for all $\varepsilon \in (0, \beta)$ there exists some $\delta > 0$ such that

$$P(|\psi - a| < 2\delta) < \beta + \varepsilon. \tag{2.2}$$

Since $Q_t \stackrel{P}{\to} \psi$ as $t \to \infty$, there exists $t' = t'(\varepsilon)$ such that

$$P(|\psi - Q_t| \geq \delta) = P(|M_t \psi_t| \geq \delta) < \varepsilon \quad \text{for all } t \geq t'. \tag{2.3}$$

Then (2.2) and (2.3) imply that, for all $t \geq t'$,

$$P(|Q_t - a| < \delta) \leq P(|Q_t - \psi| \geq \delta) + P(|\psi - a| < 2\delta) \leq \beta + 2\varepsilon. \tag{2.4}$$

Now observe that, for all $t \geq 0$,

$$\begin{aligned}
\beta = P(\psi = a) &\leq P(\psi = a, |\psi - Q_t| < \delta) + P(|\psi - Q_t| \geq \delta) \\
&= \int_{\mathbb{R}} P(Q_t + M_t s = a, |M_t s| < \delta)\, dP(\psi_t \leq s) + P(|\psi - Q_t| \geq \delta) \\
&= \sum_{s \in D_t} P(Q_t + M_t s = a, |M_t s| < \delta)\, P(\psi_t = s) \; + P(|\psi - Q_t| \geq \delta).
\end{aligned}$$

Here, the last equation follows from the fact that $P(Q_t + M_t s = a)$ can be positive for only a countable number of s, $s \in D_t$, say, since the number of atoms of any random variable is countable.

142   J. Bertoin, et al.

Since $\sum_{s\in D_t} P(\psi_t = s) \leq 1$ for all s, and since $P(|\psi - Q_t| \geq \delta) < \varepsilon$ for $t > t'$, by (2.3), it follows that for such t there is some $s_t \in \mathbb{R}$ such that

$$\beta_t := P(Q_t + M_t s_t = a, |M_t s_t| < \delta) \geq \beta - \varepsilon. \tag{2.5}$$

Observing that, for all $t \geq 0$

$$\{\psi = a\} \cup \{Q_t + M_t s_t = a, |M_t s_t| < \delta\} \subset \{|\psi - Q_t| \geq \delta\} \cup \{|Q_t - a| < \delta\},$$

we obtain for $t \geq t'$ that

$$P(|\psi - Q_t| \geq \delta) + P(|Q_t - a| < \delta)$$
$$\geq P(\psi = a) + P(Q_t + M_t s_t = a, |M_t s_t| < \delta)$$
$$-P(Q_t + M_t s_t = a, |M_t s_t| < \delta, \psi = a)$$
$$= \beta + \beta_t - \beta_t P(\psi_t = s_t).$$

We used here that $P(M_t = 0) = 0$. From (2.3) and (2.4) it now follows that

$$\beta_t P(\psi_t = s_t) \geq \beta + \beta_t - \varepsilon - (\beta + 2\varepsilon) = \beta_t - 3\varepsilon.$$

Using (2.5) and the fact that $\psi \overset{D}{=} \psi_t$, we obtain

$$P(\psi = s_t) = P(\psi_t = s_t) \geq 1 - \frac{3\varepsilon}{\beta_t} > 1 - \frac{3\varepsilon}{\beta - \varepsilon}.$$

Letting $\varepsilon \to 0$ and observing that $P(\psi = a) > 0$, it follows that $P(\psi = a) = 1$. □

As a consequence, we obtain:

Theorem 2.2 Let (ξ, η) be a bivariate Lévy process such that ξ_t converges almost surely to ∞ as $t \to \infty$, and let $I_t := \int_0^t e^{-\xi_{s-}}\, d\eta_s$. Denote the characteristic triplet of (ξ, η) by $(\gamma, \Sigma, \Pi_{\xi,\eta})$, where $\gamma = (\gamma_1, \gamma_2)$, and denote the upper diagonal element of Σ by σ_ξ^2. Then the following assertions are equivalent:

(i) I_t converges a.s. to a finite random variable I as $t \to \infty$, where I has an atom.
(ii) I_t converges a.s. to a constant random variable as $t \to \infty$.
(iii) $\exists\, k \in \mathbb{R} \setminus \{0\}$ such that $P\left(\int_0^t e^{-\xi_{s-}}\, d\eta_s = k(1 - e^{-\xi_t})\text{ for all }t > 0\right) = 1$.
(iv) $\exists\, k \in \mathbb{R} \setminus \{0\}$ such that $e^{-\xi} = \mathcal{E}(-\eta/k)$, i.e. $e^{-\xi}$ is the stochastic exponential of $-\eta/k$.
(v) $\exists\, k \in \mathbb{R} \setminus \{0\}$ such that

$$\Sigma_{\xi,\eta} = \begin{pmatrix} 1 & k \\ k & k^2 \end{pmatrix} \sigma_\xi^2,$$

the Lévy measure $\Pi_{\xi,\eta}$ of (ξ,η) is concentrated on $\{(x, k(1 - e^{-x})) : x \in \mathbb{R}\}$,
and

$$\gamma_1 - k^{-1}\gamma_2 = \sigma_\xi^2/2 + \int_{x^2+k^2(1-e^{-x})^2 \leq 1} (e^{-x} - 1 + x)\, \Pi_\xi(dx). \qquad (2.6)$$

Proof. To show the equivalence of (i) and (ii), suppose that I exists a.s. as a finite random variable and define

$$\psi := I = \int_0^\infty e^{-\xi_{s-}}\, d\eta_s, \quad Q_t := I_t = \int_0^t e^{-\xi_{s-}}\, d\eta_s \ \text{ and } \ M_t := e^{-\xi_t}, \quad t \geq 0.$$

Then

$$\psi \overset{D}{=} \int_{t+}^\infty e^{-(\xi_{s-} - \xi_t)}\, d(\eta_\cdot - \eta_t)_s =: \psi_t.$$

So we have the setup of Lemma 2.1:

$$\psi = Q_t + M_t\psi_t, \ t \geq 0, \qquad (2.7)$$

Q_t converges in probability (in fact, a.s.) to ψ as $t \to \infty$, and ψ_t is independent of (Q_t, M_t) for all $t \geq 0$. We conclude from Lemma 2.1 that $I = \psi$ is finite a.s. and has an atom if and only if it is constant, equivalently, if (ii) holds.

Now suppose that (ii) holds and that the constant value of the limit variable is k. Then it follows from (2.7) that, a.s.,

$$k = \int_0^t e^{-\xi_{s-}}\, d\eta_s + e^{-\xi_t}k, \ \text{for each } t > 0,$$

hence

$$\int_0^t e^{-\xi_{s-}}\, d\eta_s = k(1 - e^{-\xi_t}) \quad \text{for all } t > 0. \qquad (2.8)$$

Observe that $k = 0$ is impossible by uniqueness of the solution to the stochastic differential equation $d\int_0^t X_{s-}\, d\eta_s = 0$ (which implies $e^{-\xi_s} = X_s = 0$, impossible). Since Q_t and $e^{-\xi_t}$ are càdlàg functions, (2.8) holds on an event of probability 1. This shows that (ii) implies (iii). The converse is clear, since $\lim_{t\to\infty} \xi_t = \infty$ a.s. by assumption.

Dividing (2.8) by $-k$, we obtain $e^{-\xi_t} = 1 + \int_0^t e^{-\xi_{s-}}\, d(-\eta_s/k)$, which is just the defining equation for $e^{-\xi} = \mathcal{E}(-\eta/k)$, see Protter [20], p. 84, giving the equivalence of (iii) and (iv).

The equivalence of (iv) and (v) follows by straightforward but messy calculations using the Doléans-Dade formula and the Lévy-Itô decomposition (for the calculation of γ), and is relegated to the appendix. $\qquad \square$

Remarks. (i) Under stronger assumptions, Theorem 2.2 may be strengthened to conclude that I has a density or is constant. Suppose (ξ, η) is a bivariate Lévy process such that ξ has no positive jumps and drifts to ∞, i.e.

$\lim_{t\to\infty}\xi_t = \infty$ a.s. Assume further that $\int_{\mathbb{R}\setminus[-e,e]}(\log|y|)\,\Pi_\eta(dy) < \infty$. Then the condition (2.1) is fulfilled, and thus $I := \lim_{t\to\infty}\int_0^t e^{-\xi_{s-}}\,d\eta_s$ exists and is finite a.s. Applying the strong Markov property at the first passage time $T_x := \inf\{t \geq 0 : \xi_t > x\} = \inf\{t \geq 0 : \xi_t = x\}$ (since ξ has no positive jumps) yields the identity

$$I = \int_0^{T_x} e^{-\xi_{s-}}\,d\eta_s + e^{-x}I'$$

where I' has the same distribution as I and is independent of $\int_0^{T_x} e^{-\xi_{s-}}\,d\eta_s$. Thus I is a self-decomposable random variable, and as a consequence its law is infinitely divisible and unimodal and hence has a density, if it is not constant; see Theorem 53.1, p. 404, in Sato [21]. Thus I is continuous. A generalisation of this result to the case of multivariate η was recently obtained by Kondo et al. [15].

(ii) As another important special case, suppose ξ is a Brownian motion with a positive drift, and in addition that $\int_{\mathbb{R}\setminus[-e,e]}(\log|y|)\,\Pi_\eta(dy) < \infty$. Then I is finite a.s. From Condition (iii) of Theorem 2.2 we then see that $\Delta\eta_t = 0$, so the condition can hold only if η_t is also a Brownian motion. By Ito's lemma, Condition (iii) implies $d\eta_t = k(d\xi_t - \sigma_\xi^2 dt/2)$, or, equivalently, $\eta_t = k(\xi_t - \sigma_\xi^2 t/2)$. Similarly, if η is a Brownian motion, (iii) of Theorem 2.2 can only hold if ξ is a Brownian motion and the same relation is satisfied. Thus we can conclude that, apart from this degenerate case, $\int_0^\infty e^{-B_s}\,d\eta_s$ and $\int_0^\infty e^{-\xi_s}\,dB_s$, when convergent a.s., have continuous distributions, for a Brownian motion B_t.

3 Integrals with general g

We now turn our attention to the question of whether the integral $\int_0^\infty g(\xi_t)\,dt$ can have atoms, where g is a more general deterministic function, and $\xi = (\xi_t)_{t\geq 0}$ is a non-zero Lévy process. To start with, we shall discuss some natural motivating examples. Then we shall present a few criteria that ensure the absence of atoms. Finally, we shall obtain by a different technique, which is a variant of the stratification method, a sufficient condition for the absolute continuity of the integral.

3.1 Some examples

Example 3.1 Let $(\xi_t)_{t\geq 0}$ be a compound Poisson process (with no drift) and $g : \mathbb{R} \to \mathbb{R}$ a deterministic function such that $g(0) \neq 0$ and such that $\int_0^\infty g(\xi_t)\,dt$ is finite almost surely. Then $\int_0^\infty g(\xi_t)\,dt$ has a Lebesgue density.

Proof. Denote the time of the first jump of ξ by T_1. Recall that ξ is always assumed nondegenerate, so T_1 is a nondegenerate exponential random variable. We can write

$$\int_0^\infty g(\xi_t)\, dt = g(0)T_1 + \int_0^\infty g(\xi_{T_1+t})\, dt$$

(from which it is evident that the integral on the righthand side converges a.s.). Recall that the jump times in a compound Poisson process are independent of the jump sizes. By the strong Markov property of Lévy processes (see [1], Prop. 6, p. 20), the process $(\xi_{T_1+t})_{t\geq 0}$, and *a fortiori* the random variable $\int_0^\infty g(\xi_{T_1+t})\, dt$, are independent of T_1. From this follows the claim, since $g(0)T_1$ has a Lebesgue density and hence its sum with any independent random variable has also. $\qquad\square$

The following example shows that this property does not carry over to compound Poisson processes with drift, at least not if the support of g is compact.

Example 3.2 *Let $\xi = (\xi_t)_{t\geq 0} = (at + Q_t)_{t\geq 0}$ be a compound Poisson process together with a deterministic drift $a \neq 0$, such that $\lim_{t\to\infty} \xi_t = \mathrm{sgn}(a)\infty$ a.s. Suppose that g is a deterministic integrable Borel function with compact support. Then $\int_0^\infty g(\xi_t)\, dt$ is finite almost surely and its distribution has atoms.*

Proof. Since ξ drifts to $\pm\infty$ a.s., there is a random time τ after which $\xi_t \notin \mathrm{supp}\, g$ for all t; that is, if ξ enters $\mathrm{supp}\, g$ at all; if it doesn't, then $g(\xi_t) = 0$ for all $t \geq 0$. In either case, $\int_\tau^\infty g(\xi_t)\, dt = 0$, and since g is integrable and the number of jumps of Q until time τ is almost surely finite, it follows that $\int_0^\infty g(\xi_t)\, dt < \infty$ a.s.

Suppose now that $a > 0$, so that ξ drifts to $+\infty$ a.s., and let $r = \sup(\mathrm{supp}\, g)$. If $r \leq 0$ there is a positive probability that ξ does not enter $\mathrm{supp}\, g$, except, possibly, when $r = t = 0$, and then $g(\xi_t) = 0$; in either case, $\int_0^\infty g(\xi_t)\, dt = 0$ with positive probability, giving an atom at 0. If $r > 0$, let $T = r/a$. The event A that the first jump of ξ occurs at or after time T has positive probability. On A, $\xi_t = at$ for all $0 \leq t \leq T$. Also, since ξ drifts to $+\infty$ a.s., on a subset of A with positive probability ξ does not re-enter $\mathrm{supp}\, g$ after time T. On this subset, we have $\int_0^\infty g(\xi_t)\, dt = \int_0^T g(at)\, dt$, which is constant. Similarly if $a < 0$. $\qquad\square$

Our third example relies on the following classical criterion for the continuity of infinitely divisible distributions (cf. Theorem 27.4, p. 175, in Sato [21]), that we shall further use in the sequel.

Lemma 3.3 *Let μ be an infinitely divisible distribution on $\mathbb{R}$ with an infinite Lévy measure, or with a non-zero Gaussian component. Then μ is continuous.*

If ξ has infinite Lévy measure, or no drift, Example 3.2 may fail, as shown next:

Example 3.4 *Suppose that ξ is a subordinator with infinite Lévy measure, or is a non-zero subordinator with no drift. Then $\int_0^\infty 1_{[0,1]}(\xi_t)\,dt$ is finite a.s. and has no atoms.*

Proof. Since ξ_t drifts to ∞ a.s. it is clear that $\int_0^\infty 1_{[0,1]}(\xi_t)\,dt$ is finite almost surely. For $x > 0$ define

$$L_x := \inf\{t > 0 : \xi_t > x\}.$$

Then $\int_0^\infty 1_{[0,1]}(\xi_t)\,dt = L_1$, and for $a > 0$ we have

$$
\begin{aligned}
\{L_1 = a\} &= \{\inf\{u : \xi_u > 1\} = a\} \\
&= \{\xi_{a-\varepsilon} \le 1 \text{ for all } \varepsilon > 0, \quad \xi_{a+\varepsilon} > 1 \text{ for all } \varepsilon > 0\} \\
&\subseteq \{\xi_a = 1\} \cup \{\Delta\xi_a > 0\}.
\end{aligned}
$$

A Lévy process is stochastically continuous so $P(\Delta\xi_a > 0) = 0$. If ξ is a subordinator with infinite Lévy measure, then $P(\xi_a = 1) = 0$ by Lemma 3.3. Thus we get $P(L_1 = a) = 0$. If ξ is a subordinator with no drift, then $\Delta\xi_{L_1} > 0$ a.s. ([1], p. 77) (and this includes the case of a compound Poisson), so again

$$P(L_1 = a) = P(L_1 = a, \Delta\xi_{L_1} > 0) \le P(\Delta\xi_a > 0) = 0. \qquad \square$$

3.2 Some criteria for continuity

We shall now present some fairly general criteria which ensure the continuity of the distribution of the integral $\int_0^\infty g(\xi_t)dt$ whenever the latter is finite a.s. and the Lévy process ξ is transient (see Bertoin [1], Section I.4 or Sato [21], Section 35 for definitions and properties of transient and recurrent Lévy processes).

Remarks. (i) One might expect that the existence of $\int_0^\infty g(\xi_t)\,dt$ already implies the transience of ξ. That this is not true in general was shown by Erickson and Maller [8], Remark (2) after Theorem 6. As a counterexample, we may take ξ to be a compound Poisson process with Lévy measure $\Pi(dx) = \sqrt{2}\delta_1 + \delta_{-\sqrt{2}}$. Note that $\int x\Pi(dx) = 0$, so ξ is recurrent. Nonetheless ξ never returns to 0 after its first exit-time and thus $0 < \int_0^\infty 1_{\{\xi_t=0\}}\,dt < \infty$ a.s.

(ii) Sufficient conditions under which the existence of $\int_0^\infty g(\xi_t)\,dt$ implies the transience of ξ are mentioned in Remark (3) after Theorem 6 of [8]. One such sufficient condition is that there is some non-empty open interval $J \subset \mathbb{R}$ such that $\inf\{g(x) : x \in J\} > 0$.

We shall now turn to the question of atoms of $\int_0^\infty g(\xi_t)\,dt$. For the next theorem, denote by E° the set of inner points of a set E, by $\overline{E}$ its topological closure and by ∂E its boundary.

Theorem 3.5 *Let $g : \mathbb{R} \to [0, \infty)$ be a deterministic Borel function. Assume that its support, $\operatorname{supp} g$, is compact, that $g > 0$ on $(\operatorname{supp} g)^{\circ}$, and that $0 \in (\operatorname{supp} g)^{\circ}$. Write $\partial \operatorname{supp} g := \operatorname{supp} g \setminus (\operatorname{supp} g)^{\circ}$ for the boundary of $\operatorname{supp} g$. Let ξ be a transient Lévy process, and assume that $I := \int_0^{\infty} g(\xi_t) \, dt$ is almost surely finite. If either*

(i) ξ is of unbounded variation and $\partial \operatorname{supp} g$ is finite,

or

(ii) ξ is of bounded variation with zero drift and $\partial \operatorname{supp} g$ is at most countable, then the distribution of I has no atoms.

Proof. If ξ is a compound Poisson process without drift, the result follows from Example 3.1, so we will assume that ξ has unbounded variation, or is of bounded variation with zero drift such that its Lévy measure is infinite, and that g has the properties specified in the statement of the theorem. Write

$$I(x) := \int_0^x g(\xi_t) \, dt, \quad x \in (0, \infty].$$

Then $x \mapsto I(x)$ is increasing and $I = I(\infty)$ is finite a.s. by assumption, so $I(x) < \infty$ a.s. for all $x \geq 0$. Plainly $I(x)$ is a.s. continuous at each $x > 0$. Assume by way of contradiction that there is some $a \geq 0$ such that $P(I = a) > 0$, and proceed as follows. Define

$$T_s := \inf\{u \geq 0 : I(u) = s\}, \quad s \geq 0.$$

Since ξ_t is adapted to the natural filtration $\{\mathcal{F}_t\}_{t \geq 0}$ of $(\xi_t)_{t \geq 0}$, so is $g(\xi_{\cdot})$ (g is Borel), thus $\{T_s > u\} = \{\int_0^u g(\xi_t) dt < s\} \in \mathcal{F}_u$, because $I(\cdot)$ is adapted to $\{\mathcal{F}_t\}_{t \geq 0}$. Thus T_s is a stopping time for each $s \geq 0$. Further, $T_s > 0$ for all $s > 0$. Since $0 \in (\operatorname{supp} g)^{\circ}$, it is clear that $a \neq 0$. By assumption, ξ is transient, so there is a finite random time σ such that $\xi_t \notin \operatorname{supp} g$ for all $t \geq \sigma$. Then $I(\infty) = I(\sigma)$, and it follows that $P\{T_a < \infty\} > 0$.

Define the stopping times $\tau_n := T_{a-1/n} \wedge n$. Then $(\tau_n)_{n \in \mathbb{N}}$ is strictly increasing to T_a, showing that T_a is announceable; it follows that $t \mapsto \xi_t$ is continuous at $t = T_a$ on $\{T_a < \infty\}$, see e.g. Bertoin [1], p. 21 or p. 39.

Let $B = \{T_a < \infty, I(\infty) = a\}$. We restrict attention to $\omega \in B$ from now on. Since T_a is the first time $I(\cdot)$ reaches a, for every $\varepsilon > 0$ there must be a subset $J_{\varepsilon} \subset (T_a - \varepsilon, T_a)$ of positive Lebesgue measure such that $g(\xi_t) > 0$ for all $t \in J_{\varepsilon}$. Thus $\xi_t \in \operatorname{supp} g$ for all $t \in J_{\varepsilon}$, and so $\xi_{T_a} \in \operatorname{supp} g$. Since we assume that $\partial \operatorname{supp} g := \operatorname{supp} g \setminus (\operatorname{supp} g)^{\circ}$ is countable, and that ξ has infinite Lévy measure or a non-zero Gaussian component, we have by Lemma 3.3 that $P(\xi_t \in \partial \operatorname{supp} g) = 0$ for all $t > 0$. Consequently

$$E(\lambda\{t \geq 0 : \xi_t \in \partial \operatorname{supp} g\}) = \int_0^{\infty} P(\xi_t \in \partial \operatorname{supp} g) dt = 0.$$

It follows that there are times $t < T_a$ arbitrarily close to T_a with ξ_t in $(\operatorname{supp} g)^{\circ}$. By the continuity of $t \mapsto \xi_t$ at $t = T_a$, we then have $\xi_{T_a} \in \overline{(\operatorname{supp} g)^{\circ}}$

for $\omega \in B' \subseteq B$, where $P(B') = P(B) > 0$. Since $g > 0$ on $(\operatorname{supp} g)^\circ$ it follows that $\xi_{T_a} \in \partial((\operatorname{supp} g)^\circ)$ on the event $B' \subseteq \{I(\infty) = a\}$; for, if not, this would imply, by an application of the Markov property, that $I(t) > a$ for $t > T_a$, which is impossible.

Now suppose (i), so that ξ is of infinite variation. Then it follows from Shtatland's (1965) result ([22], see also Sato [21], Thm 47.1, p. 351) that 0 is regular for both $(-\infty, 0)$ and $(0, \infty)$. Since ξ_{T_a} belongs to the finite set $\partial \operatorname{supp} g$, there is an open interval $U \subset (\operatorname{supp} g)^\circ$ which has ξ_{T_a} either as left or right end point. In either case, the regularity of 0 for $(0, \infty)$ and for $(-\infty, 0)$ implies that immediately after time T_a there must be times t such that ξ_t is strictly less than ξ_{T_a} and other times t such that ξ_t is strictly greater than ξ_{T_a}. By the continuity of ξ at T_a, it follows that there must be times after T_a such that $\xi_t \in U$. Consequently, there is some $\varepsilon = \varepsilon(\omega) > 0$ such that $\xi_{T_a + \varepsilon} \in (\operatorname{supp} g)^\circ$. By the right-continuity of ξ at $T_a + \varepsilon$ it follows further that $I(\infty) > I(T_a) = a$ on B', where $P(B') > 0$ and $B' \subseteq \{I(\infty) = a\}$, a contradiction.

Alternatively, suppose (ii), so that ξ has finite variation and zero drift (and infinite Lévy measure). Then it follows that ξ almost surely does not hit single points (by Kesten's theorem [13]; see [1], p. 67). Thus, since $\partial((\operatorname{supp} g)^\circ) \subseteq \operatorname{supp} g \setminus (\operatorname{supp} g)^\circ$ and the latter is at most countable, ξ almost surely does not hit $\partial((\operatorname{supp} g)^\circ)$. But on the set B', where $P(B') > 0$ and $B' \subseteq \{T_a < \infty, I(\infty) = a\}$, we have $\xi_{T_a} \in \partial((\operatorname{supp} g)^\circ)$, contradicting $P(I(\infty) = a) > 0$. □

Remarks. (i) The assumptions on the topological structure of $\{x : g(x) > 0\}$ in the previous theorem are easy to check. That they cannot be completely relaxed can be seen from the following example: let $g(x) = 1$ for all $x \in \mathbb{Q} \cap [-1, 1]$ and $g(x) = 0$ otherwise, then $\operatorname{supp} g = [-1, 1]$, $(\operatorname{supp} g)^\circ = (-1, 1)$, but $g > 0$ on $(-1, 1)$ does not hold. And in fact, it is easy to see that in that case we have for every Lévy process of unbounded variation or infinite Lévy measure that

$$E \int_0^\infty g(\xi_t)\, dt = E \int_0^\infty 1_{\mathbb{Q} \cap [-1,1]}(\xi_t)\, dt = \int_0^\infty P(\xi_t \in \mathbb{Q} \cap [-1, 1])\, dt = 0$$

by Lemma 3.3, so that $\int_0^\infty g(\xi_t)\, dt = 0$ a.s.

(ii) Suppose g is as in Theorem 3.5, and assume $\int_0^\infty g(x)dx < \infty$. Let ξ be a Brownian motion with non-zero drift. Then $\int_0^\infty g(\xi_t)dt < \infty$ a.s. by Theorem 6 of [8] and the integral has a continuous distribution by Theorem 3.5.

Theorem 3.5 allows a wide class of transient Lévy processes (we have to exclude ξ which are of bounded variation with nonzero drift, by Ex. 3.2), but restricts us, essentially, to nonnegative g which have compact support. Another approach which combines excursion theory and Lemma 3.3 allows a much wider class of g at the expense of placing restrictions on the local behaviour of ξ. Here is the first result in this vein. We refer e.g. to Chapters IV and V in [1] for background on local time and excursion theory for Lévy processes.

Theorem 3.6 *Let $g : \mathbb{R} \to [0, \infty)$ be a measurable function such that $g > 0$ on some neighbourhood of 0. Suppose that ξ is a transient Lévy process such that 0 is regular for itself, in the sense that $\inf\{t > 0 : \xi_t = 0\} = 0$ a.s., and that the integral $I := \int_0^\infty g(\xi_t)dt$ is finite a.s. Then the distribution of I has no atoms.*

Proof. Thanks to Example 3.1, we may assume without losing generality that ξ is not a compound Poisson. Then 0 is an instantaneous point, in the sense that $\inf\{t > 0 : \xi_t \neq 0\} = 0$ a.s. The assumption that ξ is transient implies that its last-passage time at 0, defined by

$$\ell := \sup\{t \geq 0 : \xi_t = 0\},$$

is finite a.s. Since the point 0 is regular for itself, there exists a continuous nondecreasing local time process at level 0 which we denote by $L = (L_t, t > 0)$; we also introduce its right-continuous inverse

$$L^{-1}(t) := \inf\{s \geq 0 : L_s > t\}, \qquad t \geq 0$$

with the convention that $\inf \emptyset = \infty$. The largest value of L, namely, L_∞, is finite a.s.; more precisely, L_∞ has an exponential distribution, and we have $L^{-1}(L_\infty-) = \ell$ and $L^{-1}(t) = \infty$ for every $t \geq L_\infty$ ([1], Prop. 7 and Thm 8, pp. 113–115). We denote the set of discontinuity times of the inverse local time before explosion by

$$\mathcal{D} := \{t < L_\infty : L^{-1}(t-) < L^{-1}(t)\}$$

and then, following Itô, we introduce for every $t \in \mathcal{D}$ the excursion $\varepsilon(t)$ with finite lifetime $\zeta_t := L^{-1}(t) - L^{-1}(t-)$ by

$$\varepsilon_s(t) := \xi_{L^{-1}(t-)+s}, \qquad 0 \leq s < \zeta_t.$$

Itô's excursion theory shows that conditionally on L_∞, the family of finite excursions $(\varepsilon(t), t \in \mathcal{D})$ is distributed as the family of the atoms of a Poisson point process with intensity $L_\infty \mathbf{1}_{\{\zeta<\infty\}} n$, where n denotes the Itô measure of the excursions of the Lévy process ξ away from 0, and ζ the lifetime of a generic excursion ([1], Thm 10, p. 118).

Since ξ is not a compound Poisson process, the set of times t at which $\xi_t = 0$ has zero Lebesgue measure a.s., and we can express the integral in the form $I = A + B$ with

$$A := \sum_{t \in \mathcal{D}} \int_{L^{-1}(t-)}^{L^{-1}(t)} g(\xi_s)ds = \sum_{t \in \mathcal{D}} \int_0^{\zeta_t} g(\varepsilon_s(t))ds \qquad (3.1)$$

and

$$B := \int_\ell^\infty g(\xi_s)ds.$$

Excursion theory implies that A and B are independent, and hence we just need to check that A has no atom. Now, the conditional distribution of A given L_∞ is infinitely divisible, with Lévy measure Λ given by the image of $L_\infty \mathbf{1}_{\{\zeta < \infty\}} n$ under the map $\varepsilon \to \int_0^\zeta g(\varepsilon_s) ds$.

The fact that 0 is an instantaneous point implies that the measure $\mathbf{1}_{\{\zeta < \infty\}} n$ is infinite, and further that the excursions $\varepsilon(t)$ leave 0 continuously for all $t \in \mathcal{D}$ a.s. The assumption that $g > 0$ on some neighbourhood of 0 then entails that $\int_0^{\zeta_t} g(\varepsilon_s(t)) ds > 0$ for every $t \in \mathcal{D}$. Thus $\Lambda\{(0, \infty)\} = \infty$, and we conclude from Lemma 3.3 that the conditional distribution of A given L_∞ has no atoms. It follows that $P(A = a) = E(P(A = a|L_\infty)) = 0$ for every $a > 0$, completing the proof of our statement. $\qquad \square$

Remark. See Bertoin [1], Ch. V and Sato [21], Section 43, for discussions relevant to Lévy processes for which 0 is regular for itself.

An easy modification of the argument in Theorem 3.6 yields the following criterion in the special case when the Lévy process has no positive jumps. This extends the result of Theorem 3.5 by allowing a drift, as long as there is no upward jump.

Proposition 3.7 *Let* $g : \mathbb{R} \to [0, \infty)$ *be a measurable function with* $g > 0$ *on some neighbourhood of* 0. *Suppose that* $\xi_t = at - \sigma_t$, *where* $a > 0$ *and* σ *is a subordinator with infinite Lévy measure and no drift, and such that the integral* $I := \int_0^\infty g(\xi_t) dt$ *is finite a.s. Assume further that* $a \neq E\sigma_1$, *so that* ξ *is transient. Then the distribution of* I *has no atoms.*

Remark. We point out that in the case when ξ is a Lévy process with no positive jumps and infinite variation, then 0 is regular for itself ([1], Cor. 5, p. 192), and thus Theorem 3.6 applies. Recall also Example 3.2 for the case of compound Poisson processes with drift. Therefore our analysis covers entirely the situation when the Lévy process has no positive jumps and is not the negative of a subordinator.

Proof. Introduce the supremum process $\bar{\xi}_t := \sup_{0 \le s \le t} \xi_s$. We shall use the fact that the reflected process $\bar{\xi} - \xi$ is Markovian and that $\bar{\xi}$ can be viewed as its local time at 0; see Theorem VII.1 in [1], p. 189. The first-passage process $T_x := \inf\{t \ge 0 : \xi_t \ge x\}$ $(x \ge 0)$ thus plays the role of the inverse local time. It is well-known that $T.$ is a subordinator (killed at some independent exponential time when ξ drifts to $-\infty$); more precisely, the hypothesis that $\xi_t = at - \sigma_t$ has bounded variation implies that the drift coefficient of $T.$ is $a^{-1} > 0$.

Let us consider first the case when ξ drifts to ∞, so the first-passage times T_x are finite a.s. We write $\mathcal{D}$ for the set of discontinuities of $T.$ and for every $x \in \mathcal{D}$, we define the excursion of the reflected Lévy process away from 0 as

$$\varepsilon_s(x) = x - \xi_{T_{x-}+s}, \qquad 0 \le s < \zeta_x := T_x - T_{x-}.$$

According to excursion theory, the point measure

$$\sum_{x \in \mathcal{D}} \delta_{(x,\varepsilon(x))}$$

is then a Poisson random measure with intensity $dx \otimes \bar{n}$, where $\bar{n}$ denotes the Itô measure of the excursions of the reflected process $\bar{\xi} - \xi$ away from 0. Let $b > 0$ be such that $g > 0$ on $[-b, b]$. We can express

$$\int_0^\infty g(\xi_s)ds = A + B + C$$

where

$$A = a^{-1} \int_0^\infty g(x)dx \,,$$

$$B = \sum_{x \in \mathcal{D}, x \le b} \int_{T_{x^-}}^{T_x} g(\xi_s)ds = \sum_{x \in \mathcal{D}, x \le b} \int_0^{\zeta_x} g(x - \varepsilon_s(x))ds \,,$$

$$C = \sum_{x \in \mathcal{D}, x > b} \int_{T_{x^-}}^{T_x} g(\xi_s)ds = \sum_{x \in \mathcal{D}, x > b} \int_0^{\zeta_x} g(x - \varepsilon_s(x))ds \,.$$

The first term A is deterministic, and B and C are independent infinitely divisible random variables (by the superposition property of Poisson random measures). More precisely, the Lévy measure of B is the image of $\mathbf{1}_{\{0 \le x \le b\}}dx \otimes \bar{n}$ by the map

$$(x, \varepsilon) \mapsto \int_0^\zeta g(x - \varepsilon_s)ds \,.$$

Observe that the value of this map evaluated at any $x \in [0, b]$ and excursion ε is strictly positive (because excursions return continuously to 0, as ξ has no positive jumps). On the other hand, the assumption that the Lévy measure of the subordinator $\sigma_t = at - \xi_t$ is infinite ensures that 0 is an instantaneous point for the reflected process $\bar{\xi} - \xi$, and hence the Itô measure $\bar{n}$ is infinite. It thus follows from Lemma 3.3 that the infinitely divisible variable B has no atom, which establishes our claim.

The argument in case ξ drifts to $-\infty$ is similar; the only difference is that the excursion process is now stopped when an excursion with infinite lifetime arises. This occurs at time (in the local-time scale $\bar{\xi}$) $\bar{\xi}_\infty = \sup_{t \ge 0} \xi_t$, where this variable has an exponential distribution. $\square$

3.3 A criterion for absolute continuity

Next we will investigate some different sufficient conditions, and some of them also ensure the existence of Lebesgue densities. We will work with more general integrals of the form $\int_0^\infty g(\xi_t) \, dY_t$ for a process $(Y_t)_{t \ge 0}$ of bounded variation, independent of the Lévy process ξ. The method will be a variant of the strati-fication method, by conditioning on almost every quantity apart from certain jump times. Such an approach was also used by Nourdin and Simon [18] for the study of absolute continuity of solutions to certain stochastic differential equations.

We need the following lemma, which concerns only deterministic functions. Part (a) is just a rewriting of Theorem 4.2 in Davydov et al. [5], and it is this part which will be invoked when studying $\int_0^\infty g(\xi_t)\,dY_t$ for $Y_t = t$.

Lemma 3.8 *Let $Y : [0,1] \to \mathbb{R}$ be a right-continuous deterministic function of bounded variation. Let $f : [0,1] \to \mathbb{R}$ be a deterministic Borel function such that*

$$f \neq 0 \quad a.e. \tag{3.2}$$

and such that the Lebesgue-Stieltjes integral $\int_0^1 f(t)\,dY_t$ exists and is finite. Let

$$H : (0,1] \to \mathbb{R}, \quad x \mapsto \int_{0+}^x f(t)\,dY_t,$$

and denote by $\mu := H(\lambda_{|(0,1]})$ the image measure of λ under H. Then the following are sufficient conditions for (absolute) continuity of μ:
(a) Suppose the absolute continuous part of the measure induced by Y on $[0,1]$ has a density which is different from zero a.e. Then μ is absolutely continuous.
(b) Suppose that Y is strictly increasing and that f is in almost every point $t \in [0,1]$ right- or left-continuous. Then μ is continuous.

Proof. (a) Denoting the density of the absolute continuous part of Y by ϕ, it follows that H is almost everywhere differentiable with derivative $f\phi \neq 0$ a.e., and the assertion follows from Theorem 4.2 in Davydov et al. [5].

(b) Suppose that Y is strictly increasing and denote

$$K := \{t \in (0,1) : f \text{ is right- or left-continuous in } t\}.$$

By assumption, K has Lebesgue measure 1. Using the right-/left-continuity, for every $t \in K$ such that $f(t) > 0$ there exists a unique maximal interval $J_+(t) \subset (0,1)$ of positive length such that $t \in J_+(t)$ and $f(y) > 0$ for all $y \in J_+(t)$. By the axiom of choice there exists a subfamily $K_+ \subset K$ such that $(J_+(t) : t \in K_+)$ are pairwise disjoint and their union covers $K \cap \{t \in (0,1) : f(t) > 0\}$. Since each of these intervals has positive length, there can only be countably many such intervals, so K_+ must be countable.

Similarly, we obtain a countable cover $(J_-(t) : t \in K_-)$ of $K \cap \{t \in (0,1) : f(t) < 0\}$ with disjoint intervals. Now let $a \in \text{Range}(H)$. Then

$$H^{-1}(\{a\}) \subset \left(\bigcup_{t \in K_+} (H^{-1}(\{a\}) \cap J_+(t))\right) \cup \left(\bigcup_{t \in K_-} (H^{-1}(\{a\}) \cap J_-(t))\right)$$
$$\cup ([0,1] \setminus K) \cup \{t \in [0,1] : f(t) = 0\} \cup \{0,1\}.$$

Observing that

$$\lambda\left(H^{-1}(\{a\}) \cap J_\pm(t)\right) = \lambda\left((H_{|J_\pm(t)})^{-1}(\{a\})\right) = 0$$

since H is strictly increasing (decreasing) on $J_+(t)$ $(J_-(t))$ as a consequence of $f > 0$ on $J_+(t)$ $(f < 0$ on $J_-(t))$ and strict increase of Y, it follows that $\lambda(H^{-1}(\{a\})) = 0$, showing continuity of μ.   □

We now come to the main result of this subsection. Note that the case $Y_t = t$ falls under the case (i) considered in the following theorem, giving particularly simple conditions for absolute continuity of $\int_0^\infty g(\xi_t)\, dt$. In particular, part (b) shows that if ξ has infinite Lévy measure and g is strictly monotone on a neighbourhood of 0, then $\int_0^\infty g(\xi_t)\, dt$ is absolutely continuous.

Theorem 3.9 *Let* $\xi = (\xi_t)_{t\geq 0}$ *be a transient Lévy process with non-zero Lévy measure* Π_ξ. *Let* $Y = (Y_t)_{t\geq 0}$ *be a stochastic process of bounded variation on compacts which has càdlàg paths and which is independent of* ξ. *Denote the density of the absolutely continuous part of the measure induced by the paths* $t \mapsto Y_t(\omega)$ *by* ϕ_ω. *Let* $g : \mathbb{R} \to \mathbb{R}$ *be a deterministic Borel function and suppose that the integral*

$$I := \int_{(0,\infty)} g(\xi_t)\, dY_t$$

exists almost surely and is finite.
(a) [general Lévy process] Suppose that there are a compact interval $J \subset \mathbb{R} \setminus \{0\}$ *with* $\Pi_\xi(J) > 0$ *and some constant* $t_0 > 0$ *such that*

$$\lambda(\{|t| \geq t_0 : g(t) = g(t+z)\}) = 0 \quad \text{for all } z \in J. \tag{3.3}$$

Case (i): If $\lambda(\{t \in [t_0,\infty) : \phi(t) = 0\}) = 0$ *a.s., then* I *is absolutely continuous.*
Case (ii): If Y *is strictly increasing on* $[t_0,\infty)$ *and* g *has only countably many discontinuities, then* I *does not have atoms.*

(b) [infinite activity Lévy process] Suppose the Lévy measure Π_ξ *is infinite. Suppose further that there is* $\varepsilon > 0$ *such that*

$$\lambda(\{t \in (-\varepsilon,\varepsilon) : g(t) = g(t+z)\}) = 0 \quad \text{for all } z \in [-\varepsilon,\varepsilon]. \tag{3.4}$$

Case (i): If $\lambda(\{t \in (0,\varepsilon) : \phi(t) = 0\}) = 0$ *a.s., then* I *is absolutely continuous.*
Case (ii): If Y *is strictly increasing on* $(0,\varepsilon)$ *and* g *has only countably many discontinuities, then* I *does not have atoms.*

Proof. (a) Let J be an interval such that (3.3) is satisfied, and define

$$R_t := \sum_{0 < s \leq t, \Delta\xi_s \in J} \Delta\xi_s, \quad M_t := \xi_t - R_t, \quad t \geq 0.$$

Then $R = (R_t)_{t \geq 0}$ is a compound Poisson process, independent of $M = (M_t)_{t \geq 0}$. For $i \in \mathbb{N}$ denote by T_i and Z_i the time and size of the ith jump of R, respectively, and let $T_0 := 0$. Further, denote

$$
\begin{aligned}
I_i &:= \int_{(T_{2i-2}, T_{2i}]} g(\xi_t) \, dY_t \\
&= \int_{(T_{2i-2}, T_{2i-1}]} \left(g\left(M_t + \sum_{j=1}^{2i-2} Z_j \right) - g\left(M_t + \sum_{j=1}^{2i-1} Z_j \right) \right) dY_t \\
&\quad + \int_{(T_{2i-2}, T_{2i}]} g\left(M_t + \sum_{j=1}^{2i-1} Z_j \right) dY_t \\
&\quad + \left[g(\xi_{T_{2i-1}}) - g(\xi_{T_{2i-1}} - Z_{2i-1}) \right] \Delta Y_{T_{2i-1}} \\
&\quad + \left[g(\xi_{T_{2i}}) - g(\xi_{T_{2i}} - Z_{2i}) \right] \Delta Y_{T_{2i}}.
\end{aligned}
\tag{3.5}
$$

We now condition on all random quantities present except the odd numbered T_i. Thus, for every Borel set $B \subset \mathbb{R}$, we write

$$
P(I \in B) = E P \left(\sum_{i=1}^{\infty} I_i \in B \,\middle|\, Y, M, (T_{2j})_{j \in \mathbb{N}}, (Z_j)_{j \in \mathbb{N}} \right).
$$

To show that I has no atoms, it is hence sufficient to show that

$$
P \left(\sum_{i=1}^{\infty} I_i \in B \,\middle|\, Y, M, (T_{2j})_{j \in \mathbb{N}}, (Z_j)_{j \in \mathbb{N}} \right) = 0 \quad \text{a.s.}
\tag{3.6}
$$

for every Borel set B of the form $B = \{a\}$ with $a \in \mathbb{R}$. Similarly, for showing that I is absolutely continuous it is sufficient to show that (3.6) holds for every Borel set B of Lebesgue measure 0. Observe that the $(I_i)_{i \in \mathbb{N}}$ are conditionally independent given

$$
V := (Y, M, (T_{2j})_{j \in \mathbb{N}}, (Z_j)_{j \in \mathbb{N}}).
$$

Thus the conditional probability that $I = \sum_{i=1}^{\infty} I_i \in B$ is the convolution of the conditional probabilities that $I_i \in B$, $i \in \mathbb{N}$. Hence it suffices to show that there is some random integer $i_0 \in \mathbb{N}$ such that almost surely, the conditional distribution of I_{i_0} given V is absolutely continuous (case (i)) or has no atoms (case (ii)), respectively.

Define the integer i_0 as the first index i such that

$$
\min \left\{ \inf_{t \in (T_{2i-2}, T_{2i}]} \left\{ \left| M_t + \sum_{j=1}^{2i-2} Z_j \right| \right\}, T_{2i-2} \right\} \geq t_0,
\tag{3.7}
$$

with t_0 as in (3.3). Since ξ is transient i_0 is almost surely finite. As a function of V, i_0 is constant under the conditioning by V. The right hand side of (3.5) is comprised of four summands. The second and fourth summands are constant given V. The third summand is still random, after conditioning, since T_{2i-1} enters in ΔY; but here R and Y are independent, so that the third summand equals 0 a.s. Thus it is sufficient to show that, given V, the first summand, evaluated at i_0, namely

$$\widetilde{I}_{i_0} := \int_{(T_{2i_0-2}, T_{2i_0-1}]} \left(g\left(M_t + \sum_{j=1}^{2i_0-2} Z_j \right) - g\left(M_t + \sum_{j=1}^{2i_0-1} Z_j \right) \right) dY_t,$$

is almost surely absolutely continuous (case (i)) or has no atoms (case (ii)). Define the functions $f = f_V : [T_{2i_0-2}, T_{2i_0}] \to \mathbb{R}$ and $H = H_V : (T_{2i_0-2}, T_{2i_0}] \to \mathbb{R}$ by

$$f(t) = g\left(M_t + \sum_{j=1}^{2i_0-2} Z_j \right) - g\left(M_t + \sum_{j=1}^{2i_0-1} Z_j \right),$$

$$H(x) := \int_{(T_{2i_0-2}, x]} f(t) \, dY_t.$$

Observing that T_{2i_0-1} is uniformly distributed on (T_{2i_0-2}, T_{2i_0}) given V, it follows from Fubini's theorem that for any Borel set $B \subset \mathbb{R}$

$$P(\widetilde{I}_{i_0} \in B | V) = E(1_{\{H(T_{2i_0-1}) \in B\}} | V) = \int_{(T_{2i_0-2}, T_{2i_0})} 1_{\{H(x) \in B\}} \frac{dx}{T_{2i_0} - T_{2i_0-2}}$$

$$= \frac{\lambda(H^{-1}(B))}{T_{2i_0} - T_{2i_0-2}}.$$

We shall apply Lemma 3.8 to show that $\widetilde{I}_{i_0}$ given V is absolutely continuous or has no atoms, respectively. For this, observe that (3.2) is satisfied because of (3.3) and (3.7), and note that $z := Z_{2i_0-1} \in J$, since all the jumps of R are in the interval J. In case (i) this then gives absolute continuity of $\widetilde{I}_{i_0}$ conditional on V by Lemma 3.8 (a) and hence of the distribution of I. Now concentrate on case (ii), when Y is strictly increasing on $[t_0, \infty)$ and g has only countably many discontinuities. Denote this set of discontinuities of g by F. By assumption, F is countable. This then implies that almost surely, the function f is almost everywhere right-continuous. For by the a.s. right-continuity of the paths of Lévy processes, f can happen to be non-right-continuous at a point t only if $\xi_t^{(1)} := M_t + \sum_{j=1}^{2i_0-1} Z_j \in F$ or $\xi_t^{(2)} := M_t + \sum_{j=1}^{2i_0-2} Z_j \in F$. But

$$E(\lambda\{t \geq 0 : \xi_t^{(1)} \in F \text{ or } \xi_t^{(2)} \in F\}) = \int_0^\infty P(\xi_t^{(1)} \in F \text{ or } \xi_t^{(2)} \in F) \, dt,$$

and by Lemma 3.3 the last integral is zero if ξ has infinite Lévy measure, so that almost surely, f is almost everywhere right-continuous if Π_ξ is infinite.

If ξ has finite Lévy measure, then f is trivially almost everywhere right-continuous. So we see that in case (ii) our assumptions imply the conditions of Lemma 3.8 (b), which then gives the claim.

(b) The proof is similar to the proof of (a): for $0 < \delta < \varepsilon/2$, let

$$R_t^{(\delta)} := \sum_{|\Delta\xi_s|\in[\delta,\varepsilon/2]} \Delta\xi_s, \quad M_t^{(\delta)} := \xi_t - R_t^{(\delta)}, \quad t \geq 0,$$

and denote the time and size of the ith jump of $R^{(\delta)} = (R_t^{(\delta)})_{t\geq 0}$ by $T_i^{(\delta)}$ and $Z_i^{(\delta)}$, respectively. Further, define the set Ω_δ by

$$\Omega_\delta := \{T_2^{(\delta)} \leq \varepsilon, \sup_{0\leq t < T_2^{(\delta)}} |M_t^{(\delta)}| \leq \varepsilon/2\}.$$

Let $P_\delta(\cdot) := P(\cdot|\Omega_\delta)$, and denote expectation with respect to P_δ by E_δ. Since $P(\Omega_\delta) \to 1$ as $\delta \downarrow 0$ because the Lévy measure of ξ is infinite, it is sufficient to show that, given $\delta > 0$, we have $P_\delta(B) = 0$ for all Borel sets B such that $\lambda(B) = 0$ (case (i)), or such that $B = \{a\}$, $a \in \mathbb{R}$ (case (ii)), respectively. Let

$$V_\delta := (Y, M^{(\delta)}, (T_j^{(\delta)})_{j\geq 2}, (Z_j^{(\delta)})_{j\in\mathbb{N}}.$$

Then we can write

$$P_\delta(I \in B) = E_\delta P_\delta(I \in B|V_\delta),$$

and it suffices to show that $P_\delta(I \in B|V_\delta) = 0$ a.s. for the sets B under consideration. But, conditional on V_δ, I almost surely differs from

$$\widetilde{I}_1 := \int_{(0,T_2^{(\delta)}]} \left(g\left(M_t^{(\delta)}\right) - g\left(M_t^{(\delta)} + Z_1^{(\delta)}\right)\right) dY_t$$

only by a constant. It then follows in complete analogy to the proof of (a) that under P_δ, $\widetilde{I}_1$ given V_δ has no atoms or is absolutely continuous, respectively, which then transfers to I under P_δ and hence to I under P.   □

Remarks. (i) The preceding proof has shown that the independence assumption on ξ and Y can be weakened. Indeed, we need only assume that the processes $(R_t)_{t\geq 0}$ and Y are independent.

(ii) In addition to the assumptions of Theorem 3.9, assume that g is continuous. Then almost surely, $g(\xi_{t-}) = g(\xi_t)_-$ exist for all $t > 0$, and the assertions of Theorem 3.9 remain true for integrals of the form

$$\int_{(0,\infty)} g(\xi_t)_- dY_t.$$

This follows in complete analogy to the proof of Theorem 3.9.

(iii) Similar statements as in Theorem 3.9 can be made for integrals of the form $\int_0^\infty (g(\xi_t + \psi(t)) dt$, where ψ is some deterministic function behaving nicely. We omit the details.

Appendix

Proof of the equivalence of (iv) and (v) in Theorem 2.2. Assume (iv), and observe that by the Doléans-Dade formula (e.g. [20], p. 84), $e^{-\xi} = \mathcal{E}(-\eta/k)$, where $k \neq 0$, if and only if $\Pi_\eta(\{y \in \mathbb{R} : k^{-1}y \geq 1\}) = 0$ and $\xi_t = X_t$, where

$$X_t := k^{-1}\eta_t + k^{-2}\sigma_\eta^2 t/2 - \sum_{0 \leq s \leq t} \left(\log(1 - k^{-1}\Delta\eta_s) + k^{-1}\Delta\eta_s\right), \quad t \geq 0. \quad (3.8)$$

Now (X, η) is a bivariate Lévy process, whose Gaussian covariance matrix is given by $\Sigma_{X,\eta} = \begin{pmatrix} 1 & k \\ k & k^2 \end{pmatrix} \sigma_X^2$. Further, (3.8) implies $\Delta X_t = -\log(1 - k^{-1}\Delta\eta_t)$, showing that the Lévy measure $\Pi_{X,\eta}$ of (X, η) is concentrated on $\{(x, k(1 - e^{-x})) : x \in \mathbb{R}\}$.

Conversely, if (Y, η) is a bivariate Lévy process with Gaussian covariance matrix given by $\Sigma_{Y,\eta} = \Sigma_{X,\eta}$, whose Lévy measure is concentrated on $\{(x, k(1 - e^{-x})) : x \in \mathbb{R}\}$, then $\Delta Y_t = -\log(1 - k^{-1}\Delta\eta_t)$, and it follows that there is some $c \in \mathbb{R}$ such that $Y_t = X_t + ct$, so that $e^{-Y_t + ct} = (\mathcal{E}(-\eta/k))_t$. Hence we have established the equivalence of (iv) and (v) in Theorem 2.2, subject to relating γ_1 and γ_2 as in (2.6).

To do this, let X_t as in (3.8) and use the Lévy–Itô decomposition. Define

$$\begin{pmatrix} X_t^{(1)} \\ \eta_t^{(1)} \end{pmatrix} :=$$

$$\lim_{\varepsilon \downarrow 0} \left(\sum_{\substack{0 < s \leq t \\ (\Delta X_s)^2 + (\Delta\eta_s)^2 > \varepsilon^2}} \begin{pmatrix} \Delta X_s \\ \Delta\eta_s \end{pmatrix} - t \iint_{x_1^2 + x_2^2 \in (\varepsilon^2, 1]} \begin{pmatrix} x_1 \\ x_2 \end{pmatrix} \Pi_{X,\eta}(d(x_1, x_2)) \right)$$

and $(X_t^{(2)}, \eta_t^{(2)})' := (X_t, \eta_t)' - (X_t^{(1)}, \eta_t^{(1)})'$ where the limit is a.s. as $\varepsilon \downarrow 0$. (Note that the expression in big brackets on the right is not precisely the compensated sum of jumps.) Then $(X_t^{(2)}, \eta_t^{(2)})'_{t \geq 0}$ is a Lévy process with characteristic triplet $(\gamma, \Sigma, 0)$, so has the form $(X_t^{(2)}, \eta_t^{(2)})' = (\gamma_1 t, \gamma_2 t)' + \mathbf{B}_t$, $t \geq 0$, where $(\mathbf{B}_t)_{t \geq 0}$ is a Brownian motion in $\mathbb{R}^2$. From this follows that

$$X_t^{(2)} - k^{-1}\eta_t^{(2)} = (\gamma_1 - k^{-1}\gamma_2)t + \tilde{B}_t, \quad t \geq 0, \quad (3.9)$$

for some Brownian motion $(\tilde{B}_t)_{t \geq 0}$ in $\mathbb{R}^1$. We wish to determine $\gamma_1 - k^{-1}\gamma_2$. To do this, observe that from (3.8) and $\sigma_X^2 = k^{-2}\sigma_\eta^2$, we have

$$(X_t - X_t^{(1)}) - k^{-1}(\eta_t - \eta_t^{(1)})$$

$$= \sigma_X^2 t/2 + \sum_{0 \leq s \leq t} (\Delta X_s - k^{-1}\Delta\eta_s)$$

$$- \lim_{\varepsilon \downarrow 0} \left(\sum_{\substack{0 < s \leq t \\ (\Delta X_s)^2 + (\Delta\eta_s)^2 > \varepsilon^2}} (\Delta X_s - k^{-1}\Delta\eta_s) \right.$$

$$\left. - t \iint_{x_1^2 + x_2^2 \in (\varepsilon^2, 1]} (x_1 - k^{-1}x_2)\, \Pi_{X,\eta}(d(x_1, x_2)) \right).$$

Noting that $k^{-1}\Delta\eta_s = 1 - e^{-\Delta X_s}$ and that $\sum_{0 < s \leq t}(\Delta X_s - 1 + e^{-\Delta X_s})$ converges absolutely, we obtain, letting $\varepsilon \downarrow 0$, that

$$X_t^{(2)} - k^{-1}\eta_t^{(2)} = \sigma_X^2 t/2 + t \iint_{x_1^2 + x_2^2 \leq 1} (x_1 - k^{-1}x_2)\Pi_{X,\eta}(d(x_1, x_2))$$

$$= \sigma_X^2 t/2 + t \int_{x^2 + k^2(1 - e^{-x})^2 \leq 1} (x - 1 + e^{-x})\, \Pi_X(dx).$$

Comparing this with (3.9) gives (2.6). $\qquad\qquad\qquad\qquad\qquad\qquad \square$

Acknowledgements. This research was carried out while JB and AL were visiting the Centre for Mathematical Analysis and the School of Finance & Applied Statistics at ANU in Canberra. They take pleasure in thanking both for their hospitality. AL gratefully acknowledges financial support by the German Science Foundation (Deutsche Forschungsgemeinschaft), research grant number Li 1026/2-1, and RM's research was partially supported by ARC grant DP00664603.

References

1. Bertoin, J. (1996) *Lévy Processes.* Cambridge University Press, Cambridge.
2. Carmona, P., Petit, F. and Yor, M. (1997) On the distribution and asymptotic results for exponential functionals of Lévy processes. In: *Exponential Functionals and Principal Values Related to Brownian Motion* (M. Yor, Ed.), pp. 73–130. Biblioteca de la Revista Matemática Iberoamericana.
3. Carmona, P., Petit, F. and Yor, M. (2001) Exponential functionals of Lévy processes. In: *Lévy Processes, Theory and Applications* (O.E. Barndorff-Nielsen, T. Mikosch and S. Resnick, Eds.), pp. 41–55. Birkhäuser, Boston.
4. Davydov, Yu. A. (1987) Absolute continuity of the images of measures. *J. Sov. Math.* **36**, 468–473.
5. Davydov, Yu.A., Lifshits, M.A. and Smorodina, N.V. (1998) *Local Properties of Distributions of Stochastic Functionals. Translations Math. Monographs* **173**, AMS, Providence.

6. Dufresne, D. (1990) The distribution of a perpetuity, with application to risk theory and pension funding. *Scand. Actuar. J.* **9**, 39–79.

7. Erickson, K.B. and Maller, R.A. (2004) Generalised Ornstein-Uhlenbeck processes and the convergence of Lévy integrals. In: *Séminaire de Probabilités XXXVIII* (M. Émery, M. Ledoux, M. Yor, Eds.), pp. 70–94. *Lect. Notes Math.* **1857**, Springer, Berlin.

8. Erickson, K.B. and Maller, R.A. (2007) Finiteness of integrals of functions of Lévy processes. *Proc. London Math. Soc.* **94**, 386–420.

9. Gjessing, H.K. and Paulsen, J. (1997) Present value distributions with applications to ruin theory and stochastic equations. *Stoch. Proc. Appl.* **71**, 123–144.

10. Grincevičius, A. (1980) Products of random affine transformations. *Lithuanian Math. J.* **20** no. 4, 279–282. Translated from *Litovsk. Mat. Sb.* **20** no. 4, 49–53, 209 (Russian).

11. de Haan, L. and Karandikar, R.L. (1989) Embedding a stochastic difference equation into a continuous-time process. *Stoch. Proc. Appl.* **32**, 225–235.

12. Haas, B. (2003). Loss of mass in deterministic and random fragmentations. *Stochastic Process. Appl.* **106**, 245–277.

13. Kesten, H. (1969) Hitting probabilities of single points for processes with stationary independent increments. *Memoirs Amer. Math. Soc.* **93**.

14. Klüppelberg, C., Lindner, A. and Maller, R. (2004) A continuous time GARCH process driven by Lévy process: stationarity and second order behaviour. *J. Appl. Probab.* **41**, 601–622.

15. Kondo, H., Maejima, M. and Sato, K. (2006) Some properties of exponential integrals of Lévy processes and examples. *Electr. Comm. Probab.* **11**, 291–303.

16. Lifshits, M.A. (1984) An application of the stratification method to the study of functionals of processes with independent increments. *Theory Probab. Appl.* **29**, 753–764.

17. Lindner, A. and Maller, R. (2005) Lévy integrals and the stationarity of generalised Ornstein-Uhlenbeck processes. *Stoch. Proc. Appl.* **115**, 1701–1722.

18. Nourdin, I. and Simon, T. (2006) On the absolute continuity of Lévy processes with drift. *Ann. Probab.* **34**, 1035–1051.

19. Paulsen, J. (1993) Risk theory in a stochastic economic environment. *Stoch. Proc. Appl.* **46**, 327–361.

20. Protter, P.E. (2004) *Stochastic Integration and Differential Equations.* 2nd edition. Springer, Berlin.

21. Sato, K. (1999) *Lévy Processes and Infinitely Divisible Distributions.* Cambridge University Press, Cambridge.

22. Shtatland, E.S. (1965) On local properties of processes with independent increments. Theor. Prob. Appl. **10**, 317–322.

23. Yor, M. (2001) *Exponential Functionals of Brownian Motion and Related Processes.* Springer, New York.

A Law of the Iterated Logarithm for Fractional Brownian Motions

Driss Baraka and Thomas Mountford

Département de Mathématiques, École Polytechnique Fédérale
1015 Lausanne, Switzerland.
e-mail: driss.baraka@epfl.ch, thomas.mountford@epfl.ch

Summary. We show that for a class of Gaussian processes indexed by one dimensional time, the local times obey the behavior conjectured by Xiao.

Key words: Local times, Gaussian processes, fractional Brownian motion, Girsanov's theorem.

1 Introduction

The starting point for this article is the paper of Xiao [12] which considers the local times for processes $X : \mathbb{R}^N \to \mathbb{R}^d$ given by

$$X(t) = \big(X_1(t), \ldots, X_d(t)\big)$$

for X_i iid real valued Gaussian processes defined on $\mathbb{R}^N$. The class considered contained the fractional Brownian motions. Using clever Fourier analysis techniques and building on previous work of [2] and [1], [12] was able to establish very strong global and local estimates upper bound for the local times which were in line with the Brownian results of Kesten. In this note we show that the bounds of Xiao are the best possible (up to a multiplicative constant), by giving suitable lower bounds on the lim sup behavior of local times.

Let $Y = \{Y(t) : t \in \mathbb{R}^N\}$ be a fractional Brownian motion in $\mathbb{R}$ of Hurst index H ($0 < H < 1$), i.e. the centered real-valued Gaussian random field with covariance function

$$\mathrm{Cov}\big(Y(t), Y(s)\big) = \frac{1}{2}\big(|t|^{2H} + |s|^{2H} - |t - s|^{2H}\big). \tag{1}$$

We associate with Y a d-dimensional fractional Brownian motion of the same index H, $X = \{X(t) : t \in \mathbb{R}^N\}$, in $\mathbb{R}^d$ by

$$X(t) = \big(X_1(t), \ldots, X_d(t)\big), \ t \in \mathbb{R}^N, \tag{2}$$

where $X_1(t), \ldots, X_d(t)$ are independent copies of Y.

Our work concerns the case $N = 1$ and d-fractional Brownian motion. More general dimension and processes will be considered in a later paper.

The approach is similar to that used in [5] for additive Brownian motions. The present case is more complicated in that the processes considered are highly non Markov and for this reason we do not go as far as the results in [5] for the additive Brownian case.

For the process X, the local time $L(x, D)$, if it exists, is defined for $x \in \mathbb{R}^d$, $D \subset \mathbb{R}^N$, by

$$L(x, D) = \lim_{\epsilon \downarrow 0} \frac{1}{c_d \, \epsilon^d} \int_D I_{|X(t)-x| \le \epsilon} \, dt \qquad (3)$$

where c_d is the volume of the unit ball in d-dimensional Euclidean space.

It is known that for d-dimensional fractional Brownian motions of index H, the local time is a.s. well defined if and only if $N > dH$, see e.g. [12] or [4]. In the Brownian case ($H = \frac{1}{2}$) an enormous amount is known, see e.g. [2], [7] or [9], but the picture is less clear elsewhere.

Xiao showed, in a generalized context, that the local times $L(x, D)$ for the process X, satisfy a.s.

$$\limsup_{r \downarrow 0} \sup_{y \in B(0,1) \subset \mathbb{R}^N} \frac{L^*\big([y, y + r]\big)}{\phi_1(r)} < \infty \qquad (4)$$

where $B(0, 1)$ is the unit ball in $\mathbb{R}^N$,

$$L^*(D) := \sup_{x \in \mathbb{R}^d} L(x, D) , \quad \phi_1(s) = s^N \left[s / \log(\frac{1}{s}) \right]^{-dH}$$

and $[y, y + r]$ denotes the N-dimensional cube

$$\prod_{i=1}^N [y_i, y_i + r].$$

The corresponding local result was shown : a.s. for y fixed

$$\limsup_{r \downarrow 0} \frac{L\big(X(y), [y, y + r]\big)}{\phi_2(r)} < \infty, \qquad (5)$$

for

$$\phi_2(s) = s^N \left[s / \log \log(\frac{1}{s}) \right]^{-dH}.$$

We will show that these lim sup's are however strictly positive.

This will, thanks to a 0-1 law, imply

Theorem 1. *For $N = 1$ and $dH < 1$, the local time of the process X satisfies*

$$\limsup_{r\downarrow 0} \frac{L\big(0,[0,r]\big)}{\phi_2(r)} = C_H \in (0,\infty), \qquad (6)$$

for

$$\phi_2(s) = s\left[s/\log\log\left(\frac{1}{s}\right)\right]^{-dH}.$$

We are not able to give the analogous global result, we content ourselves with showing

Theorem 2. *For $N = 1$ and $dH < 1$, on any compact interval I with no empty interior,*

$$\liminf_{r\downarrow 0}\ \sup_{y,y+r\in I} \frac{L^*\big([y,y+r]\big)}{\phi_1(r)} > 0, \qquad (7)$$

for

$$\phi_1(s) = s\left[s/\log\left(\frac{1}{s}\right)\right]^{-dH}.$$

We first deal with the local question. The strategy is to consider associated Gaussian processes

$$X_i^n,\ i = 1, 2, \ldots, d,\ \ n = 1, 2, \ldots$$

so that

(i) the local time of $X_i^n = \{X_i^n(t) : t \geq 0\}$ at zero on $D = [2^{-n}/2, 2^{-n}]$ is close to that of $X_i = \{X_i(t) : t \geq 0\}$ on this interval.

(ii) $X_i^n,\ i = 1, 2, \ldots, d$ are fractional Brownian motions of index H.

(iii) for a reasonable subsequence, n_k, $X_i^{n_k}$ are independent.

This approach is similar to the one used by Xiao in [13]. In [14], Xiao examined the large deviation behavior for local times for a much wider class of Gaussian processes (Theorem 3.23) and examined the small ball probability which is strongly related to our problem.

Secondly we wish to estimate the probability that for process $X^n = \{X^n(t) : t \geq 0\}$ defined by

$$X^n(t) = \big(X_1^n(t), \ldots, X_d^n(t)\big) \text{ for } t \geq 0, \qquad (8)$$

the local time for interval $[2^{-n}/2, 2^{-n}]$ is greater than $h\, 2^{-n(1-dH)}(\log n)^{dH}$ for h small.

The key is to profit from the representation

$$X_i(t) = \int \left(\frac{1}{|x-t|^\alpha} - \frac{1}{|x|^\alpha}\right) dW_i(x), \qquad (9)$$

for a white noise $W_i(.)$, see e.g. [11]. Here $(1 - 2\alpha)/2 = H \in (0,1)$, the Hurst parameter.

In the previous work [5], we "produce" a large local time by "replacing" $W_i(x)$ with an Ornstein-Uhlenbeck process $W_i^{c,n}(x)$, where

$$W_i^{c,n}(0) = 0,$$
$$dW_i^{c,n}(x) = -c\,W_i^{c,n}dx + dB_i(x)$$

for $c = h2^n \log n$ and B_i a Brownian motion. In our case, the application of this idea becomes a little more complicated and c must be replaced by a time inhomogeneous $c(t)$.

The paper is organized as follows: first we work toward Theorem 1. We introduce independent fractional Brownian motions X_i^n, $i = 1, 2, \ldots, d$, $n = 1, 2, \ldots$, so that a rich enough subsequence of these processes is independent and so that on interval $[2^{-(n+1)}, 2^{-n}]$ the local time at zero of $X^n = \{X^n(t) : t \geq 0\}$ is close to that of the original process $X = \{X(t) : t \geq 0\}$. In the following section we introduce a comparison Gaussian process $X^{c,n}(.)$ whose law is absolutely continuous with respect to the original process but which on interval $[2^{-(n+1)}, 2^{-n}]$ has a stochastically larger local time at 0. In Section 4 we consider the Radon-Nikodym derivative of the law of this process with respect to that of our original process and are thus in a position to establish Theorem 1. Finally in the last section we detail how our approach extends to showing Theorem 2.

In the following section we denote the property: $\exists c_1 > 0, c_2 > 0$ and $n_0 > 0$ such that $\forall n > n_0$: $c_1 B_n \leq A_n \leq c_2 B_n$ by a simple notation $A_n \sim B_n$. We will use $K, K', K_1, K_2, \ldots$, to denote unspecified positive finite constants which may not be the same in each occurrence.

2 Independent Processes

Let the processes $X : \mathbb{R} \to \mathbb{R}^d$ given by $X = \{X(t) : t \geq 0\}$ be defined as in (2), and use the representation

$$X_i(t) = \int \left(\frac{1}{|x-t|^\alpha} - \frac{1}{|x|^\alpha} \right) dW_i(x), \; i = 1, \ldots, d \qquad (1)$$

where $2\alpha = 1 - 2H$ and $W_i(.)$ are independent white noises.

We introduce associated Gaussian processes $X^n = \{X^n(t) : t \geq 0\}$ defined as in (8) with

$$X_i^n(t) = \int \left(\frac{1}{|x-t|^\alpha} - \frac{1}{|x|^\alpha} \right) dZ_i^n(x), \; i = 1, \ldots, d \qquad (2)$$

and

$$Z_i^n(x) = W_i(x) \text{ for } A_n \leq |x| \leq R_n$$
$$= W_i^n(x) \text{ elsewhere,}$$

where $\left(W_i^n(.)\right)_{1\le i\le d}$ are independent white noises, also independent of $\left(W_i(.)\right)_{1\le i\le d}$ on $\left([-R_n, -A_n]\cup[A_n, R_n]\right)^c$ where A_n and R_n are two constants, to be fixed later, depending on n such that $0 < A_n << 2^{-n}$ and $R_n >> 2^{-n}$.

Let

$$X_i^p(t) = \int_{\{|x|<A_n\}\cup\{|x|>R_n\}} \left(\frac{1}{|x-t|^\alpha} - \frac{1}{|x|^\alpha}\right) dW_i(x) \qquad (3)$$

and

$$X_i^{p,n}(t) = \int_{\{|x|<A_n\}\cup\{|x|>R_n\}} \left(\frac{1}{|x-t|^\alpha} - \frac{1}{|x|^\alpha}\right) dW_i^n(x), \qquad (4)$$

for $i = 1, ..., d$.

Lemma 1. *For $X_i^p(.)$, $i = 1, ..., d$, defined in (3), all $s, s' \in [2^{-n}/2, 2^{-n}]$, $A_n \le 2^{-n}/10$ and $R_n \ge 2^{-n}.10$, we have*

$$E\left[X_i^p(s)X_i^p(s')\right] \sim \frac{ss'}{|R_n|^{2(1-H)}} + A_n^{2H}. \qquad (5)$$

Proof.

$$E\left[X_i^p(s)X_i^p(s')\right] = 2\int_{R_n}^\infty \left(\frac{1}{|x-s|^\alpha} - \frac{1}{|x|^\alpha}\right)\left(\frac{1}{|x-s'|^\alpha} - \frac{1}{|x|^\alpha}\right)dx$$

$$+2\int_0^{A_n}\left(\frac{1}{|x-s|^\alpha} - \frac{1}{|x|^\alpha}\right)\left(\frac{1}{|x-s'|^\alpha} - \frac{1}{|x|^\alpha}\right)dx$$

$$\sim \int_{R_n}^\infty \frac{\alpha^2 ss'}{|x|^{2\alpha+2}} dx + \int_0^{A_n} \frac{1}{x^{2\alpha}} dx$$

$$\sim \frac{ss'}{|R_n|^{2\alpha+1}} + A_n^{1-2\alpha}. \qquad \square$$

Lemma 2. *For $s, s' \in [2^{-n}/2, 2^{-n}]$ such that $s' < s$, let $\rho(s, s')$ and $\rho'(s, s')$ be, respectively, the correlations of $\left(X_i(s), X_i(s')\right)$ and $\left(X_i(s), X_i^n(s')\right)$, we have*

(i)

$$\rho(s, s') - \rho'(s, s') \ge 0; \qquad (6)$$

(ii)

$$\rho(s, s') - \rho'(s, s') \sim \left(\frac{1}{R_n^{2(1-H)}} + \frac{A_n^{2H}}{ss'}\right)(ss')^{1-H}; \qquad (7)$$

(iii)

$$1 - \rho^2(s, s') \ge \left|1 - \frac{s'}{s}\right|^{2H}. \qquad (8)$$

Proof. The property $\rho(s, s') - \rho'(s, s') \geq 0$ is immediate from direct calculation of $E\big[X_i(s)X_i(s')\big] - E\big[X_i(s)X_i^n(s')\big]$.

$$
\begin{aligned}
\rho(s, s') - \rho'(s, s') &= \frac{E\big[X_i(s)X_i(s')\big] - E\big[X_i(s)X_i^n(s')\big]}{\sqrt{\mathrm{Var}(X_i(s))\mathrm{Var}(X_i(s'))}}\\
&= \frac{E\big[X_i^p(s)X_i^p(s')\big]}{s^H s'^H}\\
&\sim \frac{(ss')^{1-H}}{R_n^{2(1-H)}} + \frac{A_n^{2H}}{s^H s'^H}.
\end{aligned}
$$

On the other hand, if Σ denotes the covariance matrix of $\big(X_i(s), X_i(s')\big)$, then

$$
\begin{aligned}
\det(\Sigma) &= \mathrm{Var}\big(X_i(s)\big)\,\mathrm{Var}\big(X_i(s')\big)\big(1 - \rho^2(s, s')\big)\\
&= \mathrm{Var}\big(X_i(s')\big)\,\mathrm{Var}\big(X_i(s)|X_i(s')\big),
\end{aligned}
$$

and the second result of the lemma is immediate.   □

In the following, we will be interested in

$$
A_n \leq \frac{2^{-n}}{n^{\frac{1-H}{H}\gamma}} \quad \text{and} \quad R_n \geq n^\gamma 2^{-n}, \tag{9}
$$

with $\gamma > 1$ is a constant. In this case we have

$$
\rho - \rho' \leq \frac{1}{n^{2\gamma(1-H)}}. \tag{10}
$$

Theorem 3. *Let $L[0, I_n]$ and $L^n[0, I_n]$ be, respectively, the local times of $X(.)$ and $X^n(.)$ at zero on $I_n = [2^{-n}/2, 2^{-n}]$. If*

$$
R_n \geq n^\gamma 2^{-n}, \quad A_n \leq 2^{-n}/n^{\frac{1-H}{H}\gamma} \quad \text{and} \quad \gamma > \frac{1}{2d(1-H)}, \tag{11}
$$

then

$$
E\left[\left(L[0, I_n] - L^n[0, I_n]\right)^2\right] \leq 2^{-2n(1-dH)} g(n), \tag{12}
$$

where $\sum g(n) < \infty$.

Proof. Let $P_{s,s'}$ be the joint density of $\big(X_1(s), X_1(s')\big)$ and $P_{s,s'}^n$ that of $\big(X_1(s), X_1^n(s')\big)$. Both pairs are Gaussian centered vectors, so their densities at $(0, 0)$ are simply the reciprocal of the square roots of the determinants of the covariance matrices times $1/2\pi$,

$$
P_{s,s'}(0, 0) = \frac{1}{2\pi}\frac{1}{\sigma\big(X_1(s)\big)\sigma\big(X_1(s')\big)}\frac{1}{\sqrt{1 - \rho^2\big(X_1(s), X_1(s')\big)}},
$$

$$P_{s,s'}^n(0,0) = \frac{1}{2\pi} \frac{1}{\sigma(X_1(s))\,\sigma(X_1^n(s'))} \frac{1}{\sqrt{1 - \rho^2(X_1(s), X_1^n(s'))}},$$

and

$$E\left[\left(L[0, I_n] - L^n[0, I_n]\right)^2\right]$$

$$= \int_{2^{-(n+1)}}^{2^{-n}} \int_{2^{-(n+1)}}^{2^{-n}} \left(P_{s,s'}(0,0)\right)^d - \left(P_{s,s'}^n(0,0)\right)^d ds\,ds'.$$

We note that $P_{s,s'}(0,0) \geq P_{s,s'}^n(0,0)$.

For a positive constant β to be fixed later, we split the last integral into two parts

$$\left\{|s - s'| \leq \frac{2^{-n}}{n^\beta}\right\} \quad \text{and} \quad \left\{|s - s'| \geq \frac{2^{-n}}{n^\beta}\right\}$$

which we denote by I_1 and I_2, respectively.

1)

$$I_1 = \int_{2^{-(n+1)}}^{2^{-n}} \int_{2^{-(n+1)}}^{2^{-n}} \left[\left(P_{s,s'}(0,0)\right)^d - \left(P_{s,s'}^n(0,0)\right)^d\right] I_{|s-s'| \leq \frac{2^{-n}}{n^\beta}} ds\,ds'$$

$$\leq 2 \int_{0 < s-s' \leq \frac{2^{-n}}{n^\beta}} \left(P_{s,s'}(0,0)\right)^d ds\,ds'$$

$$= 2\left(\frac{1}{2\pi}\right)^d \int_{0 < s-s' \leq \frac{2^{-n}}{n^\beta}} \left(\frac{1}{\sigma(X_1(s'))\sqrt{\mathrm{Var}(X_1(s)|X_1(s'))}}\right)^d ds\,ds'.$$

By "local nondeterminism" (LND), see [12], we find that

$$I_1 \leq 2\left(\frac{1}{2\pi}\right)^d \int_{0 < s-s' \leq \frac{2^{-n}}{n^\beta}} \left(\frac{1}{s'^H(s-s')^H}\right)^d ds\,ds'$$

$$\leq K_1 \frac{2^{-2n(1-dH)}}{n^{\beta(1-dH)}}.$$

2)

$$I_2 = \int_{|s-s'| \geq \frac{2^{-n}}{n^\beta}} \left[\left(P_{s,s'}(0,0)\right)^d - \left(P_{s,s'}^n(0,0)\right)^d\right] ds\,ds'$$

$$\leq 2 \int_{s-s' \geq \frac{2^{-n}}{n^\beta}} d\left(P_{s,s'}(0,0)\right)^{d-1} \left(P_{s,s'}(0,0) - P_{s,s'}^n(0,0)\right) ds\,ds'.$$

The last inequality follows from the fact that $x^d - x'^d \leq d\,x^{d-1}(x - x')$, for $x > x'$.

$$I_2 \leq 2d \left(\frac{1}{2\pi}\right)^d \int_{s-s' \geq \frac{2-n}{n^\beta}} \left(\frac{1}{\sigma(X(s))\,\sigma(X(s'))\sqrt{1-\rho^2}}\right)^{d-1}$$

$$\times \left[\frac{1}{\sigma(X(s))\,\sigma(X(s'))} \left(\frac{1}{\sqrt{1-\rho^2}} - \frac{1}{\sqrt{1-\rho'^2}}\right)\right] ds\,ds'$$

$$\leq 2d \left(\frac{1}{2\pi}\right)^d \int_{s-s' \geq \frac{2-n}{n^\beta}} \left(\frac{1}{\sigma(X(s))\,\sigma(X(s'))\sqrt{1-\rho^2}}\right)^{d-1}$$

$$\times \left[\frac{1}{\sigma(X(s))\,\sigma(X(s'))} \frac{\rho}{(1-\rho^2)^{3/2}}(\rho-\rho')\right] ds\,ds'$$

$$\leq 2d \,(\frac{1}{2\pi})^d \int_{s-s' \geq \frac{2-n}{n^\beta}} \left(\frac{1}{\sigma(X(s))\,\sigma(X(s'))\sqrt{1-\rho^2}}\right)^{d}$$

$$\times \left[\frac{\rho}{(1-\rho^2)}(\rho-\rho')\right] ds\,ds'.$$

Using $\rho \leq 1$ and Lemma 1 (ii),

$$I_2 \leq K_2 \frac{2^{2ndH}}{n^{2\gamma(1-H)}} \int_{s-s' \geq \frac{2-n}{n^\beta}} \left(\frac{1}{\sqrt{1-\rho^2}}\right)^{d+2} ds\,ds'$$

$$\leq K_2 \frac{2^{2ndH}}{n^{2\gamma(1-H)}} \int_{s-s' \geq \frac{2-n}{n^\beta}} \left(\frac{s}{s-s'}\right)^{H(d+2)} ds\,ds'$$

$$= K_2 \frac{2^{2ndH}}{n^{2\gamma(1-H)}} \int_{2^{-n}/2}^{2^{-n}} \int_{\frac{2-n}{n^\beta}}^{2^{-n}/2} \left(\frac{s}{u}\right)^{H(d+2)} ds\,du$$

$$\leq K_2 \frac{2^{-n(2H+1-dH)}}{n^{2\gamma(1-H)}} \int_{\frac{2-n}{n^\beta}}^{2^{-n}/2} \frac{1}{u^{H(d+2)}}\,du$$

$$= K_2 2^{-2n(1-dH)} n^{-2\gamma(1-H)} \int_{\frac{1}{n^\beta}}^{1/2} \frac{1}{t^{H(d+2)}}\,dt\,.$$

If we pose $I = I_1 + I_2$ then we will have
(i) If $H(d+2) < 1$,

$$I \leq K_3 2^{-2n(1-dH)} \left(\frac{1}{n^{\beta(1-dH)}} + \frac{1}{n^{2d\gamma(1-H)}}\left((1/2)^{1-H(d+2)} - \frac{1}{n^{\beta(1-H(d+2))}}\right)\right);$$

(ii) If $H(d+2) > 1$,

$$I \leq K_4 2^{-2n(1-dH)} \left(\frac{1}{n^{\beta(1-dH)}} + \frac{1}{n^{2d\gamma(1-H)}}\left((1/2)^{1-H(d+2)} - \frac{1}{n^{\beta(1-H(d+2))}}\right)\right);$$

(iii) If $H(d+2) = 1$,

$$I \leq K_5 2^{-2n(1-dH)} \left(\frac{1}{n^{\beta(1-dH)}} + \frac{1}{n^{2d\gamma(1-H)}}\left(\beta \log n - \log 2\right)\right).$$

We can choose β and γ such as

$$\beta(1-dH) > 1 \;,\; 2d\gamma(1-H) > 1 \text{ and } 2d\gamma(1-H) + \beta(1-H(d+2)) > 1. \quad (13)$$

$\square$

Proposition 1. *For $L[0, I_n]$ and $L^n[0, I_n]$ defined as in Theorem 3, we have for n large*

$$\left| L[0, I_n] - L^n[0, I_n] \right| < 2^{-n(1-dH)}, \; a.s. \quad (14)$$

Proof. As is well known

$$P\left(|X| \geq a\right) \leq \frac{E[X^2]}{a^2},$$

the result follows from the Borell-Cantelli theorem. $\square$

3 An Ornstein-Uhlenbeck Gaussian process

In this section we consider only one dimensional processes X.

For $s \in [0, 2^{-n}]$, let $X^R(s)$ be the random variable

$$\int_{|y| \geq R} \left(\frac{1}{|y - s|^\alpha} - \frac{1}{|y|^\alpha} \right) dW(y), \quad (1)$$

for white noise $W(.)$.

Lemma 3. *For all $s \in [0, 2^{-n}]$ and $R \geq 2.2^{-n}$, the variance of $X^R(.)$ defined by (1) verifies*

$$\mathrm{Var}\left(X^R(s)\right) \leq K' \frac{s^2}{R^{2(1-H)}} \quad (2)$$

where K' does not depend on R or n.

Proof. For $|y| \geq R$,

$$\left| \frac{1}{|y - s|^\alpha} - \frac{1}{|y|^\alpha} \right| \leq \frac{Ks}{|y|^{1+\alpha}}$$

and so

$$\begin{aligned}
\mathrm{Var}\left(X^R(s)\right) &= \int_{|y|>R} \left(\frac{1}{|y-s|^\alpha} - \frac{1}{|y|^\alpha} \right)^2 dy \\
&\leq 2K^2 s^2 \int_{|y|>R} \frac{dy}{y^{2(1+\alpha)}} \\
&= \frac{K's^2}{R^{1+2\alpha}} = \frac{K's^2}{R^{2(1-H)}}.
\end{aligned}$$

$\square$

In the following we will be interested in

$$R_0 = 2.2^{-n} \left(\log \log(\frac{1}{2^{-n}}) \right)^{\frac{2H}{2(1-H)}}. \tag{3}$$

We introduce the inhomogeneous Ornstein Uhlenbeck process $W^{c,n}$ on $[-R_0, R_0]$ by

$$W^{c,n}(0) = 0,$$
$$dW^{c,n}(t) = dB_t - f_n(t)W^{c,n}(t)\,dt \quad \text{for all } t \geq 0, \tag{4}$$

where $\{B_t : t \geq 0\}$ is a standard Brownian motion and $f_n(t)$ is defined by

a) for $t \leq 3.2^{-n}$, $f_n(t) = \dfrac{h \log \log(2^n)}{2^{-n}}$;

b) for $t \geq 2.2^{-n} \left(\log \log(2^n) \right)^{\frac{2H}{2(1-H)}}$, $f_n(t) = 0$;

c) for $3.2^{-n} \leq t \leq 2.2^{-n} \left(\log \log(2^n) \right)^{\frac{2H}{2(1-H)}}$,

$$f_n(t) = \frac{3h \log \log(2^n)}{t\big(1 + \log(2^n t/3)\big)^2}$$

and similarly for $t \leq 0$, we have $f_n(t) = -f_n(-t)$.

The next lemma considers stochastic integrals with respect to such "time inhomogeneous" Ornstein-Uhlenbeck processes.

Lemma 4. *Let $\{\beta_t : t \geq 0\}$ be a standard Brownian motion and for nonrandom piecewise smooth $r_.$, let $\{X_t : t \geq 0\}$ be the solution to*

$$X_0 = 0, \quad dX_t = d\beta_t - r_t X_t\,dt. \tag{5}$$

Then

$$\int_0^R g(x) dX_s = \int_0^R \left(g(x) - \int_x^R g(s) \exp\left\{-\int_x^s r_u\,du\right\} r_s\,ds \right) d\beta_x$$

Proof. It is elementary that X_t can be written as $\int_0^t d\beta_s \left(\exp\left\{-\int_s^t r_u\,du\right\}\right)$, So

$$\int_0^R g(x)\,dX_s = \int_0^R g(x)\,d\beta_x - \int_0^R g(x)\,r_x X_x\,dx$$

$$= \int_0^R g(x)\,d\beta_x - \int_0^R g(x)\,r_x \left(\int_0^x d\beta_s \exp\left\{-\int_s^x r_u\,du\right\} \right) dx$$

$$= d\beta_x - \int_0^R \left(\int_x^R g(s) \exp\left\{-\int_x^s r_u\,du\right\} r_s\,ds \right) d\beta_x$$

$$= \int_0^R \left(g(x) - \int_x^R g(s) \exp\left\{-\int_x^s r_u\,du\right\} r_s\,ds \right) d\beta_x.$$

the before last equation follows from the stochastic Fubini Theorem [9], p. 175.

□

Now consider for fixed t, $g(x) = g(x,t) = \dfrac{1}{|x-t|^\alpha} - \dfrac{1}{|x|^\alpha}$ for $x \geq 2t$, we have

$$g'(x) = -\alpha\left(\frac{1}{|x-t|^{\alpha+1}} - \frac{1}{|x|^{\alpha+1}}\right)$$

$$= -\alpha\frac{|x|^{\alpha+1} - |x-t|^{\alpha+1}}{|x|^{\alpha+1}|x-t|^{\alpha+1}}.$$

So $|g'(x)| \leq K\dfrac{t}{|x||x-t|^{\alpha+1}} \leq K\dfrac{t}{x^{2+\alpha}}$, and

$$g(x) - \int_x^R g(s)\exp\{-\int_x^s r_u du\}r_s ds$$

$$\leq g(x)\left(1 - \int_x^R \exp\{-\int_x^s r_u du\}r_s\,ds\right)$$

$$\times \int_x^R\left(\int_x^s \frac{t}{u^{2+\alpha}}du\right)\exp\{-\int_x^s r_u du\}r_s\,ds$$

$$= g(x)\left(1 - \exp\{-\int_x^R r_u du\}\right) - \exp\{-\int_x^R r_u du\}\int_x^R \frac{t}{u^{2+\alpha}}\,du$$

$$+ \int_x^R \frac{t}{s^{2+\alpha}}\exp\{-\int_x^s r_u du\}\,ds.$$

$\square$

Lemma 5. *Let* $W^{c,n}(.)$ *be as defined in (4). For* $t \in [2^{-(n+1)}, 2^{-n}]$ *and* $c = h\dfrac{\log\log(2^n)}{2^{-n}}$ *the random variable* $X^{c,n}(t)$ *defined by*

$$X^{c,n}(t) = \int_{-\infty}^{\infty}\left(\frac{1}{|x-t|^\alpha} - \frac{1}{|x|^\alpha}\right)dW^{c,n}(x) = \int_{-\infty}^{\infty} g(x,t)\,dW^{c,n}(x) \quad (6)$$

is a centered Gaussian random variable of variance bounded by

$$K'\left(\frac{t}{\log\log(\frac{1}{t})}\right)^{2H} \leq \mathrm{Var}\big(X^{c,n}(t)\big) \leq K\left(\frac{t}{\log\log(\frac{1}{t})}\right)^{2H} \quad (7)$$

for K' *and* K *not depending on* t.

Proof. As noted in Lemma 4, $X^{c,n}(t)$ can be written as the stochastic integral with respect to a white noise $\dot{B}(.)$ of $h(t,x)$ which equals

$$g(t,x) - \int_x^\infty \exp\{-\int_x^y r_u du\}r_y g(t,y)dy \quad \text{for } x \geq 0$$

$$g(t,x) + \int_{-\infty}^x \exp\{\int_y^x r_u du\}r_y g(t,y)dy \quad \text{for } x \leq 0$$

where $r_t = f_n(t)$ and so

$$E\left[X^{c,n}(t)\right]^2 = \int_{-\infty}^{\infty} h(t,x)^2 dx$$

$$= \int_{-R_0}^{R_0} h(t,x)^2 dx + \int_{[-R_0,R_0]^c} g(t,x)^2 dx$$

$$= \int_{R_0}^{R_0} h(t,x)^2 dx + 0\left(\frac{t}{\log\log\frac{1}{t}}\right)^{(1-2\alpha)}$$

by our choice of R_0 and Lemma 1. To bound the remaining interval we use the easily but painfully verified inequalities.

a) for $x \in (t, t+\frac{2}{c})$, $\;|h(x,t)| \le \dfrac{1}{|x-t|^\alpha}$;

b) for $x \in [t+\frac{2}{c}, 2.2^{-n}]$, $\;|h(x,t)| \le \dfrac{1}{c|x-t|^{1+\alpha}}$;

c) for $x \in \left[2.2^{-n}, 2^{-n}\left(\log\log(2^n)\right)^{\frac{1-2\alpha}{1+2\alpha}}\right]$, $\;|h(x,t)| \le \dfrac{t\left(1+\log(2^n|x|)\right)^2}{|x|^{1+\alpha}\log\log(2^n)}$;

d) for $x \in \left[2^{-n}\left(\log\log(2^n)\right)^{\frac{1-2\alpha}{1+2\alpha}}, 2.2^{-n}\left(\log\log(2^n)\right)^{\frac{1-2\alpha}{1+2\alpha}}\right], |h(x,t)| \le g(x,t)$;

e) for $x \in (t-\frac{2}{c}, t)$, $\;|h(x,t)| \le \dfrac{K}{|x-t|^\alpha} + Kc^\alpha$ where K is a universal constant depending on α ;

f) for $x \in [\frac{t}{2}, t-\frac{2}{c}]$, $\;|h(x,t)| \le \dfrac{K}{c}\dfrac{t}{|x|^{2+\alpha}} + Ke^{-(t-x)c}c^\alpha$;

g) for $x \in [\frac{2}{c}, \frac{t}{2}]$, $\;|h(x,t)| \le \dfrac{1}{c|x|^{1+\alpha}} + Ke^{-(t-x)c}c^\alpha$;

h) for $x \in [0, \frac{2}{c}]$, $\;|h(x,t)| \le \dfrac{1}{|x|^\alpha}$;

i) for $x \in (-\frac{2}{c}, 0)$, $\;|h(x,t)| \le Kc^\alpha + \dfrac{K}{|x|^\alpha}$;

j) for $x \in (-2t, -\frac{2}{c})$, $\;|h(x,t)| \le \dfrac{1}{c|x|^{1+\alpha}}$;

k) for $x \in \left(-2^{-n}\left(\log\log(2^n)\right)^{\frac{1-2\alpha}{1+2\alpha}}, -2t\right)$, $\;|h(x,t)| \le \dfrac{t\left(1+\log(2^n|x|)\right)^2}{|x|^{1+\alpha}\log\log(2^{3n})}$;

l) for $x \in \left(-2.2^{-n}\left(\log\log(2^n)\right)^{\frac{1-2\alpha}{1+2\alpha}}, -2^{-n}\left(\log\log(2^n)\right)^{\frac{1-2\alpha}{1+2\alpha}}\right)$, $\;|h(x,t)| \le g(x,t)$.

Given this we find that

$$E\left[X^{c,n}(t)^2\right] \le Kc^{-(1-2\alpha)} = Kc^{-2H}. \tag{8}$$

The lower bound is more easily verified. $\qquad\qquad\qquad\qquad\qquad\qquad\square$

Lemma 6. *For $s, t \in [2^{-(n+1)}, 2^{-n}]$ the joint density $P_{s,t}(0,0)$ of $\left(X_s^{c,n}, X_t^{c,n}\right)$ at $(0,0)$ satisfies*

$$P_{s,t}(0,0) \leq K\left(c^H \vee |t-s|^{-H}\right) c^H, \tag{9}$$

where $c = h2^n \log n$.

Proof. The pair $(X_s^{c,n}, X_t^{c,n})$ is a Gaussian centered vector, accordingly its density at $(0,0)$ is simply the reciprocal of the square root of the determinant of its covariance matrix times $1/2\pi$

$$\frac{1}{\sigma(X_s^{c,n})} \frac{1}{\sqrt{\mathrm{Var}(X_t^{c,n}|X_s^{c,n})}} \frac{1}{2\pi}. \tag{10}$$

$\frac{1}{\sigma(X_s^{c,n})} \leq Kc^H$, it remains to bound $\mathrm{Var}(X_t^{c,n}|X_s^{c,n})$. This latter quantity is simply (with the notation of the previous lemma)

$$\inf_{\rho \in M} E\left[(X_t^{c,n} - \rho X_s^{c,n})^2\right] = \inf_\rho \int_{-\infty}^{\infty} \left(h(x,t) - \rho h(x,s)\right)^2 dx$$

$$\geq \inf_\rho \int_{|x-t| \leq \frac{|t-s|}{K}} \left(h(x,t) - \rho h(x,s)\right)^2 dx$$

$$+ \int_{|x-s| \leq \frac{|t-s|}{K}} \left(h(x,t) - \rho h(x,s)\right)^2 dx.$$

Now we have for $|t-s| \geq \frac{1}{c}$ that for $|x-t| \leq \frac{c}{K}$ that

$$\left|h(x,t)\right| \geq \frac{1}{2|x-s|^{(1-2H)/2}}, \quad \left|h(x,s)\right| \leq \frac{1}{4|x-t|^{(1-2H)/2}} \tag{11}$$

for K a large positive constant not depending on s, t or n and n is fixed large enough. Similarly for $|x-s| \leq c/K$.

Thus if $|\rho| \leq 1$ we have

$$\int \left(h(x,t) - \rho h(x,s)\right)^2 ds \geq \int_{|x-t| \leq \frac{c}{K}} \left(\frac{1}{4|x-t|^{(1-2H)/2}}\right)^2 dx = K'c^{2H} \tag{12}$$

while if $|\rho| \geq 1$

$$\int \left(f(x,t) - f(x,s)\right)^2 ds \geq \int_{|x-s| \leq \frac{c}{K}} \left(\frac{1}{4|x-s|^{(1-2H)/2}}\right)^2 dx = K'c^{2H} \tag{13}$$

where K' is a universal constant.

We argue similarly for $|t-s| \leq \frac{1}{c}$. $\qquad \square$

We can apply this to the $\mathbb{R}^d$ valued process

$$\{\underline{X}^{c,n}(t) : t \geq 0\} = \{X_1^{c,n}(t), \ldots, X_d^{c,n}(t) : t \geq 0\} \tag{14}$$

where $X_i^{c,n}$ are independent copies of $\{X^{c,n}(t) : -\infty < t < \infty\}$.

Corollary 1. *There exists a constant k not depending on n so that for all n and $c = 2^n h \log \log(2^n)$ the probability that the local time at zero of the process $\{\underline{X}^{c,n}(t) : -\infty < t < \infty\}$ for the interval $[2^{-(n+1)}, 2^{-n}]$ is at least*

$$hk2^{-n(1-dH)}(\log n)^{dH} \tag{15}$$

is at least k.

Proof. This follows from the usual two moment estimate for strictly positive random variables. Let L^n be the corresponding local time. We have by Lemma 5 that

$$E[L^n] = \int_{2^{-(n+1)}}^{2^{-n}} P_s(0)^d ds \geq 2^{-(n+1)} K 2^{ndH} (\log n)^{dH}, \tag{16}$$

(for $P_s(0)$ the density of $X(s)$ at 0) and so that (see [2])

$$E[(L^n)^2] = 2 \int_{2^{-(n+1)}}^{2^{-n}} \int_t^{2^{-n}} \left(P_{t,s}(0,0)\right)^d ds\, dt \leq K'' \left(2^{-n} 2^{ndH} (\log n)^{dH}\right)^2. \tag{17}$$

As it is well known for a positive random variable X

$$P(X \geq E[X]) \geq \frac{E[X]^2}{4E[X^2]}, \qquad \text{the result holds.} \qquad \square$$

4 Radon-Nikodym derivative

In this section we consider the Radon Nikodym derivative of the law of the "white noise" $W^{c,n}(.)$ with respect to that of the white noise $W(.)$. This will give the Radon Nikodym derivative of the law of process $X^{c,n}(.)$ with respect to the law for the original process $X(.)$.

We have the following from second moment estimates

Lemma 7. *For the function $f_n(.)$ and the process $W^{c,n}(.)$ defined in (4) and $R_0 = 2.2^{-n}(\log \log 2^n)^{\frac{H}{1-H}}$*

$$\frac{1}{h \log \log(2^n)} \int_{-R_0}^{R_0} f_n(t)^2 \left(W^{c,n}(t)\right)^2 dt$$

$$= \frac{1}{h \log \log(2^n)} \int_{-\infty}^{\infty} f_n(t)^2 \left(W^{c,n}(t)\right)^2 dt \xrightarrow{pr} k \text{ as } n \to \infty \tag{1}$$

for k a strictly positive finite constant not depending on n or h.

Lemma 8. *For P the law of Brownian motion on canonical two sided path space and Q^n is the law of $\{W^{c,n}(t) : -\infty < t < \infty\}$ on this space we have*

$$\frac{dQ^n}{dP}(\omega) = \exp\left(-\int_{-R_0}^{R_0} f_n(t)w(t)\, dw(t) - \frac{1}{2}\int_{-R_0}^{R_0} f_n(t)^2 w(t)^2 dt\right) \quad (2)$$

and

$$\frac{dP}{dQ^n}(w) = \exp\left(\int_{-R_0}^{R_0} f_n(t)w(t)\, d\tilde{w}(t) - \frac{1}{2}\int_{-R_0}^{R_0} f_n(t)^2 w(t)^2 dt\right) \quad (3)$$

where $\tilde{w}(t)$ is the (Q^n) Brownian motion

$$\tilde{w}(t) = w(t) + \int_0^t f_n(s)w(s)ds. \quad (4)$$

For the proof see e.g. [9].
    We obtain as in [5],

Lemma 9. *If $Q^n(A_n) \geq c > 0$ for n large where A_n is an event in path space then*

$$P(A_n) \geq c\, e^{-2kh\log\log(2^n)} \quad (5)$$

for n sufficiently large and k the constant of (1).

Equally for d independent copies. If P^d is the Wiener measure on two sided continuous d-dimensional path space and $Q^{d,n}$ is the product measure of Q^n on this space, then

Corollary 2. *If A_n is a sequence of events on d-dimensional path space and if for some $c > 0$, $Q^{d,n}(A_n) > c$ for n large then*

$$P^d(A_n) \geq c\, e^{-2kh\log\log(2^n)d} \text{ for } n \text{ sufficiently large.} \quad (6)$$

5 Proof of Theorem 1

Define for $K > \dfrac{\gamma}{\log 2}\sup(1, \dfrac{1-H}{H})$, where γ is the constant defined in Section 2, and large integer r the sequence t_i^n, $0 \leq i \leq r$, by

$$t_0^n = 2^{-n},$$
$$t_i^n = t_{i-1}^n 2^{-3K\log n} \text{ for } i \geq 1$$

until t_r^n is first less than 2^{-2n}. Thus r is of the order $\dfrac{n}{\log n}$.

We introduce independent white noises $W_j^{n,i}$, $i = 0, 1, \ldots, r$, $j = 1, 2, \ldots, d$ also independent of $(W_j)_{1 \leq j \leq d}$ and processes

$$X_j^{n,i}(t) = \int_{-\infty}^{\infty} \left(\frac{1}{|x - t|^\alpha} - \frac{1}{|x|^\alpha} \right) dZ_j^{n,i}(x) \tag{1}$$

where

$$\dot{Z}_j^{n,i}(x) = \dot{W}_j(x) \text{ for } 3t_i^n 2^{-K \log n} \leq |x| \leq t_i^n \left(1 + \frac{2^{K \log n}}{3} \right)$$

$$= \dot{W}_j^{n,i}(x) \text{ elsewhere.}$$

Then, for $i = 0, 1, \ldots r$, the processes

$$\{ X^{n,i}(t) : t \geq 0 \} = \{ X_1^{n,i}(t), \ldots, X_d^{c,n}(t) : t \geq 0 \}$$

are independent. But by Proposition 1, the local time of $X^{n,i}$ is close to the original X at zero on interval $[t_i^n, 2 t_i^n]$ for n large, explicitly

$$\left| L^{X^{n,i}}[t_i^n, 2 t_i^n] - L^X[t_i^n, 2 t_i^n] \right| < (t_i^n)^{1-dH}, \ a.s. \tag{2}$$

We consider for $i = 0, 1, \ldots, r$ the events

$$A(i, n) = \{ L^{n,i} \geq h(t_i^n)^{1-dH} (\log n)^{dH} \}, \tag{3}$$

where $L^{n,i}$ is the local time at zero for process $\{ X^{n,i}(t) : t \geq 0 \}$ over the time interval $[t_i^n, 2t_i^n]$.

The constant h is chosen to be small enough in the sense of Corollary 1, we have by Corollary 2 that

$$P(A(i, n)) \geq \frac{1}{n^{1/2}}. \tag{4}$$

If n is sufficiently large and so by independence of the processes $X^{n,i}$ we have

$$P(\cup_{i=0}^r A(i, n)) \geq 1 - (1 - \frac{1}{n^{1/2}})^{n/K \log n} \tag{5}$$

which tends strongly to 1 as $n \to \infty$.

By Proposition 1, this and the arbitrariness of n, imply that

Theorem 4. *The* $\limsup$ *of* $\dfrac{L[0, t]}{t^{1-dH} (\log \log \frac{1}{t})^{dH}}$ *for process* $\{ X(t) : t \geq 0 \}$ *is strictly positive.*

Corollary 3. *There exists a constant* $C_\alpha \in (0, \infty)$ *so that*

$$\limsup_{t \downarrow 0} \frac{L(0, [0, t])}{t^{1-dH} (\log \log \frac{1}{t})^{dH}} = C_\alpha. \tag{6}$$

Proof. By employing Proposition 1 we have that the a.s. strictly positive value of

$$\limsup_{t \downarrow 0} \frac{L[0,t]}{t^{1-dH}(\log\log \frac{1}{t})^{dH}} \tag{7}$$

is independent of the white noise $\dot{W}$ on $(-t,t)^c$ for each $G > 0$.

The result now follows from a standard 0-1 law argument, see [8].   □

6 Proof of Theorem 2

The argument is essentially the same as in the previous sections. If we take in the third section, $R_0 = 2.2^{-n}(\log \frac{1}{2^{-n}})^{\frac{1-2\alpha}{1+2\alpha}}$ and $f_n(t)$ defined by

a) for $t \le 3.2^{-n}$, $f_n(t) = \dfrac{h\log(2^n)}{2^{-n}}$;

b) for $t \ge 2.2^{-n}\left(\log(2^n)\right)^{\frac{1-2\alpha}{1+2\alpha}}$, $f_n(t) = 0$;

c) for $3.2^{-n} \le t \le 2.2^{-n}\left(\log(2^n)\right)^{\frac{1-2\alpha}{1+2\alpha}}$,

$$f_n(t) = \frac{3h\log(2^n)}{t\left(1+\log(2^n t/3)\right)^2}. \tag{1}$$

Similarly for $t \le 0$, we have $f_n(t) = -f_n(-t)$, we can prove as with Corollary 2,

Corollary 4. *There exists a strictly positive constant k not depending on n so that for all n and $c = 2^n h\log(2^n)$ the probability that the local time at zero of the process $\{\underline{X}^{c,n}(t) : -\infty < t < \infty\}$ for the interval $[2^{-(n+1)}, 2^{-n}]$ is at least*

$$hk2^{-n(1-dH)}\, n^{dH} \tag{2}$$

is at least k.

And as with Corollary 3

Corollary 5. *If A_n is a sequence of events on d-dimensional path space and if for some $c > 0$, $Q^{d,n}(A_n) > c$ for n large then*

$$P^d(A_n) \ge c\, e^{-2kh\log(2^n)d} \quad \text{for } n \text{ sufficiently large.} \tag{3}$$

For an interval I, say for definiteness $[1, 2]$, we consider $2^{n/2}$ subintervals of length 2^{-3n} each separated from their peers by a distance $2^{-n/2} - 2^{-3n}$. We denote these intervals by I_i^n, with

$$I_i^n = \left[1+(i-1)2^{-n/2}, 1+(i-1)2^{-n/2}+2^{-3n}\right], \quad i = 1, 2, ..., 2^{n/2}, \tag{4}$$

and we consider for $1 \leq i \leq 2^{n/2}$, the processes

$$\{X^{n,i}(t);\ t \geq 0\} = \{X_1^{n,i}(t), \ldots, X_d^{n,i}(t);\ t \geq 0\},$$

with $X_j^{n,i}$ are H-fractional Brownian motions generated by white noises $(W_j^{n,i})$ for $j = 1, \ldots, d$ with formally

$$\dot{Z}_j^{n,i}(x) = \dot{W}_j(x) \text{ for } d(x, I_i^n) \leq 2^{-n/2}/3$$
$$= \dot{W}_j^{n,i}(x) \text{ otherwise}$$

where $(W_j^{n,i})_{1 \leq i \leq 2^{n/2}, 1 \leq j \leq d}$ are independent white noises also independent of $(W_j)_{1 \leq j \leq d}$. Then for $i = 1, \ldots, 2^{n/2}$, the processes $X^{n,i}$ are independent.

Let t_i^n be the center of I_i^n and apply the Proposition 1 to processes

$$Y(s) = X(t_i^n + s) - X(t_i^n), \quad |s| \leq 2^{-3n}/2$$
$$Y^{n,i}(s) = X^{n,i}(t_i^n + s) - X^{n,i}(t_i^n), \quad |s| \leq 2^{-3n}/2$$

with $A_n = 0$ and $R_n = 2^{-n/2}/3$. By the Borell-Cantelli lemma we find that the local time at $X^{n,i}(t_i^n)$ for process $\{X^{n,i}(t) : t \geq 0\}$ over the time interval I_i^n, $L\big(X^{n,i}(t_i^n), I_i^n\big)$, is close to the local time at $X(t_i^n)$ for the original process $\{X(t) : t \geq 0\}$ over the time interval I_i^n, $L\big(X(t_i^n), I_i^n\big)$, for n large, explicitly: for all $1 \leq i \leq 2^{n/2}$ and n sufficiently large

$$\left| L^{X^{n,i}}\big(X^{n,i}(t_i^n), I_i^n\big) - L^X\big(X(t_i^n), I_i^n\big) \right| < (2^{-3n}/2)^{1-dH}, \ a.s. \qquad (5)$$

We consider for $i = 1, 2, \ldots, 2^{n/2}$ the events

$$A(i, n) = \{L_i^n \geq h(2^{-3n})^{1-dH} n^{dH}\}, \qquad (6)$$

where L_i^n is the local time at $X^{n,i}(t_i^n)$ for process $\{X^{n,i}(t) : t \geq 0\}$ over the time interval I_i^n.

The constant h is chosen to be small enough in the sense of Corollary 4. So that we have, by Corollary 5, that

$$P\big(A(i, n)\big) \geq \frac{1}{2^{n/3}} \qquad (7)$$

if n is sufficiently large. Then by independence of the processes $X^{n,i}$ we have

$$P\big(\cup_{i=0}^{2^{n/2}} A(i, n)\big) \geq 1 - \big(1 - \frac{1}{2^{n/3}}\big)^{2^{n/2}} \qquad (8)$$

which tends strongly to 1 as $n \to \infty$.

By Proposition 1, this and the arbitrariness of n, imply

Theorem 5. *There exists a positive constant C, so that for each interval I non trivial*

$$\liminf_{r \downarrow 0} \sup_{y, y+r \in I} \frac{L^*\big([y, y+r]\big)}{r^{1-dH} \log\big(\frac{1}{r}\big)^{dH}} > C. \qquad (9)$$

References

1. Ehm, W. (1981), Sample function properties of multiparameter stable processes. *Zeit. Wahr. Theorie* **56**, 195-228.
2. Geman, D. and Horowitz, J. (1980), Occupation densities. *Ann. Probab.*, **8**, 1-67.
3. Geman, D., Horowitz, J. and Rosen, J. (1984), A local time analysis of intersections of Brownian paths in the plane. *Ann. Probab.*, **12**, 86-107.
4. Khoshenevisan, D., Xiao, Y. and Zhong, Y. (2003), Local times of additive Lévy processes. *Stochastic Process. Appl.*, **104**, no. 2, 193-216.
5. Mountford, T and Nualart, E (2004), Level sets of multiparameter Brownian motions. *Electron. J. Probab.*, **9**, no. 20, 594-614.
6. Mueller, C. and Tribe, R. (2002), Hitting properties of a random string. *Electron. J. Probab.*, **7**, 1-29.
7. Perkins, E. (1981), The exact Hausdorff measure of the level sets of Brownian motion. *Z. Wahrsch. verw. Gebiete*, **58**, 373-388.
8. Pitt, L.D. and Tran, L.T. (1979), Local Sample Path Properties of Gaussian Fields. *Ann. Probab.*, **7**, no. 3, 477-493.
9. Revuz, D. and Yor, M. (1999), *Continuous martingales and Brownian motion*. Third edition. Springer-Verlag, New York.
10. Rogers, C.A. (1998), *Hausdorff measures*. Cambridge University Press.
11. Samorodnitsky, G and Taqqu, M. S. (1994), *Stable non-Gaussian random processes: Stochastic models with infinite variance*. Stochastic Modeling. Chapman & Hall, New York.
12. Xiao, Y. (1997), Hölder conditions for the local times and the Hausdorff measure of the level sets of Gaussian random fields. *Probab. Th. Rel. Fields*, **109**, 129-157.
13. Xiao, Y. (2003), The packing measure of the trajectories of multiparameter fractional Brownian motion. *Math. Proc. Camb. Phil. Soc.*, **135**, 349-375.
14. Xiao, Y. (2005), Strong Local Nondeterminism and Sample Path Properties of Gaussian Random Fields. *Preprint*.

A simple theory for the study of SDEs driven by a fractional Brownian motion, in dimension one

Ivan Nourdin

Laboratoire de Probabilités et Modèles Aléatoires, Université Pierre et
Marie Curie Paris VI
Boîte courrier 188, 4 Place Jussieu, 75252 Paris Cedex 5, France
e-mail: nourdin@ccr.jussieu.fr

Summary. We will focus – in dimension one – on the SDEs of the type $dX_t = \sigma(X_t)dB_t + b(X_t)dt$ where B is a fractional Brownian motion. Our principal aim is to describe a simple theory – from our point of view – allowing to study this SDE, and this for any $H \in (0,1)$. We will consider several definitions of solutions and, for each of them, study conditions under which one has existence and/or uniqueness. Finally, we will examine whether or not the canonical scheme associated to our SDE converges, when the integral with respect to fBm is defined using the Russo-Vallois symmetric integral.

Key words: Stochastic differential equation; fractional Brownian motion; Russo-Vallois integrals; Newton-Cotes functional; Approximation schemes; Doss-Sussmann transformation.

MSC 2000: 60G18, 60H05, 60H20.

1 Introduction

The fractional Brownian motion (fBm) $B = \{B_t, t \geq 0\}$ of Hurst index $H \in (0,1)$ is a centered and continuous Gaussian process verifying $B_0 = 0$ a.s. and

$$E[(B_t - B_s)^2] = |t - s|^{2H} \tag{1}$$

for all $s, t \geq 0$. Observe that $B^{1/2}$ is nothing but standard Brownian motion. Equality (1) implies that the trajectories of B are $(H - \varepsilon)$-Hölder continuous, for any $\varepsilon > 0$ small enough. As the fBm is selfsimilar (of index H) and has stationary increments, it is used as a model in many fields (for example, in hydrology, economics, financial mathematics, etc.). In particular, the study of stochastic differential equations (SDEs) driven by a fBm is important in view of the applications. But, before raising the question of existence and/or uniqueness for this type of SDEs, the first difficulty is to give a meaning to the integral with respect to a fBm. It is indeed well-known that B is not a semimartingale when $H \neq 1/2$. Thus, the Itô or Stratonovich calculus does not apply to this case. There are several ways of building an integral with respect to the fBm and of obtaining a change of variables formula. Let us point out some of these contributions:

1. *Regularization or discretization techniques.* Since 1993, Russo and Vallois [31] have developed a regularization procedure, whose philosophy is similar to the discretization. They introduce forward (generalizing Itô), backward, symmetric (generalizing Stratonovich, see Definition 3 below) stochastic integrals and a generalized quadratic variation. The regularization, or discretization technique, for fBm and related processes have been performed by [12, 17, 32, 36], in the case of zero quadratic variation (corresponding to $H > 1/2$). Note also that Young integrals [35], which are often used in this case, coincide with the forward integral (but also with the backward or symmetric ones, since covariation between integrand and integrator is always zero). When the integrator has paths with finite p-variation for $p > 2$, forward and backward integrals cannot be used. In this case, one can use some symmetric integrals introduced by Gradinaru *et al.* in [15] (see Section 2 below). We also refer to Errami and Russo [11] for the specific case where $H \geq 1/3$.

2. *Rough paths.* An other approach was taken by Lyons [20]. His absolutely pathwise method based on Lévy stochastic areas considers integrators having p-variation for any $p > 1$, provided one can construct a canonical geometric rough path associated with the process. We refer to the survey article of Lejay [18] for more precise statements related to this theory. Note however that the case where the integrator is a fBm with index $H > 1/4$ has been studied by Coutin and Qian [7] (see also Feyel and de La Pradelle [13]). See also Nourdin and Simon [26] for a link between the regularization technique and the rough paths theory.

3. *Malliavin calculus.* Since fBm is a Gaussian process, it is natural to use a Skorohod approach. Integration with respect to fBm has been attacked by Decreusefond and Üstünel [8] for $H > 1/2$ and it has been intensively studied since (see for instance [1, 2, 6]), even when the integrator is a more general Gaussian process. We refer to Nualart's survey article [27] for precise statements related to this theory.

4. *Wick products.* A new type of integral with zero mean defined using Wick products was introduced by Duncan, Hu and Pasik-Duncan in [10], assuming $H > 1/2$. This integral turns out to coincide with the divergence operator. In [3], Bender considers the case of arbitrary Hurst index $H \in (0, 1)$ and proves an Itô formula for generalized functionals of B.

In the sequel, we will focus – in dimension one – on SDEs of the type:

$$\begin{cases} dX_t = \sigma(X_t)\,dB_t + b(X_t)dt, \ \ t \in [0,T] \\ X_0 = x_0 \in \mathbb{R} \end{cases} \tag{2}$$

where $\sigma, b : \mathbb{R} \to \mathbb{R}$ are two continuous functions and $H \in (0,1)$. Our principal motivation is to describe a simple theory – from our point of view – allowing to study the SDE (2), for *any* $H \in (0,1)$. It is linked to the regularization technique (see point 1 above). Moreover, we emphasize that it is already used and quoted in some research articles (see for example [4, 14, 21, 22, 24–26]). The aim of the current paper is, in particular, to clarify this approach.

The paper is organized as follows. In the second part, we will consider several definitions of solution to (2) and for each of them we will study under which condition one has existence and/or uniqueness. Finally, in the third part, we will examine whether or not the canonical scheme associated to (2) converges, when the integral with respect to fBm is defined using the Russo-Vallois symmetric integral.

2 Basic study of the SDE (2)

In the sequel, we denote by B a fBm of Hurst parameter $H \in (0,1)$.

Definition 2.1 *Let X, Y be two real continuous processes defined on $[0,T]$. The* symmetric integral *(in the sense of Russo-Vallois) is defined by*

$$\int_0^T Y_u d^\circ X_u = \lim_{\varepsilon \to 0} \text{in prob} \int_0^T \frac{Y_{u+\varepsilon} + Y_u}{2} \times \frac{X_{u+\varepsilon} - X_u}{\varepsilon}\, du, \tag{3}$$

provided the limit exists and with the convention that $Y_t = Y_T$ and $X_t = X_T$ when $t > T$.

Remark 2.2 If X, Y are two continuous semimartingales then $\int_0^T Y_u d^\circ X_u$ coincides with the standard Stratonovich integral, see [31].

Let us recall an important result for our study:

Theorem 2.3 *(see [15], p. 793). The symmetric integral $\int_0^T f(B_u)d^\circ B_u$ exists for any $f : \mathbb{R} \to \mathbb{R}$ of class C^5 if and only if $H \in (1/6, 1)$. In this case, we have, for any antiderivative F of f:*

$$F(B_T) = F(0) + \int_0^T f(B_u)d^\circ B_u.$$

When $H \leq 1/6$, one can consider the so-called *m-order Newton-Cotes functional*:

Definition 2.4 *Let* $f : \mathbb{R}^n \to \mathbb{R}$ *(with* $n \geq 1$*) be a continuous function,* $X : [0, T] \times \Omega \to \mathbb{R}$ *and* $Y : [0, T] \times \Omega \to \mathbb{R}^n$ *be two continuous processes and* $m \geq 1$ *be an integer. The* m*-order Newton-Cotes functional of* (f, Y, X) *is defined by*

$$\int_0^T f(Y_u) d^{\mathrm{NC}, m} X_u$$

$$= \lim_{\varepsilon \to 0} \text{ in prob} \int_0^T \left(\int_0^1 f(Y_u + \beta(Y_{u+\varepsilon} - Y_u)) \nu_m(d\beta) \right) \frac{X_{u+\varepsilon} - X_u}{\varepsilon} du,$$

provided the limit exists and with the convention that $Y_t = Y_T$ *and* $X_t = X_T$ *when* $t > T$*. Here,* $\nu_1 = \frac{1}{2}(\delta_0 + \delta_1)$ *and*

$$\nu_m = \sum_{j=0}^{2(m-1)} \left(\int_0^1 \prod_{k \neq j} \frac{2(m-1)u - k}{j - k} du \right) \delta_{j/(2m-2)}, \quad m \geq 2, \qquad (4)$$

δ_a *being the Dirac measure at point* a*.*

Remark 2.5 • The 1-order Newton-Cotes functional $\int_0^T f(Y_u) d^{\mathrm{NC}, 1} X_u$ is nothing but the symmetric integral $\int_0^T f(Y_u) d^\circ X_u$ defined by (3). On the contrary, when $m > 1$, the m-order Newton-Cotes functional $\int_0^T f(Y_u) d^{\mathrm{NC}, m} X_u$ is not *a priori* a "true" integral. Indeed, its definition could be different from $\int_0^T \tilde{f}(\tilde{Y}_u) d^{\mathrm{NC}, m} X_u$ even if $f(Y) = \tilde{f}(\tilde{Y})$. This is why we call it "*functional*" instead of "*integral*".

• The terminology "*Newton-Cotes* functional" is due to the fact that the definition of ν_m via (4) is related to the Newton-Cotes formula of numerical analysis. Indeed, ν_m is the unique discrete measure carried by the numbers $j/(2m - 2)$ which coincides with Lebesgue measure on all polynomials of degree smaller than $2m - 1$.

We have the following change of variable formula.

Theorem 2.6 *(see [15], p. 793). Let* $m \geq 1$ *be an integer. The* m*-order Newton-Cotes functional* $\int_0^T f(B_u) d^{\mathrm{NC}, m} B_u$ *exists for any* $f : \mathbb{R} \to \mathbb{R}$ *of class* C^{4m+1} *if and only if* $H \in (1/(4m + 2), 1)$*. In this case, we have, for any antiderivative* F *of* f*:*

$$F(B_T) = F(0) + \int_0^T f(B_u) d^{\mathrm{NC}, m} B_u. \qquad (5)$$

Remark 2.7 An immediate consequence of this result is that

$$\int_0^T f(B_u) d^{\mathrm{NC}, m} B_u = \int_0^T f(B_u) d^{\mathrm{NC}, n} B_u = F(B_T) - F(0)$$

when $m > n$, f is C^{4m+1} and $H \in (1/(4n+2), 1)$. Then, for f regular enough, it is possible to define the so-called *Newton-Cotes functional* $\int_0^T f(B_u)d^{NC}B_u$ without ambiguity by:

$$\int_0^T f(B_u)d^{NC}B_u := \int_0^T f(B_u)d^{NC,n}B_u \text{ if } H \in (1/(4n+2), 1). \quad (6)$$

In the sequel, we put $n_H = \inf\{n \geq 1 : H > 1/(4n+2)\}$. An immediate consequence of (5) and (6) is that, for any $H \in (0,1)$ and any $f : \mathbb{R} \to \mathbb{R}$ of class C^{4n_H+1}, we have:

$$F(B_T) = F(0) + \int_0^T f(B_u)d^{NC}B_u,$$

where F is an antiderivative of f.

To specify the sense of $\int_0^t \sigma(X_s)dB_s$ in (2), it now seems natural to try and use the Newton-Cotes functional. But for the time being we are only able to consider integrands of the form $f(B)$ with $f : \mathbb{R} \to \mathbb{R}$ regular enough, see (6). That is why we first choose the following definition for a possible solution to (2):

Definition 2.8 *Assume that $\sigma \in C^{4n_H+1}$ and that $b \in C^0$.*
i) Let $\mathfrak{C}_1$ be the class of processes $X : [0,T] \times \Omega \to \mathbb{R}$ verifying that there exists $f : \mathbb{R} \to \mathbb{R}$ belonging to C^{4n_H+1} and such that, for every $t \in [0,T]$, $X_t = f(B_t)$ a.s.
ii) A process $X : [0,T] \times \Omega \to \mathbb{R}$ is a solution to (2) if:

- $X \in \mathfrak{C}_1$,
- $\forall t \in [0,T], X_t = x_0 + \int_0^t \sigma(X_s)d^{NC}B_s + \int_0^t b(X_s)ds.$

Remark 2.9 Note that the first point of definition *ii)* allows to ensure that the integral $\int_0^t \sigma(X_s)d^{NC}B_s$ makes sense (compare with the adaptedness condition in the Itô context).

We can now state the following result.

Theorem 2.10 *Let $\sigma \in C^{4n_H+1}$ be a Lipschitz function, b be a continuous function and x_0 be a real. Then equation (2) admits a solution X in the sense of Definition 2.8 if and only if b vanishes on $\mathfrak{S}(\mathbb{R})$, where $\mathfrak{S}$ is the unique solution to $\mathfrak{S}' = \sigma \circ \mathfrak{S}$ with initial value $\mathfrak{S}(0) = x_0$. In this case, X is unique and is given by $X_t = \mathfrak{S}(B_t)$.*

Remark 2.11 As a consequence of the mean value theorem, $\mathfrak{S}(\mathbb{R})$ is an interval. Moreover, it is easy to see that either $\mathfrak{S}$ is constant or $\mathfrak{S}$ is strictly monotone, and that $\inf \mathfrak{S}(\mathbb{R})$ and $\sup \mathfrak{S}(\mathbb{R})$ are elements of $\{\sigma = 0\} \cup \{\pm\infty\}$. In particular, if σ does not vanish, then $\mathfrak{S}(\mathbb{R}) = \mathbb{R}$ and an immediate consequence of Theorem 2.10 is that (2) admits a solution in the sense of Definition 2.8 if and only if $b \equiv 0$.

Proof of Theorem 2.10. Assume that $X_t = f(B_t)$ is a solution to (2) in the sense of Definition 2.8. Then

$$f(B_t) = x_0 + \int_0^t \sigma \circ f(B_s) d^{\mathrm{NC}} B_s + \int_0^t b \circ f(B_s) ds$$

$$= G(B_t) + \int_0^t b \circ f(B_s) ds, \tag{7}$$

where G is the antiderivative of $\sigma \circ f$ verifying $G(0) = x_0$. Set $h = f - G$ and denote by Ω^* the set of $\omega \in \Omega$ such that $t \mapsto B_t(\omega)$ is differentiable at least one point $t_0 \in [0, T]$ (it is well-known that $\mathrm{P}(\Omega^*) = 0$). If $h'(B_{t_0}(\omega)) \neq 0$ for one $(\omega, t_0) \in \Omega \times [0, T]$ then h is strictly monotone in a neighborhood of $B_{t_0}(\omega)$ and, for $|t - t_0|$ sufficiently small, one has $B_t(\omega) = h^{-1}(\int_0^t b(X_s(\omega)) ds)$ and, consequently, $\omega \in \Omega^*$. Then, a.s., $h'(B_t) = 0$ for all $t \in [0, T]$, so that $h \equiv 0$. By uniqueness, one deduces $f = \mathfrak{S}$. Thus, if (2) admits a solution X in the sense of Definition 2.8, one necessarily has $X_t = \mathfrak{S}(B_t)$. Thanks to (7), one then has $b \circ \mathfrak{S}(B_t) = 0$ for all $t \in [0, T]$ a.s. and then b vanishes on $\mathfrak{S}(\mathbb{R})$. $\qquad \square$

Consequently, when the SDE (2) has no drift b, there is a natural solution. But what can we do when $b \not\equiv 0$?

Denote by $\mathscr{A}$ the set of processes $A : [0, T] \times \Omega \to \mathbb{R}$ having C^1-trajectories and verifying $\mathrm{E}\big(e^{\lambda \int_0^T A_s^2 ds}\big) < \infty$ for at least one $\lambda > 1$.

Lemma 2.12 *Let $A \in \mathscr{A}$ and $m \in \mathbb{N}^*$. Then $\int_0^T f(B_u + A_u) d^{\mathrm{NC}, m} B_u$ exists for any $f : \mathbb{R} \to \mathbb{R}$ of class C^{4m+1} if and only if $H > 1/(4m+2)$. In this case, for any antiderivative F of f, one has:*

$$F(B_T + A_T) = F(A_0) + \int_0^T f(B_u + A_u) d^{\mathrm{NC}, m} B_u + \int_0^T f(B_u + A_u) A_u' du.$$

Proof. Set $\tilde{B} = B + A$. On the one hand, using the Girsanov theorem in [28] and taking into account the assumption on A, we have that $\tilde{B}$ is a fBm of index H under some probability $\mathbb{Q}$ equivalent to the initial probability $\mathbb{P}$. On the other hand, it is easy, by going back to Definition 2.4, to prove that $\int_0^T f(B_u + A_u) d^{\mathrm{NC}, m} B_u$ exists if and only if $\int_0^T f(B_u + A_u) d^{\mathrm{NC}, m} (B_u + A_u)$ does, and in this case, one has

$$\int_0^T f(B_u + A_u) d^{\mathrm{NC}, m} (B_u + A_u) = \int_0^T f(B_u + A_u) d^{\mathrm{NC}, m} B_u + \int_0^T f(B_u + A_u) A_u' du.$$

Then, since convergence under $\mathbb{Q}$ or under $\mathbb{P}$ is equivalent, the conclusion of Lemma 2.12 is a direct consequence of Theorem 2.6. $\qquad \square$

Then, as previously, it is possible to define a functional (still called *Newton-Cotes functional*) verifying, for any $H \in (0, 1)$, for any $f : \mathbb{R} \to \mathbb{R}$ of class

C^{4n_H+1} and any process $A \in \mathscr{A}$:

$$F(B_T + A_T) = F(A_0) + \int_0^T f(B_u + A_u)d^{NC}B_u + \int_0^T f(B_u + A_u)A'_u du,$$

where F is an antiderivative of f.

Now, we can introduce an other definition of a solution to (2):

Definition 2.13 *Assume that $\sigma \in C^{4n_H+1}$ and that $b \in C^0$.*
i) Let $\mathfrak{C}_2$ be the class of processes $X : [0,T] \times \Omega \to \mathbb{R}$ such that there exist a function $f : \mathbb{R} \to \mathbb{R}$ in C^{4n_H+1} and a process $A \in \mathscr{A}$ such that $A_0 = 0$ and, for every $t \in [0,T]$, $X_t = f(B_t + A_t)$ a.s.
ii) A process $X : [0,T] \times \Omega \to \mathbb{R}$ is a solution to (2) if:

- $X \in \mathfrak{C}_2$,
- $\forall t \in [0,T], X_t = x_0 + \int_0^t \sigma(X_s)d^{NC}B_s + \int_0^t b(X_s)ds.$

Theorem 2.14 *Let $\sigma \in C^{4n_H+1}$ be a Lipschitz function, b be a continuous function and x_0 be a real.*

- *If $\sigma(x_0) = 0$ then (2) admits a solution X in the sense of Definition 2.13 if and only if $b(x_0) = 0$. In this case, X is unique and is given by $X_t \equiv x_0$.*
- *If $\sigma(x_0) \neq 0$, then (2) admits a solution. If moreover $\inf_\mathbb{R} |\sigma| > 0$ and $b \in \text{Lip}$, this solution is unique.*

Proof. Assume that $X = f(B + A)$ is a solution to (2) in the sense of Definition 2.13. Then, we have

$$f(B_t + A_t) = G(B_t + A_t) - \int_0^t \sigma(X_s)A'_s ds + \int_0^t b(X_s)ds \qquad (8)$$

where G is the antiderivative of $\sigma \circ f$ verifying $G(0) = x_0$. As in the proof of Theorem 2.10, we obtain that $f = \mathfrak{S}$ where $\mathfrak{S}$ is defined by $\mathfrak{S}' = \sigma \circ \mathfrak{S}$ with initial value $\mathfrak{S}(0) = x_0$. Thanks to (8), we deduce that, a.s., we have $b \circ \mathfrak{S}(B_t + A_t) = \sigma \circ \mathfrak{S}(B_t + A_t) A'_t$ for all $t \in [0,T]$. Consequently:

- If $\sigma(x_0) = 0$ then $\mathfrak{S} \equiv x_0$ and $b(x_0) = 0$.
- If $\sigma(x_0) \neq 0$ then $\mathfrak{S}$ is strictly monotone and the ordinary integral equation

$$A_t = \int_0^t \frac{b \circ \mathfrak{S}}{\mathfrak{S}'}(B_s + A_s)\, ds$$

admits a maximal (in fact, global since we know already that A is defined on $[0,T]$) solution by Peano's theorem. If moreover $\inf_\mathbb{R} |\sigma| > 0$ and $b \in \text{Lip}$ then $\frac{b \circ \mathfrak{S}}{\mathfrak{S}'} = \frac{b \circ \mathfrak{S}}{\sigma \circ \mathfrak{S}} \in \text{Lip}$ and A is uniquely determined. $\square$

The previous theorem is not quite satisfactory because of the prominent role played by x_0. That is why we will finally introduce a last definition for a solution to (2). We first need an analogue of Theorem 2.6 and Lemma 2.12:

Theorem 2.15 *(see [23], Chapter 4). Let A be a process having C^1-trajectories and $m \geq 1$ be an integer. If $H > 1/(2m+1)$ then the m-order Newton-Cotes functional $\int_0^T f(B_u, A_u)d^{\mathrm{NC},m}B_u$ exists for any $f : \mathbb{R}^2 \to \mathbb{R}$ of class $\mathrm{C}^{2m,1}$. In this case, we have, for any function $F : \mathbb{R}^2 \to \mathbb{R}$ verifying $F_b' = f$:*

$$F(B_T, A_T) = F(0, A_0) + \int_0^T f(B_u, A_u)d^{\mathrm{NC},m}B_u + \int_0^T F_a'(B_u, A_u)A_u' du.$$

Remark 2.16 • F_a' (resp. F_b') means the derivative of F with respect to a (resp. b).

- The condition is here $H > 1/(2m+1)$ and not $H > 1/(4m+2)$ as in Theorem 2.6 and Lemma 2.12. Thus, for instance, if $A \in \mathscr{A}$, if $g : \mathbb{R} \to \mathbb{R}$ is C^5 and if $h : \mathbb{R}^2 \to \mathbb{R}$ is $\mathrm{C}^{5,1}$ then $\int_0^T g(B_s + A_s)d^\circ B_s$ exists if (and only if) $H > 1/6$ while $\int_0^T h(B_s, A_s)d^\circ B_s$ exists *a priori* only when $H > 1/3$.
- We define $m_H = \inf\{m \geq 1 : H > 1/(2m+1)\}$. As in the Remark 2.7, it is possible to consider, for any $H \in (0,1)$ and without ambiguity, a functional (still called *Newton-Cotes functional*) which verifies, for any $f : \mathbb{R}^2 \to \mathbb{R}$ of class $\mathrm{C}^{2m_H,1}$ and any process A having C^1-trajectories:

$$F(B_T, A_T) = F(0, A_0) + \int_0^T f(B_u, A_u)d^{\mathrm{NC}}B_u + \int_0^T F_a'(B_u, A_u)A_u' du,$$

where F is such that $F_b' = f$.

Finally, we introduce our last definition for a solution to (2):

Definition 2.17 *Assume that $\sigma \in \mathrm{C}^{2m_H}$ and $b \in \mathrm{C}^0$.*
i) Let $\mathfrak{C}_3$ be the class of processes $X : [0,T] \times \Omega \to \mathbb{R}$ verifying that there exist a function $f : \mathbb{R}^2 \to \mathbb{R}$ of class $\mathrm{C}^{2m_H,1}$ and a process $A : [0,T] \times \Omega \to \mathbb{R}$ having C^1-trajectories such that $A_0 = 0$ and verifying, for every $t \in [0,T]$, $X_t = f(B_t, A_t)$ a.s.
ii) A process $X : [0,T] \times \Omega \to \mathbb{R}$ is a solution to (2) if:

- $X \in \mathfrak{C}_3$,
- $\forall t \in [0,T]$, $X_t = x_0 + \int_0^t \sigma(X_s)d^{\mathrm{NC}}B_s + \int_0^t b(X_s)ds$.

Theorem 2.18 *Let $\sigma \in \mathrm{C}_b^2$, b be a Lipschitz function and x_0 be a real. Then the equation (2) admits a solution X in the sense of Definition 2.17. Moreover, if σ is analytic, then X is the unique solution of the form $f(B, A)$ with f analytic (resp. of class C^1) in the first (resp. second) variable and A a process having C^1-trajectories and verifying $A_0 = 0$.*

Remark 2.19 • If $H > 1/3$, one can improve Theorem 2.18. Indeed, as shown in [26], uniqueness holds without any supplementary condition on σ. Moreover, in that reference, another meaning to (2) than Definition 2.17 is given, using the concept of Lévy area.

- In [25], one studies the problem of absolute continuity in equation (2), where the solution is in the sense of Definition 2.17. It is proved that, if $\sigma(x_0) \neq 0$, then $\mathcal{L}(X_t)$ is absolutely continuous with respect to the Lebesgue measure for all $t \in]0,T]$. More precisely, the Bouleau-Hirsch criterion is shown to hold: if $x_t = x_0 + \int_0^t b(x_s)ds$ and $t_x = \sup\{t \in [0,T] : x_t \notin \mathrm{Int}\,J\}$ where $J = \sigma^{-1}(\{0\})$ then $\mathcal{L}(X_t)$ is absolutely continuous if and only if $t > t_x$.
- We already said that, among the m-order Newton-Cotes functionals, only the first one (that is, the symmetric integral, defined by (3)) is a "true" integral. For this integral, the main results contained in this paper are summarized in the following table (where f denotes a regular enough function and A a process having C^1-trajectories):

Table 1. *Existence and uniqueness in SDE* $X_t = x_0 + \int_0^t \sigma(X_s)d^\circ B_s + \int_0^t b(X_s)ds$

If we use Definition	we have to choose $H \in$	X is then of the form	we have existence if	and uniqueness if moreover	See Theorem
2.8	$(1/6,1)$	$f(B)$	$\sigma \in \mathrm{C}^5 \cap \mathrm{Lip}$, $b \in \mathrm{C}^0$ and $b_{|\mathfrak{S}(\mathbb{R})} \equiv 0$	-	2.10
2.13	$(1/6,1)$	$f(B+A)$	$\sigma \in \mathrm{C}^5 \cap \mathrm{Lip}$, $b \in \mathrm{C}^0 +$ $i)\,\sigma(x_0) = 0$ $b(x_0) = 0$ or $ii)\,\sigma(x_0) \neq 0$	$i) -$ $ii)\, \inf_{\mathbb{R}} |\sigma| > 0$ and $b \in \mathrm{Lip}$	2.14
2.17	$(1/3,1)$	$f(B,A)$	$\sigma \in \mathrm{C}_b^2$ and $b \in \mathrm{Lip}$	-	2.18 and [25]

Proof of Theorem 2.18. Let us remark that the classical Doss-Sussmann [9,33] method gives a natural solution X of the form $f(B,A)$. Then, in the remainder of the proof, we will concentrate on the uniqueness. Assume that $X = f(B,A)$ is a solution to (2) in the sense of Definition 2.17. On the one hand, we have

$$X_t = x_0 + \int_0^t \sigma(X_s)d^{\mathrm{NC}}B_s + \int_0^t b(X_s)ds \qquad (9)$$

$$= x_0 + \int_0^t \sigma \circ f(B_s, A_s)d^{\mathrm{NC}}B_s + \int_0^t b \circ f(B_s, A_s)ds.$$

On the other hand, using the change of variables formula, we can write

$$X_t = x_0 + \int_0^t f_b'(B_s, A_s)d^{\mathrm{NC}}B_s + \int_0^t f_a'(B_s, A_s)A_s'ds. \qquad (10)$$

Using (9) and (10), we deduce that $t \mapsto \int_0^t \varphi(B_s, A_s) d^{NC} B_s$ has C^1-trajectories where $\varphi := f_b' - \sigma \circ f$. As in the proof of Theorem 2.14, we show that, a.s.,

$$\forall t \in]0, T[, \; \varphi(B_t, A_t) = 0. \tag{11}$$

Similarly, we can obtain that, a.s.,

$$\forall k \in \mathbb{N}, \forall t \in]0, T[, \; \frac{\partial^k \varphi}{\partial b^k}(B_t, A_t) = 0.$$

If σ and $f(., y)$ are analytic, then $\varphi(., y)$ is analytic and

$$\forall t \in]0, T[, \forall x \in \mathbb{R}, \; \varphi(x, A_t) = f_b'(x, A_t) - \sigma \circ f(x, A_t) = 0. \tag{12}$$

By uniqueness, we deduce

$$\forall t \in [0, T], \forall x \in \mathbb{R}, \; f(x, A_t) = u(x, A_t),$$

where u is the unique solution to $u_b' = \sigma(u)$ with initial value $u(0, y) = y$ for any $y \in \mathbb{R}$. In particular, we obtain a.s.

$$\forall t \in [0, T], \; X_t = f(B_t, A_t) = u(B_t, A_t). \tag{13}$$

Identity (9) can then be rewritten as:

$$X_t = x_0 + \int_0^t \sigma \circ u(B_s, A_s) d^{NC} B_s + \int_0^t b \circ u(B_s, A_s) ds,$$

while the change of variables formula yields:

$$X_t = x_0 + \int_0^t u_b'(B_s, A_s) d^{NC} B_s + \int_0^t u_a'(B_s, A_s) A_s' ds.$$

Since $u_b' = \sigma \circ u$, we obtain a.s.:

$$\forall t \in [0, T], \; b \circ u(B_t, A_t) = u_a'(B_t, A_t) A_t'. \tag{14}$$

But we have existence and uniqueness in (14). Then the proof of Theorem is done. $\qquad\square$

3 Convergence or not of the canonical approximating schemes associated to SDE (2) when $d = d^\circ$

Approximating schemes for stochastic differential equations (2) have already been studied only in few articles. The first work in that direction has been proposed by Lin [19] in 1995. When $H > 1/2$, he showed that the Euler approximation of equation (2) converges uniformly in probability–but only in the easier case when $\sigma(X_t)$ is replaced by $\sigma(t)$, that is, in the additive case. In 2005, I introduced in [24] (see also Talay [34]) some approximating schemes for the analogue of (2) where B is replaced by a Hölder continuous function of order α, for any $\alpha \in (0,1)$. I determined upper error bounds and, in particular, my results apply almost surely when the driving Hölder continuous function is a path of the fBm B, for any Hurst index $H \in (0,1)$.

Results on lower error bounds are available only since very recently: see Neuenkirch [21] for the additive case, and Neuenkirch and Nourdin [22] (see also Gradinaru and Nourdin [14]) for equation (2). In [22], it is proved that the Euler scheme $\overline{X} = \{\overline{X}^{(n)}\}_{n \in \mathbb{N}^*}$ associated to (2) verifies, under classical assumptions on σ and b and when $H \in (\frac{1}{2}, 1)$, that

$$n^{2H-1}\big[\overline{X}_1^{(n)} - X_1\big] \xrightarrow{\text{a.s.}} -\frac{1}{2}\int_0^1 \sigma'(X_s)D_sX_1 ds, \quad \text{as } n \to \infty, \qquad (15)$$

where X is the solution given by Theorem 2.18 and DX its Malliavin derivative with respect to B. Still in [22], it is proved that, for the so-called Crank-Nicholson scheme $\widehat{X} = \{\widehat{X}^{(n)}\}_{n \in \mathbb{N}^*}$ associated to (2) with $b = 0$ and defined by

$$\begin{cases} \widehat{X}_0^n = x \\ \widehat{X}_{(k+1)/n}^n = \widehat{X}_{k/n}^n + \frac{1}{2}\big(\sigma(\widehat{X}_{k/n}^n) + \sigma(\widehat{X}_{(k+1)/n}^n)\big)(B_{(k+1)/n} - B_{k/n}), \\ \hspace{6cm} k \in \{0, \ldots, n-1\}, \end{cases} \qquad (16)$$

we have, for σ regular enough and when $H \in (\frac{1}{3}, \frac{1}{2})$:

$$\text{for any } \alpha < 3H - 1/2, \quad n^\alpha\big[\widehat{X}_1^n - X_1\big] \xrightarrow{\text{Prob}} 0 \text{ as } n \to \infty, \qquad (17)$$

where X is the solution given by Theorem 2.10. Of course, this result does not give the exact rate of convergence but only an upper bound. However, when the diffusion coefficient σ verifies

$$\sigma(x)^2 = \alpha x^2 + \beta x + \gamma \text{ for some } \alpha, \beta, \gamma \in \mathbb{R}, \qquad (18)$$

the exact rate of convergence can be derived: indeed, in this case, we have

$$n^{3H-1/2}\big[\widehat{X}_1^n - X_1\big] \xrightarrow{\text{Law}} \frac{\alpha}{12}\sigma(X_1)\,G, \quad \text{as } n \to \infty, \qquad (19)$$

with G a centered Gaussian random variable independent of X_1, whose variance depends only on H. Note also that, in [14], the exact rate of convergence associated to the schemes introduced in [24] are computed and results of the type (17)–(19) are obtained.

In this section, we are interested in whether scheme (16) converges, according to the value of H and the expression of σ. First of all, this problem looks easier than computing the exact rate of convergence, as in [14,22]. But, in these two papers, no optimality is sought in the domain of validity of H. For instance, in (17), we impose that $H > 1/3$ although it seems more natural to only assume that $H > 1/6$.

Unfortunately, we were able to find the exact barrier of convergence for (16) only for particular σ, namely those which verify (18). In this case, we prove in Theorem 3.1 below that the barrier of convergence is $H = 1/6$. In the other cases, it is nevertheless possible to prove that the scheme (16) converge when $H > 1/3$ (see the proof of Theorem 3.1). But the exact barrier remains an open question.

The class (18) is quite restricted. In particular, I must acknowledge that Theorem 3.1 has a limited interest. However, its proof is instructive. Moreover it contains a useful formula for $\widehat{X}^{(n)}_{k/n}$ (see Lemma 3.4), which is the core of all the results concerning the Crank-Nicholson scheme proved in [22] (see also [14]).

Now, we state the main result of this section:

Theorem 3.1 *Assume that $\sigma \in C^1(\mathbb{R})$ verifies (18). Then the sequence $\{\widehat{X}^{(n)}_1\}$ defined by (16) converges in L^2 if and only if $H > 1/6$. In this case, the limit is the unique solution at time 1 to the SDE $X_t = x_0 + \int_0^t \sigma(X_s)d^\circ B_s$, in the sense of Definition 2.8 and given by Theorem 2.10.*

Remark 3.2 When $\sigma(x) = x$ it is easy to understand why $\widehat{X}^{(n)}_1$ converges in L^2 if and only if $H > 1/6$. Indeed, setting $\Delta_k^n = B_{(k+1)/n} - B_{k/n}$, we have

$$\widehat{X}^{(n)}_1 = x_0 \prod_{k=0}^{n-1} \frac{1 + \frac{1}{2}\Delta_k^n}{1 - \frac{1}{2}\Delta_k^n} = x_0 \exp\left\{ \sum_{k=0}^{n-1} \ln \frac{1 + \frac{1}{2}\Delta_k^n}{1 - \frac{1}{2}\Delta_k^n} \right\};$$

but

$$\ln \frac{1 + \frac{1}{2}\Delta_k^n}{1 - \frac{1}{2}\Delta_k^n} = \Delta_k^n + \frac{1}{12}(\Delta_k^n)^3 + \frac{1}{80}(\Delta_k^n)^5 + O((\Delta_k^n)^6),$$

and, because $\sum_{k=0}^{n-1} \Delta_k^n = B_1$ and by using Lemma 3.3 below, one has that $\widehat{X}^{(n)}_1$ converges if and only if $H > 1/6$ and that, in this case, the limit is $x_0 \exp(B_1)$.

As a preliminary of the proof of Theorem 3.1, we need two lemmas:

Lemma 3.3 *Let $m \geq 1$ be an integer.*

- *We have*

$$\sum_{k=0}^{n-1} (B_{(k+1)/n} - B_{k/n})^{2m} \text{ converges in } L^2 \text{ as } n \to \infty \text{ if and only if } H \geq \frac{1}{2m}.$$

  In this case, the limit is zero if $H > 1/2m$ and is $(2m)!/(2^m m!)$ if $H = 1/2m$.

- *We have*

$$\sum_{k=0}^{n-1}(B_{(k+1)/n}-B_{k/n})^{2m+1} \text{ converges in } L^2 \text{ as } n \to \infty \text{ if and only if}$$

$$H > \frac{1}{4m+2}.$$

In this case, the limit is zero.

Proof of Lemma 3.3. The first point is an obvious consequence of the well-known convergence

$$n^{2mH-1}\sum_{k=0}^{n-1}(B_{(k+1)/n}-B_{k/n})^{2m} \xrightarrow{L^2} (2m)!/(2^m m!), \text{ as } n \to \infty.$$

Let us then prove the second point. On the one hand, for $H > 1/(4m+2)$, we can prove directly that

$$\sum_{k,\ell=0}^{n-1} E[(B_{(k+1)/n}-B_{k/n})^{2m+1}(B_{(\ell+1)/n}-B_{\ell/n})^{2m+1}] \longrightarrow 0, \text{ as } n \to \infty,$$

by using a Gaussian linear regression, see for instance [16], Proposition 3.8. On the other hand, it is well known that, when $H < 1/2$,

$$n^{(2m+1)H-1/2}\sum_{k=0}^{n-1}(B_{(k+1)/n}-B_{k/n})^{2m+1} \xrightarrow{\mathcal{L}} N(0,\sigma^2_{m,H}), \text{ as } n \to \infty,$$

for some $\sigma_{m,H} > 0$

(use, for instance, the main result by Nualart and Peccati [29]). We can then deduce the non-convergence when $H \le 1/(4m+2)$ as in [15], Proof of 2(c), page 796. □

Lemma 3.4 *Assume that $\sigma \in C^5(\mathbb{R})$ is bounded together with its derivatives. Consider ϕ the flow associated to σ, that is, $\phi(x,\cdot)$ is the unique solution to $y' = \sigma(y)$ with initial value $y(0) = x$. Then we have, for any $\ell \in \{0,1,\ldots,n\}$:*

$$\widehat{X}^{(n)}_{\ell/n} = \phi\bigg(x_0, B_{\ell/n} + \sum_{k=0}^{\ell-1}f_3(\widehat{X}^{(n)}_{k/n})(\Delta^n_k)^3 + \sum_{k=0}^{\ell-1}f_4(\widehat{X}^{(n)}_{k/n})(\Delta^n_k)^4$$

$$+ \sum_{k=0}^{\ell-1}f_5(\widehat{X}^{(n)}_{k/n})(\Delta^n_k)^5 + O(n\Delta^6(B))\bigg). \tag{20}$$

Here we set

$$f_3 = \frac{(\sigma^2)''}{24}, \; f_4 = \frac{\sigma(\sigma^2)'''}{48} \; \text{and} \; f_5 = \frac{\sigma'^4}{80} + \frac{\sigma^2\sigma'\sigma'''}{15} + \frac{3\sigma\sigma'^2\sigma''}{40} + \frac{\sigma^2\sigma''^2}{20} + \frac{\sigma^3\sigma^{(4)}}{80},$$

$$\Delta_k^n = B_{(k+1)/n} - B_{k/n}, \; \text{when} \; n \in \mathbb{N} \; \text{and} \; k \in \{0, 1, \ldots, n-1\}$$

and

$$\Delta^p(B) = \max_{k=0,\ldots,n-1} |(\Delta_k^n)^p|, \; \text{when} \; p \in \mathbb{N}^*.$$

Proof of Lemma 3.4. Assume, for an instant, that σ does not vanish. In this case, $\phi(x, \cdot)$ is a bijection from $\mathbb{R}$ to himself for any x and we can consider $\varphi(x, \cdot)$ such that

$$\forall x, t \in \mathbb{R} : \; \varphi(x, \phi(x,t)) = t \; \text{and} \; \phi(x, \varphi(x,t)) = t. \tag{21}$$

On the one hand, thanks to (21), it is a little long but easy to compute that

$$\begin{aligned}
\varphi(x,x) &= 0, \\
\varphi_t'(x,x) &= 1/\sigma(x), \\
\varphi_{tt}''(x,x) &= [-\sigma'/\sigma^2](x), \\
\varphi_{ttt}^{(3)}(x,x) &= [(2\sigma'^2 - \sigma\sigma'')/\sigma^3](x), \\
\varphi_{tttt}^{(4)}(x,x) &= [(-6\sigma'^3 + 6\sigma\sigma'\sigma'' - \sigma^2\sigma''')/\sigma^4](x) \\
\varphi_{ttttt}^{(5)}(x,x) &= [(24\sigma'^4 - 36\sigma\sigma'^2\sigma'' + 8\sigma^2\sigma'\sigma''' + 6\sigma^2\sigma''^2 - \sigma^3\sigma^{(4)})/\sigma^5](x).
\end{aligned}$$

Then, for u sufficiently small, we have

$$\begin{aligned}
\varphi(x,x+u) &= \tfrac{1}{\sigma}(x)u - \tfrac{\sigma'}{2\sigma^2}(x)u^2 + \tfrac{2\sigma'^2 - \sigma\sigma''}{6\sigma^3}(x)u^3 + \tfrac{-6\sigma'^3 + 6\sigma\sigma'\sigma'' - \sigma^2\sigma'''}{24\sigma^4}(x)u^4 \\
&\quad + \tfrac{24\sigma'^4 - 36\sigma\sigma'^2\sigma'' + 8\sigma^2\sigma'\sigma''' + 6\sigma^2\sigma''^2 - \sigma^3\sigma^{(4)}}{\sigma^5}(x)u^5 + O(u^6).
\end{aligned}$$

On the other hand, using (16) and some basic Taylor expansions, one has for $k \in \{0, 1, \ldots, n-1\}$:

$$\begin{aligned}
\widehat{X}_{(k+1)/n}^{(n)} &= \widehat{X}_{k/n}^{(n)} + \sigma(\widehat{X}_{k/n}^{(n)})\Delta_k^n + \tfrac{\sigma\sigma'}{2}(\widehat{X}_{k/n}^{(n)})(\Delta_k^n)^2 + \tfrac{\sigma\sigma'^2 + \sigma^2\sigma''}{4}(\widehat{X}_{k/n}^{(n)})(\Delta_k^n)^3 \\
&\quad + \left(\tfrac{\sigma\sigma'^3}{8} + \tfrac{3\sigma^2\sigma'\sigma''}{8} + \tfrac{\sigma^3\sigma'''}{12}\right)(\widehat{X}_{k/n}^{(n)})(\Delta_k^n)^4 \\
&\quad + \left(\tfrac{\sigma\sigma'^4}{16} + \tfrac{3\sigma^2\sigma'^2\sigma''}{8} + \tfrac{\sigma^3\sigma'\sigma'''}{6} + \tfrac{\sigma^3\sigma''^2}{8} + \tfrac{\sigma^4\sigma^{(4)}}{48}\right)(\widehat{X}_{k/n}^{(n)})(\Delta_k^n)^5 \\
&\quad + O(\Delta^6(B)).
\end{aligned}$$

Then, we have

$$\begin{aligned}
&\varphi(\widehat{X}_{k/n}^{(n)}, \widehat{X}_{(k+1)/n}^{(n)}) \\
&= \varphi(\widehat{X}_{k/n}^{(n)}, \widehat{X}_{k/n}^{(n)} + [\widehat{X}_{(k+1)/n}^{(n)} - \widehat{X}_{k/n}^{(n)}]) \\
&= \Delta_k^n + \tfrac{\sigma'^2 + \sigma\sigma''}{12}(\widehat{X}_{k/n}^{(n)})(\Delta_k^n)^3 + \left(\tfrac{\sigma\sigma'\sigma''}{8} + \tfrac{\sigma^2\sigma'''}{24}\right)(\widehat{X}_{k/n}^{(n)})(\Delta_k^n)^4 \\
&\quad + \left(\tfrac{\sigma'^4}{80} + \tfrac{\sigma^2\sigma'\sigma'''}{15} + \tfrac{3\sigma\sigma'^2\sigma''}{40} + \tfrac{\sigma^2\sigma''^2}{20} + \tfrac{\sigma^3\sigma^{(4)}}{80}\right)(\widehat{X}_{k/n}^{(n)})(\Delta_k^n)^5 + O(\Delta^6(B)) \\
&= \Delta_k^n + f_3(\widehat{X}_{k/n}^{(n)})(\Delta_k^n)^3 + f_4(\widehat{X}_{k/n}^{(n)})(\Delta_k^n)^4 + f_5(\widehat{X}_{k/n}^{(n)})(\Delta_k^n)^5 + O(\Delta^6(B)).
\end{aligned}$$

We deduce, using (21):

$$\widehat{X}^{(n)}_{(k+1)/n} = \phi\big(\widehat{X}^{(n)}_{k/n}, \Delta^n_k + f_3(\widehat{X}^{(n)}_{k/n})(\Delta^n_k)^3 + f_4(\widehat{X}^{(n)}_{k/n})(\Delta^n_k)^4$$
$$+ f_5(\widehat{X}^{(n)}_{k/n})(\Delta^n_k)^5 + O(\Delta^6(B))\big).$$

Finally, by using the semi-group property verified by ϕ, namely

$$\forall x, s, t \in \mathbb{R} : \phi(\phi(x,t), s) = \phi(x, t+s).$$

we easily deduce (20).

In fact, we assumed that σ does not vanish only for having the possibility to introduce φ. But (20) is an algebraic formula then it is also valid for general σ, as soon as it is bounded together with its derivatives. $\square$

Proof of Theorem 3.1. Assume that σ verifies (18). Although σ is not bounded in general, it is easy to verify that we still have $O(n\Delta^6(B))$ as remainder in (20). Moreover, simple but tedious computations show that we can simplify in (20) to obtain

$$\widehat{X}^{(n)}_1 = \phi\big(x_0, B_1 + \tfrac{\alpha}{12} \sum_{k=0}^{n-1}(\Delta^n_k)^3 + \tfrac{\alpha^2}{80} \sum_{k=0}^{n-1}(\Delta^n_k)^5 + O(n\Delta^6(B))\big).$$

Thus, as a conclusion of Lemma 3.3, we obtain easily that $\widehat{X}^{(n)}_1$ converges to $\phi(x_0, B_1)$ if and only if $H > 1/6$. $\square$

Acknowledgement. I am indebted to the anonymous referee for the careful reading of the original manuscript and for a number of suggestions.

References

1. Alos, E., Mazet, O., Nualart, D. *Stochastic calculus with respect to fractional Brownian motion with Hurst parameter less than* $\frac{1}{2}$. Stochastic Process. Appl. **86** (2000), 121-139.
2. Alos, E., Leon, J.A., Nualart, D. *Stratonovich calculus for fractional Brownian motion with Hurst parameter less than* $\frac{1}{2}$. Taiwanese J. Math. **5** (2001), 609-632.
3. Bender, C. *An Itô formula for generalized functionals of a fractional Brownian motion with arbitrary Hurst parameter.* Stochastic Process. Appl. **104** (2003), 81-106.
4. Baudoin, F., Coutin, L. *Etude en temps petit du flot d'équations conduites par des mouvements browniens fractionnaires.* C.R. Math. Acad. Sci. Paris I **341** (2005), 39-42.
5. Boufoussi, B., Tudor, C.A. *Kramers-Smoluchowski approximation for stochastic equations with fBm.* Rev. Roumaine Math. Pures Appl. **50** (2005), 125-136.
6. Carmona, P., Coutin, L. *Intégrales stochastiques pour le mouvement brownien fractionnaire.* C.R. Math. Acad. Sci. Paris I **330** (2000), 213-236.
7. Coutin, L., Qian, Z. *Stochastic analysis, rough path analysis and fractional Brownian motions.* Probab. Theory Related Fields **122** (2002), no. 1, 108-140.

8. Decreusefond, L., Ustunel, A.S. *Stochastic analysis of the fractional Brownian motion.* Potential Anal. **10** (1998), 177-214.
9. Doss, H. *Liens entre équations différentielles stochastiques et ordinaires.* Ann. Inst. H. Poincaré Sect. B **13** (1977), 99-125.
10. Duncan, T.E., Hu, Y., Pasik-Duncan, B. *Stochastic calculus for fractional Brownian motion I. Theory.* SIAM J. Control Optim. **38** (2000), 582-612.
11. Errami, M., Russo, F. *n-covariation and symmetric SDEs driven by finite cubic variation process.* Stochastic Process. Appl. **104** (2003), 259-299.
12. Feyel, D., De La Pradelle, A. *On fractional Brownian processes.* Potential Anal. **10** (1999), no. 3, 273-288.
13. Feyel, D., De La Pradelle, A. *Curvilinear integrals along enriched paths.* Electron. J. Probab. **11** (2006), 860-892.
14. Gradinaru, M., Nourdin, I. *Weighted power variations of fractional Brownian motion and application to approximating schemes.* Preprint Paris VI.
15. Gradinaru, M., Nourdin, I., Russo, F., Vallois, P. *m-order integrals and Itô's formula for non-semimartingale processes; the case of a fractional Brownian motion with any Hurst index.* Ann. Inst. H. Poincar Probab. Statist. **41** (2005), 781-806.
16. Gradinaru, M., Russo, F., Vallois, P. *Generalized covariations, local time and Stratonovich Itô's formula for fractional Brownian motion with Hurst index $H \geq \frac{1}{4}$.* Ann. Probab. **31** (2001), 1772-1820.
17. Klingenhöfer, F., Zähle, M. *Ordinary differential equations with fractal noise.* Proc. AMS **127** (1999), 1021-1028.
18. Lejay, A. *An Introduction to Rough Paths.* Séminaire de probabilités XXXVII, vol. **1832** of Lecture Notes in Mathematics (2003), 1-59.
19. Lin, S.J. *Stochastic analysis of fractional Brownian motion.* Stochastics Stochastics Rep. **55** (1995), 121-140.
20. Lyons, T.J. *Differential equations driven by rough signals.* Rev. Math. Iberoamer. **14** (1998), 215-310.
21. Neuenkirch, A. *Optimal approximation of SDEs with additive fractional noise.* J. Complexity **22** (4), 459-475, 2006.
22. Neuenkirch, A., Nourdin, I. *Exact rate of convergence of some approximation schemes associated to SDEs driven by a fractional Brownian motion.* J. Theor. Probab., to appear.
23. Nourdin, I. *Calcul stochastique généralisé et applications au mouvement brownien fractionnaire; Estimation non-paramétrique de la volatilité et test d'adéquation.* PhD thesis, University of Nancy (2004).
24. Nourdin, I. *Schémas d'approximation associés à une équation différentielle dirigée par une fonction höldérienne; cas du mouvement brownien fractionnaire.* C.R. Math. Acad. Sci. Paris, Ser. I **340** (2005), 611-614.
25. Nourdin, I., Simon, T. *On the absolute continuity of one-dimensional SDE's driven by a fractional Brownian motion.* Statist. Probab. Lett. **76** (2006), no. 9, 907-912.
26. Nourdin, I., Simon, T. *Correcting Newton-Côtes integrals corrected by Lévy areas.* Bernoulli **13** (2007), no. 3, 695-711.
27. Nualart, D. *Stochastic calculus with respect to the fractional Brownian motion and applications.* Contemp. Math. **336** (2003), 3-39.
28. Nualart, D., Ouknine, Y. *Stochastic differential equations with additive fractional noise and locally unbounded drift.* Progr. Probab. **56** (2003), 353-365.

29. D. Nualart, D., Peccati, G. *Central limit theorems for sequences of multiple stochastic integrals.* Ann. Probab. **33** (1) (2005), 177-193.
30. Nualart, D., Răşcanu, A. *Differential equations driven by fractional Brownian motion.* Collect. Math. **53** (2002), no. 1, 55–81.
31. Russo, F., Vallois, P. *Forward, backward and symmetric stochastic integration.* Probab. Theory Related Fields **97** (1993), 403-421.
32. Russo, F., Vallois, P. *Stochastic calculus with respect to a finite quadratic variation process.* Stochastics Stochastics Rep. **70** (2000), 1-40.
33. Sussmann, H.J. *An interpretation of stochastic differential equations as ordinary differential equations which depend on a sample point.* Bull. Amer. Math. Soc. **83** (1977), 296-298.
34. Talay, D. *Résolution trajectorielle et analyse numérique des équations différentielles stochastiques.* Stochastics **9** (1983), 275-306.
35. Young, L. C. *An inequality of the Hölder type connected with Stieltjes integration.* Acta Math. **67** (1936), 251-282.
36. Zähle, M. *Integration with respect to fractal functions and stochastic calculus I.* Probab Theory Related Fields **111** (1998), 333-374.

Proof of a Tanaka-like formula
stated by J. Rosen in Séminaire XXXVIII

Greg Markowsky

1 Edgewood Dr.
Orono, ME 04473 USA
e-mail: greg@markowsky.com

Summary. Let B_t be a one dimensional Brownian motion, and let α' denote the derivative of the intersection local time of B_t as defined by J. Rosen in [2]. The object of this paper is to prove the following formula

$$\frac{1}{2}\alpha'_t(x) + \frac{1}{2}\operatorname{sgn}(x)t = \int_0^t L_s^{B_s - x}\, dB_s - \frac{1}{2}\int_0^t \operatorname{sgn}(B_t - B_u - x)\, du \qquad (1)$$

which was given as a formal identity in [2] without proof.

Let B denote Brownian motion in R^1. In [2], Rosen demonstrated the existence of a process which he termed the derivative of self intersection local time for B. That is, he showed that there is a process $\alpha_t(y)$, formally defined as

$$\alpha_t(y) = -\int_0^t \int_0^s \delta'(B_s - B_r - y)\, dr\, ds \qquad (2)$$

such that, for any C^1 function g, we have

$$\int_0^t \int_0^s g'(B_s - B_r)\, dr\, ds = -\int_R g(y)\alpha'_t(y)\, dy\ . \qquad (3)$$

In this paper we'll prove a Tanaka-style formula for α' which was given without proof by Rosen in [2]. We define

$$\operatorname{sgn}(x) = \begin{cases} -1 & \text{if } x < 0 \\ 0 & \text{if } x = 0 \\ 1 & \text{if } x > 0\ . \end{cases} \qquad (4)$$

Our result is

Theorem 1 *There exist versions of the processes in (5) such that, on a set of measure one, the following holds for all x and t:*

$$\frac{1}{2}\alpha'_t(x) + \frac{1}{2}\operatorname{sgn}(x)t = \int_0^t L_s^{B_s - x}\, dB_s - \frac{1}{2}\int_0^t \operatorname{sgn}(B_t - B_u - x)\, du\ . \qquad (5)$$

Proof: Let $f(x)$ be a regular, even, and compactly supported function with $\int f = 1$. Let $f_\varepsilon(x) = \frac{1}{\varepsilon}f(\frac{x}{\varepsilon})$, so that $f_\varepsilon \longrightarrow \delta$ weakly as $\varepsilon \longrightarrow 0$. We assume in all calculations below that $\varepsilon < 1$. Let

$$F_\varepsilon(x) = \int_0^x f_\varepsilon(t)\, dt = \int_0^{\frac{x}{\varepsilon}} f(t)\, dt . \tag{6}$$

We apply Ito's formula to F_ε to get

$$F_\varepsilon(B_t - B_u - x) - F_\varepsilon(-x)$$
$$= \int_u^t f_\varepsilon(B_s - B_u - x)\, dB_s + \frac{1}{2}\int_u^t f_\varepsilon'(B_s - B_u - x)\, ds , \tag{7}$$

which gives

$$\int_0^t F_\varepsilon(B_t - B_u - x)\, du - tF_\varepsilon(-x)$$
$$= \int_0^t \int_0^s f_\varepsilon(B_s - B_u - x)\, du\, dB_s + \frac{1}{2}\int_0^t \int_u^t f_\varepsilon'(B_s - B_u - x)\, ds\, du . \tag{8}$$

Note that $F_\varepsilon(x) \longrightarrow \frac{1}{2}\operatorname{sgn}(x)$ as $\varepsilon \longrightarrow 0$. Furthermore, $|F_\varepsilon(x)| \leq \frac{1}{2}$ for all x, ε, so by the dominated convergence theorem, the first integral on the left approaches $\frac{1}{2}\int_0^t \operatorname{sgn}(B_t - B_u - x)\, du$ as $\varepsilon \longrightarrow 0$. By Theorem 1 in [2], the rightmost integral on the right side is equal to

$$-\frac{1}{2}\int_R f_\varepsilon(y - x)\alpha_t'(y)dy . \tag{9}$$

This term approaches $-\frac{1}{2}\alpha_t'(x)$ as $\varepsilon \longrightarrow 0$ for all x at which $\alpha_t'(x)$ is continuous. In [2] it was shown that $\alpha_t'(x)$ is continuous for all $x \neq 0$. To deal with the case $x = 0$, we need another fact proved in [2], namely that $\alpha_t'(x) + \operatorname{sgn}(x)$ is continuous in x. Using this, together with the fact that $f_\varepsilon(x)\operatorname{sgn}(x)$ is an odd function, we have the following string of equalities:

$$\lim_{\varepsilon \longrightarrow 0}\int_R f_\varepsilon(y)\alpha_t'(y)dy = \lim_{\varepsilon \longrightarrow 0}\int_R f_\varepsilon(y)(\alpha_t'(y) + \operatorname{sgn}(y))dy \tag{10}$$
$$= \alpha_t'(0) + \operatorname{sgn}(0) = \alpha_t'(0) .$$

The only term which remains is the leftmost term on the right side of (8):

$$V(x, \varepsilon, t) := \int_0^t \int_0^s f_\varepsilon(B_s - B_u - x)\, du\, dB_s . \tag{11}$$

We will show that there is a set of measure one upon which, for all x, t

$$V(x, \varepsilon, t) \longrightarrow \int_0^t L_s^{B_s - x}\, dB_s . \tag{12}$$

We may choose a bounded stopping time T such that B and L are bounded on $[0, T]$. Upon proving the result for the stopped processes $B_{T \wedge t}$ and $L_{T \wedge t}$, we let T increase to ∞ to obtain the general result. To simplify notation, we will denote the stopped processes $B_{T \wedge t}$ and $L_{T \wedge t}$ by B and L, respectively. We will show first that, if we let

$$U = \sup_{\substack{a,b \\ a \neq b}} \left(\frac{\sup_t |L_t^a - L_t^b|}{|a - b|^{1/4}} \right) , \tag{13}$$

then

$$E[U^8] < \infty . \tag{14}$$

This follows from standard results on Brownian local time. For example, in the course of proving joint continuity of local times (Theorem VI.1.7) in [1] it is shown that

$$E[\sup_t |L_t^a - L_t^b|^{2p}] \leq C|a - b|^p \tag{15}$$

for any $p \geq 1$, where C depends on the bound for B but is independent of a, b. Note that $L_t^a = 0$ for $|a|$ large enough, as we are assuming that B is bounded. We may apply Kolmogorov's criterion (Theorem I.2.1 in [1]), and we get

$$E\left[\sup_{\substack{a,b \\ a \neq b}} \left(\frac{\sup_t |L_t^a - L_t^b|}{|a - b|^\alpha} \right)^{2p} \right] < \infty \tag{16}$$

for any $\alpha \in (0, \frac{p-1}{2p})$. We may of course choose $\alpha = 1/4$ and $p = 4$, and this gives (14). Let

$$M_t^x = \int_0^t L_s^{B_s - x} \, dB_s . \tag{17}$$

Let τ be an upper bound for T. Using (14), we have the following string of inequalities:

$$E[\sup_{t \geq 0} |M_t^x - M_t^y|^8] \leq CE\left[\left(\int_0^T (L_s^{B_s - x} - L_s^{B_s - y})^2 ds \right)^4 \right] \tag{18}$$

$$\leq CE[(|x - y|^{1/2} U^2 \tau)^4] = C\tau^4 |x - y|^2 E[U^8] .$$

Now we consider $(M^x)_{x \in R}$ as a process taking values in the Banach space $C([0, \tau], R)$; (18) then allows us to apply Kolmogorov's criterion in a general form (Theorem I.2.1 in [1]), to obtain a version of M_t^x which is jointly continuous in x and t. By the occupation-times formula,

$$\int_0^s f_\varepsilon(B_s - B_u - x) \, du = \int_R f_\varepsilon(y) L_s^{B_s - x - y} \, dy . \tag{19}$$

Thus,

$$V(x, \varepsilon, t) = \int f_\varepsilon(y) M_t^{x+y} \, dy \; . \tag{20}$$

Hence $V(x, \varepsilon, t)(\omega)$ approaches $M_t^x(\omega)$ for all x, t on the set of all ω where M is jointly continuous for all x and t. That is, $V(x, \varepsilon, t)$ approaches M_t^x a.s., and this completes the proof.

Acknowledgements. I owe a great deal to Jay Rosen, who suggested this problem to me and supported me throughout much of my graduate career. I am also grateful to the referee who suggested a large simplification in the above proof.

References

1. Revuz, D., Yor, M. (1986) *Continuous Martingales and Brownian Motion* Springer, Berlin.
2. Rosen, J. (2005). Derivatives of self-intersection local times, *Séminaire de Probabilités*, XXXVIII, Springer-Verlag, New York , LNM 1857, 171–184.

Une preuve simple d'un résultat de Dufresne

Ismael Bailleul

Université Paris Sud
e-mail: ismael.bailleul@math.u-psud.fr

Summary. We give a simple proof of the following result by Dufresne [Duf90]: if $\{w_s\}_{s\geq 0}$ is a linear Brownian motion and c a positive constant,

$$\mathbb{P}\Big(\int_0^\infty e^{-w_s-cs}\,ds \in [a,b[\Big) = \frac{2^{2c}}{\Gamma(2c)} \int_a^b \frac{e^{-2/u}}{u^{1+2c}}du.$$

On donne dans cette note une démonstration simple du résultat suivant de Dufresne, [Duf90].

1 Le résultat de Dufresne

Théorème 1.1 *Soit $\{w_s\}_{s\geq 0}$ un mouvement Brownien réel et c une constante > 0. On a quels que soient $0 \leq a \leq b \leq \infty$*

$$\mathbb{P}\Big(\int_0^\infty e^{-w_s-cs}\,ds \in [a,b[\Big) = \frac{2^{2c}}{\Gamma(2c)} \int_a^b \frac{e^{-2/u}}{u^{1+2c}}du.$$

Remarque $\int_0^\infty e^{-w_s-cs}\,ds$ a même loi que $\dfrac{2}{\gamma_{2c}}$, où γ_{2c} est une variable aléatoire de loi gamma, de paramètre $2c$.

$\triangleleft$ Etant donnés $y > 0$ et $\xi \in \mathbb{R}$, considérons la diffusion $\big\{(y_s, \xi_s)\big\}_{s\geq 0}$ sur $\mathbb{R}_*^+ \times \mathbb{R}$ solution des équations différentielles stochastiques

$$\begin{aligned} dy_s &= y_s dw_s + \left(c + \frac{1}{2}\right) y_s\,ds, \\ d\xi_s &= \frac{ds}{y_s}, \end{aligned} \tag{1}$$

où w est un mouvement brownien, avec pour valeurs initiales $y \in \mathbb{R}_*^+$ et $\xi \in \mathbb{R}$. L'équation de $\{y_s\}_{s\geq 0}$ s'intègre explicitement.

$$y_s = y \, e^{w_s + cs},$$

et

$$\xi_s = \xi + \frac{1}{y} \int_0^s e^{-w_r - cr} \, dr.$$

Notons $\mathbb{P}_{y,\xi}$ la loi de la diffusion $\{(y_s, \xi_s)\}_{s \geq 0}$. Le processus $\{\xi_s\}_{s \geq 0}$ admet $\mathbb{P}_{y,\xi}$-presque sûrement une limite finie

$$\xi_\infty = \xi + \frac{1}{y} \int_0^\infty e^{-w_r - cr} \, dr$$

lorsque $s \to +\infty$. Si θ_r désigne le shift usuel sur les trajectoires de la diffusion (y, ξ), quel que soit $r > 0$, on a $\mathbb{P}_{y,\xi}$-presque sûrement

$$\xi_\infty \circ \theta_r = \xi_\infty.$$

Notons $G(t) \equiv \mathbb{P}\left(\int_0^\infty e^{-w_r - cr} \, dr \geq t\right)$. Soit $A \in \mathbb{R}$. On a

$$\mathbb{P}_{y,\xi}(\xi_\infty \geq A) = \mathbb{P}_{y,\xi}\left(\xi + \frac{1}{y} \int_0^{+\infty} e^{-w_r - cr} \, dr \geq A\right) = G\big(y(A - \xi)\big).$$

Si l'on arrive à justifier par un argument a priori que la fonction

$$F_{\geq A}(y, \xi) \equiv \mathbb{P}_{y,\xi}(\xi_\infty \geq A)$$

est de classe $\mathcal{C}^2$, ou que G est de classe $\mathcal{C}^2$, la fonction $F_{\geq A}$ vérifiera l'équation différentielle

$$\left(\frac{y^2}{2} \partial_y^2 + \left(c + \frac{1}{2}\right) y \, \partial_y + \frac{\partial_\xi}{y}\right) F_{\geq A} = 0, \tag{2}$$

par la formule d'Itô, du fait que le processus $\{F_{\geq A}(y_t, \xi_t)\}_{t \geq 0}$ est une martingale.

L'équation (2) devient pour G

$$\frac{\big(y(A - \xi)\big)^2}{2} G''\big(y(A - \xi)\big) + \left(\left(c + \frac{1}{2}\right) y(A - \xi) - 1\right) G'\big(y(A - \xi)\big) = 0, \tag{3}$$

soit

$$G''(r) + \left(\frac{1 + 2c}{r} - \frac{2}{r^2}\right) G'(r) = 0.$$

On la résout explicitement:

$$G'(r) = C \frac{e^{-2/r}}{r^{1+2c}} \mathbf{1}_{r > 0},$$

pour une certaine constante C que l'on identifie avec la condition $G(0) = 1$. Cela donne la formule de l'énoncé. $\quad\triangleright$

On aura donc établi l'identité de Dufresne si l'on arrive à justifier a priori le caractère $\mathcal{C}^2$ de la fonction de répartition de la variable aléatoire $\int_0^\infty e^{-w_u-cu}\,du$, ou de la fonction $F_{\geq A}$. Ce dernier point peut s'obtenir de l'hypoellipticité du générateur L de la diffusion à l'aide du théorème de Hörmander. Il semble cependant raisonnable de se passer d'un outil si puissant dans notre cadre simple. On donne dans la section suivante une preuve directe du premier point, basée sur une formule d'intégration par parties.

2 Régularité de la loi de la variable aléatoire $\int_0^\infty e^{w_u-cu}\,du$

On démontre une formule d'intégration par parties inspirée du calcul de Malliavin. On trouvera ce calcul développé à un niveau élémentaire dans les livres de Bell et Bass, [Bel87], [Bas98] (dernier chapitre), ou dans celui de Nualart [Nua06], et dans une bien plus grande généralité, dans le livre de Malliavin [Mal97].

La formule d'intégration par parties tire ici son intérêt d'un fait élémentaire bien connu (voir par exemple le livre [Bas98] de Bass, Chap.8, proposition 3.1).

Proposition 2.1 *Soit ν une probabilité sur $\mathbb{R}$ et $k \geqslant 2$ un entier. Supposons qu'il existe une constante $C_k > 0$ telle que l'inégalité*

$$\left| \int \phi^{(k)}(x)\nu(dx) \right| \leqslant C_k \|\phi\|_\infty$$

est vérifiée quelle que soit la fonction ϕ de classe $\mathcal{C}^\infty$, bornée, ainsi que toutes ses dérivées. Alors ν admet une densité de classe $\mathcal{C}^{k-2}$ par rapport à la mesure de Lebesgue.

On va tirer partie de l'écriture suivante

$$
\begin{aligned}
\int_0^\infty e^{w_r - c\,r}\,dr &= \int_0^1 e^{(w_r - r\,w_1)+(w_1-c)r}\,dr + e^{-c}e^{w_1}\int_1^\infty e^{(w_r-w_1)-c(r-1)}\,dr \\
&= \int_0^1 e^{\mathfrak{p}_r + (w_1-c)r}\,dr + e^{-c}e^{w_1}\int_0^\infty e^{\widetilde{w}_r - cr}\,dr,
\end{aligned}
\tag{4}
$$

dans laquelle $\int_0^\infty e^{w_r - c\,r}\,dr$ apparaît comme une fonctionnelle du pont brownien

$$\{\mathfrak{p}_r\}_{0\leq r\leq 1} = \{w_r - rw_1\}_{0\leq r\leq 1},$$

de w_1, et du mouvement brownien

$$\{\widetilde{w}_r\}_{r\geq 0} = \{w_{r+1} - w_1\}_{r\geq 0}.$$

Ces trois processus sont indépendants sous $\mathbb{P}$. On notera dorénavant

$$\Omega = \mathcal{C}\big([0,1],\mathbb{R}\big) \times \mathbb{R} \times \mathcal{C}(\mathbb{R}^+,\mathbb{R}),$$

qu'on munit de la tribu produit des tribus boréliennes de chaque facteur, et sur lequel on met la probabilité

$$\mathbb{Q} = \mathbf{P}^{[0,1]} \otimes \mathbb{W} \otimes \widetilde{\mathbb{P}},$$

où $\mathbf{P}^{[0,1]}$ est la mesure du pont brownien sur $\mathcal{C}\big([0,1],\mathbb{R}\big)$, $\mathbb{W}$ une loi normale centrée, réduite, et $\widetilde{\mathbb{P}}$ la mesure de Wiener sur $\mathcal{C}(\mathbb{R}^+,\mathbb{R})$. On notera $\omega = (\mathfrak{p}, w_1, \widetilde{w})$ un élément de Ω, et w le brownien reconstruit à l'aide de $\mathfrak{p}$, w_1, $\widetilde{w}$.

Pour ne pas charger l'écriture, on prendra $c = 1$. On va montrer que la variable aléatoire $\int_0^\infty e^{w_u - u}\, du$ a une densité de classe $\mathcal{C}^2$ par rapport à la mesure de Lebesgue sur $\mathbb{R}$. Il n'y a bien entendu aucune difficulté à transposer ce qui suit à la variable aléatoire $\int_0^\infty e^{-w_u - cu}\, du$, $c > 0$, qu'on a rencontrée dans la première partie.

Les acteurs et leurs qualités

Les quantités suivantes vont nous intéresser.

- $X(\omega) = \int_0^\infty e^{w_s - s}\, ds$,
- Pour $k \geq 1$, $X^{(k)}(\omega) = \int_0^1 u^k\, e^{\mathfrak{p}_u + (w_1 - 1)u}\, du + e^{w_1 - 1} \int_0^\infty e^{\widetilde{w}_r - r}\, dr$
- $F(\omega) = \frac{1}{X^{(1)}(\omega)}$.

ϕ désignera une fonction lisse dont toutes les dérivées sont bornées.

On écrira dans la suite $\mathbb{L}^p$ pour $\mathbb{L}^p\big(\mathbb{Q}(dw)\big)$, et l'on écrira toujours $\mathbb{E}$ l'espérance sous la probabilité $\mathbb{Q}$.

Pour $\widetilde{\mathbb{P}}$-presque tout $\widetilde{w}$ et tout $\mathfrak{p}$, les fonctions $X(\mathfrak{p}, ., \widetilde{w})$ et $F(\mathfrak{p}, ., \widetilde{w})$ sont de classe $\mathcal{C}^\infty$. Les dérivées de $X(\mathfrak{p}, ., \widetilde{w})$ sont les $X^{(k)}(\mathfrak{p}, ., \widetilde{w})$ et celles de $F(\mathfrak{p}, ., \widetilde{w})$ sont explicites:

$$F^{(1)}(\mathfrak{p}, ., \widetilde{w}) = -\big(X^{(2)} F^2\big)(\mathfrak{p}, ., \widetilde{w}),$$
$$F^{(2)}(\mathfrak{p}, ., \widetilde{w}) = -\big(2 F^{(1)} X^{(2)} F + F^2 X^{(3)}\big)(\mathfrak{p}, ., \widetilde{w}),$$

et ainsi de suite. On écrira par exemple

$$\partial_{w_1}\big(\phi(X) F\big) = X^{(1)} \phi'(X) F + \phi(X) F^{(1)} = \phi'(X) + \phi(X) F^{(1)}.$$

Proposition 2.2 (a) $X \in \mathbb{L}^p$, pour tout $1 \leq p < 2$. Comme $0 \leq X^{(k)} \leq X$, on a $X^{(k)} \in \mathbb{L}^p$, pour tout $1 \leq p < 2$.

(b) $F \in \mathbb{L}^p$, quel que soit $p \geq 1$. Toutes ses dérivées par rapport à w_1 sont aussi dans tous les espaces $\mathbb{L}^p$, $p \geq 1$.

◁ 1. Soit $0 < a < 1$ une constante. La fonction $e^{w_r - r}$ est intégrable sur $(0, +\infty[$ sur un événement de probabilité 1. On peut appliquer sur cet événement l'inégalité de Jensen à la fonction $\frac{1}{a} e^{w_r - (1-a)r}$ et la probabilité $a e^{-ar} \mathbf{1}_{r>0}$; on obtient

$$\left(\int_0^\infty e^{w_r - r}\, dr\right)^p = \left(\int_0^\infty \frac{e^{w_r - (1-a)r}}{a} a e^{-ar}\, dr\right)^p$$

$$\leq a^{1-p}\int_0^\infty e^{pw_r - (1-a)pr} e^{-ar}\, dr.$$

On a ainsi

$$\mathbb{E}\left[\left(\int_0^\infty e^{w_r - r}\, dr\right)^p\right] \leq a^{1-p}\int_0^\infty e^{\frac{p^2 r}{2} - \left((1-a)p+a\right)r}\, dr < \infty,$$

si

$$\frac{p^2}{2} < (1-a)p + a. \qquad (5)$$

Comme on peut trouver pour tout $p < 2$ une constante $0 < a < 1$ assez petite pour que la condition (5) ait lieu, le résultat s'ensuit.

2. Estimons $\mathbb{Q}\big(F(w) > r\big)$.

$$\mathbb{Q}\big(F(w) > r\big) = \mathbb{Q}\left(\int_0^1 u\, e^{\mathbf{p}_u + (w_1 - 1)u}\, du + e^{w_1 - 1}\int_0^\infty e^{\widetilde{w}_u - u}\, du < \frac{1}{r}\right)$$

$$\leq \mathbb{Q}\left(e^{w_1}\int_0^\infty e^{\widetilde{w}_u - u}\, du < \frac{e}{r}\right) \leq \mathbb{Q}\left(e^{w_1}\int_0^1 e^{\widetilde{w}_u - u}\, du < \frac{e}{r}\right)$$

$$\leq \mathbb{Q}\left(e^{w_1}\int_0^1 e^{\widetilde{w}_u - u}\, du < \frac{e}{r}\; ;\; \inf_{u\in[0,1]} \widetilde{w}_u \geq -\frac{\ln r}{2}\right)$$

$$+ \mathbb{Q}\left(e^{w_1}\int_0^1 e^{\widetilde{w}_u - u}\, du < \frac{e}{r}\; ;\; \inf_{u\in[0,1]} \widetilde{w}_u < -\frac{\ln r}{2}\right)$$

$$\leq \mathbb{Q}\left(e^{w_1}\int_0^1 e^{\frac{-\ln r}{2} - u}\, du < \frac{e}{r}\; ;\; \inf_{u\in[0,1]} \widetilde{w}_u \geq -\frac{\ln r}{2}\right)$$

$$+ \mathbb{Q}\left(\inf_{u\in[0,1]} \widetilde{w}_u < -\frac{\ln r}{2}\right)$$

$$\leq \mathbb{Q}\left(w_1 \leq -\frac{\ln r}{2} + \text{Cte}\right) + 2\,\mathbb{Q}\left(\widetilde{w}_1 \geq \frac{\ln r}{2}\right). \qquad (6)$$

Ces deux termes sont du même ordre de grandeur: $\dfrac{1}{r^{\frac{\ln(r)}{8}}\ln(r)}$.

Cette fonction de r décroît assez vite pour que l'on ait quel que soit $p \geqslant 1$

$$\mathbb{E}[F^p] = \int_0^\infty r^{p-1}\mathbb{P}(F > r)\, dr < \infty.$$

Pour ce qui est des dérivées de F, on traite le cas de $F^{(1)}$ et $F^{(2)}$, les dérivées d'ordre supérieur se traitant de la même façon.
De l'inégalité presque sûre

$$0 \leq X^{(2)}F = \frac{X^{(2)}}{X^{(1)}} \leq 1,$$

on tire

$$|X^{(2)}F^2| \le F,$$

d'où il vient que $F^{(1)} = -X^{(2)}(\omega)F(\omega)^2$ est dans tous les espaces $\mathbb{L}^p$, $p \ge 1$.

De même, puisque $X^{(3)}(\omega)F(\omega) = \frac{X^{(3)}(\omega)}{X(\omega)} \le 1$, la variable aléatoire $F^2 X^{(3)}$ se trouve dans tous les espaces $\mathbb{L}^p$. On sait en outre que

$$0 \le X^{(2)} F \le 1,$$

et

$$F^{(1)} \in \mathbb{L}^p, \ \forall p \ge 1.$$

Ainsi, $F^{(2)} = -2F^{(1)}X^{(2)}F - F^2 X^{(3)}$ est dans tous les $\mathbb{L}^p$, $p \ge 1$.   ▷

Théorème 2.1 *Soit ϕ une fonction lisse, dont toutes les dérivées sont bornées. Chacun des membres de l'égalité est bien défini et l'on a*

$$\mathbb{E}\big[w_1 \, \phi(X)F\big] = \mathbb{E}\big[\partial_{w_1}\big(\phi(X)F\big)\big]. \tag{7}$$

◁ w_1 et F appartenant à tous les espaces $\mathbb{L}^p$, $w_1 F$ est intégrable. $\phi(X)$ est bornée. Le théorème de Fubini et la formule d'intégration par parties pour une loi normale sur $\mathbb{R}$, centrée, réduite, $\mathcal{N}(dw_1)$, nous permettent alors d'écrire

$$\mathbb{E}[w_1 \, \phi(X)F] = \mathbb{P}^{[0,1]} \otimes \widetilde{\mathbb{P}}\Big(\int w_1\phi(X(\mathfrak{p}, w_1, \widetilde{w}))F(X(\mathfrak{p}, w_1, \widetilde{w})) \, \mathcal{N}(dw_1)\Big)$$

$$= \mathbb{P}^{[0,1]} \otimes \widetilde{\mathbb{P}}\Big(\int \partial_{w_1}\big(\phi(X(\mathfrak{p}, w_1, \widetilde{w}))F(X(\mathfrak{p}, w_1, \widetilde{w}))\big) \, \mathcal{N}(dw_1)\Big)$$

$$= \mathbb{E}\big[\partial_{w_1}\big(\phi(X)F\big)\big]$$

▷

Comme $\partial_{w_1}\big(\phi(X)F\big) = X^{(1)}\phi'(X)F + \phi(X)F^{(1)} = \phi'(X) + \phi(X)F^{(1)}$, le théorème a le corollaire suivant, dans lequel on n'a besoin que d'une fonction de classe $\mathcal{C}^1$ pour appliquer ce qui précède.

Corollaire 2.1 *Il existe une constante $C_1 > 0$ telle que l'inégalité suivante est vraie pour toute fonction bornée $\phi \in \mathcal{C}^1$, ayant sa dérivée bornée.*

$$\big|\mathbb{E}\big[\phi'(X)\big]\big| \le C_1\|\phi\|_\infty.$$

Pour obtenir le même genre d'estimations avec la dérivée seconde $\phi^{(2)}$ de ϕ au lieu de ϕ', on applique la formule d'intégration par parties à $\phi'(=\phi^{(1)})$ pour obtenir

$$\mathbb{E}\big[\phi^{(2)}(X)\big] = \mathbb{E}\big[\phi^{(1)}(X)\{Fw_1 - F^{(1)}\}\big]. \tag{8}$$

Notons $K = Fw_1 - F^{(1)}$. Comme F et w_1 sont dans tous les espaces $\mathbb{L}^p$, Fw_1 est aussi dans tous les espaces $\mathbb{L}^p$, ainsi que K. Il en va de même de KF, qui est pour $\widetilde{\mathbb{P}}$-presque tout $\widetilde{w}$ et tout $\mathfrak{p}$ une fonction lisse de w_1, de dérivée

$$\partial_{w_1}(KF) = K^{(1)}F + KF^{(1)}.$$

Pour estimer $\mathbb{E}[\phi'(X)K]$, on applique la formule d'intégration par parties non pas à la fonctionnelle $\phi(X)F$, mais à $\phi(X)KF$.

L'utilisation de la formule est justifiée comme dans la démonstration du théorème. Cela donne

$$\mathbb{E}\big[\phi'(X)K\big] = \mathbb{E}\big[\phi(X)\big\{KF\,w_1 - \partial_{w_1}(KF)\big\}\big]. \tag{9}$$

On a donc

$$\mathbb{E}[\phi'(X)K] \le \|\phi\|_\infty \, \|KF\,w_1 - \partial_{w_1}(KF)\|_{\mathbb{L}^1}.$$

On déduit de (8) qu'on peut trouver une constante C_2 telle que

$$\big|\mathbb{E}\big[\phi^{(2)}(X)\big]\big| \le C_2\|\phi\|_\infty.$$

Pour obtenir une telle estimation avec $\phi^{(3)}(X)$, on applique la formule d'intégration par parties à la fonctionnelle $\phi(X)w_1F^2K$; puis, pour obtenir le théorème suivant, à la fonctionnelle $\phi(X)(w_1F)^2FK$.

Théorème 2.2 *Il existe d'une constante C_4 telle que l'on a pour toute fonction bornée ϕ, dont toutes les dérivées sont bornées,*

$$\big|\mathbb{E}\big[\phi^{(4)}(X)\big]\big| \le C_4\|\phi\|_\infty. \tag{10}$$

Comme on l'a vu au lemme 2.1, cette inégalité implique que la loi de $X(w)$ a une densité par rapport à la mesure de Lebesgue, de classe $\mathcal{C}^2$. L'équation $(3)(^1)$ qui permet son identification nous montre que cette densité est en fait de classe $\mathcal{C}^\infty$. On pourrait obtenir cela en itérant indéfiniment la formule d'intégration par parties.

3 Remarques

1. On trouvera dans les articles de Dufresne [Duf90], proposition 4.4.4, Yor [Yor92], [CPY01], théorème 3.1, et Matsumoto & Yor [MY05a], p.335, (ainsi que dans les références données juste avant le théorème dans l'article de Matsumoto et Yor), d'autres démonstrations de ce résultat, d'inspirations très différentes.
   • Dans [Yor92], la loi de $\int_0^{+\infty} e^{w_u - cu}\, du$ est identifiée à la loi du dernier temps de passage d'un processus de Bessel, issu de 0, de dimension $2(1+c)$;

---

[1] Où l'on prend $c = 1$

le résultat de Dufresne provient alors d'un résultat de Getoor identifiant cette loi comme celle d'un multiple de l'inverse d'une loi gamma, de paramètre $2c$.

• Dans [CPY01], la loi μ de l'intégrale de Dufresne est identifiée comme la probabilité stationnaire d'un processus d'Ornstein-Uhlenbeck de générateur explicite. La résolution d'une équation de type $L^*\mu = 0$ permet d'obtenir μ.

• Dans [MY05a], théorème 6.2, c'est l'utilisation de la transformée de Laplace et de fonctions de Bessel qui donne le résultat.

La démonstration que l'on propose a l'avantage d'être automatique et de ne demander aucun effort, dès lors que l'on sait que la densité de la loi de l'intégrale de Dufresne admet une densité de classe $\mathcal{C}^2$ par rapport à la mesure de Lebesgue. On aurait pu utiliser une formule d'intégration par parties obtenue en perturbant tout le chemin w par une fonction convenable, comme on le fait usuellement pour montrer de tels résultats. On aurait alors eu affaire à des dérivées de Fréchet de fonctionnelles à valeurs dans des espaces $\mathbb{L}^p$ là où l'on a eu des dérivées usuelles. Cette démonstration évitant le recours à des espaces de dimension infinie est peut-être plus facile à appréhender.

De nombreux résultats sur les lois des fonctionnelles $\int_0^t e^{-w_u-cu}\,du$ et $\int_0^\infty e^{-w_u-cu}\,du$ se trouvent dans [MY05a] et [MY05b].

2. Cette démonstration trouve ses origines dans l'étude d'une diffusion sur $\mathbb{H} \times \mathbb{R}^3$. Dessinons ce cadre.

Notation: $(\varepsilon_0, \varepsilon_1, \varepsilon_2)$ désigne la base canonique de $\mathbb{R} \times \mathbb{R}^2$ et ξ^0, ξ^1, ξ^2 les coordonnées sur $\mathbb{R} \times \mathbb{R}^2$.

Munissons l'espace vectoriel $\mathbb{R} \times \mathbb{R}^2$ de la forme quadratique Lorentzienne

$$q(\xi) = (\xi^0)^2 - \left((\xi^1)^2 + (\xi^2)^2\right).$$

L'ensemble $\{\xi \in \mathbb{R} \times \mathbb{R}^2 ;\ q(\xi) = 1\}$ est un hyperboloïde à deux nappes. Notons $\mathbb{H}$ celle qui correspond aux $\xi^0 > 0$. Bien que la forme quadratique q ne soit pas définie positive, sa restriction à chaque plan tangent de $\mathbb{H}$ est définie négative. Cela fait de $\mathbb{H}$ une variété riemannienne. Il s'agit là de l'un des modèles de l'espace hyperbolique de dimension 2.

Soit $\{\dot{\xi}_s\}_{s\geq 0}$ un mouvement brownien sur $\mathbb{H}$. Définissons

$$\xi_s = \xi_0 + \int_0^s \dot{\xi}_u \, du$$

et considérons la diffusion $\left\{(\dot{\xi}_s, \xi_s)\right\}_{s\geq 0}$ sur $\mathbb{H} \times \mathbb{R}^3$. Dans ce décor, on s'est intéressé au problème suivant. **Déterminer la tribu invariante** $Inv\left((\dot{\xi}, \xi)\right)$ **de la diffusion** $\{(\dot{\xi}_s, \xi_s)\}_{s\geq 0}$.

On appréhende mieux le problème en prenant pour carte sur $\mathbb{H}$ les coordonnées du demi-espace. Munissons le demi-espace $\{(y, x) \in \mathbb{R}_*^+ \times \mathbb{R}\}$ de la métrique hyperbolique $ds^2 = \frac{dx^2 + dy^2}{y^2}$. L'application suivante est une isométrie entre $(\mathbb{R}_*^+ \times \mathbb{R}, ds^2)$ et $(\mathbb{H}, q)$.

$$\psi \; : \; (y, x) \in \mathbb{R}_*^+ \times \mathbb{R} \mapsto \left(\frac{x^2 + y^2 + 1}{2y}, \frac{x^2 + y^2 - 1}{2y}, \frac{x}{y} \right) \in \mathbb{H}. \qquad (11)$$

Dans les coordonnées (y, x) du demi-espace, le mouvement brownien sur $\mathbb{H}$ est solution d'équations différentielles stochastiques simples.

$$dy_s = y_s \, dw_s^y,$$
$$dx_s = y_s \, dw_s^x,$$

où w^y et w^x sont deux mouvements browniens réels indépendants. On voit sur ces équations que $\{y_s\}_{s \geq 0}$ tend vers 0 et que $\{x_s\}_{s \geq 0}$ converge vers une variable aléatoire $x_\infty \in Inv(\dot{\xi}) \subset Inv\big((\dot{\xi}, \xi)\big)$. Il est naturel, pour trouver d'autres quantités qui convergent le long des trajectoires de la diffusion $\big\{(\dot{\xi}_s, \xi_s)\big\}_{s \geq 0}$, de conditionner par x_∞ et de regarder la diffusion conditionnée. Cette entreprise est facilitée par la présence de nombreuses symétries qui permettent de ramener cette étude à celle de la diffusion conditionnée par l'évènement (de probabilité nulle) $\{|x_\infty| = +\infty\}$. Le processus $\{(\dot{\xi}_s, \xi_s)\}_{s \geq 0}$ est alors solution du système

$$dy_s = y_s \, dw_s^y + y_s \, ds,$$
$$dx_s = y_s \, dw_s^x, \qquad (12)$$
$$d\xi_s \equiv \dot{\xi}_s \, ds = \psi\big((y_s, x_s)\big) \, ds.$$

où w^y et w^x sont deux mouvements browniens réels indépendants. Notons $\mathbb{P}_{(y,x),\xi}$ la loi de la solution du système (12), ayant pour conditions initiales $\big((y, x), \xi\big) \in (\mathbb{R}_*^+ \times \mathbb{R}) \times \mathbb{R}^3$. L'équation donnant y_s s'intègre explicitement: $y_s = y \, e^{w_s^y + \frac{s}{2}}$.

Intéressons-nous à la quantité

$$h(\xi_s) = q(\xi_s, \varepsilon_0 + \varepsilon_1) = \xi_s^0 - \xi_s^1.$$

On a, d'après la formule (11),

$$
\begin{aligned}
h(\xi_s) &= h(\xi) + \int_0^s \dot{\xi}_r^0 - \dot{\xi}_r^1 \, dr \\
&= h(\xi) + \int_0^s \left(\frac{x_r^2 + y_r^2 + 1}{2y_r} - \frac{x_r^2 + y_r^2 - 1}{2y_r} \right) dr \\
&= h(\xi) + \int_0^s \frac{dr}{y_r} \\
&= h(\xi) + \frac{1}{y} \int_0^s e^{-w_r^y - \frac{r}{2}} \, dr.
\end{aligned}
$$

On retrouve la quantité utilisée dans la démonstration du théorème de Dufresne. Si l'on note $h_\infty = \lim_{s \to +\infty} h(\xi_s)$ et $G(t) = \mathbb{P}\left(\int_0^\infty e^{w_r^y - \frac{r}{2}}\, dr\right)$, on a quel que soit $A \in \mathbb{R}$,

$$\mathbb{P}_{\big((y,x),\xi\big)}\big(h_\infty \geq A\big) = G\big(y(A - h(\xi))\big).$$

L'équation aux dérivées partielles vérifiée par la fonction $\mathbb{P}_\cdot(h_\infty \geq A)$ permet d'identifier G comme on l'a fait dans la preuve du théorème **1.1**, dont l'idée provient de ce fait.

L'identification de la tribu invariante de la diffusion $\big\{\dot\xi_s, \xi_s\big\}_{s \geq 0}$ est exposée dans l'article [Bai07], qui paraîtra ailleurs.

3. Revenons à l'équation (1) et posons

$$G(s, A) \equiv \mathbb{P}\left(\int_0^s e^{w_u - cu}\, du \geq A\right).$$

On montre, à l'aide d'une formule d'intégration par parties analogue à celle que l'on a établi pour $G(A)$, que $G(s, A)$ est une fonction de A de classe $\mathcal{C}^2$ et une fonction de s de classe $\mathcal{C}^1$ ([2]). Notons

$$F_{\geq A}^s\big((y,x),\xi\big) \equiv \mathbb{P}_{(y,x),\xi}(\xi_s \geq A).$$

On écrira $u = \big((y,x),\xi\big)$ et $u_s = \big((y_s, x_s), \xi_s\big)$ afin d'alléger l'écriture.

$$F_{\geq A}^s(u) = G\big(s, y(A - \xi)\big).$$

D'un côté, on a

$$\frac{\mathbb{E}_u\big[\mathbb{P}_{u_\varepsilon}(\xi_s \geq A)\big] - \mathbb{P}_u(\xi_s \geq A)}{\varepsilon} \xrightarrow[\varepsilon \searrow 0]{} \left(\frac{y^2}{2}\partial_y^2 + \left(c + \frac{2}{3}\right) y\, \partial_y + \frac{\partial_\xi}{y}\right) F_{\geq A}^s(u_0),$$

et de l'autre

$$\frac{\mathbb{E}_u\big[\mathbb{P}_{u_\varepsilon}(\xi_s \geq A)\big] - \mathbb{P}_{u_0}(\xi_s \geq A)}{\varepsilon}$$

$$= \frac{\mathbb{P}_{u_0}((\xi_{s+\varepsilon} \geq A)) - \mathbb{P}_{u_0}(\xi_s \geq A)}{\varepsilon} \xrightarrow[\varepsilon \searrow 0]{} \partial_s F_{\geq A}^s(u_0).$$

On a donc

$$\partial_s F_{\geq A}^s = \left(\frac{y^2}{2}\partial_y^2 + \left(c + \frac{2}{3}\right) y\, \partial_y + \frac{\partial_\xi}{y}\right) F_{\geq A}^s.$$

Cette équation devient pour $G(s, A)$

$$\partial_s\, G(s, A) = \frac{A^2}{2} G''(s, A) + \left(\left(c + \frac{1}{2}\right) A - 1\right) G'(s, A),$$

où le $'$ désigne la dérivation par rapport à A.

---

[2] On étudie pour cela la loi du couple $\big(w_s, \int_0^s e^{w_u - cu}\, du\big)$ grâce à une formule d'intégration par parties.

References

[Bai07]   Ismaël Bailleul. Poisson boundary of a relativistic diffusion. A paraitre dans *P.T.R.F.*, 2007.

[Bas98]   Richard F. Bass. *Diffusions and elliptic operators*. Probability and its Applications (New York). Springer-Verlag, New York, 1998.

[Bel87]   Denis R. Bell. *The Malliavin calculus*, volume 34 of *Pitman Monographs and Surfeys in Pure and Applied Mathematics*. Longman Scientific & Technical, Harlow, 1987.

[CPY01]   Philippe Carmona, Frédérique Petit, and Marc Yor. Exponential functionals of *Lévy processes*. In Lévy processes, pages 41–55. Birkhäuser Boston, Boston, MA, 2001.

[Duf90]   Daniel Dufresne. The distribution of a perpetuity, with applications to risk theory and pension funding. *Scand. Actuar. J.*, (1-2): 39–79, 1990.

[Mal97]   Paul Malliavin. *Stochastic analysis*, volume 313 of *Grundlehren der Mathematischen Wissenschaften [Fundamental Principles of Mathematical Sciences]*. Springer-Verlag, Berlin, 1997.

[MY05a]   Hirojuki Matsumoto and Marc Yor. Exponential functionals of Brownian motion. I. Probability laws at fixed time. *Probab. Surv.*, 2: 312–347 (electronic), 2005.

[MY05b]   Hirojuki Matsumoto and Marc Yor. Exponential functionals of Brownian motion. II. Some related diffusion processes. *Probab. Surv.*, 2: 348–384 (electronic), 2005.

[Nua06]   David Nualart. *The Malliavin calculus and related topics*. Probability and its Applications (New York). Springer-Verlag, Berlin, second edition, 2006.

[Yor92]   Marc Yor. Sur certaines fonctionnelles du mouvement brownien réel. J. Appl. Probab., 29(1): 202–208, 1992.

Creation or deletion of a drift on a Brownian trajectory

Laurent Serlet

Laboratoire de Mathématiques, Université Blaise Pascal
Campus Universitaire des Cézeaux, 63177 Aubière cedex, France
e-mail: Laurent.Serlet@math.univ-bpclermont.fr

Summary. We show that a negative drift can be created on a Brownian trajectory by cutting excursions according to a certain Poisson measure. Conversely a negative drift can be annihilated by inserting independent excursions again according to a certain Poisson measure. We first give results in discrete time by considering the random walks as contour processes of Galton-Watson trees and then pass to the limit.

Key words: Galton-Watson trees, pruning, Poisson measure, drift, approximation of diffusion, Markov chains

1 Introduction

1.1 Summary of results and methods

It is possible to create a drift on a linear Brownian motion by cutting some excursions of this trajectory. By excursion we mean a connected part of this trajectory above a certain level. These excursions are chosen according to a certain Poisson measure defined conditionally on the initial trajectory. This is explained in section 2 below but this result is essentially Proposition 4 of [AS]. However the arguments are given there in the setting of random snakes, a subject we want to avoid in the present paper in order to make it readable by a broader public. So we give anew an exposition of theses results, in a slightly more general setting, with more connections to trees and give the ideas of the proofs skipping some details. Our point of view is to approximate (reflected) Brownian motion by random walks, seen as contour processes of Galton-Watson forests. Then the cutting of (discrete) excursions on the contour process amounts to a percolation on the Galton-Watson trees which gives again Galton-Watson trees, with a new offspring law, for which the contour process is a random walk, with a higher downward bias. The version for continous time i.e. Brownian motion as stated in Theorem 3, is obtained by examining the limit of this cutting procedure.

A natural question is to ask whether this operation can be reversed, that is, if a negative drift can be annihilated by adding excursions. The answer is positive and is the subject of section 3. Again the problem is easily solved on random walks interpretating them as the contour processes of trees. The issue is to see how a Galton-Watson tree can be "decorated" by the graft of small trees on certain vertices to give a new Galton-Watson tree with higher progeny. All our Galton-Watson trees have geometric progeny law and we rely simply on a property of these laws. The next step, explained in subsection 3.3, is to consider the continuous-time limit in order to state Theorem 7, our main result. This result specifies how excursions must be added to a Brownian motion with drift to destroy the drift, a procedure that is roughly the converse of the cutting procedure specified in Theorem 3. We conclude this paper with an example of application. Other applications to Brownian snake and super-Brownian motion will be given in [Se].

1.2 Bibliographical notes

The idea of percolation on the edges of a Galton-Watson tree to retain the connected component of the root is exploited for instance by Aldous and Pitman in [AP1] to define what they call a pruning process. They describe the transition rates and give special attention to the Poisson offspring law. This idea of pruning appears also in the setting of the Continuum Random Tree (CRT) in [AP2] where Poissonian pruning leads to a description of a self-similar fragmentation process. This work is related to the results stated here because CRT can be represented by a Brownian excursion: this correspondence is used in [AS2]. Conversely the idea of grafting small trees to a forest in order to obtain a new forest having a law of similar type but with different parameters is central in [PW]. The edges of the trees and forest considered there have variable lengths; a composition rule is proved and the link with Williams decomposition for the Brownian trajectory is explained. Galton-Watson forests with random edge lengths are also studied by Duquesne and Winkel in [DW]. They define a growing family of these trees which is consistent under a Bernoulli percolation on vertices that is described there as "tree coloring". They show the existence of a limit called the Lévy tree in the topological setting of real trees . This setting avoids the coding of trees by real valued processes such as the height process. Our point of view is completely opposite since we seek results on real-valued processes by seeing them as limit of contour processes of forests. In [EW] an operation of tree pruning and regrafting is studied for real trees; for the continuous tree associated to a real valued continuous function, it consists in cutting an excursion in the graph of the function and inserting it to another place.

2 How to create a negative drift ?

In this paper we deal with random rooted trees. We refer to [AP1] for the general terminology of trees, for instance the notions of tree, vertex, edge, root,... and also the notion of Galton-Watson tree. We concentrate here on geometric Galton-Watson trees : each "vertex" has, independently of the others, an offspring distributed according to a geometric law $\mathcal{G}(\rho)$ of parameter $\rho \in [1/2, 1)$. We mean that the probability for a vertex to have k children is $\rho (1 - \rho)^k$ for every $k \in \mathbb{N} = \{0, 1, 2, \ldots\}$. The expectation of this law is $\frac{1}{\rho} - 1$ and then is smaller or equal to 1 for $\rho \in [1/2, 1)$ which implies the a.s. finitness of the tree. For basic facts on Galton-Watson trees such as the previous one the reader can refer to [AP1] Section 2.2. Moreover we let ρ depend on the height of the vertex, that is the number of generations computed from the root, but we still suppose that $\rho(\cdot) \in [1/2, 1)$. In the following we will denote $\mathcal{GW}(\rho(\cdot))$ this inhomogeneous Galton-Watson random tree. Let us perform a percolation with probability $p \in (0, 1)$ on the edges of this tree, that is, each edge is kept with probability p, independently of the others. Again the value of p may depend of the height in the tree i.e. we cut with probability $p(t)$ an edge linking two vertices of respective heights t and $t + 1$. The connected component of the root in the remaining tree is still a inhomogeneous Galton-Watson tree but now the generating function of the number $\tilde{N}$ of children of a vertex can be obtained as below, by conditioning on the number of children N that this vertex had in the original tree :

$$
\begin{aligned}
\mathbb{E}\left(s^{\tilde{N}}\right) &= \mathbb{E}\left[\mathbb{E}\left(s^{\tilde{N}}\middle| N\right)\right] \\
&= \mathbb{E}\left[(p\,s + 1 - p)^N\right] \\
&= \frac{\rho}{1 - (1 - \rho)(p\,s + 1 - p)} \\
&= \frac{\nu}{1 - (1 - \nu)\,s} \quad \text{where } \nu = \nu(\cdot) = \frac{\rho(\cdot)}{\rho(\cdot) + p(\cdot) - p(\cdot)\rho(\cdot)}.
\end{aligned}
$$

This means that this (inhomogeneous) Galton-Watson tree has a geometric offspring law of parameter $\nu(\cdot)$ given above.

A convenient way to describe a tree is to use the contour process of this tree. To our purpose it is more convenient to work with forests –in our case Galton-Watson forests– than trees. A forest is merely a sequence of independent trees. Such a forest can be seen as a sort of tree by connecting with an edge each root of the independent trees to an added vertex that we call the root of the forest. This would create an (infinite) tree with an infinite number of individuals at the first generation. We will still denote $\mathcal{GW}(\rho(\cdot))$ the forest consisting of independent (inhomogeneous) Galton-Watson trees with $\mathcal{G}(\rho(\cdot))$ offspring law where $\rho(\cdot)$ is a function of the height in the tree. The contour process of such a forest is a nearest-neighbour random walk $(X(k),\ k \in \mathbb{N})$ on $\mathbb{N}$, reflecting at 0 and with transition law given, for $j > 0$, by :

$$\mathbb{P}(X(k+1) = j - 1 | X(k) = j) = \rho(j) = 1 - \mathbb{P}(X(k+1) = j + 1 | X(k) = j)$$

and whose is denoted $\mathcal{RW}(\rho(\cdot))$. To be more specific we consider for this contour process that the roots of the independent trees belonging to the forest are at height 1 and the passage from one tree to the following consists for the contour process in a passage at 0. How can we interpret the percolation procedure of the tree on the contour process ? Cutting an edge and keeping only the part containing the root amounts to cut an excursion of the contour process. More precisely, let us denote by $(X(t), t \geq 0)$ the continuous time process which coincides with $(X(k))$ at all integer times and which is piecewise linear between those times. We denote $\mathcal{E}(X)$ the epigraph of $(X(t), t \geq 0)$ i.e. the set of points of $[0, +\infty)^2$ which are under the graph of $(X(t), t \geq 0)$:

$$\mathcal{E}(X) = \{(s,t) \in [0, +\infty)^2; \ X(s) > t\}. \tag{1}$$

For each $(s,t) \in \mathcal{E}(X)$ we denote by $[\alpha(X,s,t), \beta(X,s,t)]$ the excursion of X above level t and containing time s :

$$\alpha(X,s,t) = \sup\{s' < s; \ X(s') = t\}, \tag{2}$$
$$\beta(X,s,t) = \inf\{s' > s; \ X(s') = t\}. \tag{3}$$

For a non-negative function b on $\mathbb{N}$, we consider the Poisson point measure Λ with intensity

$$\sum_{(s,t) \in \mathbb{N}^2 \cap \mathcal{E}(X)} \frac{b(t)}{\beta(X,s,t) - \alpha(X,s,t) - 1} \delta_{(s,t)}$$

where $\delta_{(s,t)}$ denotes the Dirac measure at (s,t). Of course the law of Λ given above must be understood as a conditional law given X. The "part to cut" is defined as

$$\mathcal{C} = \bigcup_{(s,t) : \Lambda((s,t)) \neq 0} [\alpha(X,s,t), \beta(X,s,t)]. \tag{4}$$

We can now state the result of the above discussion.

Proposition 1 *The process $Y(s) = X(A(s))$ where*

$$A(s) = \inf \left\{ u; \ \int_0^u \mathbf{1}_{\{v \notin \mathcal{C}\}} \, dv > s \right\} \tag{5}$$

is the contour process of the connected component of the root after percolation at rate $p = e^{-b}$ of the forest having contour X. In other words, assuming that X is a (interpolated) random walk on $\mathbb{N}$ following the law $\mathcal{RW}(\rho(\cdot))$ as defined above then Y is a (interpolated) random walk on $\mathbb{N}$ distributed as $\mathcal{RW}(\nu(\cdot))$ where

$$\nu(\cdot) = \frac{\rho(\cdot)}{\rho(\cdot) + p(\cdot) - p(\cdot)\,\rho(\cdot)}.$$

We now want to see the counterpart of this result on diffusion processes, making such processes appear as limit of random walks. We consider a sequence of random walks on $\frac{1}{\sqrt{N}}\mathbb{N}$, at first indexed by $k \in \frac{1}{N}\mathbb{N}$, that we denote $(X_N(k); \; k \in \frac{1}{N}\mathbb{N})$, which is reflecting at 0 and has the transition law given, for every $j \in \frac{1}{\sqrt{N}}\mathbb{N} \setminus \{0\}$ and every $k \in \frac{1}{N}\mathbb{N}$, by

$$\mathbb{P}\left(X_N(k + \frac{1}{N}) = j - \frac{1}{\sqrt{N}} \,\Big|\, X_N(k) = j\right) = \frac{1}{2} + \frac{\theta_N(j)}{2\sqrt{N}}, \tag{6}$$

$$\mathbb{P}\left(X_N(k + \frac{1}{N}) = j + \frac{1}{\sqrt{N}} \,\Big|\, X_N(k) = j\right) = \frac{1}{2} - \frac{\theta_N(j)}{2\sqrt{N}} \tag{7}$$

where θ_N is a sequence of continuous non-negative functions on $\mathbb{R}_+$. We extend X_N to continuous time by linear interpolation between consecutive times of $k \in \frac{1}{N}\mathbb{N}$. Such a rescaled reflecting random walk will be denoted $\mathcal{RW}(N, \theta_N)$ from now on. In the case $\theta_N = 0$ we call it, as usual, a standard rescaled reflecting random walk. The following "classical" result, gives the limit in law when (θ_N) converges. It can be deduced from general results on interpolated Markov chains, for instance as stated in [Ku] except that it applies to a non-reflecting process. However we give a short proof in the appendix, as a corollary of Donsker Theorem, for the convenience of the reader and because this proof can easily be generalized to path-valued processes which is the setting of the applications we will develop in [Se]. We recall that a reflecting Brownian motion with non-positive drift $-\theta(\cdot)$ has the law of $(|Z_t|)$ where (Z_t) is a solution of the stochastic differential equation $dZ_t = dB_t - \text{sign}(Z_t)\,\theta(Z_t)\,dt$ where (B_t) is a standard Brownian motion.

Proposition 2 *If the sequence (θ_N) of non-negative continuous functions converges to the continuous function θ on $\mathbb{R}_+$, uniformly on compact sets of $\mathbb{R}_+$, then the law of the process $(X_N(s); \; s \geq 0)$ described above converges weakly to the law of a reflecting Brownian motion with drift $-\theta(\cdot)$.*

The procedure of cutting excursion described above on discrete random walks also makes sense on continuous time processes and we are able to state a continuous time analogue of Proposition 1. The following theorem shows that it is possible to create a negative drift on a linear Brownian motion by cutting certain excursions (case $\theta = 0$) or more generally to increase the negative drift of a Brownian motion.

Theorem 3 *Let $(X(t), \; t \geq 0)$ be a Brownian motion reflecting at 0 with continuous drift $-\theta(\cdot)$. Let b be a continuous function on $\mathbb{R}_+$ and Λ be a point measure which is, conditionally on X, a Poisson measure with intensity*

$$\frac{2\,b(t)}{\beta(X,s,t) - \alpha(X,s,t)} \, \mathbf{1}_{\mathcal{E}(X)}(s,t) \; ds\,dt$$

where $\mathcal{E}(\cdot)$, α and β are defined by (1,2,3). Also $\mathcal{C}$ is still defined by (4) and we set $Y(t) = X(A(t))$ where $A(\cdot)$ is given by (5).

Then $(Y(t),\ t \geq 0)$ *is a Brownian motion reflecting at 0 with drift* $-(\theta(\cdot) + b(\cdot))$.

Proof. We consider X_N a rescaled reflecting random walk $\mathcal{RW}(N, \theta)$ so that in particular, for $j \in \frac{1}{\sqrt{N}}\mathbb{N} \setminus \{0\}$,

$$\mathbb{P}\left(X_N(k + \frac{1}{N}) = j - \frac{1}{\sqrt{N}} \middle| X_N(k) = j\right) = \rho^N(j) = \frac{1}{2} + \frac{\theta(j)}{2\sqrt{N}}.$$

Let Λ_N be the Poisson point measure with intensity

$$\mu_N = \frac{1}{N\sqrt{N}} \sum_{(s,t) \in (\frac{1}{N}\mathbb{N} \times \frac{1}{\sqrt{N}}\mathbb{N}) \cap \mathcal{E}(X_N)} \frac{2\,b(t)}{\beta(X_N, s, t) - \alpha(X_N, s, t) - \frac{1}{N}}\, \delta_{(s,t)} \cdot$$

We set

$$\mathcal{C}_N = \bigcup_{(s,t)\,:\,\Lambda_N((s,t)) \neq 0} [\alpha(X_N, s, t), \beta(X_N, s, t)]$$

and

$$A_N(s) = \inf\left\{u;\ \int_0^u \mathbf{1}_{\{v \notin \mathcal{C}_N\}}\, dv > s\right\}$$

and, finally, $Y_N = X_N \circ A_N$. We apply Proposition 1 with a change of scale. We deduce that $(Y_N(k),\ k \in \frac{1}{N}\mathbb{N})$ is a random walk on $\frac{1}{\sqrt{N}}\mathbb{N}$ reflecting at 0 and with transition probabilities given, for $j \in \frac{1}{\sqrt{N}}\mathbb{N}$ by

$$\mathbb{P}\left(Y_N(k + \frac{1}{N}) = j - \frac{1}{\sqrt{N}} \middle| Y_N(k) = j\right) = \frac{\rho^N(j)}{1 - (1 - \rho^N(j))\left(1 - e^{-\frac{2b(j)}{\sqrt{N}}}\right)}$$

$$= \frac{1}{2}\left(1 + \frac{\theta(j) + b(j) + \varepsilon_N(j)}{\sqrt{N}}\right)$$

where ε_N is a function converging to 0 uniformly on compact sets. We now let $N \to +\infty$. By Proposition 2 we know that Y_N converges in law to a Brownian motion reflecting at 0 with drift $-(\theta(\cdot) + b(\cdot))$. Proposition 2 also applies to X_N. By Skorohod representation Theorem we may suppose that X_N converges to X uniformly on compact sets of $\mathbb{R}_+$, almost surely. Then, skipping technicalities explained in the proof of Proposition 4 of [AS], μ_N is shown to converge to the intensity given in the Theorem and we deduce that $Y_N = X_N \circ A_N$ converges to $Y = X \circ A$. We conclude that the law of Y is as stated.

3 How to create a positive drift ?

We have seen in the previous section that we can create a negative drift on a Brownian trajectory by cutting excursions. Conversely is it possible to reduce or even annihilate a negative drift by adding excursions ? The answer is affirmative as stated in Theorem 7 for total annihilation of the drift and Theorem 9 for reduction of the drift.

3.1 Graft on a Galton-Watson tree

Our first issue is the way to transform a subcritical geometric Galton-Watson tree into another one with bigger progeny expectation. We start with an elementary lemma on the geometric law whose proof is left to the reader.

Lemma 4 *Let $0 < \rho < \nu < 1$, Z_ν and Z_ρ be independent random variables distributed according to the respective laws $\mathcal{G}(\nu)$ and $\mathcal{G}(\rho)$. Let U be an independent Bernoulli variable with expectation $p = \frac{\nu-\rho}{\nu-\nu\rho}$. Then $Z_\nu + U Z_\rho$ is distributed according to the geometric law $\mathcal{G}(\rho)$.*

We will now apply this elementary result to the "decoration" of a Galton-Watson tree. Considering a forest $\mathcal{GW}(\nu(\cdot))$ we add to each vertex v, with probability $p(h(v))$ depending of the height $h(v)$ of vertex v, an independent tree $\mathcal{GW}(\rho(h(v)+\cdot))$ rooted at v. In case of addition effectively occuring at vertex v, the added tree is placed at the right of the subtrees already born at vertex v.

Proposition 5 *The forest obtained from the forest $\mathcal{GW}(\nu(\cdot))$ by "decoration" at probability $p(\cdot) = \frac{\nu(\cdot)-\rho(\cdot)}{\nu(\cdot)-\nu(\cdot)\rho(\cdot)}$ using $\mathcal{GW}(\rho(\cdot))$–trees as described above, is a $\mathcal{GW}(\rho(\cdot))$–forest.*

Proof. The independence properties being clearly satisfied, it suffices to prove that a vertex v of the decorated tree has a progeny distributed according to the $\mathcal{G}(\rho(h(v)))$–geometric law. This is obvious if v is supposed to belong to one of the added trees. Otherwise the vertex v belonged to the original forest and the number of its children is $Z_1 + U Z_2$ where Z_1 is the original number of children, distributed as $\mathcal{G}(\nu(h(v)))$, Z_2 is the number of children possibly added, and U equals 1 if decoration occurs at vertex v. But the lemma shows that this variable is $\mathcal{G}(\rho(h(v)))$–geometric. $\blacksquare$

3.2 Translation to random walks

We can translate this result in the language of random walks. The contour process of a $\mathcal{GW}(\nu(\cdot))$–forest is a random walk $(X(k), k \in \mathbb{N})$ on $\mathbb{N}$, reflecting at 0 whose law is, as denoted before, $\mathcal{RW}(\nu(\cdot))$. Moreover we prolong this random walk into a continuous time process $(X(s), s \geq 0)$ by linear interpolation between consecutive integer times. For each $(s,t) \in \mathbb{N} \times \mathbb{N}$, we define $E_{(s,t)}$ as the contour process of a $\mathcal{GW}(\rho(t+\cdot))$–tree i.e. a reflecting random walk that is going down with probability $\rho(t+\cdot)$ and stopped after a number of returns to 0 which is equal to the progeny of the first generation thus distributed as $\mathcal{G}(\rho(t))$. These processes are supposed to be independent. We set

$$p(t) = \frac{\nu(t) - \rho(t)}{\nu(t) - \nu(t)\,\rho(t)}.$$

Conditionally on $(X(s))$, for every $s \in \mathbb{N}$ such that $X(s+1) = X(s) - 1$, with probability $p(X(s))$, the walk $E_{(s,X(s))}$ is inserted in the graph of X at $(s, X(s))$. Let Y be the walk obtained after these insertions have been done altogether(they are of finite number on a bounded time interval). The reader wishing to see formulas describing this procedure can refer to the proof of Proposition 7. We let the reader check that this procedure of insertion in the graph of X to obtain Y is the translation into the language of contour process of the "decoration" procedure of a forest described in Proposition 5. Therefore we can conclude on the following result.

Proposition 6 *Y is the contour of a $\mathcal{GW}(\rho(\cdot))$–forest and as a consequence is a reflecting random walk with law $\mathcal{RW}(\rho(\cdot))$.*

3.3 From discrete time to continous time

The problem now consists in stating a continuous time analogue of Proposition 6.

Theorem 7 *Let $(X(t), \ t \geq 0)$ be a Brownian motion reflecting at 0 with continuous non-positive drift $-\theta(\cdot)$. We define, conditionally on X, a Poisson point measure Λ on $\mathbb{R}_+ \times C(\mathbb{R}_+, \mathbb{R}_+)$ with intensity*

$$2\,\theta(X(s))\,ds\,n(de) \tag{8}$$

where $n(\cdot)$ denotes the Itô measure of positive excursions of Brownian motion. Let $\sigma(e)$ denote the duration (length) of an excursion e. The function

$$A_u = u + \int_{\{s \leq u\}} \sigma(e)\,\Lambda(ds\,de)$$

is increasing right-continuous and has a jump $A_u - A_{u-} = \sigma(e_u)$ for every u such that $\Lambda(\{(u, e_u)\}) \neq 0$. We define $(Y(v))_{v \geq 0}$ by $Y(v) = X(u)$ if $v = A_u$ and $Y(v) = X(u) + e_u(v - A_{u-})$ for $A_{u-} \leq v < A_u$.
Then $(Y(t), \ t \geq 0)$ is a Brownian motion reflecting at 0.

Proof. We let $(X_N(s), \ s \in \frac{1}{N}\mathbb{N})$ be a $\mathcal{RW}(N, \theta)$ random walk. Our first goal is to apply Proposition 6 to X_N so that we now set

$$\nu^N(t) = \frac{1}{2} + \frac{\theta(t)}{2\sqrt{N}}, \quad \rho^N(t) = \frac{1}{2}$$

and

$$p^N(t) = \frac{\nu^N(t) - \rho^N(t)}{\nu^N(t) - \nu^N(t)\rho^N(t)} = \frac{2\,\theta(t)/\sqrt{N}}{1 + \theta(t)/\sqrt{N}}. \tag{9}$$

We let $(U_s, \ s \in \frac{1}{N}\mathbb{N})$ be a family of independent uniform variables on $(0,1)$ and $(B(N,s), \ s \in \frac{1}{N}\mathbb{N})$ be independent copies of a rescaled reflecting standard $\mathcal{RW}(N,0)$ random walk B^N, stopped at the time of the g-th return at 0 where

g is an independent random variable with law $\mathcal{G}(1/2)$. We consider the point measure Λ^N on $\mathbb{R}_+ \times C(\mathbb{R}_+, \mathbb{R}_+)$ given by

$$\Lambda^N = \sum_{s \in \frac{1}{N}\mathbb{N}} \mathbf{1}_{\{X_N(s+\frac{1}{N})=X_N(s)-\frac{1}{\sqrt{N}}\}} \mathbf{1}_{\{U_s \leq p^N(X_N(s))\}} \, \delta_{(s,B(N,s))}. \tag{10}$$

We set

$$A_u^N = u + \int_{\{s \leq u\}} \Lambda^N(ds\,de) \; \sigma(e)$$

where $\sigma(e)$ is the time of the last return to 0 of e. We define $(Y^N(v))_{v \geq 0}$ by $Y^N(v) = X^N(u)$ if $v = A_u^N$ and $Y^N(v) = X^N(u) + e_u(v - A_u^N)$ if

$$A_{u-}^N \leq v < A_u^N = A^N(u-) + \sigma(e_u) \text{ where } \Lambda^N(\{(u, e_u)\}) \neq 0.$$

The effect of this time change is to insert, at point $(s, X_N(s))$ preceding a descent of X_N, with probability $p^N(X_N(s))$, a rescaled reflecting standard random walk stopped after a number of return to 0 distributed according to $\mathcal{G}(1/2)$. It follows from Proposition 6 that $(Y^N(s))$ is a rescaled reflecting standard random walk. We now let $N \to +\infty$. As before, we may suppose that $(X_N(s))$ converges uniformly on every compact to $(X(s))$, almost surely, where $(X(s))$ is a Brownian motion with drift $-\theta(\cdot)$. Also Y^N converges in law to a Brownian motion, reflecting at 0.

From now on we denote $C^*(\mathbb{R}_+, \mathbb{R}_+)$ the set of the $e \in C(\mathbb{R}_+, \mathbb{R}_+)$ such that there exists $\sigma(e) = \inf\{s; \forall s' \geq s, \, e(s') = 0\}$. Let us consider an interval $p \in \mathbb{Z}_+$ and $K_1, \ldots, K_p$ be disjoint Borel subsets of $C^*(\mathbb{R}_+, \mathbb{R}_+) \cap \{\sigma \geq \eta\}$. We thus have for every $i \leq p$, $n(K_i) < +\infty$ and we suppose moreover that $n(\partial K_i) = 0$. Let $\lambda_1, \ldots, \lambda_p$ be positive real numbers. Let us define, for any interval $[a, b]$ of $\mathbb{R}_+$, the set $D_N(a, b)$ consisting in the times of descents of X_N over $[a, b]$:

$$D_N(a, b) = \left\{ s \in [a, b] \cap \frac{1}{N}\mathbb{N}; \; X_N\left(s + \frac{1}{N}\right) = X_N(s) - \frac{1}{\sqrt{N}} \right\}.$$

Then, for $0 < t_1 < t_2$, we have

$$\log \mathbb{E}\left[\exp - \left(\sum_{i=1}^{p} \lambda_i \, \Lambda^N([t_1, t_2] \times K_i) \right) \Big| X_N \right] \tag{11}$$

$$= \sum_{s \in D_N(t_1, t_2)} \log\left(1 + \sum_{i=1}^{p} (e^{-\lambda_i} - 1) \, \mathbb{P}[B(N, s) \in K_i] \, \mathbb{P}[U_s \leq p^N(X_N(s))|X_N] \right)$$

But, as $N \to +\infty$,

$$\mathbb{P}(B(N, s) \in K_i) \, \mathbb{P}(U_s \leq p^N(X_N(s))|X_N) \sim \frac{4\,\theta(X(s))\;n(K_i)}{N}. \tag{12}$$

We have used Equation (9) which shows

$$p^N(X_N(s)) \sim \frac{2}{\sqrt{N}} \, \theta(X(s))$$

and Lemma 13 below which asserts that

$$\sqrt{N} \, P(B^N \in K_i) \to 2 \, n(K_i).$$

At this point we need the following lemma.

Lemma 8 *Let $0 < t_1 < t_2$ and φ be a continuous function on $[t_1, t_2]$. Then, as $N \to +\infty$, almost surely,*

$$\frac{1}{N} \sum_{s \in D_N(t_1, t_2)} \varphi(s) \to \frac{1}{2} \int_{t_1}^{t_2} \varphi(s) \, ds$$

Proof of the lemma. For any interval $[a, b]$ of $\mathbb{R}_+$, we denote $\#D_N(a, b)$ the number of descents of X_N over $[a, b]$ i.e. the number of s in $D_N(a, b)$. We introduce a partition $[t_1 = s_0 < s_1 < \ldots < s_{k+1} = t_2]$ of the interval $[t_1, t_2]$. By immediate bounds,

$$\frac{1}{N} \sum_{s \in D_N(t_1, t_2)} \varphi(s) \le \frac{1}{N} \sum_{i=0}^{k} \#D_N(s_i, s_{i+1}) \sup_{[s_i, s_{i+1}]} \varphi. \qquad (13)$$

But an elementary count of climbs and descents gives

$$2 \, \#D_N(s_i, s_{i+1}) = N \, (s_{i+1} - s_i) - \sqrt{N}(X_N(s_{i+1}) - X_N(s_i))$$

which implies that, almost surely,

$$\#D_N(s_i, s_{i+1}) \sim N \, \frac{s_{i+1} - s_i}{2}.$$

From Inequality (13), we deduce

$$\limsup_{N \to +\infty} \frac{1}{N} \sum_{s \in D_N(t_1, t_2)} \varphi(s) \le \frac{1}{2} \sum_{i=0}^{k} (s_{i+1} - s_i) \sup_{[s_i, s_{i+1}]} \varphi.$$

When the stepsize of the considered subdivision goes to 0, the right-hand side above converges to $\int_{t_1}^{t_2} \varphi/2$. By symmetrical bounds we obtain obviously

$$\liminf_{N \to +\infty} \frac{1}{N} \sum_{s \in D_N(t_1, t_2)} \varphi(s) \ge \frac{1}{2} \int_{t_1}^{t_2} \varphi$$

and the proof of the lemma is complete.

Coming back to the proof of the theorem and more precisely Equation (11), we see, using (12) and a Taylor expansion of the logarithm, that the right-hand side in (11) has the same asymptotic behaviour as

$$\sum_{s \in D_N(t_1,t_2)} \sum_{i=1}^{p} (e^{-\lambda_i} - 1) \frac{4\,\theta(X(s))\;n(U_i)}{N}.$$

To see this, the reader can note that (12) holds uniformly for $s \in [t_1, t_2]$. But the previous lemma implies that this quantity converges, as $N \to +\infty$, to

$$2 \int_{t_1}^{t_2} ds\;\theta(X(s)) \sum_{i=1}^{p} (e^{-\lambda_i} - 1)\;n(U_i).$$

So, it follows from the preceding derivation and from obvious independence properties that, for a function $\varphi : \mathbb{R}_+ \times C^*(\mathbb{R}_+, \mathbb{R}_+) \to \mathbb{R}_+$ of the type

$$\varphi(s,e) = \sum_{i,j} \lambda_{i,j}\;\mathbf{1}_{[t_i,t_{i+1})}(s)\;\mathbf{1}_{U_j}(e)$$

with $U_j \subset \{e; \sigma(e) \le \eta\}$ for $\eta > 0$, we have

$$\lim_{N \to +\infty} \mathbb{E}\left[\exp \int_0^t \varphi(s,e)\,\Lambda^N(ds\,de) \Big| X_N\right]$$

$$= \exp\left[2 \int n(de) \int_0^t ds\;\theta(X(s)) \left(e^{\varphi(s,e)} - 1\right)\right].$$

We deduce the convergence in law of Λ^N toward a Poisson measure Λ on $\mathbb{R}_+ \times C(\mathbb{R}_+, \mathbb{R}_+)$ with intensity

$$2\,\theta(X(s))\;ds\;n(de).$$

By Skorokhod representation theorem, we may even suppose that the Λ^N are such that, almost surely, the measure Λ^N converges weakly to the measure Λ, when restricted to any set $[0,t] \times \{\sigma > \eta\}$ where $\eta, t > 0$. It follows that we can suppose that on each such set, the atoms of Λ^N converge to the atoms of Λ. This implies that

$$u \to \int_{\{s \le u\}} \Lambda^N(ds\,de)\;\mathbf{1}_{\{\sigma(e) \ge \eta\}}\;\sigma(e)$$

converges to

$$u \to \int_{\{s \le u\}} \Lambda(ds\,de)\;\mathbf{1}_{\{\sigma(e) \ge \eta\}}\;\sigma(e)$$

in the Skorokhod topology on càdlàg functions. Moreover

$$\mathbb{E}\left(\int_{\{s \le u\}} \Lambda^N(ds,de)\;\mathbf{1}_{\{\sigma(e) \le \eta\}}\;\sigma(e)\right)$$

$$\le \mathbb{E}\left(\sum_{s \le u, s \in \frac{1}{N}\mathbb{N}} \sigma(B(N,s))\;\mathbf{1}_{\{\sigma(B(N,s)) \le \eta\}}\;p(X_N(s))\right)$$

$$\le c\,\sqrt{N}\;\mathbb{E}\left(\sigma(B^N)\;\mathbf{1}_{\{\sigma(B^N) \le \eta\}}\right)$$

and this last quantity is small for all N, provided η is chosen small enough, by Lemma 14. We can thus neglect small durations up to a set of small probability. We deduce the convergence in probability of Y^N as defined above to Y as defined in the statement of the theorem. But we know that the limit in law of Y^N is Brownian motion so we can conclude on the law of Y and the proof is complete.

At the price of a complexification of the notations, the ideas of the previous proof show that we can also reduce a drift $-\theta$ to $b-\theta \leq 0$ by adding excursions of a Brownian motion (subjected itself to the drift $b - \theta$).

Theorem 9 *Let $(X(t),\ t \geq 0)$ be a Brownian motion reflecting at 0 with continuous non-positive drift $-\theta(\cdot)$. Let b be a non-negative continuous function on $\mathbb{R}_+$ such that $b \leq \theta$. We define, conditionally on X, a Poisson point measure Λ on $\mathbb{R}_+ \times C(\mathbb{R}_+, \mathbb{R}_+)$ with intensity*

$$2\, b(X(s))\ ds\, n_{(X(s))}(de) \tag{14}$$

where $n_{(t)}$ denotes the Itô measure of positive excursions of Brownian motion with drift $(b - \theta)(t + \cdot)$. The function

$$A_u = u + \int_{\{s \leq u\}} \sigma(e)\, \Lambda(ds\, de)$$

is increasing right-continuous and has a jump $A_u - A_{u-} = \sigma(e_u)$ for every u such that $\Lambda(\{(u, e_u)\}) \neq 0$. We define $(Y(v))_{v \geq 0}$ by $Y(v) = X(u)$ if $v = A_u$ and $Y(v) = X(u) + e_u(v - A_{u-})$ for $A_{u-} \leq v < A_u$.
Then $(Y(t),\ t \geq 0)$ is a Brownian motion reflecting at 0 with drift $-\theta(\cdot) + b(\cdot)$.

3.4 Extension and applications

Theorem 7 remains true if $(X(t))$ is a Brownian motion starting from $x \in \mathbb{R}$ with continuous non-positive drift $-\theta(\cdot)$ and in this case $(Y(t))$ is a Brownian motion (starting from $x \in \mathbb{R}$).

Indeed we can look at Theorem 7 when applied on $[T_x, T_0']$ where T_x is the hitting time of $x > 0$ and T_0' the following return to 0. As x is arbitrary and by translation invariance this prove the result mentioned above for a Brownian motion starting from $x \in \mathbb{R}$ up to the hitting time of any lower value. We give an application establishing a connection between Brownian motion and Brownian motion with drift.

Proposition 10 *Let $\mathbb{P}_x$ and $\mathbb{P}_x^\theta$ denote respectively the law of Brownian motion and Brownian motion with continuous non-positive drift $-\theta(\cdot)$, both starting at $x > 0$. Let T_0 denote the hitting time of 0. Then we have*

$$\mathbb{E}_x^\theta \left[\exp - \int_0^{T_0} g(B_s)\, ds \right] = \mathbb{E}_x \left[\exp - \int_0^{T_0} f(B_s)\, ds \right] \tag{15}$$

when f, g are continuous non-negative functions such that $f + \theta\, \tilde{f} = g$ where $\tilde{f}$ denotes the function

$$\tilde{f}(x) = 2 \int n(de) \left(1 - \exp - \int_0^\sigma f(x + e_r)\, dr \right)$$

which is a solution of the Ricatti differential equation $y' = -2\, f + y^2$.

Proof. The Equality (15) is a straightforward consequence of Theorem 7 and the classical exponential formula for Poisson measures. We then sketch the proof that $\tilde{f}$ satisfies the given Ricatti equation:

$$\frac{1}{2}\, \tilde{f}(x) = \int n(de) \int_0^\sigma \exp - \left(\int_0^s f(x + e_r)\, dr \right) f(x + e_s)\, ds$$

$$= \int_0^{+\infty} dy\, f(x + y)$$

$$\cdot \exp -2 \left[\int_0^y dh \int n(de) \left(1 - \exp - \int_0^\sigma f(x + h + e_u)\, du \right) \right]$$

$$= \int_0^{+\infty} dy\, f(x + y)\, \exp -2 \int_0^y dh\, \tilde{f}(x + h)$$

$$= \int_x^{+\infty} dy\, f(y)\, \exp -2 \int_x^y dh\, \tilde{f}(h)$$

The first equality is elementary calculus, the third one uses only the definition of $\tilde{f}$ and the fourth one is a change of variables. The second one involves more sophisticated arguments; first Bismut's description of the Brownian excursion under Itô measure; then we use the excursions above the future infimum of $(e(r), r \le s)$ which is a three dimensional Bessel process run up to a hitting time; these excursions have the same intensity as the excursions of a reflected Brownian motion and we finish with the exponential formula. Finally the last equality leads easily to the Ricatti equation.

As a (trivial) example consider the case of constant θ and f so that g is also constant. We obtain, using the well-known Laplace transform of T_0 under $\mathbb{P}_x$,

$$\mathbb{E}_x^\theta \left[e^{-g\, T_0} \right] = \mathbb{E}_x \left[e^{-f\, T_0} \right] = e^{-x\, \sqrt{2f}} = e^{-x \left(\sqrt{\theta^2 + 2g} - \theta \right)}$$

as could also be obtained by an application of Girsanov Theorem.

More sophisticated applications, in the setting of super-processes will be given in [Se].

4 Appendix

4.1 Proof for Proposition 2

To simplify notation we restrict ourselves to the convergence of $(X_N(s),$ $s \in [0,1])$. We denote (U_N) a reflecting and rescaled standard walk $\mathcal{RW}(N,0)$. Let F be a continuous function on $C([0,1], \mathbb{R}_+)$. By the definition of the law of X_N given by Formulas (6, 7), we have

$$\mathbb{E}\left[F(X_N)\right] = \mathbb{E}\left[F(U_N)\right.$$
$$\left. \times \prod_{k=0}^{N-1}\left(1 - \mathbf{1}_{\{U_N(\frac{k}{N}) \neq 0\}}\left(U_N(\frac{k+1}{N}) - U_N(\frac{k}{N})\right)\theta_N(U_N(\frac{k}{N}))\right)\right].$$

We introduce a reflecting Brownian motion $(B_s)_{s \in \mathbb{R}_+}$, starting from $B_0 = 0$ and the stopping times : $T_0^N = 0$,

$$T_{k+1}^N = \inf\left\{s > T_k^N, \; |B_s - B_{T_k^N}| = \frac{1}{\sqrt{N}}\right\}.$$

It is clear that $(B_{T_k^N}, \; 0 \leq k \leq N)$ is identically distributed as $(U_N(k/N), 0 \leq k \leq N)$. We set

$$B_s^N = B_{T_k^N} + (N s - k)(B_{T_{k+1}^N} - B_{T_k^N}) \text{ for } s \in [k/N, (k+1)/N].$$

We get

$$\mathbb{E}\left[F(X_N)\right] = \mathbb{E}\left[F(B^N)\,\exp(L_N)\right]$$

where

$$L_N = \sum_{k=0}^{N-1} \log\left(1 - \mathbf{1}_{\{B_{T_k^N} \neq 0\}}\,(B_{T_{k+1}^N} - B_{T_k^N})\,\theta_N(B_{T_k^N})\right)$$
$$= -\sum_{k=0}^{N-1} \mathbf{1}_{\{B_{T_k^N} \neq 0\}}\,(B_{T_{k+1}^N} - B_{T_k^N})\,\theta_N(B_{T_k^N})$$
$$-\frac{1}{2}\frac{1}{N}\sum_{k=0}^{N-1} \mathbf{1}_{\{B_{T_k^N} \neq 0\}}\,\theta_N(B_{T_k^N})^2 + R_N$$

with R_N being a remainder which converges to 0 in probability. By the Markov property for B and the scaling property of Brownian motion we can write, for $k \leq N$,

$$T_k^N = \frac{1}{N}\sum_{j=1}^{k} V_j$$

where $V_1, V_2, \ldots$ are independent and distributed as the hitting time of $\{1, -1\}$ for a Brownian motion starting from 0. In particular $\mathbb{E}(V_1) = 1$. By Kolmogorov's Lemma (or Doob's inequality) we deduce for $\varepsilon > 0$, that

$$\mathbb{P}\left[\sup_{k \leq N} |T_k^N - \frac{k}{N}| \geq \varepsilon\right] = \mathbb{P}\left[\frac{1}{N} \sup_{k \leq N} \left|\sum_{j=1}^{k}(V_j - 1)\right| \geq \varepsilon\right]$$

$$\leq \frac{1}{\varepsilon^2 N}\operatorname{Var}(V_1).$$

This shows that $\sup_{k \leq N} |T_k^N - \frac{k}{N}|$ converges to 0 in probability as $N \to +\infty$ and thus almost surely along a subsequence. Then it follows that $B_s^N \to B_s$, uniformly in s, almost surely when $N \to +\infty$ along the previous subsequence.

Noting $T_N^N \to 1$, a. s., it follows from standard arguments (see for instance [RY] Proposition IV.2.13) that, a. s., for $N \to +\infty$ along a subsequence,

$$L_N \to L = -\int_0^1 \theta(B_s) \, \mathbf{1}_{\{B_s \neq 0\}} \, dB_s - \frac{1}{2}\int_0^1 \theta(B_s)^2 \, ds.$$

Since the extraction of a converging subsequence can be made from any sequence along which N goes to infinity, we claim that $F(B^N)\exp L_N \to F(B)\exp L$ in probability. It is easy to prove, by using induction and the Markov property for B, that

$$\sup_N \mathbb{E}\left[\left(F(B^N)\exp L_N\right)^2\right] < +\infty.$$

We conclude that $\mathbb{E}[F(X_N)] \to \mathbb{E}[F(B)\exp L]$ and this is, via Girsanov Theorem, the desired result.

4.2 Convergence of discrete excursions and walks

Lemma 11 *Let $(e^N(s), \ s \geq 0)$ be an excursion of the rescaled reflecting standard random walk $\mathcal{RW}(N, 0)$. Let $\eta > 0$ and F be a bounded continuous function on $C^*(\mathbb{R}_+, \mathbb{R}_+)$ null on $\{\sigma < \eta\}$.*
Then, we have

$$\sqrt{N} \, \mathbb{E}[F(e^N)] \stackrel{N \to +\infty}{\Longrightarrow} 2\int F(e) \, n(de)$$

Proof. Let B be a standard (non-rescaled) random walk on $\mathbb{N}$ starting from 0 and stopped at its first return to 0, denoted $\sigma(B)$ so that e^N is the renormalization of B by $1/N$ in time and $1/\sqrt{N}$ in space. A classical exercise on reflection principle gives that

$$\mathbb{P}(\sigma(B) = 2n) = \binom{2n}{n}\frac{2^{-2n}}{2n - 1} \sim \frac{1}{\sqrt{\pi n}\, 2n}. \qquad (16)$$

We deduce that

$$\mathbb{P}\left(\sigma(e^N) \geq \eta\right) \sim \frac{\sqrt{2}}{\sqrt{\eta \, N \, \pi}}.$$

It is well known (see for instance [RY] Proposition XII.2.8) that

$$n\left(\sigma(e) \geq \eta\right) = \frac{1}{\sqrt{2\pi \, \eta}}.$$

So, it suffices to prove that

$$\mathbb{E}\left[F(e^N)\big|\sigma(e^N) \geq \eta\right] \overset{N \to +\infty}{\longrightarrow} \int F(e) \, n(de|\sigma(e) \geq \eta).$$

This is a conditioned version of Donsker invariance Theorem for which we refer to [Ka].

Lemma 12 Let $(B^N(s), \ s \geq 0)$ be a standard rescaled reflecting random walk $\mathcal{RW}(N,0)$, stopped at the time of the g-th return at 0. Let $\eta > 0$ and F be a bounded continuous function on $C^*(\mathbb{R}_+, \mathbb{R}_+)$ null on $\{\sigma < \eta\}$.

Then, we have

$$\sqrt{N} \, \mathbb{E}\left[F(B^N)\right] \overset{N \to +\infty}{\longrightarrow} g \, 2 \int F(e) \, n(de). \tag{17}$$

Proof. For simplicity of notations let us suppose in fact that F vanishes on $\{\sigma < g\eta\}$. We denote $e_1^N, \ldots, e_g^N$ the excursions of B^N. We have to work on the event that at least one of these excursions has a duration greater than η. From the proof of Lemma 11, we recall that

$$\mathbb{P}\left(\sigma(e_i^N) \geq \eta\right) \leq \frac{c}{\sqrt{N}}$$

so the event that two excursions are of duration larger than η is of order $1/N$ and can be asymptotically neglected. We set $H(x.) = \sup_s |x_s|$ for $x. \in C^*(\mathbb{R}_+, \mathbb{R}_+)$. The renormalization done on e^N shows that

$$\mathbb{P}\left(H(e^N) \geq \varepsilon(N)\right) \to 0 \ \text{if} \ \sqrt{N}\,\varepsilon(N) \to +\infty$$

and similarily

$$\mathbb{P}\left(\sigma(e^N) \geq \varepsilon(N)\right) \to 0 \ \text{if} \ N\,\varepsilon(N) \to +\infty.$$

From now on, we fix $\varepsilon(N) \to 0$ such that the first (hence both) of the above conditions hold. We work on one of the g events

$$\{\sigma(e_i^N) \geq \eta, \ \forall j \neq i, \ H(e_j^N) \leq \varepsilon(N), \ \sigma(e_j^N) \leq \varepsilon(N)\}$$

where $i \in \{1, \ldots, g\}$. On such an event, we have

$$|B^N(s) - e_i^N(s)| \leq \varepsilon(N) + \sup_{s \leq g\varepsilon(N)} |e_i^N(s)| + \sup_{s,\,u \leq g\varepsilon(N)} |e_i^N(s+u) - e_i^N(s)|, \tag{18}$$

because $|B^N(s) - e_i^N(s)|$ is smaller than

$$\sup_{r,\ j>i} |e_j^N(r)| \text{ if } s \geq \sum_{j \leq i} \sigma(e_j^N),$$

or is lower than

$$\sup_r \left| e_i^N \left(r + \sum_{j<i} \sigma(e_j^N) \right) - e_i^N(r) \right| \text{ if } \sum_{j<i} \sigma(e_j^N) \leq s \leq \sum_{j \leq i} \sigma(e_j^N),$$

or is lower than

$$\sup_{r,\ j<i} |e_j^N(r)| + \sup_{r \leq \sum_{j<i} \sigma(e_j^N)} |e_i^N(r)| \text{ if } s \leq \sum_{j<i} \sigma(e_j^N).$$

But under $\sqrt{N}\ \mathbb{P}$ restricted to $\{\sigma \geq \eta\}$, e_j^N converges in distribution to n restricted to $\{\sigma \geq \eta\}$. It follows that the right-hand-side of (18) converges to 0 in probability.

We deduce from these facts that the left-hand side in (17) has the same limit as $g \sqrt{N}\ \mathbb{E}\left[F(e^N)\right]$ and this one is given by Lemma 11.

Lemma 13 *Let $(B^N(s),\ s \geq 0)$ be a standard rescaled reflecting random walk $\mathcal{RW}(N,0)$, stopped at the time of the g-th return at 0 where g is an independent random variable with law $\mathcal{G}(\frac{1}{2})$. Then, for any mesurable $U \subset C^*(\mathbb{R}_+, \mathbb{R}_+) \cap \{\sigma > \eta\}$ with $\eta > 0$ such that $n(\partial U) = 0$,*

$$\sqrt{N}\ \mathbb{P}[B^N \in U] \stackrel{N \to +\infty}{\longrightarrow} 2\, n(U).$$

Proof. We first randomize g in the limit (17) according to the law $\mathcal{G}(\frac{1}{2})$ as specified here. But this law has mean 1 so (17) is now re-expressed in our new setting by replacing "g" by 1. A reformulation of this result of limit in the language of sets is the above statement.

Lemma 14 *Let $(B^N(s),\ s \geq 0)$ be a standard rescaled reflecting random walk $\mathcal{RW}(N,0)$, stopped at the time of the g-th return at 0 where g is an independent random variable with law $\mathcal{G}(\frac{1}{2})$. Let $\varepsilon > 0$.*
Then there exists $\eta > 0$ such that, for every N,

$$\sqrt{N}\ \mathbb{E}\left(\sigma(B^N)\ \mathbf{1}_{\{\sigma(B^N) \leq \eta\}}\right) \leq \varepsilon.$$

Proof. By conditioning by the value of g, it suffices to prove the same result with B^N replaced by e^N. As before, we denote by B a standard non-rescaled random walk on $\mathbb{N}$ starting from 0 and stopped at its first return to 0 i.e. e^N is the renormalization of B by $1/N$ in time and $1/\sqrt{N}$ in space. We now have to find $\eta > 0$ such that, for every N,

$$\frac{1}{\sqrt{N}}\ \mathbb{E}\left[\sigma(B)\ \mathbf{1}_{\{\sigma(B) \leq \eta N\}}\right] \leq \varepsilon.$$

From Formula (16) we deduce

$$\mathbb{E}\left(\sigma(B)\,\mathbf{1}_{\{\sigma(B)\leq\eta N\}}\right)\leq c\sum_{n=1}^{\eta N/2}\frac{1}{\sqrt{n}}\,.$$

But this quantity behaves like $\sqrt{\eta N}$ and the proof of the lemma is complete.

References

[AS1] Abraham R., Serlet L.: Representations of the Brownian snake with drift. Stochastics and stochastic Reports 73, 287-308 (2002)

[AS2] Abraham R., Serlet L.: Poisson snake and fragmentation. Elec. J. Probab. 7, 1-15 (2002)

[AP1] Aldous D., Pitman J.: Tree valued Markov chains derived from Galton-Watson processes. Ann. Inst. Henri Poincaré 34 (5), 637-686 (1998)

[AP2] Aldous D., Pitman J.: The standard additive coalescent. Ann. Probab. 26, 1703-1726 (1998)

[DW] Duquesne T., Winkel M.: Growth of Lévy trees. Available at arXiv:math.PR/0509518 (2006)

[EW] Evans S.N., Winter A.: Subtree prune and re-graft : a reversible real tree valued Markov process. Ann. Probab. 34 (3) 918-961 (2006)

[Ka] Kaigh W.D.: An invariance principle for random walk conditioned by a late return to zero. Ann. Probab. 4, 115-121 (1976)

[Ku] Kushner H.J.: On the weak convergence of interpolated Markov chains to a diffusion. Ann. Probab. 2 (1), 40-50 (1974)

[PW] Pitman J., Winkel M.: Growth of the Brownian forest. Ann. Probab. 33, 2188-2211 (2005)

[RY] Revuz D., Yor M.: Continuous martingales and Brownian motion, third edition. Springer-Verlag (1999)

[Se] Serlet L.: Survival of a snake in an hostile environment. Available at http://math.univ-bpclermont.fr/~serlet/publications.html (2006)

Extending Chacon-Walsh: Minimality and Generalised Starting Distributions

A.M.G. Cox

Department of Mathematics, University of York
York YO10 5DD, U. K.
e-mail: amgc500@york.ac.uk

Summary. In this paper we consider the Skorokhod embedding problem for general starting and target measures. In particular, we provide necessary and sufficient conditions for a stopping time to be minimal in the sense of Monroe. The resulting conditions have a nice interpretation in the graphical picture of Chacon and Walsh.

Further, we demonstrate how the construction of Chacon and Walsh can be extended to any (integrable) starting and target distributions, allowing the constructions of Azéma-Yor, Vallois and Jacka to be viewed in this context, and thus extended easily to general starting and target distributions. In particular, we describe in detail the extension of the Azéma-Yor embedding in this context, and show that it retains its optimality property.

Key words: Brownian Motion, Embedding, Azéma-Yor Embedding, Stopping Time, Minimal Stopping Time, Chacon-Walsh Construction, Balayage

MSC 2000 subject classifications. Primary: 60G40, 60J60; Secondary: 60G44, 60J65.

1 Introduction

The Skorokhod embedding problem has a long history, and was first posed (and solved) by Skorokhod in [18]. Simply stated it is the following: given a stochastic process $(X_t)_{t \geq 0}$ and a distribution μ, find a stopping time T such that $X_T \sim \mu$.

In this work we will be interested in the case where $(B_t)_{t \geq 0}$ is a Brownian motion on $\mathbb{R}$, with a given (integrable) starting distribution μ_0. Since Brownian motion on $\mathbb{R}$ is recurrent, the existence of such a stopping time is trivial: consider an independent random variable Y with distribution μ and run until the first time that the Brownian motion hits Y. Hence interest lies in the properties of the stopping time T and also of the stopped process $(B_{t \wedge T})_{t \geq 0}$.

Classically, the 1-dimensional question has been considered in the case where $B_0 = 0$, and the target distribution μ is centred. In this case many solutions are known: for example [1,2,7,9,13,17,19]. We refer the reader to [12] for an excellent recent survey of these results. A property shared by all of these embeddings is that the process $(B_{t \wedge T})_{t \geq 0}$ is uniformly integrable, and we shall call stopping times for which this is the case UI stopping times. Further, within the class of embeddings where T is UI, many of these stopping times have optimality properties: for example the Azéma-Yor embedding maximises the law of the maximum, while the Vallois construction can be used to minimise or maximise $\mathbb{E}(f(L_T))$ for a convex function f (see [20]). It is clear that either of the maximisation problems are degenerate when looked at outside this class.

The class of UI stopping times can also be characterised in the following way due to [11]. We make the following definition:

Definition 1. *A stopping time T for the process X is* minimal *if whenever $S \leq T$ is a stopping time such that X_S and X_T have the same distribution then $S = T$ a.s..*

Then the class of minimal stopping times can be shown to be equivalent to the class of UI embeddings we had before:

Theorem 2 ([11], Theorem 3). *Let S be a stopping time such that $\mathbb{E}(B_S) = 0$. Then S is minimal if and only if the process $(B_{t \wedge S})_{t \geq 0}$ is uniformly integrable.*

Such a characterisation gives a natural interpretation to the class of UI embeddings.

Our interest in this paper lies in the extension to general starting measures. In such an example, even if the means agree, there is no guarantee that there will exist a UI stopping time which has the given starting and target distributions. This can be seen by considering the example of a target distribution consisting of a point mass at zero, but with starting distribution of mass $\frac{1}{2}$ at each of -1 and 1. Clearly the only minimal stopping time is to stop the first time the process hits 0, however this stopping time is not UI.

In [6] conditions for a stopping time to be minimal were considered. When the Brownian motion starts at the origin, and the target distribution is not centred, conditions on the process can be given which are equivalent to the stopping time being minimal. One of the main results of this work is to show that the conditions have an extension to the case of a general starting distribution, however the simple example given above shows that the extension is not trivial.

It will turn out that the characterisation of minimal stopping times is closely connected to the potentials of the two measures. In this context, the relationship between the measures can be viewed graphically in the framework of Chacon & Walsh ([4]). In this paper a graphical construction is interpreted as a sequence of exit times from compact intervals, whose limit is an embedding. This is done for starting and target distributions which satisfy the relationship

$$-\mathbb{E}^{\mu_0}|X - x| = u_{\mu_0}(x) \geq u_\mu(x) = -\mathbb{E}^\mu|X - x| \qquad (1)$$

for all $x \in \mathbb{R}$. We shall show that the construction can be extended to the case where this condition fails, and that the exact method of the extension will determine whether the stopping time is minimal.

Establishing this connection will then allow us to extend several existing embeddings [1,9,20], to the more general setting (maintaining minimality) via a limiting argument.

The paper will therefore proceed as follows: in Section 2 we describe the construction of [4] and our extension of their heuristic. In Section 3 we discuss the minimality criterion, and in Section 4 connect minimality with ideas from the potential theoretic and Chacon-Walsh approach; one of the key results (Theorem 17) appears here. In Section 5 we show (under conditions) that the limit of minimal stopping times is itself minimal, and explain in Section 6 how this enables us to conclude that the Chacon-Walsh heuristic described in Section 2 is applicable, and therefore allows us to extend existing constructions. Many of the technical results are collected in an appendix.

General Comments and Notation

In general, we will work on a probability space $(\Omega, \mathcal{F}, \mathbb{P})$, on which we assume there is a Brownian motion $(B_t)_{t \geq 0}$ with respect to the filtration $(\mathcal{F}_t)_{t \geq 0}$. We will typically assume $B_0 \sim \mu_0$, so that in particular, $\mathcal{F}_0$ is not trivial. Although we work with Brownian motion, it is also clear that the results are applicable to any continuous local martingale with almost surely infinite quadratic variation via the Dambis-Dubins-Schwartz Theorem. We shall also write

$$\overline{B}_t = \sup_{0 \leq s \leq t} B_s, \quad \underline{B}_t = \inf_{0 \leq s \leq t} B_s$$

and for $A \in \mathcal{F}$, we have $\mathbb{E}(X; A) = \mathbb{E}(X \mathbf{1}_A)$. For the most part, in what follows μ will denote our target distribution and T the stopping time which we are attempting to embed with — i.e. the embedding problem is to find a stopping time T with $B_T \sim \mu$.

2 The Balayage Construction

In the theory of general Markov processes, a common definition of the potential of a stochastic process is given by

$$U\mu(x) = \int_{\mathbb{R}} \mu(dy) \int_{\mathbb{R}^+} ds \, p_s(x, y),$$

where $p_s(x, \cdot)$ is the transition density at time s of the process started at x. In the case of Brownian motion, we note that the integral is infinite. To resolve this we use the compensated definition (and introduce new notation to emphasise the fact that this is not the classical definition of potential):

$$u_\mu(x) = \int_{\mathbb{R}} \mu(dy) \int_{\mathbb{R}+} ds \, (p_s(x, y) - p_s(0, 0)).$$

This definition simplifies to the following:

$$u_\mu(x) = - \int |x - y| \, \mu(dy). \tag{2}$$

Remark 3. The function u_μ has the following properties:

(i) The measure μ is integrable if and only if the function u_μ is finite for any (and therefore all) $x \in \mathbb{R}$.

(ii) u_μ is continuous, differentiable everywhere except the set

$$\{x \in \mathbb{R} : \mu(\{x\}) > 0\}$$

and concave.

(iii) Write

$$m = \int x \, \mu(dx).$$

As $|x| \to \infty$, we have

$$u_\mu(x) + |x| \to m \, \text{sign}(x). \tag{3}$$

(iv) As a consequence of (3), if μ and ν are integrable distributions, then there exists a constant $K > 0$ such that:

$$\sup_{x \in \mathbb{R}} |u_\mu(x) - u_\nu(x)| < K.$$

(v) u_μ is almost everywhere differentiable with left and right derivatives

$$u'_{\mu,-}(x) = 1 - 2\mu((-\infty, x));$$
$$u'_{\mu,+}(x) = 1 - 2\mu((-\infty, x]).$$

[3] contains many results concerning these potentials. We will describe a balayage technique that produces a sequence of measures and corresponding stopping times, and which will have as its limit our desired embedding. The following lemma will therefore be important in concluding that the limit we obtain will indeed be the desired distribution:

Lemma 4 ([3], Lemmas 2.5, 2.6). *Suppose $\{\mu_n\}$ is a sequence of probability measures. If*

(i) μ_n converges weakly to μ and $\lim_{n\to\infty} u_{\mu_n}(x_0)$ exists for some $x_0 \in \mathbb{R}$, then $\lim_{n\to\infty} u_{\mu_n}(x)$ exists for all $x \in \mathbb{R}$ and there exists $C \geq 0$ such that

$$\lim_{n\to\infty} u_{\mu_n}(x) = u_\mu(x) - C. \tag{4}$$

(ii) $\lim_{n\to\infty} u_{\mu_n}(x)$ exists for all $x \in \mathbb{R}$ then μ_n converges weakly to μ for some measure μ and μ is uniquely determined by the limit $\lim_n u_{\mu_n}(x)$.

We consider the embedding problem where we have a Brownian motion B with $B_0 \sim \mu_0$ (an integrable starting distribution) and we wish to embed an integrable target distribution μ. This is essentially the case considered in [4], although they only consider the case where $u_{\mu_0}(x) \geq u_\mu(x)$ for all x (when (3) implies μ_0 and μ have the same mean) — we will see that this case is simpler than the general case we consider. The embedding problem is frequently considered when μ_0 is the Dirac measure at 0. One of the appealing properties of the case where $B_0 = 0$ is that (by Jensen's inequality) for all centred target distributions

$$u_\mu(x) \leq -|x| = u_{\mu_0}(x), \tag{5}$$

and the condition on the ordering of potentials is easily satisfied.

We extend the technique of [4] to allow balayage on semi-infinite intervals. This extra step in the construction allows further flexibility later when we take limits of the constructions. In particular it will make the application of subsequent results trivial. Each step in the construction is described mathematically by a simple balayage technique:

Definition 5. *Let μ be a probability measure on $\mathbb{R}$, and I a finite, open interval, $I = (a, b)$. Then define the* balayage μ_I *of μ on I by:*

$$\mu_I(A) = \mu(A) \qquad\qquad A \cap \bar{I} = \emptyset;$$
$$\mu_I(\{a\}) = \int_{\bar{I}} \frac{b-x}{b-a} \mu(dx);$$
$$\mu_I(\{b\}) = \int_{\bar{I}} \frac{x-a}{b-a} \mu(dx);$$
$$\mu_I(I) = 0.$$

Suppose now $I = (a, \infty)$ (resp. $I = (-\infty, a)$), and define the balayage μ_I *of μ by*

$$\mu_I(A) = \mu(A) \qquad\qquad A \cap \bar{I} = \emptyset;$$
$$\mu_I(\{a\}) = \int_{\bar{I}} \mu(dx);$$
$$\mu_I(I) = 0.$$

The balayage μ_I is a probability measure and if I is a finite interval the means of μ and μ_I agree. In particular, μ_I is the law of a Brownian motion started with distribution μ and run until the first exit from I. (Note that we use the term *balayage* here in the sense of [3] and [14], rather than e.g. [21].)

Our motivation for introducing the balayage technique is that the potential of μ_I is readily calculated from the potential of μ:

Lemma 6 ([3], Lemma 8.1). *Let μ be a probability measure with finite potential, $I = (a, b)$ a finite open interval and μ_I the balayage of μ with respect to I. Then*

(i) $u_\mu(x) \geq u_{\mu_I}(x)$ $x \in \mathbb{R}$;
(ii) $u_\mu(x) = u_{\mu_I}(x)$ $x \in I^C$;
(iii) u_{μ_I} *is linear for* $x \in \bar{I}$.

When I is a semi-infinite interval we may calculate the potential in a similar way:

Lemma 7. *Let μ be a probability measure with finite potential u_μ, $I = (-\infty, a)$ or $I = (a, \infty)$ a semi-infinite interval and μ_I the balayage of μ with respect to I. Then*

$$u_{\mu_I}(x) = u_\mu(x) + \Delta m \qquad x \notin I;$$
$$u_{\mu_I}(x) = u_\mu(a) + \Delta m - |a - x| \ x \in I,$$

where we have written

$$\Delta m = \int_I |x - a| \, \mu(dx).$$

The semi-infinite balayage step in Definition 5 can be recreated using the balayage steps on bounded intervals, for example by taking the sequence of intervals $(a, a + 1), (a, a + 2), (a, a + 3), \ldots$. However the more general construction will allow us an extra degree of flexibility.

Formally, we may use balayage to define an embedding as the following result shows. In the formulation of the result we assume we are given the sequence of functions we use to construct the stopping time, and from these deduce the target distribution. However we will typically use the result in situations where we have a desired target distribution and choose the sequence to fit this distribution.

Lemma 8. *Let $f_1, f_2, \ldots$ be a sequence of linear functions on $\mathbb{R}$ such that $|f_n'(x)| \leq 1$ and*

$$g(x) = \inf_{n \in \mathbb{N}} f_n(x) \wedge (u_{\mu_0}(x)). \qquad (6)$$

Set $T_0 = 0$, $g_0(x) = u_{\mu_0}(x)$ and, for $n \geq 1$, define

$$a_n = \inf\{x \in \mathbb{R} : f_n(x) < g_{n-1}(x)\};$$
$$b_n = \sup\{x \in \mathbb{R} : f_n(x) < g_{n-1}(x)\};$$
$$T_n = \inf\{t \geq T_{n-1} : B_t \notin (a_n, b_n)\};$$
$$g_n(x) = g_{n-1}(x) \wedge f_n(x).$$

Then the T_n are increasing so we define $T = \lim_{n \to \infty} T_n$. If

$$g(x) = u_\mu(x) - C \tag{7}$$

for some $C \in \mathbb{R}$ and some integrable probability measure μ then $T < \infty$ a.s. and T is an embedding of μ.

If we only consider the theorem under the condition $|f'_n(x)| < 1$ this is a formalised statement of the construction implicit in [3] and made explicit under the further condition (1) in [4]. The introduction of the balayage steps on the half-line is the novel content of the result.

Proof. The hard part is to show that if (7) holds then the stopping time T is almost surely finite. We prove in fact that $\mathbb{E}(L_T) < \infty$, where L is the local time of B at zero. By considering the martingale $|B_t| - L_t$ we must have

$$\mathbb{E}(L_{T_1}) = u_{\mu_0}(0) - f_1(0) \wedge u_{\mu_0}(0) \tag{8}$$

when the interval (a_1, b_1) is compact; by approximating the semi-infinite interval by compact intervals (and a monotone convergence argument) this will extend to all possible choices of f_1, and, by an induction argument, we deduce

$$\mathbb{E}(L_{T_n}) = u_{\mu_0}(0) - \inf_{k \leq n} f_k(0) \wedge u_{\mu_0}(0).$$

A monotone convergence argument allows us to deduce that

$$\mathbb{E}(L_T) = u_{\mu_0}(0) - \inf_{n \in \mathbb{N}} f_n(0) \wedge u_{\mu_0}(0)$$

which is finite by (7), and hence $T < \infty$ a.s..

The functions g_n correspond to a potential of a measure μ_n (μ_n being the law of B_{T_n}) via:

$$g_n(x) = u_{\mu_n}(x) - C_n$$

for some constant C_n, and hence we have

$$u_{\mu_0}(x) \geq u_{\mu_n}(x) - C_n \geq u_\mu(x) - C$$

and as $n \to \infty$ the middle two terms converge. From (3) we can deduce that

$$C_n \geq \left| \int x\, \mu_0(dx) - \int x\, \mu_n(dx) \right| \geq 0,$$

so that since $g_n(x) \geq g(x)$, we have

$$u_\mu(x) - C \leq u_{\mu_n}(x) \leq 0,$$

the second inequality coming from the definition of the potential. Consequently, taking $x = 0$, we can find a subsequence n_j for which $\lim_{j \to \infty} u_{\mu_{n_j}}(0)$

exists, and hence for which $\lim_{n\to\infty} C_{n_j}$ also exists. Since $g_n(x)$ converges pointwise to $g(x)$ we must also have pointwise convergence of $u_{\mu_{n_j}}(x)$ to $u_\mu(x) - C'$ for some constant C', and, by Lemma 4, $B_{T_{n_j}}$ converges weakly to μ. Since also $T_{n_j} \uparrow T$, by the continuity of the Brownian motion B_T has law μ.   □

The case considered by [4] has a notable property. When the starting and target measures are centred (or at least when their means agree) and

$$u_{\mu_0}(x) \geq u_\mu(x) \tag{9}$$

then we may choose a construction such that $C = 0$ in (7). In this case the process $(B_{t \wedge T})_{t \geq 0}$ is uniformly integrable [3, Lemma 5.1]. The desire to find a condition to replace uniform integrability in situations where (9) does not hold, and to construct suitable stopping times using this framework, is the motivation behind the subsequent work.

We note also that — for given μ, μ_0 — we may find a construction for any C which satisfies $C \geq \sup_x \{u_\mu(x) - u_{\mu_0}(x)\}$; as a consequence of (3) we must always have $C \geq 0$. A natural question is then to ask what might happen if C is equal to the supremum. The main result of this paper is to show that in such a case, the embeddings which are constructed are 'correct' in the sense of Definition 1. Our next section will examine this definition in further detail.

3 Minimality: Some Preliminary Results

In this and the subsequent section we discuss necessary and sufficient conditions for an embedding of an integrable target distribution to be minimal (Definition 1) when we have an integrable starting distribution. These results will extend the the conditions of Theorems 2 and the following result:

Theorem 9 ([6]). *Let T be a stopping time of Brownian motion which embeds an integrable distribution μ where $m = \int_{\mathbb{R}} x\, \mu(dx) < 0$. Then the following are equivalent:*

(i) T is minimal for μ;
(ii) for all stopping times $R \leq S \leq T$,

$$\mathbb{E}(B_S | \mathcal{F}_R) \leq B_R \quad a.s.; \tag{10}$$

In the case where $supp(\mu) \subseteq [\alpha, \infty)$ for some $\alpha < 0$ then the above conditions are also equivalent to the condition:

(iii)

$$\mathbb{P}(T \leq H_\alpha) = 1, \tag{11}$$

where $H_\alpha = \inf\{t \geq 0 : B_t = \alpha\}$ is the hitting time of α.

Remark 10. For further necessary and sufficient conditions, see also [6].

As a starting point, we recall:

Proposition 11 ([11], Proposition 2). *For any stopping time T there exists a minimal stopping time $S \leq T$ such that $B_S \sim B_T$.*

Monroe's proof does not rely on the fact that B starts at 0, and so the result extends to a general starting distribution.

It can also be seen that the argument used in [11] to show that if the process is uniformly integrable then the process is minimal does not require the starting measure to be a point mass. For completeness we state a similar result, with the proof given in [11]:

Lemma 12. *Let T be a stopping time embedding μ in $(B_t)_{t \geq 0}$, with $B_0 \sim \mu_0$ where μ and μ_0 are integrable distributions. If*

$$\mathbb{E}(B_T | \mathcal{F}_S) = B_S \text{ a.s.} \tag{12}$$

for all stopping times $S \leq T$ then T is minimal.

Note that $S \equiv 0$ implies that μ, μ_0 have the same mean.

Remark 13. We will later be interested also in necessary conditions for minimality. The condition in (12) is not necessary even when both starting and target measures are centred, as can be seen by taking $\mu_0 = \frac{1}{2}\delta_{-1} + \frac{1}{2}\delta_1$ and $\mu = \delta_0$, where it is impossible to satisfy (12) but the (only) minimal stopping time is 'stop when the process hits 0.'

The condition in (12) is equivalent to uniform integrability of the process $(B_{t \wedge T})_{t \geq 0}$. One direction follows from the Optional Stopping Theorem, the reverse implication comes from the Upward Martingale Theorem [16, Theorem II.69.5], which tells us that the process $X_t = \mathbb{E}(B_T | \mathcal{F}_t)$ is a uniformly integrable martingale on $t \leq T$. When (12) holds, $X_t = B_{t \wedge T}$, and the process $(B_{t \wedge T})_{t \geq 0}$ is a uniformly integrable martingale.

For the rest of this section we will consider minimality for general starting and target measures: particularly when the means do not agree. If this occurs when the starting measure is a point mass, necessary and sufficient conditions are given in Theorem 9. In subsequent proofs with general starting measures we will often reduce problems to the point mass case in order to apply the result.

Remark 14. The condition given in *(iii)* of Theorem 9 hints at a more general idea inherent in the study of embeddings in Brownian motion. When $B_0 = 0$, it is a well known fact that if there exist levels $\alpha < 0 < \beta$ such that $T \leq H_\alpha \wedge H_\beta$ then $(B_{t \wedge T})_{t \geq 0}$ is a uniformly integrable martingale. If $T \leq H_\alpha$ then the process is a supermartingale. In terms of embeddings, this observation has the following consequence: if the target distribution is centred and supported on a bounded interval, an embedding is minimal if and only if the process never leaves this interval. On the other hand, if the target distribution has a negative mean, but still lies on a bounded interval, any embedding must move above the interval — in fact $\mathbb{P}(\sup_{t \leq T} B_t \geq x) > 0$ for all $x \geq 0$ — otherwise the process would be bounded, and therefore uniformly integrable. In this case, Theorem 9 and Proposition 11 tell us that an embedding always exists for which $T \leq H_\alpha$ and further that all minimal embeddings satisfy this property.

Recall that there is a natural ordering on the set of (finite) measures on $\mathbb{R}$, that is $\mu \preceq \nu$ if and only if $\mu(A) \leq \nu(A)$ for all $A \in \mathcal{B}(\mathbb{R})$, in which case we say that ν dominates μ. In such instances it is possible to define a (positive, finite) measure $(\nu - \mu)(A) = \nu(A) - \mu(A)$. The notation $\nu = \mathcal{L}(B_T; T < H_\alpha)$ is used to mean the (sub-probability) measure ν such that $\nu(A) = \mathbb{P}(B_T \in A, T < H_\alpha)$.

Lemma 15. *Let $(B_t)_{t \geq 0}$ be a Brownian motion with $B_0 = 0$, T a stopping time embedding a distribution μ, $\tilde{\mu}$ a target distribution such that $\mathrm{supp}(\tilde{\mu}) \subseteq [\alpha, \infty)$ for some $\alpha < 0$ and $\int x \, \tilde{\mu}(dx) \leq 0$. Then if $\nu = \mathcal{L}(B_T; T < H_\alpha)$ is dominated by $\tilde{\mu}$, there exists a minimal stopping time $\tilde{T} \leq T \wedge H_\alpha$ which embeds $\tilde{\mu}$.*

Similarly, if $\tilde{\mu}$ is such that $\mathrm{supp}(\tilde{\mu}) \subseteq [\alpha, \beta]$ and $\int x \, \tilde{\mu}(dx) = 0$, and if $\nu = \mathcal{L}(B_T; T < H_\alpha \wedge H_\beta)$ is dominated by $\tilde{\mu}$, then there exists a minimal stopping time $\tilde{T} \leq T \wedge H_\alpha \wedge H_\beta$ which embeds $\tilde{\mu}$.

Proof. Construct a stopping time T' as follows: on $\{T < H_\alpha\}$, $T' = T$; otherwise choose T' so that $T' = H_\alpha + T'' \circ \theta_{H_\alpha}$ where T'' is chosen to embed $(\tilde{\mu} - \nu)$ on $\{T \geq H_\alpha\}$ given $B_0 = \alpha$. Then T' is an embedding of $\tilde{\mu}$ and $T' \leq T$ on $\{T < H_\alpha\}$. So by Proposition 11 and Theorem 9(*iii*), we may find a minimal embedding $\tilde{T} \leq T' \wedge H_\alpha = T \wedge H_\alpha$ which embeds $\tilde{\mu}$.

The proof in the centred case is essentially identical, but now stopping the first time the process leaves $[\alpha, \beta]$. □

We turn now to the case of interest — that is when $B_0 \sim \mu_0$ and $B_T \sim \mu$ for integrable measures μ_0 and μ. The following lemma is essentially technical in nature, but will allow us to deduce the required behaviour on letting the set A increase in density. Initially we will suppose that A is a countable set of points that contains no aggregation points, but also does not contain arbitrarily large gaps. The idea is that we will approximate our starting distribution with a distribution on A, where we are able to use Theorem 9.

Lemma 16. *Let T be a minimal stopping time, and A a countable subset of $\mathbb{R}$ such that A has finitely many elements in every compact subset of $\mathbb{R}$ and $d(x, A) < M$ for all $x \in \mathbb{R}$ and some $M > 0$. We consider the stopping time*

$$R(A) = \inf\{t \geq 0 : B_t \in A\} \wedge T$$

and we write

$$E_A(x) = \begin{cases} \mathbb{E}(B_T | T > R(A), B_{R(A)} = x) & : \mathbb{P}(T > R(A), B_{R(A)} = x) > 0; \\ x & : \mathbb{P}(T > R(A), B_{R(A)} = x) = 0. \end{cases}$$

Then there exists $a \in \bar{\mathbb{R}} = \mathbb{R} \cup \{-\infty\} \cup \{\infty\}$ such that

$$E_A(x) > x \implies x < a, \tag{13}$$
$$E_A(x) < x \implies x > a, \tag{14}$$

and $T \le H_a$ on $\{T \ge R(A)\}$.

Further, if there exists $x < y$ such that $E_A(x) > x$ and $E_A(y) < y$ then there exists $a_\infty \in [x, y]$ such that $T \le H_{a_\infty}$.

The proof of this result appears in Appendix A. The result, although technical in nature, can be thought of as beginning to describe the behaviour we shall expect from minimal embeddings in this general context. The cases considered in Theorem 9 suggest behaviour of the form: 'the process always drifts in the same direction,' if indeed it drifts at all. The example of Remark 13 suggests that this is not always possible in the general case, and the previous result suggests that this is modified by breaking the space into two sections, in each of which the process can be viewed separately. The way these sections are determined is clearly dependent on the starting and target measures, and we shall see in the next section that the potential of these measures provides an important tool in determining how this occurs.

4 Minimality and Potential

The main aim of this section is to find equivalent conditions to minimality which allow us to characterise minimality simply in terms of properties of the process $(B_{t \wedge T})_{t \ge 0}$. This is partly in order to prove the following result:

The Chacon-Walsh type embedding is minimal when constructed using the functions u_{μ_0} and $c(x) = u_\mu(x) - C$ where

$$C = \sup_x \{ u_\mu(x) - u_{\mu_0}(x) \}. \tag{15}$$

Note that the function we define here is related to the function c_μ used in [6]: in the case where μ is centered and $\mu_0 = \delta_0$, we have $c(x) = -c_\mu(x)$.

We have already shown that provided the means of our starting and target distribution match, and (9) holds (so that $C = 0$ — the solution in this case to (15)), then the process constructed using the Chacon-Walsh technique is uniformly integrable, and therefore minimal. Of course the Chacon-Walsh construction is simply an example of an embedding, and the functions u_{μ_0} and c are properties solely of the general problem — it seems reasonable however that these functions will appear in the general problem of classifying all minimal embeddings.

So consider a pair μ_0, μ of integrable measures. Remark 3(iv) tells us we we can choose $C < \infty$ such that (15) holds. We know $u_{\mu_0}(x) - c(x)$ is bounded above, and $\inf_{x \in \mathbb{R}} (u_{\mu_0}(x) - c(x)) = 0$. We consider

$$\mathcal{A} = \{x \in [-\infty, \infty] : \lim_{y \to x} [u_{\mu_0}(y) - c(y)] = 0\}. \tag{16}$$

Since both the functions are Lebesgue almost-everywhere differentiable, Remark 3(v) implies $\mathcal{A} \subseteq \mathcal{A}'$ where $\mathcal{A}'$ is the set

$$\{x \in [-\infty, \infty] : \mu((-\infty, x)) \le \mu_0((-\infty, x)) \le \mu_0((-\infty, x]) \le \mu((-\infty, x])\}. \tag{17}$$

One consequence of this is that if the starting distribution has an atom at a point of $\mathcal{A}$ then the target distribution has an atom at least as large. Also we introduce the following definition. Given a measure ν, $a \in \mathbb{R}$ and $\theta \in [\nu((-\infty, a)), \nu((-\infty, a])]$ we define the measure $\check{\nu}^{a,\theta}$ to be the measure which is ν on $(-\infty, a)$, has support on $(-\infty, a]$ and $\check{\nu}^{a,\theta}(\mathbb{R}) = \theta$. We also define $\hat{\nu}^{a,\theta} = \nu - \check{\nu}^{a,\theta}$. Then for $a \in \mathcal{A}$, by Remark 3(iv), we may find θ such that

$$\check{\mu}^{a,\theta}((-\infty, a]) = \check{\mu}_0^{a,\theta}((-\infty, a])$$
$$\hat{\mu}^{a,\theta}([a, \infty)) = \hat{\mu}_0^{a,\theta}([a, \infty)).$$

When $\mu_0((-\infty, a)) < \mu_0((-\infty, a])$ there will exist multiple θ. We will occasionally drop the θ from the notation since this is often unnecessary.

These definitions allows us to write the potential in terms of the new measures (for any suitable θ)

$$u_\mu(x) = \int_{(-\infty, x]} (y - x) \check{\mu}^x(dy) + \int_{[x, \infty)} (x - y) \hat{\mu}^x(dy) \tag{18}$$

$$= 2 \int y \, \check{\mu}^x(dy) - \int y \, \mu(dy) + x(1 - 2\check{\mu}^x(\mathbb{R})) \tag{19}$$

$$= \int y \, \mu(dy) - 2 \int y \, \hat{\mu}^x(dy) + x(2\hat{\mu}^x(\mathbb{R}) - 1) \tag{20}$$

As a consequence of this and a similar relation for u_{μ_0}, we are able to deduce the following important facts about the set $\mathcal{A}$:

• if $x < z$ are both elements of $\mathcal{A}$ (possibly $\pm\infty$), then we have

$$u_\mu(x) - u_{\mu_0}(x) = u_\mu(z) - u_{\mu_0}(z)$$

and by (19), (20),

$$\int y \, (\mu - \check{\mu}^{x,\theta} - \hat{\mu}^{z,\phi})(dy) = \int y \, (\mu_0 - \check{\mu}_0^{x,\theta} - \hat{\mu}_0^{z,\phi})(dy) \tag{21}$$

That is, we may find measures agreeing with μ and μ_0 on (x, z) and with support on $[x, z]$ which have the same mean.

- If $x \in \mathcal{A}$, by definition

$$u_\mu(x) - u_{\mu_0}(x) \geq \lim_{z \to -\infty} (u_\mu(z) - u_{\mu_0}(z)). \tag{22}$$

This can be rearranged, using (18), to deduce

$$\int_{(-\infty,x]} y\, \check{\mu}_0^x(dy) \leq \int_{(-\infty,x]} y\, \check{\mu}^x(dy)$$

with equality if and only if there is also equality in (22) — that is when $-\infty \in \mathcal{A}$.

Together these imply that the set $\mathcal{A}$ divides $\mathbb{R}$ into intervals on which the starting and target measures place the same amount of mass. Further, the means of the distributions agree on these intervals except for the first (resp. last) interval where the mean of the target distribution will be larger (resp. smaller) than that of the starting distribution unless $-\infty$ (resp. ∞) is in $\mathcal{A}$, when again they will agree. Note the connection between this idea and Lemma 16.

It is exploiting this connection that leads us to the main result of this paper: by identifying the points $a \in \mathcal{A}$ with the points a given in Lemma 16, as we let A increase in density (so that intuitively $R(A) \equiv 0$), we achieve characterisations of minimality in terms of conditions on $(B_{t \wedge T})_{t \geq 0}$. The theorem should be thought of as the extension of Theorem 9 to the setting with a general starting measure.

Theorem 17. *Let B be a Brownian motion such that $B_0 \sim \mu_0$ and T a stopping time such that $B_T \sim \mu$, where μ_0, μ are integrable. Let $\mathcal{A}$ be the set defined in (16) and $a_+ = \sup\{x \in [-\infty, \infty] : x \in \mathcal{A}\}$, $a_- = \inf\{x \in [-\infty, \infty] : x \in \mathcal{A}\}$. Then the following are equivalent:*

(i) T is minimal;

(ii) $T \leq H_\mathcal{A}$ and for all stopping times $R \leq S \leq T$

$$\mathbb{E}(B_S | \mathcal{F}_R) \leq B_R \ \text{on} \ \{B_0 \geq a_-\}$$
$$\mathbb{E}(B_S | \mathcal{F}_R) \geq B_R \ \text{on} \ \{B_0 \leq a_+\};$$

(iii) $T \leq H_\mathcal{A}$ and for all stopping times $S \leq T$

$$\mathbb{E}(B_T | \mathcal{F}_S) \leq B_S \ \text{on} \ \{B_0 \geq a_-\}$$
$$\mathbb{E}(B_T | \mathcal{F}_S) \geq B_S \ \text{on} \ \{B_0 \leq a_+\};$$

(iv) $T \leq H_\mathcal{A}$ and for all $\gamma > 0$

$$\mathbb{E}(B_T; T > H_{-\gamma}, B_0 \geq a_-) \leq -\gamma \mathbb{P}(T > H_{-\gamma}, B_0 \geq a_-)$$
$$\mathbb{E}(B_T; T > H_\gamma, B_0 \leq a_+) \geq \gamma \mathbb{P}(T > H_\gamma, B_0 \leq a_+);$$

(v) $T \leq H_A$ and as $\gamma \to \infty$

$$\gamma \mathbb{P}(T > H_{-\gamma}, B_0 \geq a_-) \to 0$$
$$\gamma \mathbb{P}(T > H_\gamma, B_0 \leq a_+) \to 0.$$

Further, if there exists $a \in \mathbb{R}$ such that $|a| < \infty$ and $T \leq H_a$, then T is minimal.

We defer the proof of this result to Appendix B. For the purposes of this paper, the key points of this theorem are firstly that we now have a mathematical interpretation of minimality, that will be easier to work with in subsequent sections; secondly, we see that the connection to the potential interpretation introduced in Section 2 is fundamental not only to the Chacon-Walsh approach, but to the general setting, appearing as it does through the set $\mathcal{A}$.

5 Minimality of the Limit

We will want to show that stopping times constructed using the techniques of Section 2 are indeed minimal when (15) is satisfied. To deduce that a stopping time T constructed using the balayage techniques is minimal, we approximate T by the sequence of stopping times T_n given in the construction (so T_1 is the exit time from the first interval we construct, and so on). Then it is clear that the stopping times T_n satisfy the conditions of Theorem 17: they are the composition of first exit times from bounded or semi-bounded intervals, where the bounded intervals must not straddle points of $\mathcal{A}$ (recalling the definitions given in Lemma 8, if the endpoints (a_n, b_n) of a step in the construction did straddle a point $a \in \mathcal{A}$, then we would have $g_n(a) > g_{n-1}(a)$, contradicting the fact that $u_{\mu_0}(a) = u_\mu(a) - C$.) Similarly, we only allow semi-infinite intervals which correspond to the first time to leave $(-\infty, \alpha]$ (resp. $[\beta, \infty)$) for some $\alpha < a_-$ (resp. $\beta > a_+$). Our aim is then to deduce that the limit is minimal.

Proposition 18. *Suppose that T_n embeds μ_n, μ_n converges weakly to μ and $\mathbb{P}(|T_n - T| > \varepsilon) \to 0$ for all $\varepsilon > 0$. Then T embeds μ.*
    If also $l_n \to l_\infty < \infty$ where $l_n = \int |x| \mu_n(dx)$ and $l_\infty = \int |x| \mu(dx)$, and T_n is minimal for μ_n, then T is minimal for μ.

Remark 19. Since $\mu_n \implies \mu$, on some probability space we are able to find random variables X_n and X with laws μ_n and μ such that $X_n \to X$ a.s.. By Scheffé's Lemma (e.g. [16]) therefore

$$\mathbb{E}|X_n - X| \to 0 \text{ if and only if } \mathbb{E}|X_n| \to \mathbb{E}|X|,$$

the second statement being equivalent to $l_n \to l_\infty$ in the statement of Proposition 18.

Before we prove Proposition 18, we will show a useful result on the distribution of the maximum. This will be used in the proof of Proposition 18, and also be important for the work in the next section, when we will show that the inequality in (23) can be attained by a class of stopping times created by balayage techniques.

Lemma 20. *Let T be a minimal embedding of μ in a Brownian motion started with distribution μ_0. Then for all $x \in \mathbb{R}$*

$$\mathbb{P}(\overline{B}_T \geq x) \leq \inf_{\lambda < x} \frac{1}{2} \left[1 + \frac{u_{\mu_0}(x) - c(\lambda)}{x - \lambda} \right]. \tag{23}$$

Proof. Define the stopping time $\bar{H}_x = \inf\{t \geq 0 : B_t \geq x\}$, the first time that B_t goes above x. Then we note the following inequality, which (by considering on a case by case basis) holds for all paths and all pairs $\lambda < x$:

$$\mathbf{1}_{\{\overline{B}_T \geq x\}} \leq \frac{1}{x - \lambda} \left[B_{T \wedge \bar{H}_x} + \frac{|B_T - \lambda| - (B_T + \lambda)}{2} - \frac{|B_0 - x| + (B_0 - x)}{2} \right]. \tag{24}$$

In particular, on $\{\overline{B}_T < x\}$, when therefore $\{B_0 < x\}$:

$$0 \leq \frac{1}{x - \lambda} \left[B_T + \begin{cases} -\lambda & : B_T > \lambda \\ -B_T & : B_T \leq \lambda \end{cases} \right]. \tag{25}$$

While on $\{\overline{B}_T \geq x\}$,

$$\begin{aligned} 1 &\leq \frac{1}{x - \lambda} \left[B_{T \wedge \bar{H}_x} + \begin{cases} -\lambda & : B_T > \lambda \\ -B_T & : B_T \leq \lambda \end{cases} - \begin{cases} B_0 - x & : B_0 > x \\ 0 & : B_0 \leq x \end{cases} \right] \\ &\leq \frac{1}{x - \lambda} \left[x + \begin{cases} -\lambda & : B_T > \lambda \\ -B_T & : B_T \leq \lambda \end{cases} \right]. \end{aligned} \tag{26}$$

So we may take expectations in (24) to get

$$\mathbb{P}(\overline{B}_T \geq x) \leq \frac{1}{2} \left[1 + \frac{2\mathbb{E}(B_{T \wedge \bar{H}_x}) + (u_{\mu_0}(x) - u_\mu(\lambda)) - (\mathbb{E}(B_T) + \mathbb{E}(B_0))}{(x - \lambda)} \right]. \tag{27}$$

We can deduce (23) provided we can show

$$C \geq 2\mathbb{E}(B_{T \wedge \bar{H}_x}) - (\mathbb{E}(B_T) + \mathbb{E}(B_0)) \tag{28}$$

since (27) holds for all $\lambda < x$.

We now consider $a \in \mathcal{A}$ possibly taking the values $\pm \infty$. Since $u_\mu(a) - u_{\mu_0}(a) = C$ for $a \in \mathcal{A}$, we can deduce (by (19), (20))

$$\begin{aligned} C &= 2\mathbb{E}(B_T; B_T < a) + 2\mathbb{E}(B_0; B_0 > a) - \mathbb{E}(B_T) - \mathbb{E}(B_0) \\ &\quad + 2a \left(1 - \mathbb{P}(B_T < a) - \mathbb{P}(B_0 > a) \right). \end{aligned}$$

Theorem 17 tells us that

$$\mathbb{E}(B_{T \wedge \bar{H}_x}; B_0 < a) \leq \mathbb{E}(B_T; B_0 < a); \tag{29}$$

$$\mathbb{E}(B_{T \wedge \bar{H}_x}; B_0 > a) \leq \mathbb{E}(B_0; B_0 > a), \tag{30}$$

and since $\{B_T < a\} \subseteq \{B_0 < a\}$, and

$$\{B_T < a\} \cup \{B_T = a, B_0 < a\} = \{B_0 < a\}$$

is a disjoint union, we have

$$\mathbb{E}(B_{T \wedge \bar{H}_x}; B_0 < a) \leq \mathbb{E}(B_T; B_T < a) + a \left(\mathbb{P}(B_0 < a) - \mathbb{P}(B_T < a)\right)$$

which we can put together to deduce (28). □

We also have the following result:

Proposition 21. *Suppose μ and $\{\mu_n\}_{n \geq 1}$ are all integrable distributions such that $\mu_n \implies \mu$ and $l_n = \int |y| \, \mu_n(dy) \to \int |y| \, \mu(dy) = l_\infty$. Then u_{μ_n} converges uniformly to u_μ.*

Proof. Fix $\varepsilon > 0$. By (18), using the fact that $\mu - \hat{\mu} = \check{\mu}$ we may write

$$u_\mu(x) = \int_{-\infty}^{\infty} (x - y) \, \mu(dy) + 2 \int_{-\infty}^{x} (y - x) \, \mu(dy)$$

$$= x - \int_{-\infty}^{\infty} y \, \mu(dy) + 2 \int_{-\infty}^{x} (y - x) \, \mu(dy),$$

and similarly for u_{μ_n}, hence

$$u_{\mu_n}(x) - u_\mu(x) = (m_\infty - m_n) + 2 \int_{-\infty}^{x} (y - x) \, (\mu_n - \mu)(dy), \tag{31}$$

where we write m_n, m_∞ for the means of μ_n and μ respectively; $m_n \to m_\infty$ as a consequence of Remark 19. Since μ is integrable, as $x \downarrow -\infty$,

$$\int_{-\infty}^{x} (x - y) \, \mu(dy) \downarrow 0.$$

By (31) and Lemma 4 (which implies u_{μ_n} converges to u_μ pointwise, the C in (4) being 0 since $l_n \to l_\infty$), for all $x \in \mathbb{R}$

$$\int_{-\infty}^{x} (x - y) \mu_n(dy) \to \int_{-\infty}^{x} (x - y) \mu(dy)$$

as $n \to \infty$. Finally we note that both sides of the above are increasing in x.
Consider

$$|u_{\mu_n}(x) - u_\mu(x)| \leq |m_\infty - m_n| + 2 \int_{-\infty}^{x} (x - y) \, \mu_n(dy) + 2 \int_{-\infty}^{x} (x - y) \, \mu(dy).$$

We may choose x_0 sufficiently small that $\int_{-\infty}^{x_0} (x_0 - y)\, \mu(dy) < \varepsilon$, and therefore such that

$$\int_{-\infty}^{x} (x - y)\, \mu(dy) \leq \int_{-\infty}^{x_0} (x_0 - y)\, \mu(dy) < \varepsilon$$

for all $x \leq x_0$. By the above and Remark 19 we may now choose $n_0(\varepsilon)$ such that for all $n \geq n_0(\varepsilon)$

$$|m_\infty - m_n| < \varepsilon \text{ and } \left| \int_{-\infty}^{x_0} (x_0 - y)\, \mu_n(dy) - \int_{-\infty}^{x_0} (x_0 - y)\, \mu(dy) \right| < \varepsilon.$$

Then for all $x \leq x_0$ and for all $n \geq n_0(\varepsilon)$,

$$|u_{\mu_n}(x) - u_\mu(x)| \leq \varepsilon + 2 \times 2\varepsilon + 2\varepsilon = 7\varepsilon.$$

Similarly we can find $x_1, n_1(\varepsilon)$ such that $|u_{\mu_n}(x) - u_\mu(x)| \leq 7\varepsilon$ for all $x \geq x_1$ and all $n \geq n_1(\varepsilon)$. Finally u_{μ_n}, u_μ are both Lipschitz and pointwise $u_{\mu_n}(x) \to u_\mu(x)$ and we must have uniform convergence on any bounded interval, and in particular on $[x_0, x_1]$.   □

Proof (Proof of Proposition 18.). Suppose first that there exists $a \in \mathcal{A} \cap \mathbb{R}$. We show that $T \leq H_a$ for all such a. As usual, we write μ_0 for the starting measure, and $c(x) = u_\mu(x) - C$. We define C_n to be the smallest value such that $u_{\mu_0}(x) \geq u_{\mu_n}(x) - C_n$ and the functions $c_n(x) = u_{\mu_n}(x) - C_n$. Note that $l_n = u_{\mu_n}(0)$, so $\lim_{n\to\infty} u_{\mu_n}(0)$ exists. Then (by Lemma 4(*i*) or equivalently [3, Lemma 2.5]) weak convergence implies

$$\lim_{n\to\infty} u_{\mu_n}(x) = u_\mu(x) - K$$

for all $x \in \mathbb{R}$ and (here) $K = 0$ since $u_{\mu_n}(0) \to u_\mu(0)$.

By Lemma 20 for $x \in \mathbb{R}$ and $\lambda < x$

$$\mathbb{P}(\overline{B}_{T_n} \geq x) \leq \frac{1}{2} \left[1 + \frac{u_{\mu_0}(x) - u_{\mu_n}(\lambda) + C}{x - \lambda} + \frac{C_n - C}{x - \lambda} \right],$$

and we take the limit as $n \to \infty$, using Proposition 21 (so that $C_n \to C$) and noting that $\mathbb{P}(\overline{B}_{T_n} \geq x) \to \mathbb{P}(\overline{B}_T \geq x)$, to get

$$\mathbb{P}(\overline{B}_T \geq x) \leq \frac{1}{2} \left[1 + \frac{u_{\mu_0}(x) - c(\lambda)}{x - \lambda} \right].$$

Suppose now $\lambda = a$. Since the above holds for all $x > a$, we may take the limit of both sides as $x \downarrow a$, in which case $u_{\mu_0}(a) = c(a)$, and by Remark 3(v)

$$\lim_{x \downarrow a} \mathbb{P}(\overline{B}_T \geq a) \leq \lim_{x \downarrow a} \frac{1}{2} \left[1 + \frac{u_{\mu_0}(x) - c(\lambda)}{x - \lambda} \right]$$

$$\mathbb{P}(\overline{B}_T > a) \leq \frac{1}{2} \left[1 + (u_{\mu_0})'_+(a) \right]$$

$$\leq \frac{1}{2} \left[1 + (1 - 2\mu_0((-\infty, a])) \right]$$

$$\leq \mu_0((a, \infty)).$$

Since also $\{\overline{B}_T > a\} \supseteq \{B_0 > a\}$, we must therefore have $\{\overline{B}_T > a\} = \{B_0 > a\}$ a.s., and similarly by considering $-B_t$ we may deduce that $\mathbb{P}(\underline{B}_T < a) \leq \mu_0((-\infty, a))$, and therefore $\{\underline{B}_T < a\} = \{B_0 < a\}$ a.s.. Hence $\mathbb{P}(T \leq H_a) = 1$, and we deduce that T is minimal.

It only remains to show (by Lemma 31) that if $\infty \in \mathcal{A}$ then

$$\mathbb{E}(B_T | \mathcal{F}_S) \geq B_S$$

for all stopping times $S \leq T$. The case where $-\infty \in \mathcal{A}$ follows from $B_t \mapsto -B_t$. In particular, for $S \leq T$ and $A \in \mathcal{F}_S$ we need to show

$$\mathbb{E}(B_T; A) \geq \mathbb{E}(B_S; A). \tag{32}$$

In fact we need only show the above for sets $A \subseteq \{S < T\}$ since it clearly holds on $\{S = T\}$. So we can define $A_n = A \cap \{S < T_n\}$ and therefore $\mathbb{P}(A \setminus A_n) \to 0$ as $n \to \infty$. Also $A_n \in \mathcal{F}_{S \wedge T_n}$. By Theorem 17 and the fact that the T_n are minimal

$$
\begin{aligned}
\mathbb{E}(B_{S \wedge T_n}; A_n) &\leq \mathbb{E}(B_{T_n}; A_n \cap \{B_0 \leq a_+^n\}) + \mathbb{E}(B_{S \wedge T_n}; B_0 > a_+^n) \\
&\quad - \mathbb{E}(B_{S \wedge T_n}; A_n^C \cap \{B_0 > a_+^n\}) \\
&\leq \mathbb{E}(B_{T_n}; A_n \cap \{B_0 \leq a_+^n\}) \mathbb{E}(B_0; B_0 > a_+^n) \\
&\quad - \mathbb{E}(B_{T_n}; A_n^C \cap \{B_0 > a_+^n\}) \\
&\leq \mathbb{E}(B_{T_n}; A_n) - \mathbb{E}(B_{T_n}; \{B_0 > a_+^n\}) + \mathbb{E}(B_0; B_0 > a_+^n)
\end{aligned}
$$

where a_+^n is the supremum of the set $\mathcal{A}_n$ (that is the corresponding set to $\mathcal{A}$ for the measures μ_0, μ_n). This is not necessarily infinite.

So it is sufficient for us to show (since $S < T_n$ on A_n) that

$$\lim_n \mathbb{E}(B_{T_n}; A_n) = \mathbb{E}(B_T; A); \tag{33}$$

$$\lim_n \mathbb{E}(B_S; A_n) = \mathbb{E}(B_S; A), \tag{34}$$

and

$$\lim_n |\mathbb{E}(B_0; B_0 > a_+^n) - \mathbb{E}(B_{T_n}; B_0 > a_+^n)| = 0. \tag{35}$$

For (33) we consider $|\mathbb{E}(B_T; A) - \mathbb{E}(B_{T_n}; A_n)|$. Then

$$|\mathbb{E}(B_T; A) - \mathbb{E}(B_{T_n}; A_n)| \leq \mathbb{E}(|B_T|; A \setminus A_n) + \mathbb{E}(|B_T - B_{T_n}|; A_n)$$

and the first term tends to zero by dominated convergence (this follows from the assumption that T_n converges to T in probability). For the second term we show $\mathbb{E}(|B_T - B_{T_n}|) \to 0$. Fix $\varepsilon > 0$. We have

$$|B_T - B_{T_n}| \leq |B_{T_n}| - |B_T| + 2|B_T| \mathbf{1}_{\{|T_n - T| \geq \varepsilon\}} + 2|B_T - B_{T_n}| \mathbf{1}_{\{|T_n - T| \leq \varepsilon\}}.$$

We take expectations and let $n \to \infty$. By the definition of μ_n the first two terms cancel each other out, while the third tends to zero by dominated convergence. For the last term, by the (strong) Markov property (so $\tilde{B}_t = B_{T \wedge T_n + t} - B_{T \wedge T_n}$ is a Brownian motion)

$$\mathbb{E}(|B_T - B_{T_n}|; |T_n - T| \leq \varepsilon) \leq \mathbb{E}(|B_{T \vee T_n} - B_{T \wedge T_n}|; (T \vee T_n - T \wedge T_n) \leq \varepsilon)$$

$$\leq \mathbb{E}\left(\sup_{0 \leq t \leq \varepsilon} |\tilde{B}_t|\right)$$

$$\leq \sqrt{\frac{2\varepsilon}{\pi}}.$$

Consequently, in the limit, $\mathbb{E}(|B_T - B_{T_n}|; |T_n - T| \leq \varepsilon) \to 0$ and (33) holds. We want to apply Lemma 31 so we can assume that $\mathbb{E}|B_S| < \infty$, and (34) follows by dominated convergence.

Finally we consider (35). Let $\theta_n = \mu_0((-\infty, a_+^n])$. Since $a_+^n \in \mathcal{A}_n$ we have

$$\mathbb{E}(B_0; B_0 > a_+^n) - \mathbb{E}(B_{T_n}; B_0 > a_+^n) = \int y \, \hat{\mu}_0^{a_+^n, \theta_n}(dy) - \int y \, \hat{\mu}_n^{a_+^n, \theta_n}(dy)$$

$$= \int (y - a_+^n) \, \hat{\mu}_0^{a_+^n, \theta_n}(dy)$$

$$- \int (y - a_+^n) \, \hat{\mu}^{a_+^n, \theta_n}(dy)$$

$$= \frac{1}{2}\left[\int y \, (\mu_0 - \mu_n)(dy)\right.$$

$$\left. + u_{\mu_n}(a_+^n) - u_{\mu_0}(a_+^n)\right]$$

$$= \frac{1}{2}\left[\int y \, (\mu_0 - \mu)(dy) - C_n\right],$$

where we have used the fact that (for a general measure ν)

$$\int (y - x) \, \hat{\nu}^x(dy) = \frac{1}{2}\left[\int y \, \nu(dy) - u_\nu(x) - x\right].$$

As $n \to \infty$, since $\infty \in \mathcal{A}$,

$$\int y \, (\mu_0 - \mu_n)(dy) \to \int y \, (\mu_0 - \mu)(dy) = C.$$

So we need only note that $C_n \to C$, which follows from the uniform convergence of u_{μ_n} to u_μ (Proposition 21). $\qquad\square$

6 Tangents and Azéma-Yor Type Embeddings

One of the motivations for this paper is to discuss generalisations of the Azéma-Yor family of embeddings (see [1, 9]) to the integrable starting/target measures we have discussed already.

The aim is therefore to find the embedding which maximises the law of the maximum, $\sup_{0 \leq t \leq T} B_t$ (or alternatively $\sup_{0 \leq t \leq T} |B_t|$). If we look for the maximum within the class of all embeddings there is no natural maximum embedding. For this reason we consider the class of minimal embeddings. Lemma 20 establishes that there is some natural limit when we consider this restriction. In fact the extended Azéma-Yor embedding will attain the limit in (23).

The idea is to use the machinery from the previous sections to show the embeddings exist as limits of the Chacon-Walsh type embeddings of Section 2. It is then possible to show that the embeddings are minimal and that they attain equality in (23).

Theorem 22. *If T is a stopping time as described in Lemma 8, where C as described in the lemma is*

$$C = \inf_{x}\{u_\mu(x) - u_{\mu_0}(x)\}, \tag{36}$$

then T is minimal.

Proof. Lemma 8 suggests a sequence T_n of stopping times for which T is the limit. We note that we can modify the definition of T_n so that T'_n is specified by the functions $f_1, f_2, \ldots, f_n, f^{-1}, f^{+1}$ without altering their limit (as a consequence of (6)), where f^{-1} is the tangent to g with gradient -1 and f^{+1} is the tangent to g with gradient 1. It is easy to see that this ensures that $\mathbb{E}(B_{T'_n}) = \mathbb{E}(B_T)$ (by (3)), and also that $u_{\mu_n}(0) \to u_\mu(0)$ and $n \to \infty$. Consequently the stopping times T'_n and T satisfy the conditions of Proposition 18, where it is clear that the T'_n are all minimal, since each step clearly satisfies the conditions of Theorem 17 as a consequence of (36) (see also the discussion at the start of Section 5). So T is minimal.   □

Define the function

$$\Phi(x) = \operatorname*{argmin}_{\lambda < x}\left\{\frac{u_{\mu_0}(x) - c(\lambda)}{x - \lambda}\right\}. \tag{37}$$

In the cases described by [1], this is the barycentre function. It can also be seen to agree with the function appearing in the generalisation of the Azéma-Yor stopping time to non-centred means which appears in [6]. A similar function is used in [8] who examines the case where starting and target means are centred and satisfy (9); the exact construction in this case is given in [12, Sec. 5.2]. $\Phi(\cdot)$ can be thought of graphically as the point (below x) at which there exists a tangent to $c(\cdot)$ meeting the function $u_{\mu_0}(\cdot)$ at x.

Lemma 23. *The Azéma-Yor stopping time*

$$T = \inf\{t \geq 0 : B_t \leq \Phi(\overline{B}_t)\} \tag{38}$$

is minimal and attains equality in (23).

We prove this lemma using an extension of an idea first suggested in [10]. We approximate T by taking tangents to c, starting with gradient -1, and increasing to $+1$. As the number of tangents we take increases, the stopping time converges to T. The general approximation sequence can be seen in Figure 1.

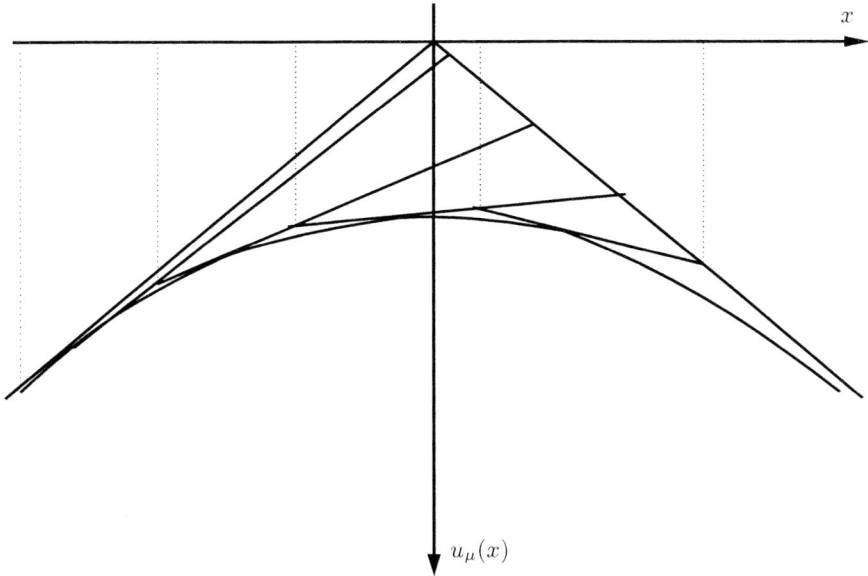

x

$u_\mu(x)$

Fig. 1. Approximating the Azéma-Yor stopping time: we take tangents to the potential from left to right. In the limit the tangents become closer. The dotted lines highlight the points at which the approximated stopping time will stop the process

Proof. We apply Lemma 8 for each n to the functions $f_1^n, f_2^n, \ldots, f_{m(n)}^n$, which are chosen as tangents to $c(\cdot)$ with increasing gradients, so that f_1^n has gradient -1, f_m^n has gradient 1, and so that the difference in the gradients of consequential tangents is less than $\frac{1}{n}$. We also choose the tangents in such a way that the points at which successive tangents intersect each other (which are B_{T_n} stops) are at most $\frac{1}{n}$ apart when they lie within $[-n, n]$ (at least as far as this is possible — if both μ_0 and μ have an interval containing no mass, it might not be possible to manage this, but this case will not be important). This defines a (minimal) stopping time T_n such that (by (3)) $\mathbb{E}(B_{T_n}) = \int x\, \mu(dx)$. Also, by considering $\mu_n = \mathcal{L}(B_{T_n})$, $|\mu_n((-\infty, x)) - \mu((-\infty, x))| \leq \frac{1}{n}$ for all $x \in \mathbb{R}$. So $\mu_n \implies \mu$. The choice of T_n also ensures that $\mathbb{P}(|T - T_n| > \varepsilon) \to 0$ for all $\varepsilon > 0$. Consequently T is minimal.

To deduce that T attains equality in (23) we note that $\Phi(x)$ is the optimal choice for λ in (23), and by the definition of $\Phi(x)$,

$$\{\overline{B}_T < x\} \subseteq \{B_T \leq \Phi(x)\}$$
$$\{\overline{B}_T \geq x\} \subseteq \{B_T \geq \Phi(x)\}.$$

This means we attain equality in (25) and (26), and so only need show that we have equality in (29) and (30) for equality in (23) to hold. But for x given, we may calculate the potential of $\mu' = \mathcal{L}(B_{T \wedge \bar{H}_x})$ — where $\bar{H}_x = \inf\{t \geq 0 : B \geq x\}$ — as:

$$u_{\mu'}(y) = \begin{cases} u_\mu(y) & : y \leq \Phi(x); \\ u_\mu(\Phi(x)) + \frac{y-\Phi(x)}{x-\Phi(x)}(u_{\mu_0}(x) - u_\mu(\Phi(x))) & : \Phi(x) \leq y \leq x; \\ u_{\mu_0}(y) & : y \geq x. \end{cases}$$

It then follows from Theorem 17 and (3) that equality holds. □

Remark 24. The embedding due to [9] can be viewed easily in this framework. Essentially the embedding can be described as follows. We wish to find an embedding which maximises $\mathbb{P}(\sup_{t\leq T}|B_t| \geq x)$ simultaneously for all x, subject to T being minimal. The optimal construction can be described thus in the Chacon-Walsh picture: choose the tangent to u_μ with gradient 0. Let (a,b) be the points at which the tangent intersects $c(x)$, where C is chosen to be the value given by (15); the first step of the construction is to run the process until it leaves the interval $I = (a,b)$. The construction will now have at least two separate halves; on the positive half we run the Azéma-Yor construction of (38), and on the negative half we run the reverse of the Azéma-Yor construction (i.e. Azéma-Yor applied to $-B_t$). Such a construction will therefore be minimal, and can be shown to be optimal — for details we refer the reader to [6].

This technique can be easily extended to maximising $\mathbb{P}(\sup_{t\leq T} f(B_t) \geq x)$, where f is a function which is increasing above some point x_0 and decreasing below x_0 (see [5] for details).

Remark 25. In a slightly different vein, the construction of [19] with decreasing functions can also be seen in this framework. Choose $\varepsilon > 0$ and construct alternate tangents; tangents of positive gradient intersecting the current potential at ε, and tangents of negative gradient intersecting the current potential at 0 (see Figure 2). This can be repeated a number of times, and results in an approximation of $c(x)$; suitable further choices of tangents can be used to construct a full embedding. If c touches u_{μ_0} at finite points, any choice of construction outside the interval containing 0 may be used. By construction this is a minimal embedding.

The Vallois construction results on taking the limit as $\varepsilon \downarrow 0$; the appearance of the local time in the construction results since the limiting embeddings are determined by the number of downcrossings of $[0,\varepsilon]$, which has as a limit the local time (see [15, VI.1.10]). An application of Proposition 18 allows us to deduce that this limit is minimal. For further details we refer the reader to [5].

Acknowledgements. The author would like to thank David Hobson and an anonymous referee for their helpful comments, which much improved the presentation of this work. He also acknowledges the financial support of the Nuffield Foundation.

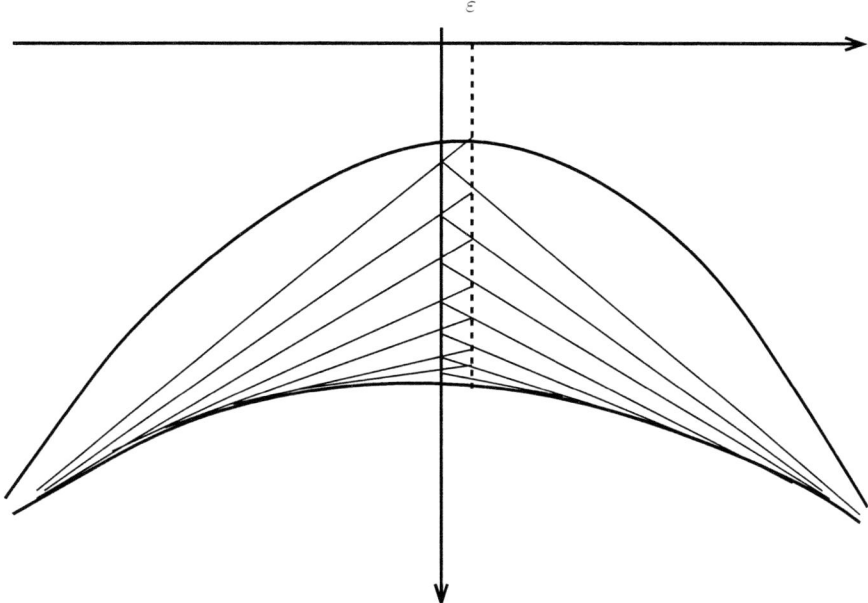

Fig. 2. The Chacon-Walsh type picture for an approximation to the Vallois stopping time, with a general starting distribution. We note that after the first two steps, there could still be mass at the extremes. This mass will have to be embedded using some suitable procedure — for example a Vallois construction using the local time at a different level

A Proof of Lemma 16

Proof (Proof of Lemma 16). Suppose that there exists $x < y$ such that $E_A(x) < x$ and $E_A(y) > y$, and note that we may suppose $E_A(w) = w$ for $x < w < y$. We show that we can construct a strictly smaller embedding, contradicting the assumption that T is minimal.

Define the stopping time $T' = R(A)\mathbf{1}_{\{B_{R(A)} \in \{x,y\}\}} + T\mathbf{1}_{\{B_{R(A)} \notin \{x,y\}\}}$ and for some $z \in (x, y)$, the stopping time

$$T'' = \inf\{t \geq T' : B_t = z\} \wedge T.$$

As a consequence of Remark 14, paths from both x and y must hit z.

Consider the set $\{T'' < T\}$. On this set we have only paths with $B_{R(A)} = x$ and $B_{R(A)} = y$. Define $\mu_x = \mathcal{L}(B_T; B_{R(A)} = x, T'' < T)$ and $\mu_y = \mathcal{L}(B_T; B_{R(A)} = y, T'' < T)$. Since Brownian motion bounded above is a submartingale,

$$\mathbb{E}(B_{T \wedge H_z}; B_{R(A)} = x, T > R(A)) \geq x\mathbb{P}(B_{R(A)} = x, T > R(A)).$$

Together with $E_A(x) < x$ this implies

$$z\mathbb{P}(B_{R(A)} = x, T'' < T) > \mathbb{E}(B_T; B_{R(A)} = x, T'' < T),$$

i.e. we must have $\frac{1}{\mu_x(\mathbb{R})} \int w\,\mu_x(dw) < z$, and similarly $\frac{1}{\mu_y(\mathbb{R})} \int w\,\mu_y(dw) > z$.
Then we apply Lemma 15 to the processes $B_{T''+t}$ on $\{B_{R(A)} = x, T'' < T\}$
and $\{B_{R(A)} = y, T'' < T\}$ with the measures

$$\tilde{\mu}_x = \mu_x|_{[a_1\infty)} + \mu_y|_{(a_2,\infty)}$$
$$\tilde{\mu}_y = \mu_x|_{(-\infty,a_1)} + \mu_y|_{(-\infty,a_2]}$$

where we choose $a_1 < z < a_2$ so that

$$\frac{1}{\tilde{\mu}_x(\mathbb{R})} \int w\,\tilde{\mu}_x(dw) \leq z \text{ and } \frac{1}{\tilde{\mu}_y(\mathbb{R})} \int w\,\tilde{\mu}_y(dw) \geq z$$

and also so that $\mu_x(\mathbb{R}) = \tilde{\mu}_x(\mathbb{R})$ and $\mu_y(\mathbb{R}) = \tilde{\mu}_y(\mathbb{R})$[1]. This will produce
a strictly smaller embedding, in contradiction to the assumption that T is
minimal.

So we have shown that there exists a such that (13) and (14) hold. We
just need to show that we can choose a so that $T \leq H_a$ on $\{T \geq R(A)\}$. Of
course, if $a = \pm\infty$, this is trivial.

Suppose that there exists $x < y$ such that $E_A(x) > x$ and $E_A(y) < y$ and
$E_A(w) = w$ for $w \in (x, y)$. If

$$\sup_{z<a} \mu_z((a, \infty)) = 0 \text{ and } \sup_{z>a} \mu_z((-\infty, a)) = 0 \text{ for some } a \in (x, y), \quad (39)$$

because T is minimal, Theorem 9 implies that $T \leq H_a$ on $\{T \geq R(A)\}$.

So suppose that (39) does not hold. We shall show that we can then find
a sequence $x_1, x_2, \ldots, x_r$ of elements of A such that we are able to transfer
mass between the x_i to produce a smaller embedding. We begin by choosing
x_1 to be the point of A satisfying $E_A(x) > x$ for which the support of μ_x
extends furthest to the right, and y_1 similarly the point satisfying $E_A(y) < y$
for which the support of μ_y extends furthest to the left. If the support of these
measures overlap we show we can exchange mass between μ_{x_1} and μ_{y_1} and
embed to find a smaller stopping time. Otherwise we look at those points for
which $E_A(x) = x$ and the support overlaps that of μ_{x_1} but extends further to
the right. In this way we can find a sequence whose supports overlap (since
(39) does not hold) and we may again perform a suitable exchange of mass
to show that we can find a smaller embedding. Then we take $x_r = y_1$ and the
points satisfy $x_2 < x_3 < \ldots < x_{r-1}$.

---

[1] It might be the case that there is in fact an atom at one or both of the points
a_1, a_2. In such a situation, it might be necessary to exchange only a part of the
atom at a_1 and/or a_2 in order to satisfy the conditions, so that the supports of
$\tilde{\mu}_x$ and/or $\tilde{\mu}_y$ might not differ from those of μ_x and μ_y. However, it is easy to
check that (provided a positive amount of mass is transferred), Lemma 15 still
holds and this will construct a strictly smaller stopping time.

There are several technical issues we need to address. Firstly, if we find at some stage there are two points which both satisfy the criterion — for example their supports have the same upper bound — then we may use either point. Secondly, if the support of all suitable points has a maximum which is not attained we may still use the same procedure but we must (and can) choose a point which approximates the bound suitably closely for subsequent steps to work. Finally we note that once we choose x_2, since there is at most one point to the right of y_1, there exists only a finite number of points left to choose from (by assumption on A) and so the sequence will be finite.

The technical construction is as follows: let x_1 be the largest value such that $E_A(x_1) > x_1$ and

$$\sup\{z : z \in \mathrm{supp}(\mu_{x_1})\} = \sup_{w: E_A(w) > w} \{\sup\{z : z \in \mathrm{supp}(\mu_w)\}\},$$

(or at least so that the left hand side approximates the right hand side sufficiently closely for the next step to work — since the support of the points to the right overlaps we shall be able to find x_1 with supremum of its support sufficiently close to the term on the left) and let y_1 be the smallest value such that $E_A(y_1) < y_1$ and

$$\inf\{z : z \in \mathrm{supp}(\mu_{y_1})\} = \inf_{w: E_A(w) < w} \{\inf\{z : z \in \mathrm{supp}(\mu_w)\}\}.$$

Then (by the assumption that (39) does not hold) we can find a sequence $x_1, x_2, \ldots, x_r$ such that $x_r = y_1$ and $x_2 < x_3 < \ldots x_{r-1}$, $E_A(x_i) = x_i$ for $1 < i < r$ and, if we define $I_i = \inf\{\text{intervals } I : \mathrm{supp}(\mu_{x_i}) \subseteq I\}$, then

$$
\begin{aligned}
Leb(I_i \cap I_{i+1}) &> 0 & k &= 1, \ldots, r-1, \\
Leb(I_i \cap I_{i+2}) &= 0 & k &= 1, \ldots, r-2.
\end{aligned}
\tag{40}
$$

This is done by choosing at each step the w with $E_A(w) = w$ which overlaps the support of the previous μ_{x_i} and whose support extends furthest to the right, until the support overlaps with the support of μ_{y_1}.

We write $\mu_i = \mu_{x_i}$. For general $1 \le i < r$ now consider μ_i^* defined by

$$\mu_i^* = \mu_i|_{(-\infty, s_i)} + \mu_{i+1}|_{(-\infty, s_i)}$$

where s_i is chosen such that $\mu_i([s_i, \infty)) = \mu_{i+1}((-\infty, s_i))^2$. Then it must be true that $\int w\, \mu_i^*(dw) < \int w\, \mu_i(dw)$. Define

$$m_i = \int w\, \mu_i(dw) - \int w\, \mu_i^*(dw) > 0$$

$$m_0 = \int w\, \mu_1(dw) - \mu_1(\mathbb{R})x_1 > 0$$

$$m_r = \mu_r(\mathbb{R})x_r - \int w\, \mu_r(dw) > 0$$

---

[2] Again, atoms at the natural break points may cause some technical problems, but the atom can again be split suitably

and set $\Delta m = \inf\{m_i : 0 \le i \le r\}$. Then for each i we can find $v_i < z_i$ such that $\mu_i([z_i, \infty)) = \mu_{i+1}((-\infty, v_i))$ and for

$$\mu_i' = \mu_i|_{(-\infty, z_i)} + \mu_{i+1}|_{(-\infty, v_i)}$$

we have

$$\int w\,\mu_i(dw) - \int w\,\mu_i'(dw) = \Delta m.$$

Set

$$\mu_1'' = \mu_1|_{(-\infty, z_1)} + \mu_2|_{(-\infty, v_1)},$$
$$\mu_i'' = \mu_{i-1}|_{[z_{i-1}, \infty)} + \mu_i|_{[v_{i-1}, z_i)} + \mu_{i+1}|_{(-\infty, v_i)} \qquad i = 2, \ldots, r-1,$$
$$\mu_r'' = \mu_{r-1}|_{[z_{r-1}, \infty))} + \mu_r|_{[v_{r-1}, \infty)}.$$

Then

$$\int x\,\mu_1'' \ge \mu_1''(\mathbb{R})x_1$$

$$\int x\,\mu_i'' = \mu_i''(\mathbb{R})x_i \qquad\qquad i = 2, \ldots, r-1$$

$$\int x\,\mu_r'' \le \mu_r''(\mathbb{R})x_r.$$

So the conditions of Lemma 15 are satisfied (due to (40)) for each μ_i'' and we can find strictly smaller stopping times on each of the sets $\{T > R(A), R(A) = x_i\}$.

It only remains to show the final statement of the lemma. Let $A' \supset A$ be another set satisfying the conditions of the lemma for some M', such that $A' \setminus A \subseteq [x, y]$. Then there exists $x', y' \in A'$ such that $x \le x' < y' \le y$, $E_{A'}(x') > x'$ and $E_{A'}(y') < y'$: to see this, suppose for a contradiction that $E_{A'}(y') \ge y'$ for all $y' \in [x, y]$. Then, by the minimality of T, and Theorem 9(ii) we must have:

$$\mathbb{E}(B_T | T > R(A), B_{R(A)} = y, B_{R(A')} = y') \ge y$$

for all $y' \in [x, y]$. However, $B_{R(A)} = y$ only if $B_{R(A')} \in [x, y]$ a.s., and we contradict $E_A(y) < y$. A similar result holds for x. Consequently, there must exist $x', y' \in [x, y]$ such that $E_{A'}(y') < y'$ and $E_{A'}(x') > x'$, and by the first part of this theorem, we must also have $x' < y'$.

Now consider a sequence $A \subset A_1 \subset A_2 \subset \ldots$ and such that $A_n \setminus A \subseteq [x, y]$ and $d(z, A_n) \le 2^{-n}$ for $z \in [x, y]$. Let

$$\Lambda = \{a \in [x, y] : T \le H_a \text{ on } \{T \ge R(A)\}\};$$
$$\Lambda_n = \{a \in [x, y] : T \le H_a \text{ on } \{T \ge R(A_n)\}\}.$$

Then the sets Λ, Λ_n are closed, $\Lambda \supseteq \Lambda_1 \supseteq \Lambda_2 \supseteq \ldots$, and each Λ_n is non-empty. So there exists $a_\infty \in \Lambda_n$ for all n. Hence $T \le H_{a_\infty}$ on $\{T \ge R(A_n)\}$ for all n. But $R(A_n) \downarrow 0$ on $\{B_0 \in [x, y]\}$ and $R(A) \le H_{a_\infty}$ on $\{B_0 \notin [x, y]\}$.   □

B Proof of Theorem 17

We prove a series of results leading to the proof of Theorem 17:

Proposition 26. *Suppose $T \leq H_{a_\infty}$ is an embedding of μ for $a_\infty \in \mathbb{R}$. Then $a_\infty \in \mathcal{A}$.*

Proof. Clearly a_∞ must lie in $\mathcal{A}'$ (see (17)). Suppose also that $a_\infty < z \in \mathcal{A}$. We may choose θ, ϕ such that $\mu_0 - \breve{\mu}_0^{a_\infty, \theta} - \hat{\mu}_0^{z, \phi}$ has no atom at either a_∞ or z.

Then

$$u_{\mu_0}(a_\infty) \geq u_\mu(a_\infty) - C, \tag{41}$$

where we note $C = u_\mu(z) - u_{\mu_0}(z)$, together with (19) and (20) imply

$$\int y \, (\mu - \breve{\mu}^{a_\infty, \theta} - \hat{\mu}^{z, \phi})(dy) \geq \int y \, (\mu_0 - \breve{\mu}_0^{a_\infty, \theta} - \hat{\mu}_0^{z, \phi})(dy), \tag{42}$$

the term on the right being equal to $\mathbb{E}(B_0; B_0 \in (a_\infty, z))$ and the term on the left less than or equal to $\mathbb{E}(B_T; B_0 \in (a_\infty, z))$ (by the choice of θ and ϕ, since $T \leq H_{a_\infty}$, $\mathcal{L}(B_T; B_0 \in (a_\infty, z))$ is dominated by $(\mu - \breve{\mu}^{a_\infty, \theta} - \hat{\mu}^{z, \phi})$ on $[a_\infty, z)$, with the same total mass, but supported on $[a_\infty, \infty)$). However $(B_{t \wedge T})_{t \geq 0} = (B_{t \wedge T \wedge H_{a_\infty}})_{t \geq 0}$ is a supermartingale on $\{B_0 \geq a_\infty\}$, so we must have equality in (42) and hence in (41). So $a_\infty \in \mathcal{A}$. A similar proof can be used for $z < a_\infty$. If there does not exists any such z, $a_\infty \in \mathcal{A}$, since $\mathcal{A} \neq \emptyset$. □

Proposition 27. *Suppose T is minimal and A is a countable subset of $\mathbb{R}$ such that A has finitely many elements in every compact subset of $\mathbb{R}$ and $d(x, A) < M$ for all $x \in \mathbb{R}$ and some $M > 0$. Suppose also that $S \leq T$ is a stopping time and $I \subseteq \mathbb{R}$ is an interval such that $\partial I \subseteq A$. If*

$$\mathbb{E}(B_T; F \cap \{B_0 \in I\}) > \mathbb{E}(B_S; F \cap \{B_0 \in I\}) \tag{43}$$

for some $F \in \mathcal{F}_S$ then $E_A(x) > x$ for some $x \in A \cap \bar{I}$.

Proof. Let $R(A)$ be as defined in Lemma 16. We may assume $F \subseteq \{B_0 \in I\}$ and we note that therefore $B_{R(A)} \in A \cap \bar{I}$ on $\{R(A) < T\} \cap F$. Since $B_{t \wedge R(A)}$ is uniformly integrable,

$$\mathbb{E}(B_S; F) = \mathbb{E}(B_{R(A)}; F \cap \{S \leq R(A)\}) + \mathbb{E}(B_S; F \cap \{R(A) < S\})$$
$$\mathbb{E}(B_T; F) = \mathbb{E}(B_{R(A)}; F \cap \{T = R(A)\}) + \mathbb{E}(B_T; F \cap \{R(A) < T\}).$$

So (43) and the above identities imply

$$\mathbb{E}(B_T; F \cap \{R(A) < T\}) > \mathbb{E}(B_{R(A)}; F \cap \{S \leq R(A) < T\})$$
$$+ \mathbb{E}(B_S; F \cap \{R(A) < S\}).$$

However if $E_A(x) \leq x$ for all $x \in A \cap \bar{I}$ and T is minimal, by the Strong Markov property and Theorem 9:

$$\mathbb{E}(B_T; F \cap \{R(A) < S\}) \leq \mathbb{E}(B_S; F \cap \{R(A) < S\})$$
$$\mathbb{E}(B_T; F \cap \{S \leq R(A) < T\}) \leq \mathbb{E}(B_{R(A)}; F \cap \{S \leq R(A) < T\})$$

and we deduce a contradiction. □

Proposition 28. *Suppose* $F \in \mathcal{F}_0$, $\mathbb{E}(B_T; F) = \mathbb{E}(B_0; F)$ *and*

$$\mathbb{E}(B_T | \mathcal{F}_S) \leq B_S \text{ a.s. on } F \tag{44}$$

for all stopping times S. Then in fact we have equality — that is

$$\mathbb{E}(B_T | \mathcal{F}_S) = B_S$$

almost surely on F.

Proof. If $\mathbb{P}(F) = 0$ there is nothing to prove. Otherwise we may condition on F to reduce to showing the result when $F = \Omega$.

By the Upward Martingale Theorem [16, Theorem II.69.5], the process

$$X_t = \mathbb{E}(B_T | \mathcal{F}_t)$$

is uniformly integrable. Also $\mathbb{E}(B_T | \mathcal{F}_0) \leq B_0$ and $\mathbb{E}(B_T) = \mathbb{E}(B_0)$ implies $\mathbb{E}(B_T | \mathcal{F}_0) = B_0$. Let $Y_t = B_{T \wedge t} - X_{T \wedge t}$. By (44), $(Y_t)_{t \geq 0}$ is a non-negative local martingale such that $Y_0 = Y_T = 0$. Hence $Y \equiv 0$. □

Lemma 29. *If T is minimal and $a \in \mathcal{A}$ then $T \leq H_a$ and for stopping times $S \leq T$,*

$$\mathbb{E}(B_T | \mathcal{F}_S) \leq B_S \text{ on } \{B_0 \geq a\}; \tag{45}$$
$$\mathbb{E}(B_T | \mathcal{F}_S) \geq B_S \text{ on } \{B_0 \leq a\}. \tag{46}$$

Proof. Suppose initially $a \in \mathbb{R}$, and let $\theta = \mu_0((-\infty, a))$. If $\{B_0 < a\} \not\subseteq \{B_T \leq a\}$ a.s. then

$$\mathbb{E}(B_0; B_0 < a) = \int y \, \check{\mu}_0^{a,\theta}(dy) \leq \int y \, \check{\mu}^{a,\theta}(dy) < \mathbb{E}(B_T; B_0 < a);$$

$$\mathbb{E}(B_0; B_0 \geq a) = \int y \, \hat{\mu}_0^{a,\theta}(dy) \geq \int y \, \hat{\mu}^{a,\theta}(dy) > \mathbb{E}(B_T; B_0 \geq a).$$

So there exists $x_1 \leq a$ and $x_2 \geq a$ such that (by Proposition 27)

$$E_A(x_1) < x_1 \text{ and } E_A(x_2) > x_2$$

for a suitable choice of A — a contradiction to Lemma 16. By symmetry, we will also have $\{B_0 > a\} \subseteq \{B_T \geq a\}$ a.s..

A similar argument can be used with $\theta = \mu_0((-\infty, a])$ to deduce that $\{B_0 \leq a\} \subseteq \{B_T \leq a\}$ a.s. and $\{B_0 \geq a\} \subseteq \{B_T \geq a\}$ a.s.. Consequently, if there is an atom of μ_0 at a then paths starting at a must also stop at a, and hence (by the minimality of T) must stop immediately — i.e. $T = 0$ on $\{B_0 = a\}$.

So consider paths for which $\{B_0 < a\}$. For almost all these paths, for some choice of A, $B_{R(A)} < a$. We suppose (46) fails and deduce a contradiction.

By Proposition 27 there exists $x < a$ such that $E_A(x) < x$. Since (for $\theta = \mu_0((-\infty, a)))$)

$$\int y \, \check{\mu}_0^{a,\theta}(dy) \leq \int y \, \check{\mu}^{a,\theta}(dy),$$

Lemma 16 implies there must also exist y such that $E_A(y) > y$, and since T is minimal, the lemma further implies $y < x$, and therefore that there exists $a' < a$ such that $T \leq H_{a'}$.

Consequently, $B_{t \wedge T}$ is a supermartingale on $\{B_0 > a'\}$ (and a submartingale on $\{B_0 \leq a'\}$). But Proposition 26 and (21) imply $\mathbb{E}(B_0; a' < B_0 < a) = \mathbb{E}(B_T; a' < B_0 < a)$ and therefore (by Proposition 28) $B_{t \wedge T}$ is a true martingale on $\{a' < B_0 < a\}$ — contradicting $E_A(x) < x$. Similarly (45) can be shown to hold.

So suppose now that $a = \infty$ (the case $a = -\infty$ is similar) and there exists $a' < \infty$ also in $\mathcal{A}$. By the above, $T \leq H_{a'}$ and so $(B_{t \wedge T})_{t \geq 0}$ is a supermartingale on $\{B_0 > a'\}$, while by (21) $\mathbb{E}(B_0; B_0 > a') = \mathbb{E}(B_T; B_0 > a')$, and hence $(B_{t \wedge T})_{t \geq 0}$ satisfies (46) by Proposition 28.

Finally suppose $\mathcal{A} = \{\infty\}$. By Lemma 16 $E_A(x) \geq x$ for all suitable choices of A and all x. Hence, by Proposition 27,

$$\mathbb{E}(B_T | \mathcal{F}_S) \geq B_S. \qquad \square$$

Remark 30. We note that some of the above arguments, particularly the use of Proposition 28, allow us to deduce that if there exists $a \in \mathcal{A}$, $|a| < \infty$ for which $T \leq H_a$ then (45) and (46) hold and $T \leq H_{a'}$ for all $a' \in \mathcal{A}$.

Lemma 31. *Suppose that for all stopping times S with $S \leq T$ and $\mathbb{E}|B_S| < \infty$ we have*

$$\mathbb{E}(B_T | \mathcal{F}_S) \leq B_S \quad a.s.. \tag{47}$$

Then T is minimal.

We refer the reader to Lemma 8 of [6], the proof of which is still valid in the more general case.

Of course we may replace the '$\leq$' in (47) with '$\geq$' or '$=$' without altering the conclusion.

Lemma 32. *If $T \leq H_{\mathcal{A}} = \inf\{t \geq 0 : B_t \in \mathcal{A}\}$ is a stopping time of the Brownian motion $(B_t)_{t \geq 0}$ where $B_0 \sim \mu_0$ and $B_T \sim \mu$, and (for stopping times $S \leq T$)*

$$\mathbb{E}(B_T | \mathcal{F}_S) \leq B_S : \text{ on } \{B_0 \geq a_-\} \tag{48}$$

$$\mathbb{E}(B_T | \mathcal{F}_S) \geq B_S : \text{ on } \{B_0 \leq a_+\}, \tag{49}$$

where $a_- = \inf \mathcal{A}$ and $a_+ = \sup \mathcal{A}$, then T is minimal.

Proof. Choose $a \in \mathcal{A}$. By assumption $T \leq H_a$ and, by Lemma 31, T is minimal for $\check{\mu}^a$ on $\{B_0 \leq a\}$ and for $\hat{\mu}^a$ on $\{B_0 \geq a\}$. It must then be minimal for μ. $\square$

Proposition 33. *If (v) holds and $S \leq T$ then $\mathbb{E}|B_S| < \infty$.*

Proof. We show that $\mathbb{E}(|B_S|; B_0 \geq a_-) < \infty$. Since $(B_{t \wedge H_{-k}})_{t \geq 0}$ is a supermartingale on $\{B_0 \geq -k\}$,

$$\mathbb{E}(B_{T \wedge H_{-k}}; B_S < 0, S < H_{-k}, B_0 \geq a_- \wedge (-k))$$
$$\leq \mathbb{E}(B_{S \wedge H_{-k}}; B_S < 0, S < H_{-k}, B_0 \geq a_- \wedge (-k)).$$

The term on the left hand side is equal to:

$$\mathbb{E}(B_T; B_S < 0, T < H_{-k}, B_0 \geq a_- \wedge (-k))$$
$$- k\mathbb{P}(B_S < 0, S \leq H_{-k} < T, B_0 \geq a_- \wedge (-k)).$$

The first term converges (by dominated convergence) to $\mathbb{E}(B_T; B_S < 0, B_0 \geq a_-)$ and the second term vanishes by the assumption. By monotone convergence

$$\mathbb{E}(B_S; B_S < 0, B_0 \geq a_-) = \lim_k \mathbb{E}(B_S; B_S < 0, S < H_{-k}, B_0 \geq a_- \wedge (-k))$$
$$\geq \lim_k \mathbb{E}(B_T; B_S < 0, S < H_{-k}, B_0 \geq a_- \wedge (-k))$$
$$\geq \mathbb{E}(B_T; B_S < 0, B_0 \geq a_-) \geq -\mathbb{E}(B_T^-) > -\infty.$$

Also

$$\mathbb{E}(B_0; B_0 \geq a_- \wedge (-k)) \geq \mathbb{E}(B_{S \wedge H_{-k}}; B_0 \geq a_- \wedge (-k))$$
$$= \mathbb{E}(B_S; B_0 \geq a_- \wedge (-k), S < H_{-k})$$
$$- k\mathbb{P}(H_{-k} \leq S, B_0 \geq a_- \wedge (-k)),$$

and

$$\mathbb{E}(B_S; B_0 \geq a_- \wedge (-k), S < H_{-k}) = \mathbb{E}(B_S^+; B_0 \geq a_- \wedge (-k), S < H_{-k})$$
$$- \mathbb{E}(B_S^-; B_0 \geq a_- \wedge (-k), S < H_{-k}),$$

so

$$\mathbb{E}(B_S^+; B_0 \geq a_- \wedge (-k), S < H_{-k}) \leq \mathbb{E}(B_0; B_0 \geq a_- \wedge (-k))$$
$$+ \mathbb{E}(B_S^-; B_0 \geq a_- \wedge (-k), S < H_{-k})$$
$$+ k\mathbb{P}(H_{-k} \leq S, B_0 \geq a_- \wedge (-k)).$$

By monotone and dominated convergence, in the limit we have

$$\mathbb{E}(B_S^+; B_0 \geq a_-) \leq \mathbb{E}(B_0; B_0 \geq a_-) + \mathbb{E}(B_S^-; B_0 \geq a_-)$$
$$< \infty.$$

So $\mathbb{E}(|B_S|; B_0 \geq a_-) < \infty$. Similarly $\mathbb{E}(|B_S|; B_0 \leq a_+) < \infty$, and together these imply $\mathbb{E}(B_S) < \infty$. $\square$

Proof (Proof of Theorem 17.). Clearly *(ii)* $\implies$ *(iii)* $\implies$ *(iv)* $\implies$ *(v)* (the final implication following from dominated convergence). We also know *(i)* $\iff$ *(iii)*. We show *(v)* $\implies$ *(ii)*.

Suppose $A \in \mathcal{F}_R$, $A \subseteq \{B_0 \geq a_-\}$ and set $A_k = A \cap \{R < H_{-k}\} \cap \{B_0 \geq -k\}$. Then

$$\mathbb{E}(B_{S \wedge H_{-k}}; A_k) \leq \mathbb{E}(B_{R \wedge H_{-k}}; A_k).$$

By Proposition 33 we may apply dominated convergence to deduce that in the limit as $k \to \infty$ the right-hand side converges to $\mathbb{E}(B_R; A)$. Also

$$
\begin{aligned}
\mathbb{E}(B_{S \wedge H_{-k}}; A_k) =& \mathbb{E}(B_S; A \cap \{B_0 \geq -k\} \cap \{S \leq H_{-k}\}) \\
& + k\mathbb{P}(A, R < H_{-k} < S, B_0 \geq -k),
\end{aligned}
$$

where the second term converges to zero by assumption and the first converges to $\mathbb{E}(B_S; A)$ by dominated convergence. $\qquad\square$

References

1. J. Azéma and M. Yor. Une solution simple au problème de Skorokhod. In *Séminaire de Probabilités, XIII (Univ. Strasbourg, Strasbourg, 1977/78)*, volume 721 of *Lecture Notes in Math.*, pages 90–115. Springer, Berlin, 1979.
2. J. Bertoin and Y. Le Jan. Representation of measures by balayage from a regular recurrent point. *Ann. Probab.*, 20(1):538–548, 1992.
3. R. V. Chacon. Potential processes. *Trans. Amer. Math. Soc.*, 226:39–58, 1977.
4. R. V. Chacon and J. B. Walsh. One-dimensional potential embedding. In *Séminaire de Probabilités, X (Prèmiere partie, Univ. Strasbourg, Strasbourg, année universitaire 1974/1975)*, pages 19–23. Lecture Notes in Math., Vol. 511. Springer, Berlin, 1976.
5. A. M. G. Cox. *Skorokhod embeddings: non-centred target distributions, diffusions and minimality*. PhD thesis, University of Bath, 2004.
6. A. M. G. Cox and D. G. Hobson. Skorokhod embeddings, minimality and non-centred target distributions. *Probab. Theory Related Fields*, To Appear.
7. L. E. Dubins. On a theorem of Skorohod. *Ann. Math. Statist.*, 39:2094–2097, 1968.
8. D. G. Hobson. The maximum maximum of a martingale. In *Séminaire de Probabilités, XXXII*, volume 1686 of *Lecture Notes in Math.*, pages 250–263. Springer, Berlin, 1998.
9. S. D. Jacka. Doob's inequalities revisited: a maximal H^1-embedding. *Stochastic Process. Appl.*, 29(2):281–290, 1988.
10. I. Meilijson. On the Azéma-Yor stopping time. In *Seminar on probability, XVII*, volume 986 of *Lecture Notes in Math.*, pages 225–226. Springer, Berlin, 1983.
11. I. Monroe. On embedding right continuous martingales in Brownian motion. *Ann. Math. Statist.*, 43:1293–1311, 1972.
12. Jan Obłój. The Skorokhod embedding problem and its offspring. *Probability Surveys*, 1:321–392, 2004.

13. E. Perkins. The Cereteli-Davis solution to the H^1-embedding problem and an optimal embedding in Brownian motion. In *Seminar on stochastic processes, 1985 (Gainesville, Fla., 1985)*, pages 172–223. Birkhäuser Boston, Boston, MA, 1986.

14. Sidney C. Port and Charles J. Stone. *Brownian motion and classical potential theory*. Academic Press [Harcourt Brace Jovanovich Publishers], New York, 1978. Probability and Mathematical Statistics.

15. D. Revuz and M. Yor. *Continuous martingales and Brownian motion*, volume 293 of *Grundlehren der Mathematischen Wissenschaften [Fundamental Principles of Mathematical Sciences]*. Springer-Verlag, Berlin, third edition, 1999.

16. L. C. G. Rogers and D. Williams. *Diffusions, Markov processes, and martingales. Vol. 1*. Cambridge University Press, Cambridge, 2000. Foundations, Reprint of the second (1994) edition.

17. D. H. Root. The existence of certain stopping times on Brownian motion. *Ann. Math. Statist.*, 40:715–718, 1969.

18. A. V. Skorokhod. *Studies in the theory of random processes*. Translated from the Russian by Scripta Technica, Inc. Addison-Wesley Publishing Co., Inc., Reading, Mass., 1965.

19. P. Vallois. Le problème de Skorokhod sur **R**: une approche avec le temps local. In *Seminar on probability, XVII*, volume 986 of *Lecture Notes in Math.*, pages 227–239. Springer, Berlin, 1983.

20. P. Vallois. Quelques inégalités avec le temps local en zero du mouvement brownien. *Stochastic Process. Appl.*, 41(1):117–155, 1992.

21. Marc Yor. Sur le balayage des semi-martingales continues. In *Séminaire de Probabilités, XIII (Univ. Strasbourg, Strasbourg, 1977/78)*, volume 721 of *Lecture Notes in Math.*, pages 453–471. Springer, Berlin, 1979.

Transformations browniennes et compléments indépendants : résultats et problèmes ouverts

Jean Brossard et Christophe Leuridan

Institut Fourier, Laboratoire de Mathématiques
UMR5582 (UJF-CNRS), BP 74, 38402 St Martin d'Hères Cedex (France)
e-mail: Christophe.Leuridan@ujf-grenoble.fr

Summary. In the first part, the natural filtration of an $\mathbf{R}^d$-valued Brownian motion B is compared to that of the BM $B' = \int_0^{\cdot} H dB$, where H is previsible in the filtration of B and valued in $O_d(\mathbf{R})$. We show that there exists a r.v. U, independent of B' and uniformly distributed on $[0, 1]$ or on some finite set, such that $\sigma(B) = \sigma(B') \vee \sigma(U)$, provided either of the following two conditions holds:
– when the transform $B \mapsto B'$ is "subordinated" to some subdivision of $\mathbf{R}_+$;
– when this transform commutes with Brownian scaling.
The r.v. U encodes the information lost by the transform $B \mapsto B'$. We show that all kinds of information loss are possible: U may take infinitely many or any finite number of values.

In the second part, we study a related question: suppose given a linear BM X' in the filtration generated by some planar BM (X, Y). Can one find another linear BM independent from X' and such that the planar BMs (X', Y') and (X, Y) generate the same filtration ? We give a necessary condition for X' to have such an independent Brownian complement, and we study some examples.

Classification mathématique : 60J65,60H20.

Mots-clés : filtrations, transformations browniennes, compléments indépendants.

1 Transformations intégrales du mouvement brownien

1.1 Introduction

Soient B un mouvement brownien dans $\mathbf{R}^d$ et $\mathcal{F}_{\cdot}^B$ la filtration naturelle (complétée) de B. Si H est un processus prévisible dans $\mathcal{F}_{\cdot}^B$ à valeurs dans $O_d(\mathbf{R})$, l'intégrale stochastique $B' = \int_0^{\cdot} H dB$ définit un nouveau mouvement brownien dans $\mathbf{R}^d$, dont la filtration naturelle est évidemment contenue dans celle de B (au sens où $\mathcal{F}_t^{B'} \subset \mathcal{F}_t^B$ pour tout t). A l'aide de la propriété de représentation prévisible, ou bien en utilisant l'indépendance de $\mathcal{F}_t^B$ et des accroissements $B'_{t+} - B'_t$, on montre facilement que la filtration naturelle de B' est *immergée* dans celle de B, au sens où toute martingale dans la filtration $\mathcal{F}^{B'}$ est encore une martingale dans $\mathcal{F}^B$ (voir par exemple [1], proposition 1).

L'inclusion $\mathcal{F}_t^{B'} \subset \mathcal{F}_t^B$ peut être stricte comme le montre l'exemple de la transformation de Lévy : si B est un mouvement brownien dans $\mathbf{R}$ et $B' = \int_0^\cdot \mathrm{sgn}(B_s)dB_s$, il est bien connu que la filtration naturelle de B' est celle de $|B|$. L'information perdue par la transformation $B \mapsto \int_0^\cdot \mathrm{sgn}(B_s)dB_s$ est donc la famille des signes des excursions de B. On peut coder cette information par une suite de signes indépendants et uniformément distribués, indépendante de B', ou, en utilisant les développements dyadiques, par une variable aléatoire de loi uniforme sur $[0,1]$, indépendante de B'.

Nous allons voir que l'existence d'un complément indépendant et de loi uniforme est une situation courante, moyennant quelques restrictions sur le processus prévisible H. Pour se convaincre qu'une filtration brownienne immergée dans une autre ne possède pas toujours de complément indépendant, posons $H_t = \mathrm{sgn}(B_t - B_1)$ si $t > 1$ et $B_1 > 0$ et $H_t = 1$ dans tous les autres cas. Autrement dit, on applique la transformation de Lévy aux accroissements à partir de l'instant 1 lorsque $B_1 > 0$, et on ne change rien lorsque $B_1 \leq 0$. Dans ce cas, pour $t > 1$ fixé, la loi de $(B_s)_{0 \leq s \leq t}$ connaissant $(B'_s)_{0 \leq s \leq t}$ est diffuse si $B'_1 > 0$ et est une masse de Dirac si $B'_1 \leq 0$. Il n'existe pas de complément indépendant à $\mathcal{F}_t^{B'}$ qui redonne la tribu $\mathcal{F}_t^B$.

Les deux paragraphes suivants présentent deux situations où une hypothèse supplémentaire assure l'existence d'un complément indépendant et de loi uniforme sur $[0,1]$ ou sur un ensemble fini, cet ensemble fini étant un singleton lorsque la transformation $B \mapsto B'$ ne perd pas d'information.

1.2 Transformations assujetties à une subdivision de $\mathbf{R}_+$.

Dans ce paragraphe, nous appellerons subdivision de $\mathbf{R}_+$ toute suite $(t_n)_{n \in \mathbf{Z}}$ strictement croissante telle que $t_n \to 0$ quand $n \to -\infty$ et $t_n \to +\infty$ quand $n \to +\infty$. Grosso modo, nous dirons que la transformation $B \mapsto \int_0^\cdot H_s dB_s$ est assujettie à une subdivision $(t_n)_{n \in \mathbf{Z}}$ lorsque le processus H restreint à un intervalle $[0, t_n]$ est une fonction de B restreint à l'intervalle $[0, t_{n-1}]$. Énonçons le résultat précis.

Théorème 1 *Soit B un mouvement brownien défini sur $(\Omega, \mathcal{A}, P)$ et à valeurs dans $\mathbf{R}^d$. Soit $(t_n)_{n \in \mathbf{Z}}$ une subdivision de $\mathbf{R}_+$. Soit H un processus à valeurs dans $O_d(\mathbf{R})$ tel que, pour tout $n \in \mathbf{Z}$, l'application $(s, \omega) \mapsto H_s(\omega)$ de $[0, t_n] \times \Omega$ soit mesurable pour la tribu $\mathcal{B}([0, t_n]) \otimes \mathcal{F}_{t_{n-1}}^B$. Soit B' le mouvement brownien défini par $B' = \int H dB$. Alors pour tout $t > 0$, il existe une variable aléatoire U_t indépendante de $\mathcal{F}_t^{B'}$ et de loi uniforme sur $[0,1]$ ou sur un ensemble fini telle que $\mathcal{F}_t^B = \mathcal{F}_t^{B'} \vee \sigma(U_t)$.*

Démonstration.   Nous montrons le résultat pour $t = +\infty$, la démonstration s'adaptant facilement au cas où $t \in \mathbf{R}_+^*$. L'idée est de se ramener à des processus à temps discret et d'appliquer les résultats sur les suites récurrentes stochastiques ou chaînes de Markov constructives indexées par $\mathbf{Z}$. L'hypothèse sur le processus H assure que la transformation $B \mapsto \int_0^{\cdot} H_s dB_s$ ne peut perdre de l'information qu'à l'instant 0, ce qui permet d'utiliser des lois du zéro-un.

Pour tout $n \in \mathbf{Z}$, notons X_n et V_{n+1} les portions de trajectoires définies par

$$X_n = (B_s)_{0 \le s \le t_n} \quad \text{et} \quad V_{n+1} = (B'_s - B'_{t_n})_{t_n \le s \le t_{n+1}}.$$

Les filtrations naturelles des processus $(X_n)_{n \in \mathbf{Z}}$, $(V_n)_{n \in \mathbf{Z}}$ et $((X_n, V_n))_{n \in \mathbf{Z}}$ sont données par $\mathcal{F}_n^{(X,V)} = \mathcal{F}_n^X = \mathcal{F}_{t_n}^B$ et $\mathcal{F}_n^V = \mathcal{F}_{t_n}^{B'}$. Pour tout $n \in \mathbf{Z}$, la variable aléatoire V_{n+1} est indépendante de $\mathcal{F}_n^{(X,V)} = \mathcal{F}_{t_n}^B$ et l'égalité $B_t - B_{t_n} = \int_{t_n}^t H_s^{-1} dB'_s$ pour $t_n \le t \le t_{n+1}$ montre que la connaissance de X_n et de V_{n+1} permet de reconstituer $(H_s)_{t_n \le s \le t_{n+1}}$ et donc $X_{n+1} = (B_s)_{0 \le s \le t_{n+1}}$. Autrement dit, $(X_n)_{n \in \mathbf{Z}}$ est une chaîne de Markov inhomogène gouvernée par une relation de récurrence de la forme $X_{n+1} = f_n(X_n, V_{n+1})$. Comme la tribu asymptotique $\mathcal{F}_{-\infty}^X = \mathcal{F}_0^B$ est triviale, il suffit d'appliquer le théorème que nous rappelons ci-dessous (corollaire 1.2 de [4]).

Théorème 2 *Soit $X = (X_n)_{n \in \mathbf{Z}}$ une chaîne de Markov inhomogène gouvernée par une suite de variables $V = (V_n)_{n \in \mathbf{Z}}$ et une suite d'applications $(f_n)_{n \in \mathbf{Z}}$, autrement dit :*

X_n est une variable aléatoire à valeurs dans un espace mesurable $(E_n, \mathcal{E}_n)$;

V_{n+1} est une variable aléatoire à valeurs dans un espace mesurable $(G_{n+1}, \mathcal{G}_{n+1})$ indépendante de la suite $((X_k, V_k))_{k \le n}$

$X_{n+1} = f_n(X_n, V_{n+1})$ où f_n est une application mesurable de $(E_n \times G_{n+1}, \mathcal{E}_n \otimes \mathcal{G}_{n+1})$ dans $(E_{n+1}, \mathcal{E}_{n+1})$.

Si les espaces $(E_n, \mathcal{E}_n)$ sont lusiniens (isomorphes à un borélien d'un espace polonais) et si la tribu $\mathcal{F}_{-\infty}^X$ est triviale, alors il existe une variable aléatoire U indépendante de V et de loi uniforme sur $[0,1]$ ou sur un ensemble fini telle que $\sigma(X) = \sigma(V) \vee \sigma(U)$.

Un exemple simple en dimension 1, étudié dans [1], est celui où $H_s = \operatorname{sgn}(B_{t_n} - B_{t_{n-1}})$ pour $t_n < s \le t_{n+1}$: la filtration $\mathcal{F}^{B'}$ est alors la filtration de Goswami-Rao. Autrement dit $\mathcal{F}_t^{B'}$ est engendrée par les fonctionnelles paires du mouvement brownien jusqu'à l'instant t. On perd exactement un bit d'information en passant de $\mathcal{F}_t^B$ à $\mathcal{F}_t^{B'}$.

Plus généralement, M. Malric [7, 8] utilise des transformations du même type pour construire un mouvement brownien qui engendre la filtration quotient de la filtration d'un mouvement brownien de dimension d par n'importe quel sous-groupe du groupe orthogonal. Par un choix convenable du sous-groupe, on peut donc perdre une information codée par une variable aléatoire uniforme sur $[0,1]$ ou sur un ensemble fini fixé.

Déterminer l'information perdue peut être difficile, même pour une transformation donnée par une formule simple en dimension 1. Par exemple, la transformation $B \mapsto B' = \int_0^{\cdot} \mathrm{sgn}(B_{s/2}) dB_s$ est assujettie à la subdivision $(2^n)_{n \in \mathbf{Z}}$ et perd au moins un bit d'information, puisque B' est une fonction paire de B. Perd-elle seulement un bit d'information ?

Cette dernière transformation appartient à l'autre classe intéressante que nous allons voir, constituée des transformations qui commutent avec les changements d'échelle.

1.3 Transformations commutant avec les changements d'échelle

Le résultat s'énonce de la façon suivante :

Théorème 3 *Soit B un mouvement brownien défini sur $(\Omega, \mathcal{A}, P)$ et à valeurs dans $\mathbf{R}^d$. Soit B' le mouvement brownien défini par $B' = \int H dB$, où H est un processus prévisible de la forme*

$$H_t = h\left((t^{-1/2} B_{tu})_{0 \leq u \leq 1}\right)$$

pour une certaine application h de $\mathcal{C}([0,1], \mathbf{R}^d)$ dans $O_d(\mathbf{R})$ ne dépendant pas de t, alors pour tout $t > 0$, il existe une variable aléatoire U_t indépendante de $\mathcal{F}_t^{B'}$ et de loi uniforme sur $[0,1]$ ou sur un ensemble fini telle que $\mathcal{F}_t^B = \mathcal{F}_t^{B'} \vee \sigma(U_t)$.

Remarque : on peut montrer – mais ce n'est pas notre propos – que la forme de H est équivalente au fait que la transformation $B \mapsto B'$ commute avec les dilatations $B \mapsto (\frac{1}{\lambda} B_{\lambda^2 t})_{t \geq 0}$. Dans la démonstration, nous utiliserons simplement le fait que la loi du couple (B, B') est invariante par changement d'échelle, ce qui est encore équivalent.

Démonstration. Cette fois-ci, on ne peut pas appliquer directement les résultats sur les chaînes de Markov constructives, mais on peut adapter la démonstration du théorème 1. Pour $t > 0$, notons $(\nu_t(\omega, \cdot))_{\omega \in \Omega}$ la loi conditionnelle de $(B_s)_{0 \leq s \leq t}$ sachant $(B'_s)_{0 \leq s \leq t}$, qui est aussi la loi conditionnelle de $(B_s)_{0 \leq s \leq t}$ sachant B' et posons

$$A_t(\omega) = \nu_t(\omega, \{(B_s(\omega))_{0 \leq s \leq t}\}).$$

Si $s \leq t$, la loi $\nu_s(\omega, \cdot)$ est l'image de de $\nu_t(\omega, \cdot)$ par la projection canonique de $\mathcal{C}([0,t], \mathbf{R}^d)$ dans $\mathcal{C}([0,s], \mathbf{R}^d)$. Cela entraîne que le processus $(A_t)_{t \geq 0}$ est décroissant.

Mais par hypothèse, la loi du couple (B, B') est invariante par changement d'échelle, ce qui entraîne que la loi de A_t ne dépend pas de t. Donc presque sûrement A_t ne dépend pas de t. Les variables aléatoires A_t sont presque sûrement égales à une variable aléatoire $\mathcal{F}_0^B$-mesurable, donc presque sûrement constante. Soit a cette constante. En conditionnant par rapport à $\sigma((B'_s)_{0 \leq s \leq t})$ l'égalité presque sûre

$$\mathbf{I}_{[\nu_t(\omega, \{(B_s(\omega))_{0 \leq s \leq t}\}) = a]} = 1,$$

on obtient pour presque tout $\omega \in \Omega$,

$$\int_{\mathcal{C}([0,t],\mathbf{R}^d)} \mathbf{I}_{[\nu_t(\omega,\{b\})=a]}\nu_t(\omega, db) = 1.$$

Donc $\nu_t(\omega, \cdot)$ est formée de $N = 1/a$ atomes de masse a si $a > 0$, et diffuse si $a = 0$. Dans les deux cas, la structure atomique de $\nu_t(\omega, \cdot)$ ne dépend pas de ω, ce qui permet de construire un complément indépendant U_t.

Si $a > 0$, il suffit de numéroter de façon $\sigma(B')$-mesurable les atomes de $\nu_t(\omega, \cdot)$; on définit $U_t(\omega)$ comme le numéro de l'atome $(B_s(\omega))_{0 \leq s \leq t}$.

Si $a = 0$, on choisit une application $F(\omega, \cdot)$ de $\mathcal{C}([0,t], \mathbf{R})$ dans $[0,1]$, dépendant de façon $\sigma(B')$-mesurable de ω, qui envoie $\nu_t(\omega, \cdot)$ sur la loi uniforme sur $[0,1]$; on pose alors $U_t(\omega) = F(\omega, (B_s(\omega))_{0 \leq s \leq t})$.

1.4 Perte d'information pour les transformations commutant avec les changements d'échelle

Voyons quelques exemples en dimension 1, montrant que tous les types de perte d'information sont possibles.

La transformation de Lévy, définie par $H_t = \text{sgn}(B_t)$ perd une information infinie.

Lorsque H_t est le signe de la plus longue excursion achevée avant l'instant t, la transformation $B \mapsto B' = \int_0^{\cdot} H dB$ perd un bit d'information et le mouvement brownien B' engendre la filtration de Goswami-Rao (voir [1]).

Pour construire une transformation perdant une information à p valeurs possibles, on peut se limiter au cas où p est impair (ou même premier ≥ 3) et composer les transformations. La construction dont nous allons donner les étapes s'inspire de la construction surprenante d'une transformation brownienne d'ordre p et commutant avec les changements d'échelle donnée dans [1].

L'idée est d'inclure dans le mouvement brownien un jeu de pile ou face (c'est-à-dire une suite i.i.d. indexée par $\mathbf{Z}$ de variables aléatoires uniformes sur $\{-1, 1\}$) qui soit invariant à translation temporelle près par changement d'échelle du mouvement brownien. Un exemple est fourni par les signes des *excursions longues* du mouvement brownien, c'est-à-dire des excursions plus longues que toutes les excursions antérieures.

On utilise alors une transformation *homogène* (c'est-à-dire qui commute avec les translations temporelles) du jeu de pile ou face perdant une information à p valeurs. Comme dans l'article [3], on obtient une telle transformation en observant l'effet sur les développements dyadiques de la transformation $x \mapsto px$ modulo 1. Plus précisément, pour tout $x \in \mathbf{R}$,

$$x \equiv \sum_{n \leq -1} a_n(x)2^n \quad \text{et} \quad px \equiv \sum_{n \leq -1} b_n(x)2^n \text{ modulo } 1,$$

avec $a_n(x) = \text{Ent}(2^{-n}x) - 2\text{Ent}(2^{-n-1}x) = \text{Ent}(2^{-n}x) \mod 2$ et $b_n(x) = a_n(px)$. Mais a_n et b_n sont des fonctions 2^{n+1}-périodiques et p est impair donc

$$b_n = \text{Ent}(2^{-n}p \sum_{k \leq n} a_k 2^k) \mod 2 = a_n + \text{Ent}(p \sum_{k \geq 1} a_{n-k} 2^{-k}) \mod 2.$$

Cette dernière formule fournit une transformation sur les suites de $\{0,1\}^{-\mathbf{N}}$, ou même de $\{0,1\}^{\mathbf{Z}}$, qui a les propriétés voulues. Il reste à changer $\{0,1\}^{\mathbf{Z}}$ en $\{-1,1\}^{\mathbf{Z}}$ pour obtenir une transformation homogène du jeu de pile ou face, c'est-à-dire de la forme $(x_n)_{n \in \mathbf{Z}} \mapsto (x_n \phi((x_{n-k})_{k \geq 1}))_{n \in \mathbf{Z}}$ avec ϕ mesurable de $\{-1,1\}^{\infty}$ dans $\{-1,1\}$.

L'homogénéité temporelle de cette transformation du jeu de pile ou face permet de l'appliquer aux signes des excursions longues sans avoir besoin de numéroter les excursions. De plus, la transformation brownienne obtenue commute avec les changements d'échelle. Mais si on ne modifie que les signes des excursions longues en conservant telles quelles les autres excursions, le mouvement brownien obtenu ne peut pas se représenter par l'intégrale stochastique d'un processus *prévisible*.

Pour obtenir un intégrande H prévisible, introduisons la suite ordonnée $(V_n)_{n \in \mathbf{Z}}$ des fins d'excursions longues du mouvement brownien B et notons ϵ_n le signe de l'excursion longue finissant à l'instant V_n. Il suffit de poser $H_t = \phi((\epsilon_{n-k})_{k \geq 1})$ sur l'événement $[V_{n-1} < t \leq V_n]$. Cette définition ne dépend pas de la façon de numéroter les excursions, pourvu que l'ordre chronologique soit respecté. Bien qu'on ne sache pas énumérer les fins d'excursions longues par une suite strictement croissante de temps d'arrêt indexée par $\mathbf{Z}$, le processus H est prévisible. En effet, pour $\epsilon > 0$ fixé, on peut choisir la numérotation de telle sorte que $(V_n)_{n \in \mathbf{N}}$ soit la suite ordonnée des fins d'excursions longues postérieures à l'instant $\epsilon > 0$, ce qui montre que la restriction de H à $[\epsilon, +\infty[$ est prévisible.

Le processus H est constant entre deux fins d'excursions longues consécutives de B, donc $\int_0^{\cdot} H dB = HB$. Autrement dit, les mouvements browniens B et $B' = \int_0^{\cdot} H dB$ ont les mêmes excursions à un signe près et ce signe ne peut varier qu'aux fins d'excursions longues. La transformation $B \mapsto B' = \int_0^{\cdot} H dB$ perd donc une information à p valeurs possibles, tout comme la transformation induite sur les signes des excursions longues.

Comparaison avec la construction de Attal, Burdzy, Émery et Hu

La construction d'une transformation d'ordre p par Attal, Burdzy, Émery et Hu présente une difficulté supplémentaire du fait que l'application $x \mapsto x + \frac{1}{p}$ modulo 1 vue sur les développements dyadiques fournit une transformation *inhomogène* du jeu de pile ou face, d'ordre p. L'inhomogénéité pose une difficulté supplémentaire pour inclure la transformation du jeu de pile ou face dans le mouvement brownien. Mais la période q du développement dyadique de $\frac{1}{p}$ leur permet de mener à bien la démonstration au prix d'une étape supplémentaire : la construction d'une partie de $\{V_n ; n \in \mathbf{Z}\}$ invariante par changement d'échelle et numérotable cycliquement par $\mathbf{Z}/q\mathbf{Z}$ de façon optionnelle. Pour les problèmes de numérotation des records d'un processus de Poisson ponctuel tel que le processus des longueurs d'excursions, nous renvoyons le lecteur à l'article [3].

La construction de Attal, Burdzy, Émery et Hu, on le voit, est loin d'être intuitive et nous amène naturellement à poser trois questions :

1. Existe-t-il des transformations homogènes du jeu de pile ou face d'ordre p impair ?
2. Existe-t-il des transformations homogènes du jeu de pile ou face d'ordre $4, 8, \ldots$?
3. Existe-t-il des transformations intégrales du mouvement brownien linéaire, d'ordre $4, 8, \ldots$ et commutant avec les changements d'échelle ?

Une réponse positive à la question 1 permettrait de simplifier la construction de Attal, Burdzy, Émery et Hu. Une réponse positive à la question 2 entraînerait une réponse positive à la question 3.

2 Compléments browniens indépendants

2.1 Mouvements browniens complémentables et maximaux

Définition 4 *Soit (X, Y) un mouvement brownien plan et X' un mouvement brownien linéaire dans la filtration naturelle complétée de (X, Y), notée $\mathcal{F}.^{(X,Y)}$.*

On dit que X' est complémentable dans $\mathcal{F}.^{(X,Y)}$ s'il existe un brownien linéaire Y' dans $\mathcal{F}.^{(X,Y)}$, indépendant de X' et tel que (X', Y') engendre $\mathcal{F}.^{(X,Y)}$.

On dit que X' est maximal dans $\mathcal{F}.^{(X,Y)}$ si pour tout brownien linéaire X'' dans $\mathcal{F}.^{(X,Y)}$, l'inclusion $\mathcal{F}.^{X'} \subset \mathcal{F}.^{X''}$ entraîne l'égalité $\mathcal{F}.^{X'} = \mathcal{F}.^{X''}$.

Commençons par deux remarques immédiates :

- Deux mouvements browniens X' et Y' dans $\mathcal{F}.^{(X,Y)}$ sont indépendants si et seulement si $\langle X', Y' \rangle = 0$.
- Si deux mouvements browniens X' et X'' dans $\mathcal{F}.^{(X,Y)}$ vérifient $\mathcal{F}.^{X'} \subset \mathcal{F}.^{X''}$, alors il existe un processus H à valeurs dans $\{-1, 1\}$, prévisible dans $\mathcal{F}.^{X''}$ (et dans $\mathcal{F}.^{(X,Y)}$), tel que $X' = \int_0^\cdot H dX''$; on a donc $\langle X', Y' \rangle = \int_0^\cdot H d\langle X'', Y' \rangle$ pour tout mouvement brownien Y' dans $\mathcal{F}.^{(X,Y)}$.

De ces remarques, on déduit les résultats suivants.

Lemme 5 *Si X', X'' et Y' sont des mouvements browniens dans $\mathcal{F}.^{(X,Y)}$ et $\mathcal{F}.^{X'} \subset \mathcal{F}.^{X''}$, alors Y' est indépendant de X' si et seulement si Y' est indépendant de X''.*

Corollaire 6 *Si X' est un mouvement brownien complémentable dans $\mathcal{F}_{\cdot}^{(X,Y)}$, alors X' est maximal dans $\mathcal{F}_{\cdot}^{(X,Y)}$.*

Démonstration.   Supposons que X' possède un complément brownien indépendant Y' dans $\mathcal{F}_{\cdot}^{(X,Y)}$. Montrons que X' est maximal dans $\mathcal{F}_{\cdot}^{(X,Y)}$. Soit X'' un mouvement brownien dans la filtration naturelle de (X,Y) vérifiant $\mathcal{F}_{\cdot}^{X'} \subset \mathcal{F}_{\cdot}^{X''}$. Pour tout $t \in \mathbf{R}_+$,

$$\mathcal{F}_t^{(X,Y)} = \mathcal{F}_t^{(X',Y')} \subset \mathcal{F}_t^{(X'',Y')} \subset \mathcal{F}_t^{(X,Y)}$$

donc $\mathcal{F}_t^{(X',Y')} = \mathcal{F}_t^{(X'',Y')}$. Par indépendance de X'' et de Y' (lemme 5), on a donc pour tout $A \in \mathcal{F}_t^{X''}$,

$$P[A|\mathcal{F}_t^{X'}] = P[A|\mathcal{F}_t^{(X',Y')}] = P[A|\mathcal{F}_t^{(X'',Y')}] = \mathbf{I}_A,$$

d'où $A \in \mathcal{F}_t^{X'}$, ce qui montre l'inclusion réciproque $\mathcal{F}_{\cdot}^{X''} \subset \mathcal{F}_{\cdot}^{X'}$.

Exemple. Le corollaire 6 montre que le mouvement brownien

$$X' = \int_0^{\cdot} \mathrm{sgn}(X_s)dX_s$$

n'est pas complémentable dans $\mathcal{F}_{\cdot}^{(X,Y)}$ puisque sa filtration naturelle est strictement incluse dans celle de X. Remarquons cependant qu'il serait complémentable si l'on acceptait des compléments non indépendants. En effet, les mouvements browniens X' et $(X+Y)/\sqrt{2}$ engendrent la filtration naturelle de (X,Y) puisque la connaissance de $\langle X', X+Y \rangle = \int_0^{\cdot} \mathrm{sgn}(X_s)ds$ permet de reconstituer le signe de X, tandis que la connaissance de X' détermine la valeur absolue de X.

Question ouverte. La réciproque du corollaire 6 est-elle vraie ?

Savoir si un mouvement brownien est maximal, complémentable ou non est une question délicate, à moins de savoir exhiber soit un mouvement brownien dont la filtration est strictement plus grosse, soit un complément brownien indépendant. L'exemple qui suit montre qu'il faut se méfier des intuitions simplistes.

Exemple. Les mouvements browniens

$$X' = \int_0^{\cdot} \mathrm{sgn}(Y_s)dX_s \qquad \text{et} \qquad Y' = \int_0^{\cdot} \mathrm{sgn}(X_s)dY_s$$

sont indépendants et complémentables (donc maximaux), mais ils n'engendrent pas la filtration naturelle de (X,Y). En effet, l'égalité

$$X = \int_0^{\cdot} \mathrm{sgn}(Y_s)dX'_s$$

permet de voir que Y est un complément brownien indépendant de X'. De même, X est un complément brownien indépendant de Y'. Cependant, la filtration naturelle de (X', Y') est strictement incluse dans celle (X, Y) car (X', Y') est inchangé lorsqu'on change (X, Y) en $(-X, -Y)$.

Question ouverte. Dans l'exemple ci-dessus, la filtration $\mathcal{F}^{(X',Y')}$ est-elle égale à la filtration $\mathcal{F}^{(X,Y)}$ quotientée par $\{-1, 1\}$?

Nous renvoyons le lecteur à [7,8] pour la définition des filtrations quotients.

2.2 Mouvements browniens associés à la décomposition polaire d'un mouvement brownien complexe

Soit $Z = X + iY$ un mouvement brownien complexe issu de 0. Notons $S = X^2 + Y^2$ et $R = \sqrt{S}$ le module de Z. Presque sûrement, le mouvement brownien Z ne repasse pas en 0, ce qui permet de poser $U_s = \frac{Z_s}{R_s}$ pour $s > 0$ et de définir un nouveau mouvement brownien complexe Z' par

$$Z'_t = \int_0^t \frac{R_s}{Z_s} dZ_s = \int_0^t \overline{U_s} dZ_s.$$

Les parties réelle et imaginaire de Z' sont les mouvements browniens donnés par

$$X'_t = \int_0^t \frac{X_s dX_s + Y_s dY_s}{R_s} \quad \text{et} \quad Y'_t = \int_0^t \frac{X_s dY_s - Y_s dX_s}{R_s}.$$

Ces mouvements browniens sont indépendants et gouvernent respectivement la partie radiale et la partie angulaire de Z. En effet, une application de la formule d'Itô montre que les processus S et U satisfont les équations différentielles stochastiques

$$dS_t = 2\sqrt{S_t} dX'_t + 2dt \quad \text{et} \quad dU_t = \frac{iU_t}{R_t} dY'_t - \frac{U_t}{2R_t^2} dt.$$

L'étude de ces équations permet de déterminer les filtrations naturelles de plusieurs processus. Ces résultats classiques, que nous rappelons dans la proposition ci-dessous, se trouvent déjà dans la proposition 3.1 et le théorème 3.4 de [10]. Le point 2 figure également dans le théorème V.2.11 de [9].

Proposition 7 *(Comparaison de plusieurs filtrations)*

1. *La filtration naturelle de U est aussi celle de $Z = X + iY$.*
2. *La filtration naturelle de X' est aussi celle de R.*
3. *La filtration naturelle de (X', Y') est la filtration de (X, Y) quotientée par $SO_2(\mathbf{R})$. Plus précisément pour tout $t > 0$, $\mathcal{F}_t^{(X',Y')} = \sigma((\frac{Z_s}{U_t})_{0 \leq s \leq t})$*
4. *La filtration naturelle de (X', Y') est strictement incluse dans celle de (X, Y). Plus précisément, pour tout $t > 0$, la variable U_t est indépendante de (X', Y') et de loi uniforme sur le cercle unité.*

Démonstration.   Montrons les différents points.

1. La filtration naturelle contient celle de $\langle U, U \rangle = \int_0^{\cdot} \frac{ds}{R_s^2}$, donc celle de R et par conséquent celle de $Z = RU$. L'inclusion réciproque est évidente.
2. Les solutions de l'équation de Bessel vérifiée par S sont fortes, donc $\mathcal{F}^R = \mathcal{F}^S \subset \mathcal{F}^{X'}$. L'inclusion réciproque découle de l'égalité $X' = \int_0^{\cdot} \frac{dS_t - 2dt}{2\sqrt{S_t}}$.
3. Notons $\mathcal{G}_t = \sigma((\frac{Z_s}{U_t})_{0 \leq s \leq t})$. L'inclusion $\mathcal{F}_t^{(X',Y')} \subset \mathcal{G}_t$ vient du fait que Z' est une fonction de Z invariante par rotation. L'inclusion réciproque vient de l'inclusion $\mathcal{F}^R \subset \mathcal{F}^{X'}$ et de l'égalité $U_t = U_s \exp(i \int_s^t \frac{dY_r'}{R_r})$ pour $t > s > 0$.
4. Soit v un complexe de module 1. On vérifie facilement que changer Z en vZ change U_t en vU_t mais ne modifie pas Z'. Comme la loi de Z est invariante par rotation, cela entraîne que la variable U_t est indépendante de Z' et de loi uniforme sur le cercle unité.

Comme la filtration naturelle de (X', Y') est strictement incluse dans celle de (X, Y), il est naturel de se demander si les mouvements browniens X' et Y' sont complémentables. Nous connaissons la réponse pour X' mais pas pour Y'.

Théorème 8 *Le mouvement brownien $X' = \int_0^{\cdot} \frac{X_s dX_s + Y_s dY_s}{R_s}$ est complémentable (donc maximal) dans $\mathcal{F}^{(X,Y)}$.*

La construction d'un complément brownien indépendant que nous allons présenter s'inspire de la construction de Émery et Schachermayer dans [6] montrant que la filtration d'un mouvement brownien circulaire indexé par $\mathbf{R}$ est brownienne après changement de temps $s = e^t$.

Démonstration.   Soit $(t_n)_{n \in \mathbf{Z}}$ une subdivision de $\mathbf{R}_+$. Pour tout $n \in \mathbf{Z}$, notons Δ_n la médiatrice du segment $[|Z_{t_n}|, Z_{t_n}]$ et Σ_n la symétrie orthogonale par rapport à Δ_n, autrement dit $\Sigma_n(z) = U_{t_n}\overline{z}$ pour tout $z \in \mathbf{C}$. Introduisons le temps d'arrêt
$$T_n = \inf\{t \in [t_n, t_{n+1}[\, Z_t \in \Delta_n\},$$
avec la convention $T_n = t_{n+1}$ si Z_t ne rencontre pas Δ_n entre les instants t_n et t_{n+1}.

Pour $t > 0$, notons $H_t = -1$ sur $[t_n \leq t < T_n]$ et $H_t = 1$ sur $[T_n \leq t < t_{n+1}]$. Le processus H, à valeurs dans $\{-1, 1\}$, est prévisible dans $\mathcal{F}^{(X,Y)}$, ce qui permet de définir un nouveau mouvement brownien dans la filtration $\mathcal{F}^{(X,Y)}$ par $Y'' = \int_0^{\cdot} H dY'$. Comme $\langle X', Y'' \rangle = 0$, Y'' est indépendant de X'. Nous allons montrer que Y'' est un complément brownien indépendant de X'.

Pour cela, introduisons le mouvement brownien par morceaux $\tilde{Z}$ défini par $\tilde{Z}_t = \Sigma_n(Z_t)$ sur $[t_n \leq t < T_n]$ et $\tilde{Z}_t = Z_t$ sur $[T_n \leq t < t_{n+1}]$. La décomposition polaire de $\tilde{Z}$ s'écrit $\tilde{Z} = R\tilde{U}$, avec $\tilde{U}_t = \frac{U_{t_n}}{U_t} = U_{t_n}\overline{U_t}$ sur

$[t_n \leq t < T_n]$ et $\tilde{U}_t = U_t$ sur $[T_n \leq t < t_{n+1}]$. On vérifie facilement que pour $t \in [t_n, t_{n+1}[$,

$$\tilde{U}_t = \exp(i \int_{t_n}^{t} \frac{dY_s''}{R_s}).$$

Mais d'après la proposition 7, la filtration naturelle de R est aussi celle de X'. Par conséquent, les processus $\tilde{U}$ et $\tilde{Z}$ sont adaptés à la filtration naturelle du couple (X', Y''). On en déduit que pour $t \in [t_n, t_{n+1}[$,

$$\sigma((Z_s)_{t_n \leq s \leq t}) = \sigma(Z_{t_n}) \vee \sigma((\tilde{Z}_s)_{t_n \leq s \leq t}) \subset \sigma(Z_{t_n}) \vee \mathcal{F}_t^{(X', Y'')},$$

et une récurrence immédiate montre que cette inclusion reste vraie pour tout $t \geq t_n$.

Pour $n \in \mathbf{Z}$, notons A_n l'événement « le mouvement brownien Z fait au moins deux tours autour de 0 entre les instants t_n et t_{n+1} » et $\tilde{A}_n$ l'événement « le processus $\tilde{Z}$ fait au moins un tour autour de 0 entre les instants t_n et t_{n+1} ». L'événement $\tilde{A}_n$ contient A_n. Mais pour un bon choix de la subdivision $(t_n)_{n \in \mathbf{Z}}$, l'événement A_n est réalisé presque sûrement pour une infinité de $n \leq 0$. Le point-clé consiste alors à remarquer que $Z_{t_n} = \tilde{Z}_{t_n-}$ sur l'événement $\tilde{A}_{n-1}$, qui appartient à $\mathcal{F}_{t_n}^{\tilde{Z}}$. Pour $t > 0$ fixé, Z_t est donc une fonction de $((X_s', Y_s''))_{0 \leq s \leq t}$ sur l'événement presque sûr $\limsup \tilde{A}_n$, ce qui montre que $\mathcal{F}_t^Z \subset \mathcal{F}_t^{(X', Y'')}$. L'inclusion réciproque est évidente.

2.3 Mouvements browniens associés au carré d'un mouvement brownien complexe

Dans ce paragraphe, on modifie la définition du mouvement brownien Z' en posant

$$Z_t' = \int_0^t \frac{Z_s}{R_s} dZ_s = \int_0^t U_s dZ_s.$$

Les parties réelle et imaginaire de Z' sont les mouvements browniens donnés par

$$X_t' = \int_0^t \frac{X_s dX_s - Y_s dY_s}{R_s} \ \text{ et } \ Y_t' = \int_0^t \frac{X_s dY_s + Y_s dX_s}{R_s}.$$

Notons $M = Z^2$. Alors $|M_t| = R_t^2$ et $dM_t = 2Z_t dZ_t = 2R_t dZ_t'$ donc M est solution de l'équation différentielle stochastique

$$dM_t = 2\sqrt{|M_t|} dZ_t'.$$

Cette équation est identique à celle qui définit un carré de Bessel de dimension 0 à la différence près que M et Z' sont complexes. Pour les équations différentielles stochastiques dans $\mathbf{R}^2$ à coefficients höldériens, les théorèmes d'unicité ne s'appliquent pas.

D'ailleurs M et le processus identiquement nul sont deux solutions issues de 0 pour cette équation. Plusieurs questions se posent donc naturellement sur les solutions qui quittent 0 immédiatement :

1. Y a-t-il des solutions fortes ?
2. Y a-t-il unicité trajectorielle ?
3. La solution M est-elle forte ?

Signalons que ce type de question est nouveau à notre connaissance et n'apparaît pas dans l'ouvrage récent [5] spécialisé sur les équations différentielles stochastiques singulières, dont le chapitre 1 fait pourtant le point sur les liens entre existence de solutions fortes, existence de solutions faibles, unicité trajectorielle, unicité en loi.

Une réponse positive aux questions 1 et 2 entraînerait une réponse positive à la question 3. Une autre façon de reformuler la question 3 est de demander si la transformation brownienne $Z \mapsto Z'$ perd seulement un bit d'information (elle en perd au moins un puisque Z' est une fonction paire de Z). Remarquons que cette transformation commute avec les changements d'échelle, donc le théorème 3 s'applique.

La question de savoir si les mouvements browniens X' et Y' sont complémentables ou non, maximaux ou non, semble encore plus délicate. On peut toutefois noter que les réponses sont les mêmes pour les deux browniens. En effet, si l'on change Z en $\exp(i\pi/4)Z$, on change $Z' = X + iY$ en $iZ' = -Y + iX$.

2.4 Addendum : une généralisation du théorème 8 par Michel Émery

Bien que notre but ne soit pas de rechercher la plus grande généralité possible, il nous semble intéressant de donner la démonstration suggérée par Michel Émery, qui permet d'étendre le théorème 8 à toute dimension.

Dans ce qui suit, on note $\langle \cdot, \cdot \rangle$ le produit scalaire, $|\cdot|$ la norme euclidienne et $(e_1, \ldots, e_m)$ la base canonique de $\mathbf{R}^m$. On suppose que $m \geq 3$.

Théorème 9 *Soit Z un mouvement brownien dans $\mathbf{R}^m$, issu de 0. Le mouvement brownien $\beta = \int_0^{\cdot} \frac{\langle Z_s, dZ_s \rangle}{|Z_s|}$ est complémentable (donc maximal) dans $\mathcal{F}^Z$.*

Démonstration.   On commence par démontrer le fait suivant : pour tout $t > 0$, la probabilité pour que Z coupe tous les hyperplans vectoriels entre les instants ϵ et t tend vers 1 quand $\epsilon \to 0$. Ce lemme et le lemme de Borel-Cantelli permettent de construire une subdivision $(t_n)_{n \in \mathbf{Z}}$ de $\mathbf{R}_+$ telle que presque sûrement, pour tout n assez proche de $-\infty$, le mouvement brownien Z coupe tous les hyperplans vectoriels entre les instants t_n et t_{n+1}.

Pour tout $n \in \mathbf{Z}$, notons H_n l'hyperplan médiateur du segment $[|Z_{t_n}|e_1, Z_{t_n}]$ et Σ_n la réflexion par rapport à H_n. Introduisons le temps d'arrêt

$$T_n = \inf\{t \in [t_n, t_{n+1}[\, Z_t \in H_n\},$$

avec la convention $T_n = t_{n+1}$ si Z_t ne rencontre pas H_n entre les instants t_n et t_{n+1}. Pour $t > 0$, notons $\tilde{Z}_t = \Sigma_n(Z_t)$ sur $[t_n \leq t < T_n]$ et $\tilde{Z}_t = Z_t$ sur $[T_n \leq t < t_{n+1}]$.

En remarquant que $Z_{t_n} = \tilde{Z}_{t_n-}$ pour tout n assez proche de $-\infty$, on démontre que la filtration de Z est aussi celle du mouvement brownien par morceaux $\tilde{Z}$, qui est encore celle du mouvement brownien $\widehat{Z}$ obtenu en recollant les morceaux , défini par

$$\widehat{Z}_t = \sum_{k=-\infty}^{n-1} (\tilde{Z}_{t_{k+1}} - \tilde{Z}_{t_k}) + \tilde{Z}_t - \tilde{Z}_{t_n} \text{ pour } t_n \leq t < t_{n+1}.$$

Prenons sur $\mathbf{R}^m \setminus \mathbf{R}_+ e_m$ un champ Q de matrices orthogonales, de classe $\mathcal{C}^\infty$ et tel que pour tout $z \in \mathbf{R}^m \setminus \mathbf{R}_+ e_m$, la dernière ligne de $Q(z)$ soit la transposée de $\frac{z}{|z|}$. On peut prendre par exemple la matrice de la réflexion par rapport à l'hyperplan médiateur du segment $[|z|e_m, z]$. Comme $m \geq 3$, le mouvement brownien $\tilde{Z}$ évite presque sûrement la demi-droite $\mathbf{R}_+ e_m$. On peut définir un nouveau mouvement brownien B en posant $B = \int_0^\cdot Q(\tilde{Z}) d\widehat{Z}$. La dernière composante de B est le mouvement brownien β. En effet, sur chaque intervalle $[t_n, t_{n+1}[$, on peut écrire

$$dB_t^{(m)} = \left\langle \frac{\tilde{Z}_t}{|\tilde{Z}_t|}, d\tilde{Z}_t \right\rangle = \left\langle \frac{Z_t}{|Z_t|}, dZ_t \right\rangle$$

puisque Σ_n est une isométrie.

Pour montrer que $(B^{(1)}, \ldots, B^{(m-1)})$ est un complément indépendant de $B^{(m)} = \beta$, il reste à vérifier que B engendre la filtration de Z. Cela se voit en remarquant que β engendre la filtration de $|Z|$ (le carré de Bessel $|Z|^2$ est une solution forte de l'équation différentielle stochastique $dX_t = 2\sqrt{|X_t|}d\beta_t + m dt$) et que sur chaque intervalle $[t_n, t_{n+1}[$, $\tilde{Z}$ est solution de l'équation différentielle stochastique $d\tilde{Z}_t = Q(\tilde{Z})^{-1} dB_t$, avec condition initiale $\tilde{Z}_{t_n} = |Z_{t_n}|e_d$ déterminée par β. Les solutions de ces équations sont fortes car Q est de classe $\mathcal{C}^\infty$ sur $\mathbf{R}^m \setminus \mathbf{R}_+ e_m$ et car $\tilde{Z}$ évite la demi-droite $\mathbf{R}_+ e_m$.

References

1. Attal S., Burdzy K., Émery M., Hu Y., *Sur quelques filtrations et transformations browniennes*, Séminaire de Probabilités XXIX, Lecture Notes in Mathematics **1613**, Springer (1995), 56–68.
2. Brossard J., Leuridan C., *Numérotations des records d'un processus de Poisson ponctuel*, Annals of Probability, **27 - 3** (1999), 1304–1323.
3. Brossard J., Leuridan C., *Perte d'information dans les transformations du jeu de pile ou face*, Annals of Probability 34-4, (2006), 1550–1588.
4. Brossard J., Leuridan C., *Chaînes de Markov constructives indexées par* **Z**, Annals of Probability 35-2, (2007), 715 – 731.

5. Cherny, A.S., Engelbert, H.J., *Singular stochastic differential equations*, Lecture Notes in Mathematics **1858**, Springer (2005).
6. Émery M., Schachermayer W., *A remark on Tsirelson's differential equation*, Séminaire de Probabilités XXXIII, Lecture Notes in Mathematics **1709**, Springer (1999), 291–303.
7. Malric M., *Filtrations quotients de la filtration brownienne*, Séminaire de Probabilités XXXV, Lecture Notes in Mathematics **1755**, Springer (2001), 260–264.
8. Malric M., *Correction au volume XXXV*, Séminaire de Probabilités XXXVI, Lecture Notes in Mathematics **1801**, Springer (2002), 492.
9. Revuz D., Yor M., *Continuous martingales and Brownian motion*, Springer (1991).
10. Stroock D.W., Yor M., *Some remarkable martingales*, Séminaire de Probabilités XV, Lecture Notes in Mathematics **850**, Springer (1980), 590–603.

Hyperbolic random walks

Jean-Claude Gruet

Laboratoire Nicolas Oresme, Université de Caen
BP 5186, F-14032 Caen Cedex, France
e-mail: gruet@math.unicaen.fr

Summary. Although the hyperbolic r.w. defined on a regular hyperbolic planar grid satisfies an invariance principle, as we shall see, the picture radically differs from the Euclidean setting: the infinite grid is the whole space when the step is too small. We also give a radial discretization of Bochner's subordinated hyperbolic Brownian motions.

Key words: Hyperbolic plane, random walk, invariance principle, non Fuchsian group, hypergroup, stable process

1 Introduction

In hyperbolic space, we construct a family of discrete (both in time and in space) random planar walks which functionally converge towards hyperbolic Brownian motion (Theorem 1). The previously known construction (e.g. Gangolli [21]) only dealt with the isotropic case. Our method is to proceed by fixed geodesic steps and independent random choices of directions, the change of direction being chosen inside a regular polygon, then parallelly transported.

The steps to which we apply the normalization are completely described by the length of the geodesic jumps. Let us stress that, unlike the nilpotent case, when the steps are sufficiently small, the subgroup generated by the associated isometries is dense in $PSL_2(\mathbb{R})$. Hence the strong analogy between Poincaré half-plane and discrete skeletons such as the regular trees is definitively not the whole story.

In the last part, assuming the process to be radial, we construct random walks which converge towards a subordinated hyperbolic Brownian motion, alias a pseudo stable hyperbolic process of the second kind.

2 The isotropic random walk

Let M be a complete Riemannian manifold with Ricci curvature bounded from below by a constant. The isotropic random walk $(W_k^{(\sqrt{\frac{2}{n}})}, k \in \mathbb{N})$ with step $d := \sqrt{\frac{2}{n}}$ is inductively constructed by requiring $W_{k+1}^{(\sqrt{\frac{2}{n}})}$ to be uniformly distributed on the geodesic sphere with radius $\sqrt{\frac{2}{n}}$ centered at $W_k^{(\sqrt{\frac{2}{n}})}$. Then, a continuous interpolation $(\xi_n(t), t \geq 0)$ of this process is defined by letting ξ_n move with constant speed along the geodesic $W_k^{(\sqrt{\frac{2}{n}})} \to W_{k+1}^{(\sqrt{\frac{2}{n}})}$ during the time interval $[k, k+1]$.

Proposition 1. *(Jørgensen [26], Pinsky [33] and Blum [11]). The processes $(\xi_n(nt), t \geq 0)$ converge in law towards the M-Brownian motion.*

Remark. Note that these authors consider the Poissonized (discontinuous) jump process (with independent exponential holding times) to simplify their proofs. The name isotropic transport process is due to Pinsky. There have been some misunderstandings about Jørgensen's work since its third part introduces in the greatest generality laws which are invariant by parallel transport.

Moreover, when M is a symmetric space of rank one (such as the Poincaré half-plane $\mathbb{H}_2$), the distance process $\big(R_n = d(W_n, o), n \in \mathbb{N}\big)$ from any given origin o is a Markov chain. This reduction to a single real process was exploited by the hypergroup approach [10,40,41]. In hypergroup theory, the convolution of two probabilities is conventionally defined by linearity from the convolution of two Dirac measures, which is a mixture of Dirac measures. Calling δ_d the uniform law on the geodesic sphere $S(o, d)$, hyperbolic trigonometry in $\mathbb{H}_2$ produces the well-known convolution formula: if $k(d_1, d_1, \phi)$ stands for $\cosh(d_1)\cosh(d_2) + \sinh(d_1)\sinh(d_2)\cos(\phi)$,

$$\delta_{d_1} \star \delta_{d_2} = \frac{1}{\pi} \int_0^\pi \delta_{\mathrm{argcosh}(k(d_1, d_2, \phi))} \, d\phi \,, \tag{1}$$

which defines the hyperbolic hypergroup with index $\alpha = 0$. This identity involves the distribution of $d(o, M)$ if M is uniformly distributed on the random sphere $S(M, d_2)$ where M is chosen uniformly on $S(o, d_1)$.

A standard fact in hypergroup theory (Zeuner [42]) is the coexistence of two different limit theorems for Chébli-Trimèche hypergroups of exponential growth, such as the hyperbolic one. The $\frac{1}{\sqrt{n}}$ outer scaling induces as functional limit the Euclidean Brownian motion, but this is the second limit result which fulfills our needs and gives a hypergroup interpretation of prop. 1: without any renormalization, the convolution deduced from (1) of many small variables converges in law towards the radial part of hyperbolic Brownian motion.

3 The directional discretization

We discretize the geodesic sphere, in such a manner that the directions of successive displacements may connect each other. More specifically, let M be a complete connected Riemannian surface. Fix a positive value d and an integer $N \geq 2$. The walker's location $X(k)$ at time k is defined recursively as follows. The walker selects at random an angle θ_k in the finite set $E_N := \{l\frac{\pi}{N},\ l = 1, 2, .., 2N\}$. Then he rotates his last speed vector w_k by θ_k to get $R_{\theta_k}(w_k)$. Then he lands at $X(k+1)$ on the geodesic issued from $(X(k), R_{\theta_k}(w_k))$, at distance d from $X(k)$. Remark that if $l = N$, he turns back. Hence we say that these successive choices of angles preserve a connectivity property. Of course this discretization destroys Markovianity.

The only reference which allows the random walk approximation to live in an arbitrary set seems to be Jørgensen [26]. But the second part of this work develops a general approach relying upon arbitrary centered laws on the tangent space. This is not adequate to handle the non Markovian case, for semi-group techniques require the Markov property.

Recall standard facts about hyperbolic models [7, 9, 36]. In the Poincaré half-plane model, $\mathbb{H}_2 = \{z \in \mathbb{C}, \Im m(z) > 0\}$ endowed with the metrics $ds^2 = \frac{dx^2 + dy^2}{y^2}$, the fractional linear transforms $z \mapsto g(z) = \frac{az+b}{cz+d}$ with real coefficients such that $ad - bc = 1$ form the direct (orientation preserving) isometry group. The full Möbius group which is obtained when complex coefficients are used is linked to three dimensional hyperbolic space. We identify g and $M = \pm \begin{pmatrix} a & b \\ c & d \end{pmatrix}$, hence the isometry group of $\mathbb{H}_2$ is $PSL_2(\mathbb{R}) = SL_2(\mathbb{R})/\{I, -I\}$.

The disk model in $D = \{z \in \mathbb{C} : |z| < 1\}$ is obtained after transporting the Riemannian structure by $z \mapsto \mathcal{I}(z) = \frac{z-i}{z+i}$ from $\mathbb{H}_2$ to D. Hence $ds^2 = \frac{4}{(1-|z|^2)^2}(dx^2 + dy^2)$. The tangent space at $z \in D$ is $\mathbb{C}$ endowed with the scalar product $\langle v, v \rangle_z = \frac{4}{(1-|z|^2)^2}|v|^2$. The Riemannian distance is always denoted by the same symbol $d(.,.)$. Isometries of the disk are the transforms $z \mapsto g(z) = \frac{\alpha z + \beta}{\bar{\beta} z + \bar{\alpha}}$ for complex α, β such that $|\beta| < |\alpha|$. The isometry g is characterized by the unimodular matrix $M = \frac{\pm 1}{\sqrt{|\alpha|^2 - |\beta|^2}} \begin{pmatrix} \alpha & \beta \\ \bar{\beta} & \bar{\alpha} \end{pmatrix}$ in $PSU(1,1) \subset PSL_2(\mathbb{C})$. We use the convenient notation $M(x + iy) = g(x + iy)$. Note that $SU(1,1)$ is a Lorentz group whose elements are all unimodular transforms which preserve the Hermitian form $|z_1|^2 - |z_2|^2$. We will need later ref. [14] about Lorentz groups. We use the same symbol g to denote the extension defined on $\overline{D}$. Points on the boundary ∂D are said to lie at infinity. For computational purposes, we often switch from one model to the other. The disk model is more frequently used in complex analysis but the Poincaré model $\mathbb{H}_2$ is preferred when geodesic flows are needed. We use tilded symbols to distinguish objects related to $\mathbb{H}_2$. The disk model enjoys a pleasant symmetry, but matrices are complex valued.

The hyperbolic analogues of the Euclidean translations are those isometries which only fix two points on the boundary ∂D and no point in D. They are called the hyperbolic elements. In the half-plane model, they are conjugated to $Z \mapsto cZ$ with $c > 0$, or equivalently their squared trace $\mathrm{tr}^2(M) = (a+d)^2$ is strictly greater than 4. This property becomes $|\Im m(\alpha)| < |\beta|$ in the disk model. The geodesic $\gamma(g)$ between its two fixed points (at infinity) is called the axis because when restricted to its axis, a hyperbolic element behaves like a translation. Let $T > 0$ denotes its step (with respect to hyperbolic distance). The remaining isometries are elliptic or parabolic whether $(a+d)^2 \leq 4$ or $(a+d)^2 = 4$. Contrary to the hyperbolic elements, no geodesic line is globally invariant.

If $\xi = \exp(i\frac{\pi}{N}) = \exp(i\theta)$ with $N \geq 2$ and $0 < r < 1$ are given, define the hyperbolic elements T_l with $0 \leq l \leq 2N - 1$:

$$z \in D \mapsto T_l(z) = \frac{z + r\xi^l}{r\overline{\xi}^l z + 1} \tag{2}$$

associated to the matrices $M(r, \xi^l) := \pm \dfrac{1}{\sqrt{1 - r^2}} \begin{pmatrix} 1 & r\xi^l \\ r\overline{\xi}^l & 1 \end{pmatrix}$. The points at infinity on the axis $\gamma(T_l)$ are the vertices of a regular Euclidean $2N$-gon. We only consider the even case because we keep in mind Cayley graphs of free groups. We will select r in a sequence (r_n) decreasing to 0. Observe that $d(n) := d(0, T_l(0)) = d(0, r_n) = \log(\frac{1+r_n}{1-r_n}) \sim 2r_n$ when n goes to infinity.

Proposition 2. *The hyperbolic random walk $(X_k^{(r)}, k \in \mathbb{N})$ in D is the projection $(P_k(0), k \in \mathbb{N})$ of a matrix random walk $(P_k, k \in \mathbb{N})$ such that P_k is a right product $M_1 M_2 ... M_k$ of i.i.d. matrices in $SL_2(\mathbb{C})$, started at identity and M_1 uniformly distributed in the generating set defined by (2).*

Remark 1: Of course, a price to pay for the right multiplication is the loss of the Markovian character for the D valued process $k \mapsto P_k(0)$. Unfortunately our geometric construction requires these unpleasant right products. Observe also that $(P_k(0), k \in \mathbb{N})$ is Markovian of order two since there is a unique geodesic connecting two points, hence the last vector speed is known from $P_{k-1}(0)$ and $P_k(0)$. But this remark is useless because the state space would depend from the distance $d(P_{k-1}(0), P_k(0)) = d(n)$ which decreases to 0 with n.

Remark 2: At the end of this section, we show that $(X_k^{(r)})$ is also the projection of the geodesic flow (with step d) interlaced with independent rotations of angle uniformly distributed on multiples of $\frac{\pi}{N}$. This flow is also constructed via right products. Although a Donsker type theorem applies for the matrix valued process (P_k), the similar result for the matrix valued interlaced flow (K_k) does not hold. As a partial explanation, note that in the limit case $d = 0$ we get $P_k = Id$ whereas the interlaced flow becomes the usual rotation flow.

Proof: Consider $g_k := \phi_1 \circ \phi_2 \circ ... \circ \phi_k$ if ϕ_k denotes the isometry associated to M_k. Thus $d(g_{k+1}(0), g_k(0)) = d(g_k(\phi_{k+1}(0)), g_k(0)) = d(\phi_{k+1}(0), 0) = d(n)$, hence $g_{k+1}(0)$ lies on a circle centered at $g_k(0)$. Let us stress the fact that this argument only works for right products: indeed $p \mapsto d(p, \phi_{k+1}(p))$ depends upon the point p. Recall that the displacement functions D_g of isometries g: $z \mapsto D_g(z) := \sinh(\frac{1}{2}d(z, g(z)))$ are computed in [7]. If g is hyperbolic with axis $\gamma(g)$ and length T (analogous to an Euclidean translation), $D_g(z) = \cosh(d(z, \gamma(g)) \sinh(T/2)$.

Using once more the fact that g_k is an isometry, adjacent points $g_{k+1}(0)$ on the circle centered at $g_k(0)$ are equally spaced because the vertices $T_l(0)$ share this property. Define now $T_{l*} = \phi_k^{-1}$, the inverse of the last factor in g_k. The connectivity property of the walk is secured by the point $g_k(T_{l*}(0))$ on the circle centered at $g_k(0)$. $\qquad \square$

For future reference, we need the following computation in the half-space model:

Lemma 1. *Define $d = \log(\frac{1+r}{1-r}) = d(0, r)$ and $\theta = \frac{\pi}{N}$. Then the transport on $\mathbb{H}_2$ of $M(r, \xi^l)$ is described by the matrix*

$$\begin{pmatrix} \cosh(d/2) + \cos(l\theta)\sinh(d/2) & -\sin(l\theta)\sinh(d/2) \\ -\sin(l\theta)\sinh(d/2) & \cosh(d/2) - \cos(l\theta)\sinh(d/2) \end{pmatrix}. \quad (3)$$

Proof: The computation becomes more transparent after a conjugation of $M(\tanh(d/2), 1)$:

$$M(\tanh(d/2), \xi^l) = \Delta(l\theta) M(\tanh(d/2), 1) \Delta(-l\theta)$$

if $\Delta(\alpha) := \begin{pmatrix} e^{i\alpha/2} & 0 \\ 0 & e^{-i\alpha/2} \end{pmatrix}$, associated to the rotation in D: $z \mapsto \exp(i\alpha) z$. Moreover the transport of the isometry represented by $\pm M(\tanh(d/2), 1) =$

$\pm \begin{pmatrix} \cosh(d/2) \sinh(d/2) \\ \sinh(d/2) \cosh(d/2) \end{pmatrix}$ is precisely $\pm G(d/2) = \pm \begin{pmatrix} \exp(d/2) & 0 \\ 0 & \exp(-d/2) \end{pmatrix}$

which is the familiar matrix associated to the normalized geodesic flow on the unit tangent bundle $T^1(\mathbb{H}_2)$. On the other hand, the rotation in $\mathbb{H}_2$ is represented by $\pm R_{-\alpha/2} := \pm \begin{pmatrix} \cos(\alpha/2) & \sin(\alpha/2) \\ -\sin(\alpha/2) & \cos(\alpha/2) \end{pmatrix}$ (beware of the minus sign). $\qquad \square$

Proposition 3. *Let $(\tilde{M}_k^{(\sqrt{\frac{2}{n}})}, k \geq 0)$ be sequences of i.i.d. matrices uniformly distributed on the set (3) with $d = \sqrt{\frac{2}{n}}$. When n goes to infinity,*

$$(\tilde{M}_1^{(\sqrt{\frac{2}{n}})} \ldots \tilde{M}_{[nt]}^{(\sqrt{\frac{2}{n}})}, t \geq 0) \xrightarrow{(d)} (\tilde{M}_t, t \geq 0)$$

such that

$$d\tilde{M}_t = \tilde{M}_t E_1 \circ d\alpha_t + \tilde{M}_t E_2 \circ d\beta_t \quad and \quad \tilde{M}_0 = Id \quad (4)$$

written in Stratonovitch form with two independent linear Brownian motions
α *and* β. *Here* $E_1 = \frac{1}{2}\begin{pmatrix} 1 & 0 \\ 0 & -1 \end{pmatrix}$ *and* $E_2 = \frac{1}{2}\begin{pmatrix} 0 & 1 \\ 1 & 0 \end{pmatrix}$ *generate the Lie algebra*
$sl(2)$. *Moreover, the linear system*

$$dZ_1(t) = Z_2(t) \circ d\overline{Z}_t \quad dZ_2(t) = Z_1(t) \circ dZ_t \tag{5}$$

directed by a single complex Brownian motion (Z_t) *describes the limit diffusion
in the disk model*
$$\left(M(t) = \begin{pmatrix} Z_1(t) & Z_2(t) \\ \overline{Z_2}(t) & \overline{Z_1}(t) \end{pmatrix}, t \geq 0 \right).$$

Remark 3. No explicit solutions of (5) in closed form are available. Matrix analogues and the infinite dimensional case are studied in H. Airault's [1] recent paper.

Remark 4. At fixed time t, proposition 3 is simply a consequence of Wehn's central limit theorem in Lie groups (neatly proved by E. Breuillard in his survey [13]). We find useful to provide in the proof below some computational details in order to avoid mistakes about the different normalizations of hyperbolic metrics or of the local coordinates associated to a Lie basis.

Proof: We apply in the group $SL_2(\mathbb{R})$ the functional convergence theorem of Stroock and Varadhan [34] for random walks in Lie groups. We use the notations from Kunita [30]. The vector space spanned by E_1 and E_2 is $\mathfrak{p}$ in the Cartan decomposition of $sl(2)$. Denote by x_1, x_2, x_3 the local coordinates near $e = Id$ such that $g = (g_{i,j}) = \exp(\sum_{k=1}^{3} x_k E_k)$ with $E_3 := [E_1, E_2]$. They satisfy when g goes to e

$$x_1 \sim g_{1,1} - g_{2,2}, \quad x_2 \sim g_{1,2} + g_{2,1} \quad \text{and} \quad x_3 \sim g_{1,2} - g_{2,1}.$$

If μ_n is the law of $\tilde{M}_1^{(\sqrt{\frac{2}{n}})}$, the measures $n\mu_n(dg)\mathbb{1}_{\{g \neq e\}}$ on $SL_2(\mathbb{R}) - \{e\}$ converge vaguely to 0. Let U_δ be an arbitrary neighborhood of e. Since $\sum_{l=1}^{2N} \exp(l\frac{\pi}{N}) = 0$, $n \int_{U_\delta} x_k(g)\mu_n(dg)$ makes sense for n large enough and goes to 0 as $n \to \infty$. Moreover, $\frac{1}{2N}\sum_{l=1}^{2N} \cos^2(l\frac{\pi}{N}) = \frac{1}{2}$ and some similar identities imply that $n \int_{U_\delta} x_i(g) \, x_j(g)\mu_n(dg)$ converges to 0 unless $i = j = 1$ or $i = j = 2$. These two integrals go to 1 since they are equivalent to $n\frac{d(n)^2}{2}$. Denote by $\mathbf{X}_k$ the left invariant field generated by E_k. By Stroock and Varadhan's result, the sequence of Markov chains $(\tilde{M}_1^{(\sqrt{\frac{2}{n}})}...\tilde{M}_{[nt]}^{(\sqrt{\frac{2}{n}})}, t \geq 0)$ functionally converges to the diffusion with infinitesimal generator $Af = \frac{1}{2}(\mathbf{X}_1^2 + \mathbf{X}_2^2)f$ acting on a domain of C^2 functions on the group. □

Remark 5. Let us sketch a direct proof relying on the well-known sufficient conditions (e.g. Ethier and Kurtz [34] corollary 7.4.2) deduced from Aldous' tightness criteria. They imply functional convergence of a sequence of $\mathbb{R}^d$ valued Markov chains $(Y_{[nt]}^{(n)}, t \geq 0)$ towards a diffusion process associated to a well-posed martingale problem. If the transition of $Y^{(n)}$ is denoted by $\mu_n(x, dy)$, the generator of the diffusion involves the limits

$$b(x) = \lim_{n \to +\infty} n \int_{|y-x| \le 1} (y - x) \, \mu_n(x, dy)$$

$$\text{and} \quad a(x) = \lim_{n \to +\infty} n \int_{|y-x| \le 1} (y - x)^T (y - x) \, \mu_n(x, dy).$$

Define the Markov chain $(Y_k^{(n)}, k \in \mathbb{N})$ started at $(1, 0, 0, 0)$ with state space $E \subset \mathbb{R}^4$ given by

$$E = \{(x_1, y_1, x_2, y_2) = (\mathfrak{Re}(\alpha), \mathfrak{Im}(\alpha), \mathfrak{Re}(\beta), \mathfrak{Im}(\beta)) \text{ such that } |\beta| < |\alpha|\}.$$

Here $Y_k = (\mathfrak{Re}(\alpha_k), \mathfrak{Im}(\alpha_k), \mathfrak{Re}(\beta_k), \mathfrak{Im}(\beta_k))$ is deduced from the upper row of the complex matrix $M_1 \cdots M_k$ with (M_l) i.i.d. uniformly distributed in the set (2).

Although the standard interpolation in $\mathbb{R}^d$ is linear, convergence is not altered by our different geodesic interpolation. We would obtain the singular covariance matrix

$$\Sigma = (a_{i,j}(x)) = \begin{pmatrix} x_2^2 + y_2^2 & 0 & x_1 x_2 - y_1 y_2 & x_2 y_1 + x_1 y_2 \\ 0 & x_2^2 + y_2^2 & x_1 y_2 + x_2 y_1 & x_1 x_2 - y_1 y_2 \\ x_1 x_2 - y_1 y_2 & x_1 y_2 + x_2 y_1 & x_1^2 + y_1^2 & 0 \\ x_2 y_1 + x_1 y_2 & x_1 x_2 - y_1 y_2 & 0 & x_1^2 + y_1^2 \end{pmatrix}$$

in accordance with the quadratic variations in (5).

Theorem 1. *The sequence of discontinuous processes $(X_{[nt]}^{(\frac{1}{\sqrt{2n}})}, t \ge 0)$ defined in proposition 2 functionally converges in law towards hyperbolic Brownian motion. Thus the sequence of geodesic interpolated processes shares the same property.*

Proof. The theorem follows from proposition 3 since the limit process $(\tilde{M}_t, t \ge 0)$ turns out to be the horizontal Brownian motion on $SL_2(\mathbb{R})$. Hence its projection $\tilde{M}_t(i)$ defines a Brownian motion on $\mathbb{H}_2$. The counterpart $(M_t(0), t \ge 0)$ in the disk model is the hyperbolic Brownian motion. $\square$

Remark 6. An analogue of equation (5) appears for instance in Baxendale's study [6] of the $SO(n, 1)$ horizontal Brownian motion. The conversion from the $SO(2, 1)$ model is given with Baxendale's notations by means of the transform $T : y \mapsto T(y) = {}^t\left(\frac{2y}{1-|y|^2}, \frac{1+|y|^2}{1-|y|^2}\right)$ from D to the upper hyperboloid $\{x_1^2 + x_2^2 - x_3^2 = -1, x_3 > 0\}$. For instance, the counterpart of E_1 (resp. E_2) is

$$E_1' = \begin{pmatrix} 0 & 0 & 1 \\ 0 & 0 & 0 \\ 1 & 0 & 0 \end{pmatrix} \quad \left(\text{resp. } E_2' = \begin{pmatrix} 0 & 0 & 0 \\ 0 & 0 & 1 \\ 0 & 1 & 0 \end{pmatrix}\right).$$

The end of this section is devoted to the link with the geodesic flow. Let T_p^1 denote the set of unit tangent vectors at $p \in D$. For instance T_0^1 is the unit circle in $\mathbb{C}$. Recall that $PSU(1,1)$ acts on the unit tangent bundle $T^1(D) := \bigcup_{p \in D} \{p\} \times T_p^1$ (e.g. [29] written in terms of the half plane model with respect to the base point $[i, i]$): if g in $PSU(1,1)$, $g([p, v_p]) = [g(p), g'(p)v_p]$. Also note that $T^1(D)$ can be identified with $PSU(1,1)$ since for any $[p, v_p] \in T^1(D)$, there is a unique $h \in PSU(1,1)$ such that $h(0) = p$ and $h'(0) = v_p$, summarized as $h([0, 1]) = [p, v_p]$.

Rotating the second component of $[p, v_p] = [h(0), h'(0)]$ by an angle α amounts to performing $h \circ \Delta_\alpha([0, 1]) = [h(0), v_p \exp(i\alpha)]$. Hence the continuous flow at time $t \in \mathbb{R}$ is obtained from the right multiplication by $\Delta_{t\alpha}$. Similarly the geodesic flow with step d is obtained from the right multiplication by $F(d/2) := M(\tanh(d/2), 1)$. In the more familiar half-space model,

$$t \mapsto \begin{pmatrix} a & b \\ c & d \end{pmatrix} \begin{pmatrix} \exp(t/2) & 0 \\ 0 & \exp(-t/2) \end{pmatrix} [i, i] \text{ represents in } T^1(\mathbb{H}_2) \text{ the geodesic}$$

started at 0 from the element $[q, u_q] = [\frac{ai+b}{ci+d}, \frac{i}{(ci+d)^2}]$.

Thus the geodesic flow in $T^1(D)$ interlaced with independent rotations is $K_k([0, 1])$ where the right walk $K := (K_k, k \in \mathbb{N})$ is recursively defined by $K_{k+1} = K_k \Delta(\alpha_{k+1}) F(d/2)$ and $K_0 = Id$.

Proposition 4. *The two projected processes $(K_k(0), k \in \mathbb{N})$ and $(P_k(0), k \in \mathbb{N})$ have the same law. But no Donsker theorem is available for the interlaced matrix flow (K_k).*

Proof. Equality in distribution comes from the fact that

$$(\Delta(\alpha_1), \Delta(\alpha_1)^{-1}\Delta(\alpha_2), \Delta(\alpha_2)^{-1}\Delta(\alpha_3), ...) \stackrel{(d)}{=} (\Delta(\alpha_1), \Delta(\alpha_2), \Delta(\alpha_3), ...)$$

for a sequence $(\alpha_1, \alpha_2, ...)$ of independent variables uniformly distributed in E_N.

There is no hope to get an analogue of Proposition 3 since the law of $\Delta(\alpha_1) F((2n)^{-\frac{1}{2}})$ does not converge towards δ_e. $\square$

4 Oddities of the hyperbolic grid

Numerous works are devoted to the link between Brownian motion on a manifold M and random walks on discrete skeletons (e.g. [27, 28, 37]). But the discretization procedure of Lie groups bears some intrinsic limitations. For instance, let us say that a topological group G is strongly approximable by an increasing sequence (G_n) of discrete subgroups if and only if any open subset U of G contains a point in almost all G_n. M. Kuranishi [31] showed that the class of nilpotent Lie groups contains all strongly approximable Lie groups. Thus our hyperbolic walks give rise to non discrete subgroups of $PSL_2(\mathbb{R})$.

A subgroup G of $PSL_2(\mathbb{R})$ is discrete if and only if it contains no convergent sequence of distinct isometries (it is not required that the sequence converges to a element of G). In the two dimensional case, a discrete group of isometries is called a Fuchsian group [7, 9]. A fundamental result is that G is Fuchsian if and only if G acts discontinuously on $\mathbb{H}_2$, i.e. if there exists $x \in \mathbb{H}_2$ and a neighborhood U of x such that $\{g \in G : g(U) \cap U \neq \emptyset\}$ is finite. Let us stress the fact that non Fuchsian groups have rarely been considered in the literature. Beardon's paper [8] (motivated by Riley's work) has given a new impulse to this rather unexplored field.

Theorem 2. (i) If $r \leq \cos(\frac{\pi}{2N})$, the set $\{T_l, 1 \leq l \leq N\}$ in (2) generates a non-Fuchsian group $G = \langle T_1, ..., T_N \rangle$.

(ii) In this case, G is dense in $PSL_2(\mathbb{R})$, hence any point in the closed half-plane is a limit point.

(iii) Yet for any given $r > 0$, when $k \to \infty$, the walk $k \mapsto g_k(i)$ a.s. goes to the boundary $\partial H_2 = \mathbb{R} \cup \{\infty\}$. Hence the walker is almost surely unable to enjoy the wildness of the hyperbolic grid.

Remark 7. Hyperbolic transforms close to the identity usually generate non discrete groups. Several theorems confirm this well-known fact. For instance Jørgensen's inequality (theorem 5.4.1 in Beardon's book [7]; note that this mathematician has nothing to do with E. Jørgensen cited above) states that if two Möbius transforms a, b in $SL_2(\mathbb{C})$ generate a non elementary discrete group then they cannot be too close to identity in the following sense: $\max(\|a - Id\|, \|b - Id\|) > 0.146$. Since any subgroup of a discrete group is discrete, this inequality provides a crude universal bound. Recall that a subgroup G of Möbius transforms is elementary if and only there is a finite G orbit.

In the same vein, Margulis' lemma [9] also implies such results. Indeed, this lemma gives (for instance in dimension two) the existence of a universal constant $\epsilon_2 > 0$ such that the group $\langle G_{\epsilon_2} \rangle$ generated by $G_{\epsilon_2} := \{g \in G : d(i, g(i)) \leq \epsilon_2\}$ is almost nilpotent when G is any discrete subgroup of $SL_2(\mathbb{R})$. Let us recall that a group G_ϵ is almost nilpotent if and only if there is a nilpotent subgroup H of finite index (ie G/H finite). By Tits' alternative [18], almost nilpotency implies that $\langle G_{\epsilon_2} \rangle$ does not contain free non Abelian groups. But there are plenty of free sub-groups in $SL_2(\mathbb{R})$, since for instance (see e.g. [18]), if a and b such that $a^2 \neq Id$ and $b^2 \neq Id$ do not fix a common point in $\mathbb{H}_2 \cup \mathbb{R} \cup \{\infty\}$, the subgroup $\langle a, b \rangle$ generated by the pair $\{a, b\}$ contains a finitely generated non Abelian free group. Thus if $\log(\frac{1+r}{1-r}) \leq \epsilon_2$, $G = \langle T_1, ..., T_N \rangle$ is nondiscrete.

Moreover if the generators a and b are randomly chosen in a neighborhood of Id, almost surely with respect to the Haar measure $\langle a, b \rangle$ is free and dense [13]. More specifically, the problem to determine whether a two-generator non elementary subgroup of $SL_2(\mathbb{R})$ is discrete is totally solved in Gilman [23] (theorem 3.1.1), including some difficult cases left out in earlier works. For instance, putting $N = 2$ in (3) leads to her difficult case (VII): two hyperbolic generators $A = \begin{pmatrix} \exp(d/2) & 0 \\ 0 & \exp(-d/2) \end{pmatrix}$ and $B = \begin{pmatrix} \cosh(d/2) & -\sinh(d/2) \\ -\sinh(d/2) & \cosh(d/2) \end{pmatrix}$

with intersecting axes. The trace of the commutator $[A, B] = ABA^{-1}B^{-1}$ is $-2(2\cosh^4(d/2) - 4\cosh^2(d/2) + 1)$. Thus if $r \geq \cos(\frac{\pi}{4})$, by an easy computation the commutator is parabolic or hyperbolic, and Gilman's classification results imply that G is discrete. If $r < \cos(\frac{\pi}{4})$, the commutator is elliptic and provided that we could check that its order is infinite, non-discreteness of G would follow from Gilman's results. But for infinitely many r the eigenvalues of the commutator are roots of unity. Hence we do not follow this route in the following proof:

Proof of Theorem 2: Cohen and Colonna have proved in [16] that $G = \langle T_1, ..., T_N \rangle$ is discrete if and only if $r \geq r_1 := \cos(\frac{\pi}{2N})$. If you connect adjacent vertices $\{g(0), g \in G\}$ of the Cayley graph, they establish that when the above inequality is violated, distinct branches can cross each other; hence the tree is very different from what should be expected. Basically they obtain infinite paths by choosing the next edge furthest to the right (resp. to the left) at each step. They prove that these paths land to a limit in the disk interior when $r < r_1$. If $r = r_1$, G is a Fuchsian group of first kind: this means that $\Lambda(G) = \partial D$.

The density of G comes from Chen & Greenberg's [14] Theorem 4.4.2, which classifies subgroups of Lorentz groups, here $SU(1, 1)$ if the complex field is used.

Since the elements of G do not have a common fixed point in $\overline{\mathbb{H}}_2$ and do not leave invariant a proper totally geodesic manifold (here this is simply a geodesic), $PSL(2, \mathbb{C}) \subset \overline{G}$. Note that D. Sullivan also gives a quick proof of the classification theorem in the first lines of the paper [35].

The third part of the theorem is proved with the help of Theorem 1 of ref [2] which provides an easier way to apply Furstenberg's theory of left matrix random walks. Since there is no common fixed point in $\overline{\mathbb{H}}_2$ and no common invariant geodesic line, the left matrix r.w. $k \mapsto \tilde{P}_k = \tilde{M}_1 \cdots \tilde{M}_k$ (or after a transposition the right walk) with generators (3) a.s. goes to infinity: $\|\tilde{P}_k\|_{HS} \to +\infty$ if $\| \cdot \|_{HS} = \mathrm{tr}(\tilde{M}^T\tilde{M}) = \sqrt{a^2 + b^2 + c^2 + d^2}$ denotes the Hilbert-Schmidt norm. Indeed $^T\tilde{M}_1$ and $\tilde{M}_1$ have the same law, hence a.s. $\frac{1}{k}\log(\|^T\tilde{P}_k\|_{HS}) \to \gamma > 0$. Thus $\tilde{P}_k(i)$ goes to infinity since $\|g\|_{HS}^2 = 2\cosh(i, g(i))$ if g belongs to $SL_2(\mathbb{R})$. Note that the usual Furstenberg statement (for instance Bougerol & Lacroix [12] Theorem 4.4) would require to check that no finite union of linear spaces is invariant by any element of the generating set.   □

Remark 8. *(i)* Let us come to the freeness of G. From Cohen and Colonna, we already know that G is free when $r \geq r_1 := \cos(\frac{\pi}{2N})$. But there is a dense set of small values of r such that G is free (there is a vast literature about some simpler cases eg. [32]), however there are infinitely many small values such that G contains elements of finite order (e.g. commutators as above).

(ii) Let us add a different feature of hyperbolic geometry: the abundance of closed polygonal convex loops. For simplicity put $N = 2$. Recall that for any integer $m \geq 5$, there is a regular hyperbolic convex m polygon with right angles. By elementary trigonometry in an isosceles hyperbolic triangle, the

edge length is $r(m) = \mathrm{argcosh}(1 + 2\cos(\frac{2\pi}{m}))$. Hence plenty of non trivial closed paths are available when $d(n)$ is a fraction $\frac{r(m)}{k}$ of some $r(m)$: the walker with unit speed begins by k steps eastward, then a quarter of turn is performed before a succession of k steps and so on. At time km the starting point is reached again. But the walk does not see these loops infinitely often.

5 Discretizations of radial stable processes

We first recall known facts about Lévy processes in the hyperbolic plane. We follow Applebaum's survey [5], where general symmetric spaces are discussed. We put more emphasis on the role of hypergroup theory in the radial case.

A radial hyperbolic process is originally the radial part $R_t = d(o, X_t)$ of some isotropic $\mathbb{H}_2$ valued process (X_t) when some origin o is chosen. If X is a Lévy process on a symmetric space, isotropy is equivalent to sphericity (Applebaum and Estrade [4]). A random variable M on $PSL(2, \mathbb{R})$ is radial (or $O(2)$ bi-invariant) when its law only depends on $\mathrm{tr}(M^t M)$. Of course, the $SL_2(\mathbb{R})$ lift of a spherical hyperbolic Lévy processes is radial. In the hyperbolic plane, the convolution of general laws is meaningless. But the convolution of radial laws is defined for instance via the hypergroup structure (1), which enables to work on the positive line: at any fixed time $t > 0$, the radial law μ_t of R_t (or equivalently of X_t) is characterized by its Fourier transform $\hat{\mu}_t(p) = \int_0^{+\infty} \phi_p(s) \, d\mu_t(s)$ given by the Lévy-Khintchine formula [10]:

$$\hat{\mu}_t(p) = \exp(-t\,\Psi(p)) = \exp\left(-t\left[\left(p^2 + \frac{1}{4}\right) + \int_0^{\infty} (1 - \phi_p(r))\, d\nu(r)\right]\right).$$
(6)

Here ν is the Lévy measure and (ϕ_p) the family of spherical functions indexed by a nonnegative real parameter p:

$$\phi_p(r) = P_{-\frac{1}{2}+ip}(\cosh(r)) = 1 - \frac{1}{2}\left(\frac{1}{4} + p^2\right)r^2 + o(r^2) \quad \text{when } r \to 0 \quad (7)$$

where P_ν denotes the Legendre function of the first kind with order $\mu = 0$ and complex degree ν. Note that the quantity $\exp(-\frac{t}{2}(p^2 + \frac{1}{4}))$ is the Fourier transform of the Brownian semi-group, generated by $\frac{1}{2}\Delta_{\mathbb{H}_2}$.

As a matter of fact, $\mathbb{H}_2$ does not carry stable laws, since non trivial endomorphisms are lacking [42]. There is a scaling operation θ_c which induces a central limit theorem. The dilation θ_c is defined for $c > 0$ via the Fourier transform of radial functions on $\mathbb{H}$ by $\widehat{\theta_c(f)}(p) = \hat{f}(cp)$. Several constructions of pseudo stable laws have been proposed. Note that in the literature the word pseudo is often omitted (e.g. in the recent [24]). From S. Cohen's paper [15], we have at hand two different kinds of Markovian hyperbolic pseudo stable processes. The first one is constructed from the stochastic development of a subordinated Brownian motion in the tangent bundle. The hypergroup counterpart was independently considered by H.M. Zeuner [42]. He generalized the radial central limit theorem: for any α in $]0,2[$, the hyperbolic infinitely divisible radial laws μ_α with Lévy measures proportional to $\frac{1}{r^{\alpha+1}}$ have a domain of attraction given by a classical condition on the tail at infinity. Of course, one says that μ is attracted by μ_α when there exists some sequence (c_n) growing to infinity such that $(\theta_{c_n} \mu)^{*n} \to \mu_\alpha$. The form $\frac{const}{r^{1+\alpha}}$ could be deduced from Feller's lemma VIII.8.2. The law of a candidate to be a stable like radial process is the convolution semi-group generated by μ_α. The choice $c_n \sim \frac{1}{n^{1/\alpha}}$ characterizes the normal domain of attraction. Since the Lévy measure of μ_α is explicit and very simple, the discretizing procedure which relies on radial compound Poisson approximations is straightforward [17]. A different method would start from Applebaum's [3] interlacing construction of spherical Lévy processes on the semisimple Lie group $SL_2(\mathbb{R})$.

Getoor [22] constructed the second kind of pseudo stable hyperbolic process $(S_t^{(2\alpha)}, t \geq 0)$ with index 2α in $]0,2[$ by Bochner's subordination: here $S_t^{(2\alpha)} = \Pi(\tau_t)$ if $(\tau_t, t \geq 0)$ is a stable subordinator on $\mathbb{R}^+$ with index α, independent from the hyperbolic Brownian motion Π started at o. Thus the stable-like radial process $(R_t^{(2\alpha)})$ is defined by $(d(0, S_t^{(2\alpha)}), t \geq 0)$, such that $\hat{\mu}_t(p) = \mathbb{E}(\phi_p(R_t^{(2\alpha)})) = \exp(-(\frac{1/4+p^2}{2})^\alpha t/2)$. This process has no Brownian part since $\lim_{p \to +\infty} \frac{\Psi(p)}{p^2}$ vanishes. The Lévy measure ν_α of $\Pi(\tau_1)$ is complicated. Accordingly we recover in the hyperbolic case the non coincidence of the two notions of pseudo stable processes established by S. Cohen. In the paper [25], we derived from Fubini's theorem the equality

$$\nu_\alpha(r) = \frac{2\pi\alpha}{\Gamma(1-\alpha)} \int_0^\infty \frac{q_s(r)}{s^{1+\alpha}} \, ds \tag{8}$$

where $q_s(r)$ denotes the radial hyperbolic heat kernel. The factor 2π comes from the fact that we used in [25] the reference measure $\sinh(r)\, dr$. Since we showed that $\nu_\alpha(r) \sim \frac{const}{r^{2\alpha+1}}$ when r goes to 0, this second kind process differs from the first one. Contrary to the three dimensional case, the formula

$$q_s(r) = \frac{1}{2(\pi s)^{3/2}} \exp(-s/8) \int_r^\infty \frac{u \exp(-u^2/2s)}{\sqrt{\cosh(u) - \cosh(r)}} \, du$$

cannot be simplified. Integrating it with respect to ds in (8) we obtain the integral

$$\nu_\alpha(r) = const \int_r^\infty \frac{u K_{\alpha+5/2}(u/4)\, du}{\sqrt{\cosh(u) - \cosh(r)}}$$

which contains a modified Bessel function.

We are unable to obtain the Getoor "stable" hyperbolic processes as a limit of subordinated random walks on the hyperbolic grid. The subordinator would be the limit of a sequence of discrete approximations. However note that if the walks are replaced by interpolated continuous processes, convergence is obvious. Indeed, the composition of a continuous function with an increasing right continuous change of time defines a continuous operation with respect to the Skorokhod topology [39]. But the interpolation looks unnatural in our discrete context. Thus the fully discrete program remains out of reach. The main difficulty comes from the fact that approximating the complicated Lévy measure ν_α seems a difficult task. Secondly, up to now, non commutative harmonic analysis on $SL_2(\mathbb{R})$ remains an obstacle to derive non radial probabilistic results.

Accordingly, we only consider hyperbolic radial processes. Since Applebaum's method is inadequate here in terms of computational simplicity, the discretizations are obtained in theorem 3 below by means of a subordinated radial random walk (with respect to the convolution (1)). The random changes of time are partial sums $C_t := \sum_{j=1}^{[nt]} N_j$ of i.i.d. integer valued r.v. $(N_j, j \in \mathbb{N})$ in the domain of attraction of a totally asymmetric stable law of index α. A convenient choice, simpler than a discrete Pareto-Zipf law relies on an idea attributed to F. Spitzer (Durrett [19] p. 106): if $0 < \alpha < 1$, let (c_k) be the Taylor coefficients at 0 of

$$F_\alpha(t) = 1 - (1-t)^\alpha = \sum_{k \geq 1} c_k t^k \quad \text{for } |t| < 1 \tag{9}$$

and put $\mathbb{P}(N_1 = k) = c_k$ if $k \geq 1$. Basically we generalize the known fact that $\sum_{j=1}^{n} \sum_{k=1}^{N_j} \frac{\epsilon_{k,j}}{n^{1/\alpha}}$ converges in law towards a symmetric stable law of index 2α when $(\epsilon_{k,j}, (k,j) \in \mathbb{N}^2)$ is a double sequence of i.i.d. symmetric Bernoulli r.v., independent from the sequence (N_j). Now the $\overset{+}{\Lambda}$ operation takes the place of the ordinary sum: the hyperbolic convolution (1) of two non-negative r.v. X and Y is concretized [10] as the randomized sum $X \overset{+}{\Lambda} Y = \Lambda(X+Y) = \Phi(X, Y, U)$ where

$$\Phi(x, y, u) = \operatorname{argcosh}\left(\cosh(x)\cosh(y) + \sinh(x)\sinh(y)\cos(\pi u)\right)$$

and U is an independent r.v. uniformly distributed on the unit interval. The associated Fourier analysis is based on the formula

$$\mathbb{E}(\phi_p(X \overset{+}{\Lambda} Y)) = \mathbb{E}(\phi_p(X)\phi_p(Y)). \tag{10}$$

Theorem 3. *Let (N_l) be a sequence of i.i.d. r.v. such that for k in $\mathbb{N} - \{0\}$, $\mathbb{P}(N_1 = k) = c_k$ defined by (9). Assume (N_l) to be independent of a double array of uniformly distributed r.v. used to define the randomized sums*

$$Z_t^{(n)} = \Lambda\left(\sum_{j=1}^{[nt]} \sum_{k=1}^{N_j} \frac{1}{n^{1/2\alpha}}\right).$$

Then the processes $(Z_t^{(n)}, t \geq 0)$ converge in distribution (for the Skorokhod J_1 topology) towards the radial part $R_t^{(2\alpha)} := d(o, S_t^{(2\alpha)})$ of the Getoor stable radial hyperbolic process.

The proof generalizes Zeuner's approach to get his Theorem 3.1. in [41]; but observe that now the steps $n^{-1/2\alpha}$ are deterministic contrary to the lengths N_j of the sums. The following lemma readily implies the finite dimensional convergences:

Lemma 2. *At any fixed time t, when n grows to infinity, $Z_t^{(n)} \xrightarrow{(d)} R_t^{(2\alpha)}$.*

Proof: Just modify in the radial case the standard proof based on radial characteristic functions; now the Bessel family of functions $r \mapsto J_0(pr)$ is replaced by the ϕ_p family recalled in eq. (7). Hence for any fixed t, if $C_t := \sum_{j=1}^{[nt]} N_j$, conditioning w.r.t. the N_j, we get from (10) with deterministic $\phi_p(\frac{1}{n^{1/2\alpha}})$,

$$\mathbb{E}(\phi_p(Z_t^{(n)})) = \mathbb{E}\left(\phi_p(\frac{1}{n^{1/2\alpha}})^{C_t}\right) = F_\alpha\left(\phi_p(\frac{1}{n^{1/2\alpha}})\right)^{[nt]} \text{ which converges to}$$

$$\exp\left(\left(\frac{\frac{1}{4}+p^2}{2}\right)^\alpha t\right).$$
□

We are now in a position to complete the proof of Theorem 3: from Zeuner's method, the unique point is to check Aldous' condition.

If $\epsilon > 0$ and $\eta > 0$ are given, we have to show that there exists $\delta_0 > 0$ and n_0 such that if $0 \le \delta \le \delta_0$ and $n \ge n_0$, $\mathbb{P}(|Z_{t+\delta}^{(n)} - Z_t^{(n)}| \ge \eta / Z_t^{(n)}) \le \epsilon$ a.s.

Define $m = [n(t+\delta)] - [nt]$ and $M(n,t,\delta) = \sum_{l=1}^m N_l$. We also denote by $\mathbb{P}^{*r}_{\frac{1}{n^{1/2\alpha}}}$ the r fold convolution (in the $\overset{+}{\Lambda}$ sense) of deterministic steps $n^{-1/2\alpha}$. Following Zeuner, if $(\mathcal{F}_t^{(n)}, t \ge 0)$ denotes the filtration generated by the process $Z_t^{(n)}$, we get a.s.

$$\mathbb{P}(|Z_{t+\delta}^{(n)} - Z_t^{(n)}| \ge \eta / \mathcal{F}_t^{(n)} \vee \sigma(N_l, l \in \mathbb{N})) \le \mathbb{P}^{*M(n,t,\delta)}_{\frac{1}{n^{1/2\alpha}}}([\eta, \infty[).$$

Therefore

$$\mathbb{P}(|Z_{t+\delta}^{(n)} - Z_t^{(n)}| \ge \eta) \le \mathbb{E}(\mathbb{P}^{*M(n,t,\delta)}_{\frac{1}{n^{1/2\alpha}}}([\eta, \infty[)).$$

Recall that the first moment function of a probability measure on a hypergroup K preserves the additivity property of K valued r.v. [10]. The first moment function m_1 of the hyperbolic hypergroup of index 0 is $m_1(x) = 2\log(\cosh(\frac{x}{2}))$ [10]; accordingly, like any Sturm-Liouville hypergroups of exponential growth, m_1 is bounded by some constant c (here $c = \frac{1}{4}$ works).

Observe that a.s. $\mathbb{P}^{*r}_{\frac{1}{n^{1/2\alpha}}}([\eta, \infty[)$ is dominated by $\frac{r}{m_1(\eta)} m_1(n^{-1/2\alpha})$. Hence if N is an integer valued r.v. independent from the $\frac{1}{n^{1/2\alpha}}$ walk and $C = \frac{m_1(\eta)}{c}$,

$$\mathbb{E}(\mathbb{P}^{*N}_{\frac{1}{n^{1/2\alpha}}}([\eta, \infty[)) \le \epsilon + \mathbb{P}(N \ge \epsilon C n^{1/\alpha}).$$

Also notice that $\mathbb{P}(\sum_{l=1}^m N_l \ge h) \le \mathbb{P}(\sum_{l=1}^{n\delta_0+1} N_l \ge h)$ for any $\delta \le \delta_0$ and $h \ge 0$. Moreover $n^{-1/\alpha} \sum_{l=1}^{n\delta_0+1} N_l$ converges in law towards $c_\alpha \tau_{\delta_0}$ where $(\tau_t, t \ge 0)$ is a stable subordinator such that if $y \ge 0$, $\mathbb{E}(e^{-y\tau_t}) = \exp(-t y^\alpha)$ and c_α is a tedious positive constant. Therefore if δ_0 is small enough, for all $n \ge n_0$, $\mathbb{P}(\sum_{l=1}^{n\delta_0+1} N_l \ge \epsilon C n^{1/\alpha}) \le \epsilon$ and Aldous' condition follows.
□

References

1. Airault H. (2004). Stochastic analysis on finite dimensional Siegel disks, approach to the infinite dimensional Siegel disk and upper half plane. *Bull. Sci. Math.* **128** (7) 605–659.
2. Ambroladze A., Wallin H. (2000). Random iteration of Möbius transformations and Furstenberg's theorem. *Ergod. Th. Dynam. Sys.* **20** (4) 953–962.
3. Applebaum D. (2000). Compound Poisson processes and Lévy processes in groups and symmetric spaces. *J Theor. Probab.* **13** (2) 383–425.
4. Applebaum D., Estrade A. (2000). Isotropic Lévy processes on Riemannian manifolds. *Ann. Probab.* **28** (1) 166–184.
5. Applebaum D. (2001). Lévy processes in stochastic differential geometry., in Lévy processes, theory and applications, Barndorff-Nielsen O.E. et al. editors, Birkhäuser, Basel.
6. Baxendale P. H. (1986). Asymptotic behavior of the stochastic flows of diffeomorphisms: two case studies. *Probab.Th. Rel. Fields* **73** 51–85.
7. Beardon A.F. (1983). *Geometry of discrete groups.* Springer, Berlin.
8. Beardon A.F. (1996). Some remarks on non-discrete Möbius groups. *Ann. Acad. Sci. Fenn.* **21** 69–79.
9. Benedetti R., Petronio C. (1992). *Lectures on hyperbolic geometry.* Springer, Berlin.
10. Bloom W.R., Heyer H. (1995). *Harmonic analysis on probability measures on hypergroups*, W. de Gruyter, Berlin-New-York.
11. Blum G. (1984). A note on the central limit theorem for geodesic random walks. *Bull. Austral. Math. Soc.* **30** (2) 169–173.
12. Bougerol Ph., Lacroix J. (1985). *Products of random matrices with applications to Schrödinger operators*, Progress in Probability and Statistics **8**, Birkhäuser, Basel.
13. Breuillard E. Random walks on Lie groups, preprint available at http://www.dma.ens.fr/~breuilla/part0gb.ps.
14. Chen S.S., Greenberg L. (1974). Hyperbolic spaces in *Contributions to analysis, a collection of papers dedicated to Lipman Bers*, 49–87, Academic Press, New York.
15. Cohen S. (1995). Some Markov properties of stochastic differential equations with jumps. Séminaire de Probabilités XXIX, 181–193, Springer, Berlin.
16. Cohen J. M., Colonna F. (1984). Embeddings of trees in the hyperbolic disk. *Compl. Var. Th. Appl.* **24** 311-335 and corrigendum (2004) in vol **49** (3) 227–228.
17. Davydov Yu., Nagaev A.V. (2002). Theoretical aspects of simulation of random vectors having a symmetric stable distribution. *J. Multivar. Anal.* **82** (1) 210–235.
18. De La Harpe P. (1983). Free groups in linear groups. *Enseignement math.* **29** 129–144.
19. Durrett R. (1996). *Probability: Theory and examples*, second edition, Duxburry press, Belmont.
20. Ethier S.N., Kurtz Th.G. (1986). *Markov processes, characterization and convergence*, Wiley, New-York.
21. Gangolli R. (1964). On the construction of certain diffusions on a differentiable manifold. *Zeit. f. Warsch.* **2** 406–419.

22. Getoor R.K. (1961). Infinitely divisible probabilities on the hyperbolic plane. *Pacific J. Math.* **11** 1287–1308.
23. Gilman J. (1995). *Two-generators discrete subgroups of PSL(2, ℝ). Memoirs of the A.M.S.* **561**.
24. Graczyck P., Stós A. (2004). Transition density for stable processes on symmetric spaces. *Pacific J. Math.* **217** (1) 87–100.
25. Gruet J-C. (1998). Jacobi radial stable processes. *Ann. Math. Blaise Pascal* **5** (2) 39–48.
26. Jørgensen E. (1975). The central limit problem for geodesic random walks. *Zeit. f. Warsch.* **32** (1) 1–64.
27. Kanai M. (1985). Rough isometries and combinatorial approximations of noncompact Riemannian manifolds. *J. Math. Soc. Japan* **37** 391–413.
28. Kanai M. (1986). Rough isometries and the parabolicity of Riemannian manifolds. *J. Math. Soc. Japan* **38** 277–288.
29. Katok A., Haselblatt B. (1995). *Introduction to the modern theory of dynamical systems, Encyclopedia of Math. vol 54.* Cambridge University Press, Cambridge.
30. Kunita H. (1995). Some problems concerning Lévy processes on Lie groups. *Proceedings of Symposia in Pure Mathematics* **57**, 323–341. A.M.S.
31. Kuranishi M. (1949). Two elements generations on semisimple Lie groups. *Kodai Math. Sem. Reports* **1** (5-6) 9–10.
32. Lyndon R., Ullman J.L. (1969). Groups generated by 2 parabolic linear fractional transforms. *Canad. J. Math.* **21** 1388–1403.
33. Pinsky M.A. (1975). Isotropic transport process on Riemannian manifold. *Trans. Amer. Math. Soc.* **218** 353–360.
34. Stroock D.W., Varadhan S.R.S. (1973). Limit theorems for random walks on Lie groups. *Sankhyā Ser. A* **35** 277–294.
35. Sullivan D. (1985). Quasiconformal homeomorphisms and dynamics II: structural stability implies hyperbolicity for Kleinian groups. *Acta Math.* **155** 253–289.
36. Terras A. (1985). *Harmonic analysis on symmetric spaces and applications,* Springer, Berlin.
37. Varopoulos. N. Th. (1984). Brownian motion and random walks on a Riemannian manifold. *Ann. Inst. Fourier* **34** (2) 243–269.
38. Wehn D. (1960). Limit distributions on Lie groups, Yale thesis.
39. Whitt W. (1980). Some useful functions for functional limit theorems. *Math. Oper. Res.* **5** 67–85.
40. Zeuner H.M. (1989). The central limit theorem for Chébli-Trimèche hypergroups, *J. Theor. Probab.* **2** (1) 51–63.
41. Zeuner H.M. (1993). Invariance principles for random walks on hypergroups on ℝ₊ and ℕ, *J. Theor. Probab.* **7** (2) 225–245.
42. Zeuner H.M. (1995). Domains of attraction with inner norming on Sturm-Liouville hypergroups. *J. Appl. Analysis* **1** 213–221.

The Hypergroup Property and Representation of Markov Kernels

D. Bakry and N. Huet

[1] IUF and Institut de Mathématiques, Université de Toulouse and CNRS
118, route de Narbonne, 31400 Toulouse, France
e-mail: dominique.bakry@math.univ-toulouse.fr
[2] Institut de Mathématiques, Université de Toulouse and CNRS
118, route de Narbonne, 31400 Toulouse, France
e-mail: nolwen.huet@math.univ-toulouse.fr

Summary. For a given orthonormal basis (f_n) on a probability measure space, we want to describe all Markov operators which have the f_n as eigenvectors. We introduce for that what we call the *hypergroup property*. We study this property in three different cases.

On finite sets, this property appears as the dual of the GKS property linked with correlation inequalities in statistical mechanics. The representation theory of groups provides generic examples where these two properties are satisfied, although this group structure is not necessary in general.

The *hypergroup property* also holds for Sturm–Liouville bases associated with log-concave symmetric measures on a compact interval, as stated in Achour–Trimèche's theorem. We give some criteria to relax this symmetry condition in view of extensions to a more general context.

In the case of Jacobi polynomials with non-symmetric parameters, the *hypergroup property* is nothing else than Gasper's theorem. The proof we present is based on a natural interpretation of these polynomials as harmonic functions and is related to analysis on spheres. The proof relies on the representation of the polynomials as the moments of a complex variable.

Contents

1 Introduction

In a number of situations, Markov operators appear to be a wonderful tool to provide useful information on a given measured space. Let us for example mention heat kernel methods to prove functional inequalities like Sobolev or Log-Sobolev inequalities, or Cauchy kernels to prove boundness results on Riesz transforms in L^p. Heat kernels are widely used in Riemannian geometry and statistical mechanics, while Poisson, Cauchy and other kernels had been proved useful in other contexts related to classical harmonic analysis (see [9, 5, 6, 7, 11, 10, 20, 28, 29, 2, 38] for example to see the action of different families of semigroups in various contexts).

It seems therefore interesting to describe all Markov kernels associated with a given structure. In what follows, we shall consider a probability space $(E, \mathcal{E}, \mu)$ on which is given an orthonormal $L^2(\mu)$ basis $\mathcal{F} = (f_0, \ldots, f_n, \ldots)$, where we impose $f_0 = 1$. Such a basis shall be called a unitary orthonormal basis (UOB in short). In general our basis $\mathcal{F}$ shall be real, but we do not exclude to consider complex bases.

We may then try to describe the Markov operators defined from a family of probability measures $k(x, dy)$ by

$$f \mapsto K(f) = \int_E f(y) \, k(x, dy)$$

which are symmetric in $L^2(\mu)$ and have the functions f_n as eigenvectors. In other words, we want to define the linear operator K from

$$K(f_n) = \lambda_n f_n,$$

and try to describe for which sequences (λ_n) this operator is a Markov kernel. We shall call these sequences Markov sequences (MS's in short) associated with the UOB $\mathcal{F}$.

For a general basis $\mathcal{F}$, this is quite impossible. But many bases which appear in natural examples have a special property, which we call the hypergroup property, under which one is able to describe all Markov sequences associated with the UOB $\mathcal{F}$.

This expository paper is not intended to be a complete account of the general theory of hypergroups, for which we may for example refer to the complete treatise [13]. In fact, we just extracted from this theory what is useful for our purpose. More precisely, we concentrated on the fundamental aspect which we are interested in, that is the possibility of describing all Markov kernels associated with our basis $\mathcal{F}$.

The paper is organized as follows.

In the first part, we present the case of finite sets, where the hypergroup property appears as the dual property of a more natural condition on the basis $\mathcal{F}$, namely the positivity of the multiplication coefficients. This property is called the GKS property in [8] because of its links with some famous correlation inequalities in statistical mechanics, and we keep this notation. These correlation inequalities did in fact motivate our efforts in this direction (see Paragraph 2.6). Many examples come from the representation theory of groups, but we propose a systematic exposition. The hypergroup property provides a convolution operation on the set of probability measures on the space, and the Markov kernels may be represented as the convolution with some given measure. In this situation, the hypergroup property appears as a special property of an orthogonal matrix. We shall see that in fact there are many situations where no group structure holds and where nevertheless the hypergroup property holds.

The second part is devoted to the presentation of Achour–Trimèche's theorem, which states this property for the basis of eigenvectors of a Sturm–Liouville operator with Neumann boundary conditions, associated with a log-concave symmetric measure on a compact interval. The original Achour–Trimèche's theorem was not stated exactly in the same way (see [1]), but his argument carries over very easily to our context. We give a complete proof of this result, since to our knowledge this proof was never published. We tried to relax the symmetry condition on the measure, and provided various rather technical extensions of the theorem. But we did not succeed to extend Achour–Trimèche's theorem to a wider class of measures which would include the case of Gasper's theorem on Jacobi polynomials studied in the next chapter. The real motivation of this effort is that in general, log-concave measures on $\mathbb{R}$ or on an interval serve as a baby model in Riemannian geometry for manifolds with non-negative Ricci curvature. Unfortunately, the symmetry condition does not seem to have any natural interpretation.

In the last part, we present Gasper's theorem, which states hypergroup property for the Jacobi polynomials. We present a proof which relies on geometric considerations on the spheres when the parameters are integers, and which easily extends to the general case. We follow Koornwinder's proof of the result, and give a natural interpretation of Koornwinder's formula (Lemma 5.2) which represents those polynomials as the moments of some complex random variable. We found after the redaction of this part that the interpretation of Jacobi polynomials as harmonic functions was already known from specialists (see [14, 23, 35]) but it seems that it was not directly used to prove this integral representation formula. We hope that this simple interpretation may provide other examples for similar representation in other contexts.

2 The finite case

In this section, we restrict ourselves to the case of finite sets, since in this context most of the ideas underlying the general setting are present, and we so avoid the analytic complexity of the more general cases that we shall study later on.

2.1 The GKS property

In what follows, we assume that our space is a finite set

$$E = \{x_0, \ldots, x_n\},$$

endowed with a probability measure

$$\mu = (\mu(x_0), \ldots, \mu(x_n)).$$

We denote by $L^2(\mu)$ the space of real functions on E, and we assume the existence of a real basis

$$\mathcal{F} = (f_0, \ldots, f_n)$$

with $f_0 = 1$. We suppose here that for any $x \in E$, $\mu(x) > 0$.

We shall write $\langle f \rangle$ for $\int f \, d\mu$ and $\langle f, g \rangle$ for $\int fg \, d\mu$.

The algebra structure of the set of functions is reflected in the multiplication tensor (a_{ijk}) for which

$$f_i f_j = \sum_k a_{ijk} f_k.$$

We therefore have

$$a_{ijk} = \langle f_i f_j f_k \rangle,$$

and we see that the tensor (a_{ijk}) is symmetric in (ijk). It has also another property which reflects the fact that the multiplication is associative.

Definition 2.1. *We shall say that $\mathcal{F}$ has the property GKS if all coefficients (a_{ijk}) are non-negative.*

This notation comes from the GKS inequality in statistical mechanics that we shall describe at the end of this section.

Observe that $a_{ij0} = \delta_{ij}$.

Many natural bases $\mathcal{F}$ share this property. For example, consider the hypercube $E = \{-1, 1\}^N$, with the uniform measure on it. Let ω_i denote the i-th coordinate

$$\omega = (\omega_1, \ldots, \omega_n) \mapsto \omega_i,$$

and, for $A \subset \{1, \ldots, N\}$

$$\omega_A = \prod_{i \in A} \omega_i, \ \omega_\emptyset = 1.$$

Then,

$$\mathcal{F} = \{\omega \mapsto \omega_A, A \subset \{1, \ldots, N\}\}$$

is a UOB of (E, μ). Since

$$\omega_A \omega_B = \omega_{A \triangle B},$$

it has the GKS property.

(We shall see later that this is a special case of a generic situation in finite groups).

Although we are here mainly interested in the case of a real basis, there are many natural complex GKS bases, issued in general from the representation theory of finite groups (see Paragraph 2.5 later). If the basis is complex, we shall still require that the multiplication coefficients are non-negative real numbers, which means that

$$a_{ijk} = \langle f_i f_j \bar{f}_k \rangle \geq 0,$$

for any (i, j, k).

In what follows, we only consider real GKS bases, although the next result remains probably true in the complex setting.

Proposition 2.1. *If a UOB has the GKS property, then there exists a unique point x_0 on which every $f_i(x_0)$ is maximal. Moreover, for any i and any x, $|f_i|(x) \leq f_i(x_0)$, and at this point x_0, $\mu(x_0)$ is minimal.*

Proof. Let us say that a function $f : E \mapsto \mathbb{R}$ is GKS if for any $i = 0, \ldots, n$, $\langle f f_i \rangle \geq 0$. In other words, f is written with non-negative coefficients in the basis $\mathcal{F}$.

We shall say that a set $K \subset E$ is GKS if $\mathbb{1}_K$ is a GKS function. We shall say that a point x is GKS if $\{x\}$ is a GKS set.

We shall see that there is only one GKS point.

Remark first that the sum of two GKS functions is GKS and that, thanks to the GKS property of $\mathcal{F}$, the product of two GKS functions is GKS. Moreover a limit of GKS functions is GKS. Observe also that a GKS function has always a non-negative integral with respect to μ since $f_0 = 1$.

Let us consider a non-zero GKS function f and consider $m = \max_{x\in E}|f|$. We see first that $\{f = m\} \neq \emptyset$.

For this, assume the contrary, that is that $f = -m$ on $|f| = m$. Since $m > 0$, we see that f^{2p+1}/m^{2p+1} is a GKS function, and converges to $-\mathbb{1}_{\{f=-m\}}$. Since a GKS function has a non-negative integral, this is impossible. Using the same argument, we see that

$$\frac{1}{2}\left(\lim_n \frac{f^{2n}}{m^{2n}} + \lim_n \frac{f^{2n+1}}{m^{2n+1}}\right) = \mathbb{1}_{\{f=m\}}$$

is a GKS function.

Therefore, the set $\{f = m\}$ is a GKS set, and there are non-trivial GKS sets.

Moreover, for any GKS function, the set $\{f = \max|f|\}$ is GKS.

Let E_1 be a nonempty GKS set, minimal for the inclusion. Then, for any GKS function f, $g = \mathbb{1}_{E_1}f$ is GKS. If g is not 0, then its maximum is attained on a subset E_2 of E_1 which is again GKS. Since E_1 is minimal, we have $E_2 = E_1$.

Therefore, for any GKS function, its restriction g to E_1 is either 0 on E_1 or constant (and equal to the maximum of g). In any case, f is constant on E_1. Since this applies to every function f_i in $\mathcal{F}$, and since $\mathcal{F}$ is a basis, every function is constant on E_1 and therefore E_1 is reduced to a single point $\{x_0\}$.

The same proof shows that any GKS set contains a GKS point.

For a GKS point x_0, $f_i(x_0) = \langle f_i \mathbb{1}_{x_0}\rangle/\mu(x_0) \geq 0$.

Then, consider two distinct points x_0 and x_1, and write

$$\frac{\mathbb{1}_{x_0}}{\mu(x_0)} = \sum_k f_k(x_0)f_k, \quad \frac{\mathbb{1}_{x_1}}{\mu(x_1)} = \sum_k f_k(x_1)f_k.$$

Writing the product, we see that

$$0 = \frac{\mathbb{1}_{x_0}\mathbb{1}_{x_1}}{\mu(x_0)\mu(x_1)} = \sum_{ijk} a_{ijk}f_i(x_0)f_j(x_1)f_k,$$

with the multiplication coefficients a_{ijk}.

So we see that for any pair of distinct points, and for any k,

$$\sum_{ij} f_i(x_0)f_j(x_1)a_{ijk} = 0.$$

Suppose then that x_0 and x_1 are GKS points. In the previous sum, all coefficients are non-negative. Therefore, for any (i,j,k)

$$a_{ijk}f_i(x_0)f_j(x_1) = 0.$$

If we apply that with $i = 0$ and $j = k$, we see that $f_j(x_1) = 0$, for any i. This is impossible since then any function would take the value 0 in x_1. So there is a unique GKS point.

Let x_0 be this unique GKS point. Any GKS set contains x_0. Since for any GKS function, the set where f is maximum is GKS, any GKS function attains its maximum at x_0.

It remains to show that μ is minimal at x_0. For this, observe that for any point x, the function

$$f_x = \mu(x)\mathbb{1}_{x_0} - \mu(x_0)\mathbb{1}_x$$

is GKS (this comes from the fact that each f_i is maximal at x_0). Therefore, the maximum value of $|f_x|$ is attained in x_0, which gives the result.   □

2.2 Orthogonal matrix representation

Consider the matrix

$$(O_{ij}) = \left(\sqrt{\mu(x_i)} f_j(x_i) \right),$$

we see easily that the matrix (O_{ij}) is a $(n+1) \times (n+1)$ orthogonal matrix with positive first column. Conversely, any such matrix may be associated with a UOB on a finite set with measure μ given by

$$\mu(x_i) = O_{i0}^2.$$

Therefore, there is a one to one correspondence between the set of orthogonal matrices with positive first column, and the set of finite probability spaces, whose probability has everywhere positive weight, endowed with a UOB. (In fact, this is not completely true, since we would not distinguish between bases given in different orders, provided that the first element is 1, which identifies the set of UOBs with a quotient of the set of orthogonal matrices through a permutation of rows and columns.)

The GKS property may be translated into the following property on such an orthogonal matrix:

$$\forall j, k, l, \quad \sum_i \frac{O_{ij} O_{ik} O_{il}}{O_{i0}} \geq 0. \tag{1}$$

The transposed of an orthogonal matrix is orthogonal, and we just saw that an orthogonal matrix which has the GKS property also has a non-negative row (corresponding to the row where the first column is minimal according to Proposition 2.1). We may of course rearrange the labelling of the points in such a way that this row is the first one. Then, the situation is completely symmetric.

We shall then consider the squares of terms in the first row as a probability measure on the dual set $\{0, 1, \ldots, n\}$:

$$\nu(i) = \mu(x_0) f_i^2(x_0).$$

Thanks to the fact that the functions f_i are maximal at x_0 and that this maximum must be larger than 1 (since $\int f_i^2 \, d\mu = 1$), the dual measure is also minimum at 0.

As an application, we have the following.

Proposition 2.2. *If a real GKS basis exists for the uniform measure on some finite set E, then the cardinal of E must be 2^k for some k, and this basis is the canonical basis (ω_A) of the characters of the group $(\mathbb{Z}/2\mathbb{Z})^k$.*

To see this, we first observe that the dual measure is uniform too. In fact, for the matrix (O_{ij}), we have $O_{00} = 1/\sqrt{n+1}$, where $n+1$ is the number of points in the space, and since the first row is positive and has minimum value $1/\sqrt{n+1}$, it must be constant since the sum of the squares of its coefficients is 1.

Now, if we multiply the matrix O by $\sqrt{n+1}$, then we see that in each column, the maximum value of the coefficients is attained on the first row and is equal to 1. Since the sum of all the squares of the coefficients in a given column must add to $n+1$, this shows that in any column, the coefficients must take only the values ± 1.

Those matrices (with entries ± 1 and orthogonal lines) are called Hadamard matrices (cf [30, 33, 37]). If $n+1 = 2^k$ for some k, such matrices are given by the basis (ω_A) on $\{-1, 1\}^k$ and satisfy the GKS property. It is known that the order of a Hadamard matrix must be 1, 2, or $4k$, and it is an open problem to find such matrices for all k (the lowest k for which no Hadamard matrix of order $4k$ is known is $k = 167$ since [33]). Nevertheless, the following proposition will prove the result of Proposition 2.2.

Proposition 2.3. *If a Hadamard matrix has the GKS property, then it must be of order 2^k for some k, and, up to permutation, it is the matrix of the canonical basis (ω_A) of the group $\{-1, 1\}^k$.*

Proof. The case of a set of size 2 is trivial, and we therefore assume that the size of the matrix is at least 3. To fix the idea, consider a matrix M of order $n+1$ with entries ± 1, with orthogonal columns. Call f_i the column vectors and $\hat{f}_i = f_i/\sqrt{n+1}$ the normalized ones, so that $M = \left(f_i(x_j)\right)_{0 \le i,j \le n}$. As usual, let us denote by $\langle f \rangle$ the mean value of a function f with respect to the normalized uniform measure. We suppose, which is possible up to reordering, that all the entries of the first line and the first column are $+1$. The GKS property says that

$$\langle f_i f_j f_k \rangle \ge 0,$$

for any (i, j, k).

First, since $\langle f_i \rangle = 0$ for any $i \ge 1$, there must be as many 1 and -1 in each column, and therefore $n+1$ is even. Let A be the set of points where $f_1 = 1$. Let $i \ge 2$, and let q be the number of points in $x \in A$ such that $f_i(x) = 1$, and r be the number of points in A^c where $f_i(x) = 1$. Writing

$\langle f_i \rangle = 0$ and $\langle f_1 f_i \rangle = 0$, we get

$$q + r = p, \; q = r,$$

which shows that p is even and also that, when $i \geq 2$,

$$\langle \mathbb{1}_A f_i \rangle = 0.$$

This is the generic argument which shows that Hadamard matrices have order $4k$. We shall now make use of the GKS property. Write

$$\mathbb{1}_A f_i = \sum_l a_{il} f_l.$$

We have

$$a_{il} = \langle \mathbb{1}_A \hat{f}_i \hat{f}_l \rangle = \frac{1}{n+1} \langle \mathbb{1}_A f_i f_l \rangle,$$

and therefore the matrix $A = (a_{il})$ is symmetric. As it is the matrix of a projector, its eigenvalue are 0 or 1. The dimension of the eigenspace associated to 1 is p, since the eigenspace is generated by $(\mathbb{1}_{\{x\}}, x \in A)$. Also, A is a GKS set because it is the set where f_1 attains it's maximum (cf proof of Proposition 2.1, Page 300). This implies that all the entries of A are not negative. Moreover, if we look at the values of the functions at x_0, it holds $\sum_j a_{ij} = 1$.

Therefore, the matrix A is Markovian. There are no transitory points since A is symmetric. The number of recurrence classes for such a matrix is the multiplicity of 1 as eigenvector, here p, so there are exactly p recurrence classes, and no recurrence class is reduced to a single point, since $a_{ii} = 1/2 < 1$. So every recurrence class has exactly two points. For example, $\{0, 1\}$ form a recurrence class. On each line of the matrix, there are exactly two places where $a_{ij} \neq 0$. The values of those entries are then $1/2$, since $a_{ii} = 1/2$. If we choose two distinct indices i and j in two different recurrent classes, then $a_{ij} = 0$. This means that $\langle \mathbb{1}_A f_i f_j \rangle = 0$.

Choose now one index in every recurrence class (say the even indices to fix the ideas, which is possible up to reordering of the columns). Then, the functions $g_i = \mathbb{1}_A f_{2i}$ form an orthogonal Hadamard GKS matrix of order $p = (n+1)/2$, and we may now use induction to see that the order must be 2^k for some k.

To see that the unique basis such basis in dimension 2^k is given by the canonical basis, it is enough to observe that if i and $\sigma(i)$ are in the same recurrence class for the matrix A, then we have

$$\mathbb{1}_A f_i = \frac{1}{2}(f_i + f_{\sigma(i)}) = \mathbb{1}_A f_{\sigma(i)},$$

and also

$$\mathbb{1}_{A^c} f_i = -\mathbb{1}_{A^c} f_{\sigma(i)}.$$

Then an easy induction leads to the result.   □

Nevertheless, unlike real bases, we shall see in Paragraph 2.5 that there always exists a complex GKS basis on any finite set with uniform measure.

On two points, an easy computation shows that a two dimensional orthogonal matrix having the GKS property must be

$$\begin{pmatrix} \cos\theta & \sin\theta \\ \sin\theta & -\cos\theta \end{pmatrix}$$

with $\theta \in [\pi/4, \pi/2]$. In fact, in dimension 2, given the measure (giving two distinct positive masses on the two points), there are exactly two UOB, and only one such that the unique non-constant function is maximal on the point with minimal mass (a necessary condition to have the GKS property as we saw). In this situation, any GKS matrix is symmetric, and the set of orthogonal matrices having the GKS property is connected.

On three points, the situation is more complicated. We saw for example that there are no real GKS basis when the measure is uniform. The set of real GKS UOBs on three points is connected, and one may see that the maximum value of $\mu(x_0)$ for which there exists a real GKS basis is $\mu(x_0) = 1/4$, and in this case the probability measure is $(1/4, 1/4, 1/2)$ and the unique GKS basis is obtained taking the real parts of the characters in the group $\mathbb{Z}/4\mathbb{Z}$ (we shall see later in Paragraph 2.5 how to associate complex or real GKS bases with any finite group). There is also a complex GKS basis with the uniform measure (the characters of the group $\mathbb{Z}/3\mathbb{Z}$).

2.3 The hypergroup property

As we saw before, when we have the GKS property, the situation is completely symmetric and we may consider the dual property. This is the hypergroup property.

Definition 2.2. *We shall say that the UOB $\mathcal{F}$ of $L^2(\mu)$ satisfies the hypergroup property (HGP in short) at point x_0 if for any $(x, y, z) \in E^3$,*

$$K(x, y, z) = \sum_i \frac{f_i(x)f_i(y)f_i(z)}{f_i(x_0)} \geq 0.$$

Of course, this supposes that for any i, $f_i(x_0) \neq 0$.

We do not require the GKS property to hold in the definition of the HGP property. We shall see later (at the end of Paragraph 2.5, Page 312) that we can find bases with HGP property and not GKS property, and the reverse. On the other hand, we shall see that in groups, the natural basis always share both property.

Remark. Since we may always change for any $i \geq 1$ $f_i \in \mathcal{F}$ into $-f_i$, and still get a UOB, and that this operation does not change the hypergroup property, we see that we may always assume that $f_i(x_0) > 0$.

To see the duality with the previous situation, let us enumerate the points in E starting from x_0, and recall our orthogonal matrix O with $O_{ij} = \sqrt{\mu(x_i)} f_j(x_i)$.

Then, under the GKS condition, O has non-negative first line and first column. Recall that the GKS property may be written as

$$\forall j, k, l, \quad \sum_i \frac{O_{ij} O_{ik} O_{il}}{O_{i0}} \geq 0.$$

while the HGP property writes

$$\forall j, k, l, \quad \sum_i \frac{O_{ji} O_{ki} O_{li}}{O_{0i}} \geq 0,$$

Notice also that if both properties occur, then the point x_0 must be the unique point where all functions f_i are non-negative, and where all f_i and $|f_i|$ are maximal.

From the symmetry of the situation, we may consider the functions

$$g_x(i) = \frac{O_{xi}}{O_{0i}}$$

to be a UOB on the set $\{0, \ldots, n\}$ endowed with the measure $\nu(i) = O_{0i}^2$.

From this we deduce that a basis has the HGP property if and only if this new basis has the GKS property, where of course the point 0 plays the role of the point x_0 in the previous paragraph. Therefore, the $g_x(i)$ are maximal at $i = 0$, and moreover 0_{i0} is minimal at $i = 0$. This means that, for the HGP property also, one has

Proposition 2.4. *If the UOB $\mathcal{F}$ has the HGP property at point x_0, then*

$$\forall i, x, \quad |f_i(x)| \leq f_i(x_0); \quad \forall x, \quad \mu(x) \geq \mu(x_0).$$

(Recall that we assume that $f_i(x_0) > 0$.)

We may reformulate the HGP property in the following way, which we shall use later in a different context, since it is more tractable.

Proposition 2.5. *The UOB $\mathcal{F}$ has the HGP property if and only if there is a probability kernel $k(x, y, dz)$ such that, for any $i = 0, \ldots, n$*

$$\frac{f_i(x) f_i(y)}{f_i(x_0)} = \int \frac{f_i(z)}{f_i(x_0)} \, k(x, y, dz).$$

Proof. The proof is straightforward. If the hypergroup property holds, then the kernel

$$k(x, y, dz) = \left(\sum_i \frac{f_i(x) f_i(y) f_i(z)}{f_i(x_0)} \right) \mu(dz)$$

is a probability kernel satisfying our conditions.

On the other hand, if such a probability kernel $k(x, y, dz)$ exists, then writing

$$k(x, y, dz) = K_1(x, y, z)\mu(dz)$$

and

$$K_1(x, y, z) = \sum_i a_i(x, y) f_i(z),$$

one sees that

$$a_i(x, y) = \frac{f_i(x) f_i(y)}{f_i(x_0)}.$$

$\square$

The link with the Markov operators is the following.

Definition 2.3. *A Markov operator is just an operator K which satisfies $K(1) = 1$ and which preserves non-negative functions.*

Given a sequence $\lambda = (\lambda_i, \ i = 0, \ldots, n)$, we define the associated linear operator K_λ by

$$K_\lambda(f_i) = \lambda_i f_i.$$

A Markov sequence (MS in short) is a sequence $\lambda = (\lambda_i)$ such that the associated operator K_λ is a Markov operator.

Remark that for any MS λ, one has $\lambda_0 = 1$. Remark also that

$$K_\lambda(f)(x) = \int f(y) \, k_\lambda(x, dy),$$

where

$$k_\lambda(x, dy) = \left(\sum_i \lambda_i f_i(x) f_i(y) \right) \mu(dy).$$

Therefore, the set of MS's is just the set of sequences λ such that the matrices $k_\lambda(x, y)$ are Markov matrices.

The HGP property asserts that, for any x, the sequence

$$\lambda(x) = \left(\frac{f_i(x)}{f_i(x_0)} \right)_{i=0,\ldots,n}$$

is a MS.

It is quite standard to see that any eigenvalue λ_i of a Markov operator must satisfy $|\lambda_i| \leq 1$. The set of Markov sequences is a convex compact set, which is stable under pointwise multiplication. Now the main interest of this property relies in the following theorem.

Theorem 2.1. *If the basis $\mathcal{F}$ has the HGP property, then the sequences*

$$\left(\lambda_i(x) = \frac{f_i(x)}{f_i(x_0)} \right)_i$$

are the extremal points in the convex set of all Markov sequences.
   More precisely, every Markov sequence may be written uniquely as

$$\left(\lambda_i = \int \frac{f_i(x)}{f_i(x_0)} \, d\nu(x) \right)_i$$

for some probability measure ν on E. Conversely, every probability measure can be associated in the same way with a Markov sequence.

The main interest of this result is that there exist numerous natural L^2 bases with the HGP property, as we shall see later.

Proof. The representation formula is straightforward. Indeed, writing $\nu(dy) = k_\lambda(x_0, dy)$, one has

$$\lambda_i f_i(x_0) = \int f_i(y) \, \nu(dy),$$

which gives the representation.
   From this, it is easy to see that if the sequences $(\lambda_i(x))_i$ are Markov sequences, then they are extremal. Indeed any representation

$$\frac{f_i(x)}{f_i(x_0)} = \theta \lambda_i^1 + (1 - \theta) \lambda_i^2$$

with MS's λ^1 and λ^2 leads to

$$f_i(x) = \int f_i(y) \, \nu(dy),$$

with

$$\nu(dy) = \theta K_{\lambda^1}(x_0, dy) + (1 - \theta) K_{\lambda^2}(x_0, dy).$$

From this we deduce that for any function f

$$f(x) = \int f(y) \, \nu(dy),$$

and therefore $\nu = \delta_x$, which gives the extremality.   □

Remark. Remark that the representation formula is still true for any MS when the basis does not verify the HGP property. As we can always embed the convex set of Markov sequences in the n-dimensional affine space $H_1 = \{(\lambda_i), \lambda_0 = 1\}$, the latter fact means that this set is actually contained in the n-simplex generated by the $n + 1$ points $\lambda^{x/x_0} = \left(f_i(x)/f_i(x_0) \right)_i$.

Then, when the hypergroup property holds, the set of Markov sequences is a n-simplex and the representation of a point in this set as affine combination of extremal points is unique.

We may ask for which kind of L^2 basis on a finite space this still happens. It is quite clear that the cardinal of the set of extremal Markov sequences is finite. Indeed, the set of Markov sequences is delimited by a finite number of $(n-1)$-hyperplanes in H_1. Namely, for any pair (x,y) of points in E, one considers the half space defined by $\{\langle \lambda, F^{x,y} \rangle \geq 0\}$, where $F^{x,y} = (f_i(x)f_i(y))$. Then the set of Markov sequences is the intersection of all these half spaces. Therefore, every extremal point lies in the finite set E_1 of possible intersections of n hyperplanes $H^{x,y} = H_1 \cap \{\langle \lambda, F^{x,y} \rangle = 0\}$. Now, consider any point x_0 such that for any index i, $f_i(x_0) \neq 0$. The point $\lambda^{x/x_0} = (f_i(x)/f_i(x_0))_i$ belongs to $H^{x_0,y}$ for any $y \neq x$, thanks to the orthogonality relations of the basis. Therefore, those points λ^{x/x_0} belong to the set E_1. When x_0 is fixed and x varies in E, those points describe a simplex S_{x_0} for which we know that every Markov sequence belongs to it. The hypergroup property holds at some point x_0 exactly when no other point in E_1 lie in the interior of S_{x_0}.

On three points, one may check directly that the hypergroup property holds at some point x_0 exactly when the set of Markov sequences is a simplex (that means no other simplex is possible than the simplices S_x, $x \in E$). We may wonder if this situation is general, that is if the hypergroup property is equivalent to the fact that the set of Markov sequences is a simplex.

2.4 Markov operators as convolutions

When the hypergroup property holds, we may introduce a convolution on the space of measures.

Indeed, consider the kernel

$$k(x,y,z) = \sum_i \frac{f_i(x)f_i(y)f_i(z)}{f_i(x_0)}.$$

We observe that, for any (x,y)

$$\int k(x,y,z)\, \mu(dz) = 1,$$

and therefore the measures

$$\mu_{x,y}(dz) = k(x,y,z)\, \mu(dz)$$

are probability measures.

We may decide that the convolution is defined from this kernel by

$$\delta_x * \delta_y = \mu_{x,y},$$

and extending it to any measure by bilinearity.

Moreover, we extend the convolution to functions by identifying a function f with the measure $f d\mu$. This gives

$$f * g(z) = \int f(x)g(y)k(x,y,z)\,d\mu(x)d\mu(y).$$

Observe that

$$f_i * f_j = \delta_{ij}\frac{f_i}{f_i(x_0)}, \tag{2}$$

and that this property again completely determines the convolution.

It is easy to verify that this convolution is commutative and that $\delta_{x_0}*\nu = \nu$ for any ν. Moreover, if an operator K satisfies $K(f_i) = \lambda_i f_i$, then

$$K(f * g) = K(f) * g = f * K(g),$$

as may be verified directly when $f = f_i$ and $g = f_j$ using (2).

On the other hand, if ν is a probability measure, then the operator $K_\nu(f) = f * \nu$ is a Markov operator which satisfies

$$K_\nu(f_i) = \lambda_i f_i,$$

with

$$\lambda_i = \frac{\int f_i\,d\nu}{f_i(x_0)}.$$

This is straightforward using (2) if we write $\nu(dx) = h(x)\mu(dx)$ and the decomposition of h along the basis $\mathcal{F}$.

Therefore, if K is a Markov operator, then we have

$$K(f) = K(f * \delta_{x_0}) = f * K(\delta_{x_0}).$$

This representation is exactly the representation of Markov sequences, with $\nu = K(\delta_{x_0})$, and every Markov operator K_λ may be defined from $K_\lambda(f) = f * \nu$, for some probability measure ν.

2.5 The case of finite groups

Many natural examples of finite sets endowed with a probability measure and a UOB which satisfies both GKS and HGP properties come from finite groups.

Since perhaps not every reader of these notes is familiar with this setting, let us summarize briefly the basic elements of the analysis on groups. We refer to [22] or [31] for more details.

Given a finite group G, one may consider linear representations $\rho : G \mapsto U(V)$, that is group homomorphisms between G and some $U(V)$, for some finite dimensional Hermitian space V (where $U(V)$ denotes the unitary group of V). Such a representation is irreducible if there is no non-trivial proper subspace of V which is invariant under $\rho(G)$. Any representation may be split into a sum of irreducible representations, acting on orthogonal subspaces of V. Two representations (ρ_1, V_1) and (ρ_2, V_2) are equivalent if there exists a linear unitary isomorphism $h : V_1 \mapsto V_2$ such that $\rho_1(g) = h^{-1}\rho_2(g)h$ for any $g \in G$. There are only a finite number of non-equivalent irreducible representations, that we denote (ρ_i, V_i), $i \in I = \{0, \ldots, n\}$.

Let $\hat{G}$ the set of the equivalence classes of G under the conjugacy relation (g_1 is conjugate to g_2 means $g_2 = g^{-1}g_1 g$ for some $g \in G$). We endow $\hat{G}$ with the probability ν which is the image measure of the uniform measure on G, which means that the measure of any class is proportional to the number of points in this class. A function on G which is constant on conjugacy classes (that we call a class function) can be seen as a function on $\hat{G}$. It is just a function which is stable under conjugacy.

For any irreducible representation (ρ_i, V_i), let us define the function χ_i on G by $\chi_i(g) = \text{trace}(\rho_i(g))$. This is a class function, that is to say constant on any conjugacy class. The function χ_i is called the character of the representation. By convention, we take $\chi_0 = 1$, that is the trace of the constant representation into the space $V = \{\mathbb{C}\}$.

Proposition 2.6. *The set $\{\chi_i, i \in I\}$ is a (complex) UOB for $(\hat{G}, \nu)$. Moreover, it has the GKS and HGP properties.*

Proof. We shall not enter in the details here. We refer to any introduction book on the representation theory of finite groups for the first fact. We shall detail a bit more the HGP and GKS properties, which are perhaps less standard.

For the GKS property, for any pair of irreducible representations (ρ_i, V_i) and (ρ_j, V_j), one may consider the representation $\rho_i \otimes \rho_j$ in the tensor product $V_i \otimes V_j$. If we split this representation into irreducible representations and take the trace, and if we notice that $\text{trace}(\rho_1 \otimes \rho_2) = \text{trace}(\rho_1)\text{trace}(\rho_2)$, then we get that

$$\chi_i \chi_j = \sum_k m_{ijk}\chi_k,$$

where m_{ijk} is the number of times that the representation ρ_k appears in this decomposition. Here we may see that not only the multiplication coefficients are non-negative, but they are integers.

We shall see next that this basis has the HGP property at the point $x_0 = e$ (which forms a conjugacy class by itself). For that, we require a bit more material.

First define the convolution on the group G itself by

$$\phi * \psi(g) = \frac{1}{|G|} \sum_{g' \in G} \phi(gg'^{-1})\psi(g').$$

The Fourier transform is defined on the set I of irreducible representation as

$$\hat{\phi}(i) = \sum_{g \in G} \phi(g)\rho_i(g).$$

(It takes values in the set of linear operators on V_i.)

One has an inversion formula

$$\phi(g) = \frac{1}{|G|} \sum_i d_i \text{trace} \left(\rho_i(g^{-1})\hat{\phi}(i) \right),$$

where d_i is the dimension of V_i (the degree of the representation).

One has

$$(\phi * \psi)\hat{} = \hat{\phi}\hat{\psi},$$

and

$$\hat{\chi}_j(i) = \delta_{ij} \frac{|G|}{d_i}.$$

Now, the convolution of two class functions is again a class function, as seen directly from the definition.

We want to show that this convolution is exactly the convolution that we defined in the previous section from the HGP property, that is

$$\chi_i * \chi_j = \delta_{ij} \frac{\chi_i.}{\chi_i(e)}.$$

For that, we look at the Fourier transform and the result is straightforward, since $\chi_i(e) = d_i$.

This convolution is then the convolution defined from the χ_i, and we have

$$\delta_x * \delta_y = k(x, y, z)d\mu(z),$$

where

$$k(x, y, z) = \sum_i \frac{\chi_i(x)\chi_i(y)\chi_i(z)}{\chi_i(e)}.$$

Since by construction in this case the convolution of two probability measures is a probability measure, the kernel $k(x, y, z)$ is non-negative, which proves the HGP property.

Observe that here the kernel $k(x, y, z)$ has a simple interpretation. Given 3 classes (x, y, z), then

$$k(x, y, z) = \frac{|G|}{|x|\,|y|} m(x, y, z),$$

where $m(x, y, z)$ is, for any point $g \in z$, the number of ways of writing $g = g_1 g_2$ with $g_1 \in x$ and $g_2 \in y$, this number being independent of the choice of $g \in z$. $\qquad \square$

If we want to stick to real bases as we did before (and as we shall do in the next chapters), we may restrict ourselves to real groups (that is groups where g and g^{-1} are always in the same class), or we may agglomerate the class of g with the class of g^{-1}. We get a new probability space, where the functions $\Re(\chi_i)$ form a UOB which again satisfies the GKS and HGP properties.

It is certainly worth noticing that, unlike the convolution on G itself, the convolution on $\hat{G}$ is always commutative.

Observe that taking the group $\mathbb{Z}/n\mathbb{Z}$, one gets a complex GKS and HGP UOB on the set of finite points with the uniform measure (with $f_l(x) = \exp(2i\pi lx)$), and that the unique real case where the measure is uniform and is GKS (the hypercube) is nothing else that the group $(\mathbb{Z}/2\mathbb{Z})^n$.

Remark. Unlike what happens for finite groups, it is not true in general that a basis $\mathcal{F}$ which has the GKS property has the dual property HGP. This is the case on two points spaces, since any orthogonal GKS matrix is symmetric (cf Page 304). If we look at the sets with three points, one may construct examples of an orthogonal matrix having the GKS property without the HGP property (and conversely, of course). In fact, consider an orthogonal matrix (O_{ij}), $0 \leq i, j \leq 2$, with positive first row and columns. O_{00}, O_{01} and 0_{10} determine entirely the first rows and columns, and then it is easy to see that there are only 2 orthogonal matrices with given O_{00}, O_{01}, O_{10}. Then, it is not hard (using a computer algebra program) to produce orthogonal matrices which have the GKS and not the HGP property, or which have neither, or both.

2.6 On the GKS inequalities

We conclude this section with some remarks on the correlation inequalities in statistical mechanics.

In this context, one is interested in the space of configurations of some system. We have a set of positions $i \in K$, K being a finite set, and at each point $i \in K$ there is some random variable x_i with values in E, where E is some finite set, endowed with a probability measure μ. One is then interested in the set E^K of configurations, which is equipped with a measure μ_H, where

$$\mu_H(dx) = \exp(H(x))\frac{\mu_0(dx)}{Z_H},$$

where μ_0 is the product measure $\mu^{\otimes K}$ on E^K, H is some function on E^K (the Hamiltonian), and Z_H is the normalizing constant.

One of the basic example of spin systems is when $E = \{-1, 1\}$, and $H = \sum_A c_A \omega_A$, where the functions ω_A are the canonical GKS basis on $\{-1, 1\}^K$ described before.

To study such systems (and more precisely their asymptotics when K enlarges), one uses some structural inequalities. We present here two fundamental such inequalities, known as GKS inequalities, from Griffiths [27], Kelly and Sherman [32]. The GKS property for a basis has been introduced in [8], in an attempt to generalize the GKS inequality to a more general context.

The classical GKS inequalities are settled in the context of $(\mathbb{Z}/2\mathbb{Z})^K$. As before, we say that F is a GKS function if $F = \sum_{A \subset K} f_A \omega_A$, where $\forall A$, $f_A \geq 0$.

Then we have

Proposition 2.7. *(i) (GKS1 inequality). Assume that F and H are GKS. Then*

$$\int F \, d\mu_H \geq 0.$$

(ii) (GKS2 inequality). Assume that F, G and H are GKS functions. Then

$$\int FG \, d\mu_H \geq \int F \, d\mu_H \int G \, d\mu_H.$$

The main advantage of the GKS and HGP properties is that they are stable under tensorization. That is, if one considers two sets (E_i, μ_i) with UOB bases $\mathcal{F}_i$ $(i = 1, 2)$, then, on the set $(E_1 \times E_2, \mu_1 \otimes \mu_2)$ one has a natural UOB basis $\mathcal{F}_1 \otimes \mathcal{F}_2 = (f_i \otimes f_j)$. Then, if both $\mathcal{F}_i$ are GKS or HGP, the same is true for $\mathcal{F}_1 \otimes \mathcal{F}_2$. This is straightforward from the definitions.

This allows us to consider a set (E, μ) with a given GKS basis $\mathcal{F}$, and then the basis $\mathcal{F}^{\otimes K}$ on E^K is again GKS.

One has the following ([8])

Proposition 2.8. *If (E, μ) has a UOB $\mathcal{F}$ which is GKS, then the GKS1 inequality is true.*

Proof. The previous statement just means that if we define a GKS function F as a function which may be written as $F = \sum_i F_i f_i$, where (f_i) are the elements of $\mathcal{F}$ and $\forall i$, $F_i \geq 0$, then if F and G are GKS functions, one has

$$\int F \, d\mu_H \geq 0.$$

The statement is straightforward, since $\exp(H)$ is again GKS, being the sum of a series with non-negative coefficients, and so $F \exp(H)$ is itself GKS. Since any GKS function has a non-negative integral, the conclusion follows.   □

The GKS2 inequality is much harder. It has only be obtained in some restricted settings, like products of abelian groups, and when the basis comes from G^N for G elementary groups like dihedral groups, and some for other few groups. Nevertheless, in any example, one has both the GKS and the HGP property.

There is no example of a GKS basis where the GKS2 inequality is not satisfied. But we may restrict ourselves to a simpler setting.

Here is one conjecture that we had been unable to prove, and which motivated most of the material of this section:

Conjecture. *If the UOB $\mathcal{F}$ has the GKS and the HGP property, then the GKS2 inequality is true.*

3 The hypergroup property in the infinite setting

In the first section, we described the hypergroup property in the context of finite sets. In what follows, we consider a general probability space $(E, \mathcal{E}, \mu)$, together with a L^2 basis $\mathcal{F} = \{1 = f_0, f_1, \ldots, f_n, \ldots\}$. Very soon, we shall restrict ourselves to the case of a topological space (in fact an interval in the basic examples of Sections 4.3 and 5), where the functions f_i of the basis will be continuous bounded functions. But some general properties may be stated in a more general context.

3.1 Markov sequences associated with a UOB

Let $(E, \mathcal{E}, \mu)$ a general probability space. In this subsection, we shall ask $(E, \mathcal{E})$ to be at least a "nice" measurable space in the context of measure theory. For us, it shall be enough to suppose that E is a separable complete metric space (a polish space) and that $\mathcal{E}$ is the σ-algebra of its σ-field. Then $L^2(\mu)$ is separable.

We suppose that some $L^2(\mu)$ orthonormal basis $\mathcal{F} = (f_0, \ldots, f_n, \ldots)$ is given, with $f_0 = 1$. In what follows, we shall assume that this is a basis of the real Hilbert space $L^2(\mu)$, although we may as well assume that the functions f_n may have complex values and be a basis of the complex Hilbert space. Such a unitary basis $\mathcal{F}$ will be called a Unitary Orthonormal Basis (UOB) associated with the measure μ.

We are interested in bounded linear operators K on $L^2(\mu)$, for which the functions f_n are eigenvectors. They are uniquely determined by

$$K(f_n) = \lambda_n f_n,$$

for some bounded sequence (λ_n). The central question we address here is to determine for which sequences (λ_n) one has

$$K(f)(x) = \int_E f(y) \, k(x, dy),$$

for some Markov kernel $k(x, dy)$ of probability measures on E.

As before, we shall call such a sequence (λ_n) a Markov Sequence (MS in short) associated with the UOB $\mathcal{F}$. We shall say that the kernel k (or rather K with a slight abuse of notation) is associated with the MS (λ_n).

In general, this is not an easy question, but as before the hypergroup property of the basis will be a way of describing all Markov sequences.

Let us start with some basic remarks.

First, for any Markov operator, $K(f_0) = f_0$, since f_0 is the constant function, and therefore $\lambda_0 = 1$.

Also, any such Markov operator is symmetric in $L^2(\mu)$, since we already know its spectral decomposition which is discrete and given by the basis $\mathcal{F}$.

That means that for any pair (f, g) of functions in $L^2(\mu)$, one has

$$\int K(f)(x)g(x)\,\mu(dx) = \int f(x)K(g)(x)\,\mu(dx).$$

Therefore, the measure $K(x, dy)\mu(dx)$ is symmetric in (x, y).

But any Markov operator is a contraction in $L^\infty(\mu)$, and any symmetric Markov kernel is a contraction in $L^1(\mu)$, since, for any $f \in L^2(\mu)$,

$$\int |K(f)|\;d\mu \le \int K(|f|)\;d\mu = \int |f|\,K(f_0)\;d\mu = \int |f|\;d\mu.$$

Therefore, by interpolation, K is a contraction in $L^p(\mu)$ for any $p \ge 1$, and in particular in L^2.

We deduce from that that any MS (λ_n) satisfies

$$\forall n, \; |\lambda_n| \le 1.$$

Also, if (λ_n) and (μ_n) are MS's, with associated kernels K and K_1, for any $\theta \in [0, 1]$, $(\theta\lambda_n + (1-\theta)\mu_n)$ is a MS, associated with the kernel $\theta K + (1-\theta)K_1$.

Therefore, the set of Markov Sequences is convex, and compact (for the product topology on $\mathbb{R}^\mathbb{N}$). This shows that describing all Markov sequences amounts to describe the extremal points of this convex set.

Notice also that the set of all Markov sequences is stable under pointwise multiplication, which corresponds to the composition of operators. In other words, if Θ is the set of extremal points in the compact set of Markov sequences, and if $(\lambda_i(\theta))$ is the MS associated with the point $\theta \in \Theta$, then one has

$$\lambda_i(\theta)\lambda_i(\theta') = \int_\Theta \lambda_i(\theta_1)\;R(\theta, \theta', d\theta_1),$$

for some probability kernel $R(\theta, \theta', d\theta_1)$ on the space Θ.

To determine that an operator K is a Markov operator starting from its spectral decomposition, we shall need the following proposition.

Proposition 3.1. *A bounded symmetric operator K on $L^2(\mu)$ is a Markov operator if and only if*

$$K(f_0) = f_0, \; f \ge 0 \implies K(f) \ge 0.$$

Proof. This is where we need the fact that the measure space $(E, \mathcal{E}, \mu)$ is a nice space. The conditions on K are obviously necessary. To see the reverse, we apply the bi-measure theorem ([21], Page 129). We consider the map

$$\mathcal{E} \times \mathcal{E} \mapsto [0, 1] \; : \; (A, B) \mapsto \int \mathbb{1}_A K(\mathbb{1}_B)\;d\mu(x).$$

For any fixed B, this is a measure in A, and by symmetry, it is also a measure in B. Since our spaces are Polish spaces, theses measures are tight, and therefore we may extend this operation into a measure $\mu_K(dx, dy)$ on the σ-algebra $\mathcal{E} \times \mathcal{E}$. The measure is symmetric, and any of its marginal is μ.

Then, we apply the measure decomposition theorem to write

$$\mu_K(dx, dy) = K(x, dy)\mu(dy).$$

The kernel $K(x, dy)$ is exactly the kernel we are looking for.   □

3.2 The hypergroup property

The GKS property is relatively easy to state in a general context, as soon as the functions f_i of the basis are in $L^3(\mu)$, since then we may just ask that

$$\forall i, j, k, \quad \int_E f_i f_j f_k \, d\mu \geq 0.$$

But for the dual hypergroup property, one has to be a bit more cautious. In general, functions in L^2 are defined up to a set of μ-measure 0. Therefore, the meaning of $f_i(x)/f_i(x_0)$ is not so clear. In order to avoid difficulties, and since this shall correspond to the examples we are going to describe below, we restrict ourselves to the following setting: E is a compact separable Hausdorff space, and the functions f_i are continuous on E.

We may then set the following definition.

Definition 3.1. *We shall say that the UOB $\mathcal{F}$ has the hypergroup property (HGP in short) at some point $x_0 \in E$, if, for any $x \in E$, the operator defined on $\mathcal{F}$ by*

$$K_x(f_i) = \frac{f_i(x)}{f_i(x_0)} f_i$$

is a Markov operator.

In other words, we require the sequences $\left(f_i(x)/f_i(x_0)\right)_i$ to be Markov sequences.

Observe that this implies that $|f_i|$ is maximal at x_0, since the eigenvalues of a Markov operator must be bounded by 1. Since the functions f_i are normalized in $L^2(\mu)$, then for any $i \in \mathbb{N}$, one has $|f_i(x_0)| \geq 1$.

As before, this definition is equivalent to the following

Proposition 3.2. *The UOB $\mathcal{F}$ has the HGP property at the point x_0 if and only if there exists a probability kernel $K(x, y, dz)$ such that, for any $i \in \mathbb{N}$*

$$\frac{f_i(x)f_i(y)}{f_i(x_0)} = \int f_i(z) \, K(x, y, dz).$$

Proof. We shall mainly use this in the obvious way: if there is a probability kernel $K(x, y, dz)$ satisfying the hypothesis of the proposition, then the HGP property holds. In fact, if such a probability kernel $K(x, y, dz)$ exists, for any $x \in E$, the Markov kernel $K(x, y, dz)$ defines a Markov operator with Markov sequence $\left(f_i(x)/f_i(x_0)\right)_i$.

For the reverse, if the HGP property holds, there exists for any x a Markov kernel $k_x(y, dz)$ which satisfies

$$\frac{f_i(x)f_i(y)}{f_i(x_0)} = \int f_i(z)\, k_x(y, dz).$$

It remains to turn this family of kernels into a two parameters kernel $K(x, y, dz)$. □

Sometimes, it is easier to see the HGP property in the reverse way.

Proposition 3.3. *If there exists a non-negative kernel $k(x, dy, dz)$ such that for any i, j*

$$\int f_i(y)f_j(z)\, k(x, dy, dz) = \delta_{ij}\frac{f_i(x)}{f_i(x_0)}$$

then the HGP property holds at the point x_0.

Proof. Remark first that from our hypotheses, the kernel $k(x, dy, dz)$ is a probability kernel (taking $i = j = 0$ in the definition).

Now, the two marginals of the kernel $k(x, dy, dz)$ are equal to μ, since

$$\int f_i(y)\, k(x, dy, dz) = \delta_{0i},$$

which shows that those marginals and μ give the same integral to any f_i, and therefore to any L^2 function.

We may then decompose the kernel $k(x, dy, dz) = k_1(x, y, dz)\mu(dy)$.

Since the functions f_i are bounded, we may consider the bounded functions

$$H_i(x, y) = \int f_i(z)\, k_1(x, y, dz).$$

From the definition of k_1 and the hypothesis on k, it is straightforward to check that

$$\int H_i(x, y)f_k(x)f_l(y)\, \mu(dx)\mu(dy) = \delta_{il}\delta_{ik}\frac{1}{f_i(x_0)},$$

and hence

$$H_i(x, y) = \frac{f_i(x)f_i(y)}{f_i(x_0)}.$$

Therefore k_1 satisfies the hypotheses of Proposition 3.2 and the proof is completed. □

The representation of Markov sequences is then the same than in the previous section.

Theorem 3.1. *If the UOB $\mathcal{F}$ has the hypergroup property, then any Markov sequence has the representation*

$$\lambda_i = \int_E \frac{f_i(x)}{f_i(x_0)} \, \nu(dx),$$

for some probability measure ν on E. Moreover, the Markov sequences

$$\left(\frac{f_i(x)}{f_i(x_0)} \right)_i$$

are the extremal Markov sequences.

Proof. The proof is exactly similar to the finite case (see Theorem 2.1).   □

Remark that the series

$$K(x, y, z) = \sum_i \frac{f_i(x) f_i(y) f_i(z)}{f_i(x_0)}$$

does not converge in general. We shall see in the examples developed in the next section that the formal measure $K(x, y, z)\mu(dz)$ is not absolutely continuous with respect to μ, and may have Dirac masses at some points.

But we may still define a convolution structure from $f_i * f_j = \delta_{ij} f_i / f_i(x_0)$, which maps probability measures onto probability measures, and all Markov kernels associated with $\mathcal{F}$ would be represented as $K(f) = f * \nu$, for some probability measure on E. We give no details here since this will not be used in the sequel.

4 Sturm–Liouville bases and Achour–Trimèche's theorem

4.1 The natural UOB associated with a measure on a compact interval

In this section, we shall consider some natural infinite UOB coming from the spectral decomposition of Sturm–Liouville operators on a compact interval of the real line.

Let us first describe the context. Consider a probability measure $\mu(dx) = \rho(x)dx$ on some compact interval $[a, b] \subset \mathbb{R}$. In what follows, we shall assume for simplicity that ρ is smooth, bounded above and away from 0 on $[a, b]$. The density ρ is associated with a canonical differential operator

$$L(f)(x) = f''(x) + \frac{\rho'}{\rho}(x)f'(x),$$

which is symmetric in $L^2(\mu)$. We shall consider here L acting on functions on $[a, b]$ with derivative 0 at the boundaries a and b (Neumann boundary conditions).

In this context, L is essentially self-adjoint on the space of smooth functions with $f'(a) = f'(b) = 0$ and there is an orthonormal basis

$$\mathcal{F} = (1 = f_0, f_1, \ldots, f_n, \ldots)$$

of $L^2(\mu)$ which is given by eigenvectors of L satisfying the boundary conditions. This means that there is an increasing sequence of real numbers

$$0 = \lambda_0 < \lambda_1 < \cdots < \lambda_n < \cdots$$

such that

$$Lf_i = -\lambda_i f_i, \quad f_i'(a) = f_i'(b) = 0.$$

From the standard theory of Sturm–Liouville operators, the eigenvalues λ_i are non-negative and simple. Therefore, there is for any λ_i a unique solution f_i of the previous equation, which has norm 1 in $L^2(\mu)$ and which satisfies $f_i(a) > 0$. We refer to any standard text book for details (see [15] or [40] for example).

This basis shall be called the canonical UOB associated with μ on $[a, b]$.

The fact that we chose to deal with the Neumann boundary conditions and not with the Dirichlet boundary conditions $(f(a) = f(b) = 0)$ comes from the fact that we require the function 1 to be an eigenvector of the operator.

It will be much more convenient in what follows, essentially for notations, to extend our functions by symmetry in a and b, (and the same for μ). In this way we may consider that we are working on functions on the real line, which are symmetric under $x \mapsto 2a - x$, and are $2(b - a)$-periodic.

The eigenvectors are perhaps not smooth then at the boundaries a and b, but they are at least C^2 (since they are solutions of the equation $Lf_i = -\lambda_i f_i$ at the boundaries).

The hypergroup property is stated at some point in $[a, b]$. In the finite case, we know at which point we may expect the hypergroup property to hold: this is a point of minimal mass. In the general case, such a reasoning does not hold, since one may choose a point with minimal density to a given reference measure, but this depends on this choice.

Here, the basis $\mathcal{F}$ is the sequence of eigenvectors of an elliptic second order differential operator L symmetric in $L^2(\mu)$. In this setting, there is a natural distance associated with the operator L (in this precise example of Sturm–Liouville operators, this is the natural distance on $\mathbb{R}$). In any example we know, the point x_0 is minimal in the following sense

$$\lim_{r \to 0} \frac{\mu\big(B(x_0, r)\big)}{\mu\big(B(x, r)\big)} \leq 1. \tag{3}$$

We did not try to prove this in a more general context. However, it is not clear how the properties of the operator L must be reflected in the properties of $\mathcal{F}$ to insure for example that the maximal values of the eigenvectors are attained at the same point, and that this point is of minimal mass in the sense of (3).

4.2 Wave equations

In this context, one has some other interpretation of the hypergroup property.

On $D = [a, b]^2$, we shall consider the following differential equation

$$L_x F(x, y) = L_y F(x, y), \qquad (4)$$

for a function F which has Neumann boundary conditions on the boundary of D. We shall say that such a function is a solution of the (modified) wave equation.

We have to be careful here with the regularity assumption on the function $F(x, y)$ that we require. We shall see later that given any smooth function $f(x)$ at the level $x = x_0$, with Neumann boundary conditions, there is exactly one smooth function $F(x, y)$ on D which is solution of Equation (4) and satisfies $F(x, y_0) = f(x)$.

In fact, if $f(x) = \sum_i a_i f_i(x)$ is the L^2 orthogonal decomposition of f, then

$$F(x, y) = \sum_i \frac{a_i}{f_i(y_0)} f_i(x) f_i(y)$$

is a formal L^2 solution of the wave equation, since

$$L_x F = \sum_i \lambda_i \frac{a_i}{f_i(y_0)} f_i(x) f_i(y) = L_y F.$$

But we do not even know (for the moment) that this solution is such that $L_x F$ is in $L^2(\mu \otimes \mu)$.

Therefore, we shall say that F is a weak L^2 solution of (4) if for any smooth function $G(x, y)$ with Neumann boundary conditions on ∂D, one has

$$\int [(L_x - L_y)G(x, y)] F(x, y) \, \mu(dx)\mu(dy) = 0.$$

Since

$$\int L_x(F)G \, \mu(dx)\mu(dy) = \int L_x(G)F \, \mu(dx)\mu(dy)$$

for any pair of smooth functions F and G satisfying the Neumann boundary conditions, then any ordinary solution is a weak one.

Now, given any $L^2(\mu)$ function $f(x)$, the above construction produces a weak $L^2(\mu \otimes \mu)$ solution $F(x, y)$ satisfying the wave equation (4), and we claim immediately that this solution is unique. In fact, writing the function $F(x, y)$ as

$$F(x,y) = \sum_{ij} a_{ij} f_i(x) f_j(x),$$

and using the fact that the eigenvalues are simple, one may check that if F satisfies weakly (4), then $a_{ij} = 0$ if $i \neq j$, from which we deduce our claim.

Observe moreover that $a_{00} = \int F(x,y)\,\mu(dx)\mu(dy)$, and that for almost every y_0, $F(x,y_0) \in L^2(\mu)$ and that

$$\int F(x,y_0)\,\mu(dx) = \int F(x,y)\,\mu(dx)\mu(dy).$$

It is not clear however that if $F(x,y_0)$ is smooth, then $F(x,y)$ is smooth. This shall be done later at least when $y_0 = a$ or $y_0 = b$.

The link between solutions of the wave equation and Markov kernels is the following.

If a Markov kernel is Hilbert–Schmidt (that is if its eigenvalues λ_i satisfy $\sum_i \lambda_i^2 < \infty$), then it may be represented as a

$$K(f)(x) = \int f(y)k(x,y)\,\mu(dy),$$

where

$$k(x,y) = \sum_i \lambda_i f_i(x) f_i(y).$$

Therefore, there is a one-to-one correspondence between Hilbert–Schmidt Markov kernels and non-negative weak L^2 solutions of the wave equation which satisfy

$$\int F(x,y)\,\mu(dx)\mu(dy) = 1.$$

We then have the following

Theorem 4.1. *Assume that for any function $f(x)$ on the interval $[a,b]$ with Neumann boundary conditions, there exists a unique C^2 solution $H_f(x,y)$ of the wave equation (4) on $[a,b]^2$ such that $H(x,y_0) = f(x)$. Then, the HGP property holds at the point y_0 for the natural UOB associated with μ if and only if whenever $f \geq 0$ one has $H_f \geq 0$ on I^2.*

In other words, the hypergroup property is equivalent to the fact that the wave equation is positivity preserving.

Proof. Assume first that the hypergroup property holds at the point y_0. Take any smooth solution $F(x,y)$ of the wave equation with $F(x,y_0) = f(x) \geq 0$. Then, from what we just saw, one has

$$F(x,y) = K_y(f)(x),$$

where K_y is the Markov kernel with eigenvalues $f_i(y)/f_i(y_0)$. Therefore, $F(x,y)$ is everywhere non-negative.

On the other hand, assume that any smooth solution of the wave equation which is non-negative on $\{y = y_0\}$ is non-negative everywhere. Consider the heat kernel

$$p_t(x,z) = \sum_i \exp(-\lambda_i t) f_i(x) f_i(z).$$

We know that it is a smooth function on D, which is everywhere positive. Then,

$$F_{t,z}(x,y) = \sum_i \frac{\exp(-\lambda_i t)}{f_i(y_0)} f_i(z) f_i(x) f_i(y)$$

is the unique L^2 solution of the wave equation with $F_{t,z}(x,y_0) = p_t(x,z)$.

Therefore, this function is non-negative, and this shows that for any $t > 0$, the sequence

$$\exp(-\lambda_i t)\frac{f_n(z)}{f_n(y_0)}$$

is a Markov sequence. It remains to let t go to 0 to get the result, since a limit of Markov sequences is a Markov sequence.   □

4.3 Achour–Trimèche's theorem and wave equations

In what follows, we consider the case of a symmetric interval $[-b, b]$. Then we have

Theorem 4.2 (Achour–Trimèche). *Let ρ be a log-concave and symmetric density on $[-b, b]$. Then, the natural UOB associated with μ has the HGP property at the point $-b$. In this case, we may as well choose $x_0 = b$. The same is true on any interval with any log-concave increasing density ρ.*

This result is one of the very few cases when one may produce hypergroup bases without any kind of group structure on the space E. We shall see in the next chapter that this property holds for Jacobi polynomials, but in this case, there are at least for the integer values of the coefficients some interpretations of the convolution which reflects the group action of some orthogonal group. There is absolutely no such interpretation in this context.

In general, Achour–Trimèche's result is stated with a density ρ which vanishes on the boundary. Under the conditions usually stated in Achour–Trimèche's theorem, there are then no difference between Neumann and Dirichlet boundary conditions. The series

$$\sum_i \frac{f_i(x) f_i(y) f_i(z)}{f_i(x_0)} \mu(dz)$$

is absolutely continuous with respect to the measure μ, which is not the case here.

We chose to present this result in the case where the density ρ is bounded from below because it seemed to us to be more natural.

Apparently, the proof of Achour–Trimèche's theorem had never been published. We found a mention of it in the reference book [13] and the result is announced in [1], with no proof. Most of the ideas presented here come from Achour's thesis. The idea follows a previous result of Chebli [16], which works on $[0, \infty)$ and is somehow simpler (It corresponds to the case of a concave decreasing density).

Proof. To prove this result, we shall make use of the characterization of the hypergroup property in terms of the wave equation given in Theorem 4.1. We shall see in the next paragraph that any smooth bounded function satisfying Neumann boundary conditions on $[a, b]$ has a unique extension as a smooth solution of the wave equation (4), when x_0 is one of the boundary points (see Paragraph 4.5). (This has nothing to do with the log-concavity of the measure or with the symmetry: this is just a consequence of the fact that $\log(\rho)$ is smooth and bounded.)

We first treat the case where the density ρ is log-concave symmetric.

First we make use of the symmetry assumption. Then, any eigenvector of the operator L on $[-b, b]$ with Neumann boundary conditions is either even or odd, since $f_i(-x)$ is also an eigenvector with the same eigenvalue.

Then, any L^2 solution $F(x, y)$ of the wave equation (4), written as

$$F(x, y) = \sum_i a_i f_i(x) f_i(y)$$

is symmetric under the change $(x, y) \mapsto (y, x)$ and under $(x, y) \mapsto (-x, -y)$.

We want to show that if $F(x, -b) \geq 0$, then $F(x, y) \geq 0$ everywhere. For this, it is enough to show this on the domain $D_1 = \{x + y \leq 0, \ x \geq y\}$.

Also, we may change F into $F + \varepsilon$ for any $\varepsilon > 0$, and we are thus reduced to prove that the result is true when the function f on the boundary is bounded below by some positive constant.

Then, a point $M \in D_1$, let Δ_M be the triangle delimited by the lines $x + y = c$ and $x - y = c'$ passing through M and the line $y = -c$. Let M_- and M_+ be the points of this triangle which lie on the line $\{y = -b\}$, M_- being the left point and M_+ the right one (see Figure 1). Let F be a smooth solution of the wave equation (4), and let $G(x, y) = F(x, y)\rho(x)\rho(y)$.

$$2G(M) = G(M_-) + G(M_+) + \int_{[M_- M]} G(s) a_+(s) \, ds + \int_{[MM_+]} G(s) a_-(s) \, ds \tag{5}$$

where

$$a_+(x, y) = \frac{1}{\sqrt{2}} \left(\frac{\rho'}{\rho}(x) + \frac{\rho'}{\rho}(y) \right), \ a_-(x, y) = \frac{1}{\sqrt{2}} \left(\frac{\rho'}{\rho}(y) - \frac{\rho'}{\rho}(x) \right),$$

and the integral $\int_{[M_- M]} H(s) \, ds$ and $\int_{[MM_+]} H(s) \, ds$ denote the one dimensional integrals along the segments $[M_- M]$ and $[MM_+]$ against the (euclidean) length measure on those lines.

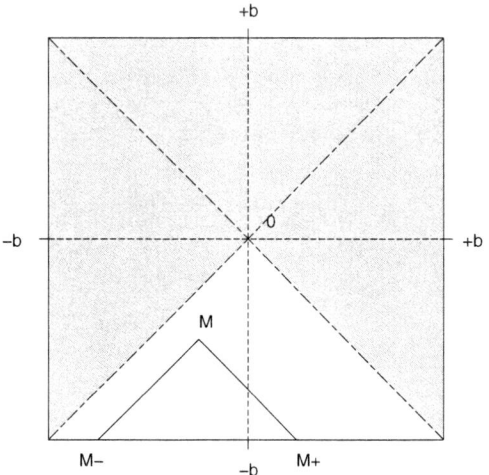

Fig. 1. Triangle Δ_M.

This formula relies on an integration by parts formula. Even though we shall use it only on domains like Δ_M, it is perhaps of some interest to state it in general. So we set it as a lemma. We shall not give too much details here, since a more general formula will be derived in the next paragraph.

Lemma 4.1. *Let H a smooth function on D and $\Omega \subset D$ a domain with a piece-wise C^1 boundary $\partial\Omega$. Then*

$$\int_\Omega (L_x - L_y)H(x,y)\ \rho(x)\rho(y)\ dxdy = \int_{\partial\Omega} \nabla H \odot n\ \rho(x)\rho(y)\ ds,$$

where $n = (n_x, n_y)$ denotes the exterior normal derivative of the domain and

$$\nabla H \odot n = \partial_x H n_x - \partial_y H n_y,$$

ds designing the length measure on the boundary $\partial\Omega$.

We shall not prove this lemma. It is the analogue of the classical Stokes formula, where the elliptic operator Δ is replaced by the hyperbolic operator $L_x - L_y$, its invariant measure being $\rho(x)\rho(y)dxdy$.

One may see this as a particular case of the general integration by parts formula

$$\int_D H_1(L_x - L_y)H_1\ \rho(x)\rho(y)\ dxdy = -\int \nabla H_1 \odot \nabla H_2\ \rho(x)\rho(y)\ dxdy,$$

applied with $H_1 = \mathbb{1}_{\Delta_M}$.

From the previous formula, applied on $\Omega = \Delta_M$ for a function F which is solution of the wave equation (4) and has normal derivative vanishing on the boundary $[M_-, M_+]$, one has

$$-\int_{[M-M]} (\partial_x F + \partial_y F)\rho(x)\rho(y)\ ds + \int_{[MM_+]} (\partial_x F - \partial_y F)\rho(x)\rho(y)\ ds = 0.$$

We may then perform a next integration by parts on both integrals to find

$$G(M_-) - G(M) + \int_{[M_- M]} G(s)a_+(s)\ ds$$

$$+ G(M_+) - G(M) + \int_{[MM_+]} G(s)a_-(s)\ ds = 0,$$

which gives (5).

Under our assumptions of ρ, both a_+ and a_- are non-negative on the subdomain D_1: under the log-concavity assumption a_- is non-negative on $\{y \leq x\}$, and $a(x) + a(y) = a(y) - a(-x) \geq 0$ if $x + y \leq 0$.

Now, consider the smallest $y \in (-b, 0)$ such that there exists some point in $(x, y) \in D_1$ with $G(x, y) = 0$. On this point, we have

$$2G(M) = 0 \geq G(M_-) + G(M_+),$$

which gives a contradiction.

For the case where the density is log-concave increasing, we may use the same argument on the domain $\{x \geq y\}$, since we still have the solution of the wave equation symmetric under the change $(x, y) \mapsto (y, x)$. Then we extend F by symmetry around the axes $x = -b$ and $x = b$, and then by periodicity, into a function defined on $\mathbb{R} \times [-b, b]$. The same argument of integration by parts remains valid, and, by means of the symmetrization, the domain of integration (Mm_+M_-) that should be used is replaced by the same triangle Δ_M as before, as shown in Figure 2. Then we use the fact that the function a is decreasing and non-negative.

$\square$

Notice that the second case (when a is non-negative) may be reduced to the first one if we extend ρ by symmetry around b, into a log-concave function on the interval $[-b, 3b]$, symmetric around the point $x = b$. (The function ρ may not be C^2 at the point $x = b$, but this causes no problem). Then one has to apply the previous result on symmetric functions on the interval $[-b, 3b]$.

4.4 Other representations of the solutions of the wave equation

In this paragraph, we shall consider an operator $L(f) = f'' + a(x)f'$ in $I = [0, 1]$, with Neumann boundary conditions, and look at different representations of the solutions of the wave equation (4).

We shall consider the probability measure $\rho(x)dx$ in I which satisfies $\rho'/\rho = a$.

Let $F(x, y)$ be a solution of the wave equation $(L_x - L_y)F = 0$ on I^2 with Neumann boundary conditions. As before, it is easier to extend F to

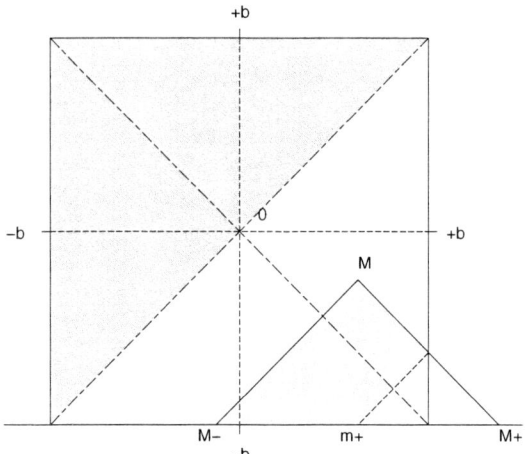

Fig. 2. Triangle Δ_M and (Mm_+M_-).

$\mathbb{R}^2$ by imposing symmetry conditions at the boundaries $x \in \mathbb{Z}$ or $y \in \mathbb{Z}$, and to extend a by imposing antisymmetry conditions on these lines (or if one prefers, symmetry conditions on ρ).

We have seen before that such an equation has an integral representation (5). Our first task shall be to change it into a new one.

As before, for $M = (X, Y)$ with $y > 0$, we denote by Δ_M the triangle delimited by the lines $\{x + y = X + Y\}$, $\{y - x = Y - X\}$ and $\{y = 0\}$. M_+ and M_- denote the edges of this triangle which lie on the line $\{y = 0\}$.

For $S \in \Delta_M$, we denote by S^{-M} the unique point U on the interval $[M_-, M]$ such that $S \in [U, U_+]$ and S^{+M} the unique point U on $[M, M_+]$ such that $S \in [U_- U]$ (see Figure 3).

Recall that

$$a_+(x, y) = \frac{1}{\sqrt{2}}\big(a(x) + a(y)\big), \ a_-(x, y) = \frac{1}{\sqrt{2}}\big(a(y) - a(x)\big),$$

and let $R(x, y) = \rho(x)\rho(y)$.

Proposition 4.1. *If a continuous function G satisfies* (5), *then the function*

$$H(x, y) = \frac{G(x, y)}{\sqrt{R(x, y)}}$$

satisfies

$$2H(M) = H(M_-) + H(M_+)$$
$$+ \int_{[M_-, M_+]} H(S)a_0(M, S) \, dS + \int_{\Delta_M} H(S)a(M, S) \, dS,$$

where

$$a(M,S) = \frac{1}{2}\sqrt{R(S)}\left(\frac{a_+(S^{-M})}{\sqrt{R(S^{-M})}}a_-(S) + \frac{a_-(S^{+M})}{\sqrt{R(S^{+M})}}a_+(S)\right),$$

and

$$a_0(M,S) = \frac{1}{2\sqrt{2}}\sqrt{R(S)}\left(\frac{a_+(S^{-M})}{\sqrt{R(S^{-M})}} + \frac{a_-(S^{+M})}{\sqrt{R(S^{+M})}}\right).$$

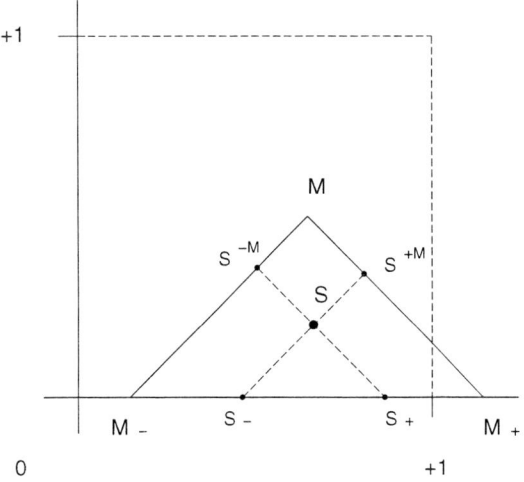

Fig. 3. S, S_+, S_-, S^{+M} et S^{-M}.

Proof. We first start by considering the function

$$\theta_M(S) = \sqrt{\frac{R(M)}{R(S)}},$$

and we notice that, for $S \in [M_-, M]$,

$$\theta_M(S) = 1 + \frac{1}{2}\int_{[S,M]} \theta_M(U)a_+(U)\,dU,$$

and similarly that, for $S \in [M, M_+]$,

$$\theta_M(S) = 1 + \frac{1}{2}\int_{[M,S]} \theta_M(U)a_-(U)\,dU.$$

Then starting from Equation (5), we replace the term

$$\int_{[M_- M]} G(S)a_+(S)\, dS$$

by

$$\int_{[M_- M]} (1 - \theta_M(S))G(S)a_+(S)\, dS + \int_{[M_- M]} \theta_M(S)G(S)a_+(S)\, dS.$$

In the last integral, replace $G(S)$ by

$$\frac{1}{2}\left(G(M_-) + G(S_+) + \int_{[M_- S]} G(U)a_+(U)\, dU + \int_{[SS_+]} G(U)a_-(U)\, dU \right).$$

Then we have

$$\int_{S \in [M_- M]} \theta_M(S)a_+(S) \left(\int_{U \in [M_- S]} G(U)a_+(U)\, dU \right) dS =$$

$$\int_{[M_- M]} G(S)a_+(S) \left(\int_{U \in [S,M]} \theta_M(U)a_+(U)\, dU \right) dS,$$

and this last expression cancels with

$$\int_{[M_- M]} (1 - \theta_M(S))G(S)a_+(S)\, dS.$$

We do the same computation on the other side $[M, M_+]$, and we collect the results. Observe that the term

$$\int_{[M_- M]} \theta_M(S)a_+(S)G(S_-)\, dS$$

gives rise to one part of the integral

$$\int_{[M_- M_+]} H(U)a_0(M, U)\, dU,$$

while the term

$$\int_{[M_- M]} \int_{[SS_+]} G(U)\theta_M(S)a_+(S)a_-(U)\, dSdU$$

produces one part of the integral

$$\int_{\Delta_M} H(U)a(M, U)\, dU. \qquad \square$$

We shall see in what follows that one may find many other integral representations of the wave equation.

One is the following

Proposition 4.2. *Let F be a solution of the wave equation* (4), *and let ψ be a smooth positive function on I^2 satisfying the Neumann boundary conditions. Then, if we set $G(x, y) = F(x, y)\psi(x, y)\rho(x)\rho(y)$, we have*

$$2G(M) = G(M_+) + G(M_-) + \int_{[M_- M]} G(s)K_+(s)\, ds$$

$$+ \int_{[MM_+]} G(s)K_-(s)\, ds - \int_{\Delta_M} G(s)\frac{(L_x - L_y)(\psi)}{\psi}\, ds,$$

where

$$K_+(x, y) = \sqrt{2}(\partial_x + \partial_y) \log\left(\psi\sqrt{\rho(x)\rho(y)}\right)$$

and

$$K_-(x, y) = \sqrt{2}(\partial_y - \partial_x) \log\left(\psi\sqrt{\rho(x)\rho(y)}\right).$$

Proof. The proof is the same as before, but we have to consider the equation satisfied by G instead of the equation satisfied by F. It is perhaps easier to make the computations under a change of variables

$$x = \frac{u + v}{\sqrt{2}}, \quad y = \frac{u - v}{\sqrt{2}},$$

in which case the operator $L_x - L_y$ becomes

$$2\partial_{uv}^2 - \frac{a_-(u, v)}{\sqrt{2}}\partial_u + \frac{a_+(u, v)}{\sqrt{2}}\partial_v.$$

We extend our functions by symmetry to the set $\{y \leq 0\}$, and then our functions become for the new variables symmetric under the symmetry $(u, v) \mapsto (v, u)$. Then the result is obtained through the integration on a square $\{u_0 \leq u \leq u_1,\ u_0 \leq v \leq u_1\}$. Since we shall not use these representations here, the details are left to the reader. □

As a consequence, if we set $U \leq V$ when $U \in \Delta_V$, and if there exists a positive function ψ satisfying the Neumann boundary conditions with $(L_x - L_y)(\psi) \leq 0$ and such that $\sqrt{\rho(x)\rho(y)}\,\psi$ is increasing for this partial order, then any continuous solution of the wave equation (4) which is non-negative on $\{y = 0\}$ is non-negative everywhere. In particular, if there is a solution of the wave equation which is increasing for this order, then the property holds.

If we are looking for the hypergroup property at the point 0 for the Neumann basis on $[0, 1]$, a good candidate for the function ψ in Proposition 4.2

seems to be

$$\psi(x, y) = 1 - \frac{f(x)f(y)}{f(1)^2},$$

where f is the (increasing) eigenvector associated with the first non-0 eigenvalue, provided that $|f(1)| \leq |f(0)|$, which is a necessary condition for the hypergroup property to hold at 0. But we were unable to derive reasonable conditions on a which would insure that for this particular case the function $\sqrt{\rho(x)\rho(y)}\psi$ is increasing for the partial order on $[0, 1]^2$.

4.5 More about the solutions of the wave equation (4)

As we saw in the previous section, there are many integral representations of the solutions of the wave equation on $[0, 1]^2$.

Most of them appear as

$$F(M) = \int F(S) \, V_0(M, dS) + \int F(S) \, V_1(M, dS),$$

where $V_0(M, dS)$ is a continuous family of bounded measures whose support is the interval $[M_-, M_+]$ on the boundary $\{y = 0\}$, and $V_1(M, dS)$ is a continuous family of bounded measures with support Δ_M.

In general, those representations lead to a unique representation

$$F(M) = \int_{[M_-, M_+]} F(S) \, W(M, dS),$$

for a continuous family of bounded measures with support in $[M_-, M_+]$. The crucial point is that in some situations the measure $W(M, dS)$ may be positive even if $V_1(M, dS)$ is not. In the case of Achour–Trimèche's theorem however, the measure W is positive only on some symmetric functions.

To understand these representations, we shall consider a more general setting.

Consider a separable compact Hausdorff space E, and two continuous families $V_0(M, ds)$ and $V_1(M, ds)$ on E (two kernels). We shall identify such a family with the operator

$$F \mapsto V_i(F)(M) = \int_E F(y) \, V_i(M, dS),$$

which maps the Banach space $C(E)$ of continuous functions into itself. The identity operator corresponds to the kernel $I(M, dS) = \delta_M(dS)$, and the composition of kernels

$$V \odot W(M, dS) = \int_U V(M, dU)W(U, dS)$$

corresponds then to the operator composition.

We set

$$\|V\| = \sup_M \int_E |V(M, dS)|,$$

which is the operator norm.

Then we have

Lemma 4.2. *Consider some continuous function $F \in C(E)$ satisfies*

$$\forall M \in E, \ F(M) = \int F(S) \, V_0(M, dS) + \int_E F(S) \, V_1(M, dS). \qquad (6)$$

If the series

$$\sum_n \|V_1^{\odot n}\| \qquad (7)$$

converges, then setting

$$\mathcal{E}(V_1) = W_1 = \sum_n V_1^{\odot n},$$

one has

$$F = W_1 \odot V_0(F).$$

In particular, if V_0 is supported by some closed subset E_0 of E, and if the condition (7) is satisfied for V_1, then there exists a unique solution F to the equation (6) given the restriction F_0 of F on E_0.

Moreover, if the kernels V_0 and V_1 are non-negative and if F_0 is non-negative, so is F.

Proof. The proof is straightforward and is just the classical representation of $(I - V_1)^{-1}$ as $\sum_n V_1^{\odot n}$. Observe moreover that if V_1 is non-negative, so is $W_1 = \mathcal{E}(V_1)$, and that the representation of the solution is then given by a non-negative kernel. □

In what follows, and to apply these lemmas, we shall consider the case where E is a compact subset of $\mathbb{R} \times [0, \infty[$, and where, for some point $M \in E$, the measure $V_0(M, dS)$ is supported by $[M_-, M_+]$, $V_1(M, dS)$ is supported by Δ_M, and has a bounded density $a(M, S)$ with respect to the Lebesgue measure on the product, in which case we write

$$F(M) = \int_{[M_-, M_+]} F(S) \, V_0(M, dS) + \int_{\Delta_M} F(S) a(M, S) \, dS. \qquad (8)$$

Then we have

Proposition 4.3. *Let κ a uniform bound on $|a(S, U)|$, $(S, U) \in \Delta_M$ in (8). Then, on Δ_M, we have*

$$\|V_1^{\odot n}\| \le \frac{\kappa^n |\Delta_M|^n}{(n!)^2},$$

where $|\Delta_M|$ denotes the area of the triangle Δ_M.

Proof. Recall the partial order $(S_1 \leq S_2) \iff S_1 \in \Delta_{S_2}$.
    Then one has

$$V_1^{\odot n}(M, dS) = \mathbb{1}_{\Delta_M}(S)a_n(M, S)dS,$$

where

$$a_n(M, S) =$$

$$\int_{S \leq S_1 \leq \cdots \leq S_{n-1} \leq M} a(M, S_{n-1})a(S_{n-1}, S_{n-2}) \ldots a(S_1, S) \, dS_1 \ldots \, dS_{n-1}.$$

It is easy to see by induction that

$$|a_n(M, S)| \leq \kappa^n \frac{|[S, M]|^n}{n!^2},$$

where $|[S, M]|$ denotes the area of the rectangle $\{U \mid S \leq U \leq M\}$.
    The conclusion follows easily from this estimate.   □

Considering the representation of the solutions of the wave equation given in Proposition 4.1, we see that any continuous solution may be represented as

$$F(M) = \int_{[M_- M_+]} F(S) \, V(M, dS)$$

where $V(M, dS)$ has two Dirac masses at the points M_- and M_+ and has a bounded density on (M_-, M_+). The smoothness of this density depends of course on the smoothness of the function a itself (and may be analyzed through the convergence of the series that we just described). For example, if a has k bounded derivative, then so has the density.
    We may observe the following.

Corollary 4.1. *Consider a solution F of Equation (6). Assume that V_0 is non-negative and supported by $E_0 \subset E$, and that $V_1 \geq V_2$, in the sense that $V_3 = V_1 - V_2$ is a non-negative kernel. Suppose that V_2 and V_3 satisfy the growth condition (7) and moreover that $\mathcal{E}(V_2) = W_2$ is non-negative.*
    Then, if the restriction F_0 of F on E_0 is non-negative, then F is non-negative everywhere on E.

Proof. Once again, this is straightforward. Setting $W_2 = (I - V_2)^{-1}$, one has

$$F = W_2 \odot V_0(F) + W_2 \odot V_3(F),$$

which is an equation of the same type, but with non-negative kernels.   □

We may apply this for example for the solutions of (8) :

Corollary 4.2. *If* $V_0(M, dS)$ *is non-negative and* $a(U, S) \geq -C$, *on* $S \leq U \leq M$, *where*

$$C = \frac{\mu_0^2}{2 |\Delta_M|}$$

and μ_0 *is the first 0 of the Bessel G function which is solution on* $(0, \infty)$ *of*

$$G'' + \frac{G'}{x} = -G, \ G(0) = 1, \ G'(0) = 0,$$

then any continuous solution of Equation (8) which is non-negative on the boundary $\{y = 0\}$ *is non-negative on* Δ_M.

Proof. It is a simple application of the previous Corollary 4.1 with $V_2(M, dS) = -C \mathbb{1}_{\Delta_M}(S) dS$.

In this case, it is not hard to see that

$$\mathcal{E}(V_2)(M, dS) = \mathbb{1}_{\Delta_M}(S) F(\|[S, M]\|) \ dS,$$

where $\|[S, M]\|$ denotes the Lebesgue measure of the rectangle $[S, M] = \{U \mid S \leq U \leq M\}$ and

$$F(x) = \sum_n \frac{(-Cx)^n}{(n!)^2}.$$

The function F is the solution of

$$xF'' + F' = -CF, \ F(0) = 1, F'(0) = -C,$$

which is related to the function G through the change of variable $x = z^2/4C$. The function G is non-negative on $[0, \mu_0)$ and this gives the result provided one observes that $\|[S, M]\| \leq \frac{|\Delta_M|}{2}$.   □

Remark. One may also observe that $1/\mu_0^2$ is the fundamental eigenvalue of the Laplace operator on the unit ball of $\mathbb{R}^2$ with the Dirichlet boundary conditions, the function $G(\|x\|/\mu_0)$ being the corresponding eigenvector.

All these considerations provide many criteria on the function a such that the associated Neumann basis on I has the hypergroup property at the left end point of the interval. In the next section, we shall deal with Gasper's theorem, where $I = [0, \pi/2]$ and $a(x) = \alpha \tan x - \beta \cot x$, with $\alpha \geq \beta > -1$. The reader should check that no one of these criteria may apply on this example. Achour–Trimèche's theorem shows that the hypergroup property holds for this example in the symmetric case $\alpha = \beta$, even on any symmetric (around $\pi/4$) subinterval of $[0, \pi/2]$. But we do not even know for the moment if the hypergroup property holds in the general case on any symmetric subinterval of $[0, \pi/2]$. (Yet it is true for small subintervals and also provided that the parameters α and β belong to some specific domains that we shall not describe here).

5 The case of Jacobi polynomials: Gasper's theorem

Gasper's theorem states the hypergroup property for the family of Jacobi polynomials. The case of Jacobi polynomials may be considered as a special case of a Sturm–Liouville basis on $[0, \pi/2]$. In this situation, both the GKS and the HGP property hold [24, 25, 26]. Actually, it is a unique situation for orthogonal polynomials, since they are the only ones, up to a linear change of variables, for which the HGP property holds (see [18, 17, 19]) (under some mild extra condition on the support of the measure which represents the product formula). In the case of symmetric Jacobi polynomials (known as Gegenbauer or ultraspherical polynomials), the HGP property may be seen as a particular example of Achour–Trimèche's theorem (although in this case the measure has a density which vanishes on the boundary). But in the general case, as we already mentioned, none of the extension we gave of Achour–Trimèche's theorem covers this result. Even worse, we do not know if the HGP property holds for any symmetric subinterval of $[0, \pi/2]$.

The Jacobi polynomials are a quite universal object, since they are basically the unique examples of a family of orthogonal polynomials which are also eigenvectors of Sturm–Liouville operators (together with their limiting cases the Hermite and Laguerre polynomials, see [36]). On the other hand, for special values of the parameters, they may be considered as eigenvectors of rank-one symmetric compact spaces (here, with our notations, it is for the parameters $(1, p)$, $(2, p)$, $(4, p)$ and (p, p), with $p \in \mathbb{N}$). But for a wider range of parameters ($p, q \in \mathbb{N}$), they may be seen as eigenvectors of a Laplace operator on a $p + q - 1$ dimensional sphere. The special case where $p = q$ is much simpler, since then one may consider a p-dimensional sphere.

In this section, after a short introduction on Jacobi polynomials and the statement of the hypergroup property for these polynomials, we present the simpler case of symmetric Jacobi polynomials, where the convolution structure has a nice geometric interpretation for $p \in \mathbb{N}$. This interpretation is for example described in [12]. For the dissymmetric case, although the Jacobi polynomials still have a simple geometric interpretation too when the parameters are integers, the convolution structure is far less obvious.

5.1 Jacobi polynomials

This polynomial family is defined for some positive parameters p and q as the family of orthogonal polynomials associated with the measure

$$\mu_{p,q}(dx) = C_{p,q}(1 - x)^{\frac{q-2}{2}}(1 + x)^{\frac{p-2}{2}}\,dx$$

on $[-1, 1]$, $C_{p,q}$ being a normalizing constant such that $\mu_{p,q}$ is a probability measure.

These polynomials are also the eigenvectors of the operator

$$L_{p,q}f(x) = (1 - x^2)f''(x) - \left(q\frac{x+1}{2} + p\frac{x-1}{2}\right)f'(x)$$

on $[-1,1]$. If $P_k^{p,q}$ is the polynomial of degree k, one has

$$L_{p,q}P_k^{p,q} = -k\left(\frac{p+q}{2} + k - 1\right)P_k^{p,q}.$$

Remark. These polynomials are traditionally parametrized by $\alpha = \frac{q-2}{2}$ and $\beta = \frac{p-2}{2}$ with $\alpha, \beta > -1$, from [39] or [24, 25, 26].

If we change $x = \cos(2\theta)$, $\theta \in [0, \frac{\pi}{2}]$, then this operator is turned into

$$L_{p,q}f(\theta) = \frac{1}{4}\left[f''(\theta) + ((q-1)\cot(\theta) - (p-1)\tan(\theta))f'(\theta)\right].$$

We see then that Jacobi polynomials is one example of a Neumann basis associated with a Sturm–Liouville operator (except that the density of the measure vanishes on the boundary points, for parameters larger than 2). We may also observe that the measure is log-concave as soon as p and q are in $[1, \infty)$.

When p and q are integers, one may see the operator $L_{p,q}$ as the action of some spherical laplacian on a quotient of the sphere.

More explicitly, we set $N = p + q$. Let us denote by $|X|$ the euclidean norm of a point X in $\mathbb{R}^N$, and let $\mathbb{S}^{N-1}$ be the unit sphere. We consider the Laplace operator $\Delta_{\mathbb{S}}$ on the unit sphere $\mathbb{S}^{N-1}$ in $\mathbb{R}^N$: this is the restriction to the sphere of the usual Laplace operator on $\mathbb{R}^N$ acting on function which are defined in a neighborhood of the sphere and do not depend on the radius of the point.

We parametrize $\mathbb{S}^{N-1}$ as

$$X = \left(\sqrt{\frac{1+x}{2}}X_1, \sqrt{\frac{1-x}{2}}X_2\right), \tag{9}$$

where $X_1 \in \mathbb{S}^{p-1}$ $X_2 \in \mathbb{S}^{q-1}$, and $x \in [-1, 1]$. The action of $\Delta_{\mathbb{S}}$ on a function which depends only on x gives again a function of x, and we have

$$\Delta_{\mathbb{S}}(h)(x) = 4L_{p,q}(h)(x).$$

We shall say that such a function on the sphere which depends only on x (that is which depends only on the norm of the projection of X onto $\mathbb{R}^p$) has the invariance $SO(p) \times SO(q)$, where the action of $SO(p) \times SO(q)$ is obtained by the action of the first component on X_1 and of the second on X_2.

The measure $\mu_{p,q}$ is the invariant measure for the operator $L_{p,q}$ and the uniform measure is the invariant measure for the Laplace operator on the sphere. This shows that $\mu_{p,q}$ is the image of the uniform measure on the

sphere (normalized as to be a probability measure) under the map $X \mapsto x$ of
Formula (9).

In fact, under this map $X \mapsto (X_1, X_2, x)$, it is straightforward to see that
the uniform measure σ_{p+q-1} on $\mathbb{S}^{p+q-1}$ is transformed into $\sigma_{p-1} \otimes \sigma_{q-1} \otimes \mu_{p,q}$.

Thanks to this remark, consider $N \geq p$ and look at the projection $\pi(X)$
from $\mathbb{S}^{N-1}$ onto the unit ball in $\mathbb{R}^p$. (That is the orthogonal projection when
the sphere is imbedded into $\mathbb{R}^N$). If we set $x = 2|\pi(X)| - 1$ and $X_1 = \frac{\pi(X)}{|\pi(X)|}$,
we see that the image measure of σ_{N-1} under $X \mapsto (x, X_1)$ is $\mu_{p,N-p} \otimes \sigma_{p-1}$.
This remark shall be used in Paragraph 5.4.

5.2 Gasper's result

Gasper proved the following product formula which gives the HGP property
for Jacobi polynomials, applying Proposition 3.2.

Theorem 5.1 (Gasper). *Let $p, q > 0$ and $-1 < x, y < 1$. Then*

- *we have the following product formula:*

$$\forall k, \quad \frac{P_k^{p,q}(x) P_k^{p,q}(y)}{P_k^{p,q}(1)} = \int P_k^{p,q}(z) \, m_{p,q}(x, y, dz)$$

  where $m_{p,q}(x, y, dz)$ is a Borel measure on $[-1; 1]$;
- *the measure $m_{p,q}$ is positive (and then is a probability measure) if and only if*

$$(p, q) \in \{q \geq p\} \cap \{p \geq 1 \text{ or } p + q \geq 4\};$$

- *moreover, if $q > p > 1$, $m_{p,q}(x, y, dz)$ is absolutely continuous with respect to $\mu_{p,q}$, with density in $L^2(\mu_{p,q})$, so that*

$$K_{p,q}(x, y, z) = \sum_k \frac{P_k^{p,q}(x) P_k^{p,q}(y) P_k^{p,q}(z)}{P_k^{p,q}(1)} \geq 0,$$

  with convergence of the sum for almost every z.

The original Gasper's proof (see [25, 26]) consisted in the explicit computation
of the sum $K_{p,q}(x, y, z)$ using formulae on special functions like Bessel's and
hypergeometric functions. There had been many other proofs of this property.
For example, Koornwinder derived it in [35] from the addition formula of
Jacobi polynomials and he found an other proof in [34], that we discuss next.

Here we restrict ourself to prove the HGP property in the symmetric case
($p = q$) and in the case when $q > p > 1$. In the latter case, we follow a proof
given by Koornwinder in [34]. However, his argument was based on an integral
representation formula of the polynomials (our Lemma 5.2), whose proof, as
we found in literature (see [3] together with [4]), relies on computational considerations on hypergeometric functions. In Section 5.4, we shall give a more
geometric interpretation of this formula, at least when p and q are integers
(it appears finally that the interpretation of Jacobi polynomials as harmonic
functions was already known – see [14, 23, 35] – but it seems that it was not
yet directly used to derive Koornwinder's representation formula).

5.3 The special case of ultraspherical polynomials ($p = q \geq 1$)

In the case when $p = q$, Jacobi polynomials are called ultraspherical polynomials. In this case, there is a much simpler representation of $L_{p,p}$ when $p \in \mathbb{N}^*$ as the action of the Laplace operator on the sphere $\mathbb{S}^p$ (and not on $\mathbb{S}^{2p-1}$ as before).

When p is a positive integer, then the hypergroup property has a simple geometric interpretation, and thus the property is quite easy to establish. This easily extends to the case when $p \notin \mathbb{N}$, by a simple extension of the formulae. This is what we are going to see in this paragraph.

Consider a smooth function $F : \mathbb{S}^p \mapsto \mathbb{R}$ which depends only on the first coordinate. To fix the ideas, let $F(X) = f(X \cdot e_1)$, where e_1 is the first unit vector in $\mathbb{R}^{p+1}$, and $Y \cdot X$ denotes the standard scalar product in $\mathbb{R}^{p+1}$. Then, if $\Delta_{\mathbb{S}^p}$ is the Laplace operator on $\mathbb{S}^p$, we have

$$\Delta_{\mathbb{S}^p} = L_{p,p}(f)(X \cdot e_1).$$

As before, the image measure of the uniform measure on the sphere through the map $X \mapsto x = X \cdot e_1$ is the invariant measure for $L_{p,p}$, that is $\mu_{p,p}$. Moreover, we may parametrize $\mathbb{S}^{p*} = \mathbb{S}^p \setminus \{e_1, -e_1\}$ by

$$X = \left(x, \sqrt{1 - x^2} X_1 \right), \tag{10}$$

where $x \in (-1, 1)$ is the first coordinate of the point $x \in \mathbb{S}^p \subset \mathbb{R}^{p+1}$, and $X_1 \in \mathbb{S}^{p-1}$. Through this map $\mathbb{S}^{p*} \mapsto (-1, 1) \times \mathbb{S}^{p-1}$, the image measure of σ_p is $\mu_{p,p} \otimes \sigma_{p-1}$.

From that, we see that if $P_k^{p,p}$ is the k-th ultraspherical polynomial, then $P_k^{p,p}(X.e_1)$ is an eigenvector of $\Delta_{\mathbb{S}^p}$, with eigenvalue $\lambda_k^p = -k(k + p - 1)$.

Observe that for any point Y on the sphere, $P_k^{p,p}(Y.X)$ is again an eigenvector on the sphere with the same eigenvalue. (This comes from the fact that the Laplace operator on the sphere is invariant under rotations.)

Now, if we take two points Y and Z on the sphere, $P_k^{p,p}(Y.X)$ and $P_k^{p,p}(Z.X)$ are two eigenvectors of $\Delta_{\mathbb{S}^p}$, with the same eigenvalue. Let us compute their scalar product in $L^2(\mathbb{S}^p)$

$$H(Y, Z) = \int_{\mathbb{S}^p} P_k^{p,p}(Y.X) P_k^{p,p}(Z.X) \, \sigma_p(dX).$$

Obviously, $H(Y, Z)$ is a smooth function, taking values in $[-1, 1]$, and if R is any rotation, $H(Y, Z) = H(RY, RZ)$. From this we see that $H(Y, Z) = h(Y \cdot Z)$, and we may write the function h in terms of ultraspherical polynomials

$$h = \sum_r a_r P_r^{p,p}.$$

We have

$$a_r = \int h(x)P_r(x)\,\mu_{p,p}(dx) = \int_{\mathbb{S}^p} h(Y.Z)P_r^{p,p}(Y.Z)\,\sigma_p(dZ)$$

$$= \int_{\mathbb{S}^p}\int_{\mathbb{S}^p} P_r^{p,p}(Y.Z)P_k^{p,p}(Y.X)P_k^{p,p}(Z.X)\,\sigma_p(dX)\sigma_p(dZ).$$

Using Fubini's theorem and the orthogonality of eigenvectors associated with different eigenvalues, we see that $a_r = 0$ unless $r = k$. We therefore see that

$$\int_{\mathbb{S}^p} P_k^{p,p}(Y.X)P_k^{p,p}(Z.X)\,\sigma_p(dX) = a_k P_k^{p,p}(Y.Z).$$

To compute a_k we choose $Y = Z$, from which we get, if we remember that the polynomials $P_k^{p,p}$ have norm 1 in L^2, that

$$a_k P_k^{p,p}(1) = 1.$$

Now, if we rewrite this formula for $Y = e_1$ and through the parametrization described above in (10), then we get, for $Z = (z, \sqrt{1-z^2}Z_1)$,

$$\frac{P_k^{p,p}(z)}{P_k^{p,p}(1)} = \int_{x,X_1} P_k^{p,p}(x)P_k^{p,p}(zx + \sqrt{1-z^2}\sqrt{1-x^2}X_1Z_1)\,\mu_{p,p}(dx)\sigma_{p-1}(dX),$$

while, for $k \neq l$,

$$\int_{x,X_1} P_k^{p,p}(x)P_l^{p,p}(zx + \sqrt{1-z^2}\sqrt{1-x^2}X_1Z_1)\,\mu_{p,p}(dx)\sigma_{p-1}(dX) = 0.$$

This may be rewritten as

$$\int_{x,t} P_k^{p,p}(x)P_l^{p,p}(zx + \sqrt{1-z^2}\sqrt{1-x^2}t)\,\mu_{p,p}(dx)\mu_{p-1,p-1}(dt) = \delta_{kl}\frac{P_k^{p,p}(z)}{P_k^{p,p}(1)}.$$

This last formula may be turned into an explicit representation

$$\int_{x,t} P_k^{p,p}(x)P_l^{p,p}(y)\,k_p(z,dx,dy) = \delta_{kl}\frac{P_k^{p,p}(z)}{P_k^{p,p}(1)}$$

for some probability kernel $k_p(z,dx,dy)$, which gives the hypergroup property thanks to Proposition 3.3.

When p is not an integer, since we have an explicit representation of the kernel $k_p(z,dx,dy) = K_p(x,y,z)\mu(dx)\mu(dy)$, it is a simple verification to check that the function $K_p(x,y,z)$ satisfies $L_x K_p = L_y K_p$ together with

$$\lim_{y\to 1} K_p(x,y,z)\,\mu_{p,p}(dz) = \delta_x(dz),$$

which is enough to get the HGP property at the point 1.

Moreover, the convolution associated with this hypergroup structure is quite easy to understand when p is an integer.

Let us say that a probability measure μ on the sphere $\mathbb{S}^p$ is zonal around $X \in \mathbb{S}^p$ if it is invariant under any rotation $R \in SO(p+1)$ such that $RX = X$.

Given any probability measure μ on $[-1, 1]$, and any $X \in \mathbb{S}^p$, we may lift μ into a unique probability measure $\hat{\mu}$ which is zonal around X such that the image measure of $\hat{\mu}$ under the projection $\pi(Y) = Y \cdot X$ from $\mathbb{S}^p$ onto $[-1, 1]$ is μ.

Now, let us choose $e_1 \in \mathbb{S}^p$, and consider two probability measures ν_1 and ν_2 on $[-1, 1]$. We may lift ν_1 into a probability measure $\hat{\nu}_1$ on $\mathbb{S}^p$, which is zonal around e_1. Then, we choose a random point in $\mathbb{S}^p$ according to $\hat{\nu}_1$. Then, given X, we consider the lift of ν_2 which is zonal around X and choose a random point Y according to this measure. Then, the resulting law of Y is zonal around e_1, and we project this measure into a new measure $\nu_1 * \nu_2$. It is an exercise to show that this convolution is the convolution associated with the hypergroup structure in this case.

5.4 The case of dissymmetric Jacobi polynomials ($q > p > 1$)

Although the dissymmetric Jacobi polynomials may be interpreted as eigenvectors of the Laplace operator on the sphere $\mathbb{S}^{p+q-1}$, it is far from trivial to prove the hypergroup property even in the case where p and q are integers. Nevertheless, the proof that we present below for completeness and which is due essentially to Koornwinder [34] has also some simple interpretation when p and q are integers in terms of harmonic analysis in $\mathbb{R}^{p+q}$.

Koornwinder's proof relies on two facts, given in the following Lemmas 5.1 and 5.2. In what follows, and to lighten the notations, we remove the indices p and q from the definitions of the polynomials $P_k^{p,q}$.

Lemma 5.1. *(Bateman's formula) Let $b_{k,r}$ the the coefficients such that*

$$\frac{P_k(s)}{P_k(1)} = \sum_{r=0}^{k} b_{k,r}(s+1)^r.$$

Then

$$\frac{P_k(s)P_k(t)}{P_k(1)^2} = \sum_{r=0}^{k} b_{k,r} \frac{(s+t)^r}{P_r(1)} P_r\left(\frac{1+st}{s+t}\right).$$

Lemma 5.2. *(Koornwinder's formula)*

$$\frac{P_k(x)}{P_k(1)} = \int_{[-1,1]^2} \left[\frac{2(1+x) - (1-x)(1+u)}{4} \right.$$

$$\left. + \frac{i\sqrt{2}\sqrt{1-x^2}\sqrt{1+u}\,v}{4} \right]^k \mu_{p,q-p}(du)\mu_{p-1,p-1}(dv).$$

Before going further, let us show that this implies the HGP property at the point $x_0 = 1$. In fact, we shall use the characterization of the hypergroup property given by Proposition 3.2.

For that, we replace in Bateman's formula of 5.1 the representation given by Koornwinder's formula. For this, we observe that, if $(s,t) \in (0,1)^2$, then

$$\left(\frac{1+st}{s+t}\right)^2 > 1,$$

and therefore if we set $x = (1+st)/(s+t)$, we may replace by analytic continuation $i\sqrt{1-x^2}$ by $\sqrt{x^2-1}$.

Then, if we set

$$\psi(s,t,u,v) = \frac{2(1+x) - (1-x)(1-u) + \sqrt{2}\sqrt{x^2-1}\sqrt{1+u}\,v}{4},$$

one has

$$\frac{P_k(s)P_k(t)}{P_k(1)^2} = \int \sum_{r=0}^{k} b_{k,r}[(s+t)\psi(s,t,u,v)]^k\,\mu_{p,q-p}(du)\mu_{p-1,p-1}(dv).$$

From the definition of the coefficients $b_{k,r}$, we get then

$$\frac{P_k(s)P_k(t)}{P_k(1)^2} = \int \frac{P_k\Big((s+t)\psi(s,t,u,v) - 1\Big)}{P_k(1)}\,\mu_{p,q-p}(du)\mu_{p-1,p-1}(dv).$$

If we define $m_{p,q}(s,t,dz)$ to be the image measure of $\mu_{p,q-p}(du)\mu_{p-1,p-1}(dv)$ under the map

$$(u,v) \mapsto (s+t)\psi(s,t,u,v) - 1,$$

one gets

$$\frac{P_k(s)P_k(t)}{P_k(1)} = \int P_k(z)\,m_{p,q}(s,t,dz),$$

which is the announced result.

Of course, one has to check that the image measure is indeed supported by $[-1,1]$, but this point is left to the reader.

We now give the proof of Lemmas 5.1 and 5.2. As it shall turn out, they rely on elementary considerations on the interpretations of the operator $L_{p,q}$. For the moment, we restrict ourselves to the case where p and q are positive integers, and we shall interpret those formulae in term of the Laplace operator on $\mathbb{R}^{p+q}$.

First observe that, given any function f on $[-1,1]$, we may lift this function on the sphere $\mathbb{S}^{p+q-1}$ into a function which has the $SO(p) \times SO(q)$ invariance. Namely, using the parametrization of the sphere given in (9), we set

$$F\left(\sqrt{\frac{1+x}{2}}X_1, \sqrt{\frac{1-x}{2}}X_2\right) = f(x).$$

Notice that in this formula,

$$x = |\pi_1 X|^2 - |\pi_2(X)|^2,$$

where π_1 and π_2 are the orthogonal projections on $\mathbb{R}^p$ and $\mathbb{R}^q$ when the sphere is imbedded into $\mathbb{R}^{p+q}$. Let us call $U(f)$ such a lift of a function from $[-1, 1]$ onto the sphere.

Now, if P_k is the Jacobi polynomial of degree k, the corresponding function $F_k = U(P_k)$ is an eigenvector of the Laplace operator on the sphere, and therefore the restriction to the sphere of a harmonic polynomial of degree $2k$. Therefore, if we parametrize a point Z in $\mathbb{R}^{p+q}$ by $R = |Z|^2$ and $X = \frac{Z}{|Z|}$, we may see that the function $R^k F_k(X)$ is harmonic in $\mathbb{R}^{p+q}$.

This may be seen in another way as we may write the Laplace operator in those coordinates

$$\Delta = \frac{\partial^2}{\partial_R^2} + \frac{N}{2R}\frac{\partial}{\partial R} + \frac{1}{4R^2}\Delta_{\mathbb{S}},$$

where $N = p + q$ and $\Delta_{\mathbb{S}}$ is the Laplace operator on $\mathbb{S}^{N-1}$. (It does not look as usual because of the change of $r = |x|$ into $r^2 = R$.) Since

$$\Delta_{\mathbb{S}} F_k = 4U\left(L_{p,q} P_k\right) = -4k\left(k + \frac{N}{2} - 1\right)F_k,$$

one may check directly that $H(R, X) = R^k F_k(X)$ is a solution of $\Delta H = 0$.

In other words, the solutions of

$$\left(\frac{\partial^2}{\partial_R^2} + \frac{N}{2R}\frac{\partial}{\partial R} + \frac{1}{R^2}L_{p,q}\right)F = 0$$

correspond to harmonic functions in $\mathbb{R}^p \times \mathbb{R}^q$ which are radial in both components (bi-radial harmonic functions). If (X, Y) are the two component of a point in $\mathbb{R}^{p+q}$, then this harmonic function is

$$(|X|^2 + |Y|^2)^k P_k\left(\frac{|X|^2 - |Y|^2}{|X|^2 + |Y|^2}\right).$$

Proof. (Of Bateman's formula 5.1.)

Let $L = L_{p,q}$. The function $K(s, t) = P_k(s)P_k(t)$ is a solution of the wave equation

$$(L_s - L_t)K = 0.$$

In order to prove the assertion, which amounts to verify the identity of two polynomials, it is enough to check it on an open set. We shall choose to prove it on the set $\{s \in (-1, 1),\ t > 1\}$, on which the wave equation $(L_s - L_t)K = 0$ becomes an elliptic equation.

On the other hand, consider a solution $G(R,x)$ of

$$\left(\frac{\partial^2}{\partial_R^2} + \frac{N}{2R}\frac{\partial}{\partial_R} + \frac{1}{R^2}L_{p,q}\right)G = 0,$$

and perform the change of variable

$$R = s + t; \quad x = \frac{1+st}{s+t}.$$

This equation becomes $(L_s - L_t)G = 0$. (We shall leave the computation to the reader, since it is just brute calculus.)

This strange (and miraculous) change of variables may be much understood if we first operate a change of variables to reduce the leading terms to $\partial_x^2 + \partial_y^2$ in both equations, and then observe that the transformation we made is conformal in $\mathbb{R}^2$, and thus preserves the leading terms. But we could find no simple geometric transformation, even in the case where p and q are positive integers, to understand this change of a bi-Jacobi equation into a bi-radial harmonic function.

Therefore, the right-hand side of Bateman's formula is a solution of the wave equation $(L_s - L_t)F = 0$. The coefficients $b_{k,r}$ are computed in such a way that the two polynomials coincide on $t = 1$.

To see that they must coincide everywhere, it is enough to remark that if two polynomials $A(s,t)$ and $B(s,t)$ in (s,t) are solutions of the wave equation which coincide on $t = 1$, they coincide everywhere. Indeed, we may write

$$A(s,t) = \sum_{r=0}^{k} a_r P_r(s)P_r(t), \quad B(s,t) = \sum_{r=0}^{k} b_r P_k(s)P_k(t),$$

and identifying the values in $t = 1$ produces $a_r = b_r$, $r = 0, \ldots, k$.   □

We now turn to the proof of Koornwinder's formula.

We begin with a lemma. Here, we shall use for the first time that $q > p$.

Lemma 5.3. *Consider a bi-radial analytic function H on $\mathbb{R}^{p+q}$, that is to say $H(X,Y) = h\left(|X|^2, |Y|^2\right)$ where h is an analytic function on $\mathbb{R}^2$. If moreover H is harmonic on $\mathbb{R}^{p+q}$, then it holds*

$$H(X,Y) = \int_{SO(q)} H(X + i\pi(RY), 0)\, \nu(dR) \tag{11}$$

where $\nu(dR)$ is the Haar measure on the group $SO(q)$, and π is the orthogonal projection from $\mathbb{R}^q$ onto $\mathbb{R}^p$.

As a consequence, if $H(X,Y) = h(|X|^2, |Y|^2)$ and $g(x) = h(x,0)$, then

$$h(|X|^2, |Y|^2) =$$

$$\int_{[-1,1]^2} g\left(|X|^2 - \frac{1+u}{2}|Y|^2 + i\sqrt{2}\,|X|\,|Y|\,\sqrt{1+us}\right) d\mu_{p,q-p}(u)d\mu_{p-1,p-1}(s). \tag{12}$$

In practice, we shall just apply this lemma with polynomials functions h.

Proof. (Of Lemma 5.3.)

The proof comes from the following remark. We observe that if F is any analytic radial function in $\mathbb{R}^p$, namely $F(X) = f(|X|^2)$ where f is real analytic, then $F(X + iY)$ is a solution in $\mathbb{R}^p \times \mathbb{R}^p$ of $\Delta_X F + \Delta_Y F = 0$, that is to say that this function is harmonic in $\mathbb{R}^{2p}$. This is clear if we consider that $F(X + Y)$ is a solution of $\Delta_X F = \Delta_Y F$.

Remark here that in this analytic continuation, we consider some functions $f(|X + iY|^2)$, where

$$|X + iY|^2 = |X|^2 - |Y|^2 + 2iX \cdot Y.$$

This is not the norm of $X + iY$ considered as a point in $\mathbb{C}^p$.

Then, $F(X + i\pi(Y))$ is harmonic in $\mathbb{R}^{p+q}$, since the projection of the Laplace operator on $\mathbb{R}^q$ is the Laplace operator on $\mathbb{R}^p$. Hence, for any element $R \in SO(q)$, $F(X + i\pi(RY))$ is harmonic in $\mathbb{R}^{p+q}$ since the Laplace operator on $\mathbb{R}^q$ is invariant under rotations.

From this, we see that

$$\tilde{H}(X, Y) = \int_{SO(q)} H(X + i\pi(RY), 0) \, \nu(dR)$$

is harmonic. Observe also that it is bi-radial. It is obviously radial in Y, since we averaged using the Haar measure on $SO(q)$. To see that it is radial in X, we just observe that, if $R_1 \in SO(p)$, one has

$$H(R_1 X + i\pi(Z), 0) = H\left(X + iR_1^{-1}\pi(Z), 0\right)$$

since $H(X, 0)$ is radial. Moreover, for any $R_1 \in SO(p)$, there exists $R_2 \in SO(q)$ such that $R_1\pi(Z) = \pi(R_2 Z)$.

Now, let us remark that we will get (12) from (11) just by expliciting the latter formula. For that, we write $X = |X| e_1$, where $e_1 \in \mathbb{S}^{p-1}$ and $\pi(RY) = |Y| \sqrt{\frac{1+u}{2}} Y_1$, where $Y_1 \in \mathbb{S}^{p-1}$.

We know that if R is chosen according to the Haar measure on $SO(q)$, the law of $R\frac{Y}{|Y|}$ is uniform on $\mathbb{S}^q$, and therefore, writing $\pi\left(R\frac{Y}{|Y|}\right) = |Y| \sqrt{\frac{1+u}{2}} Y_1$, the law of (x, Y_1) is $\mu_{q-p,p}(dx) \otimes \sigma_{p-1}(dY_1)$, as we saw at the end of paragraph 5.1. Therefore, if we set $s = e_1 \cdot Y_1$, the law of (u, s) is $\mu_{p,q-p}(du) \otimes \mu_{p-1,p-1}(ds)$.

Then,

$$|X + i\pi(RY)|^2 = |X|^2 - \frac{1+u}{2}|Y|^2 + i\sqrt{2}|X||Y|\sqrt{1+u}e_1 \cdot Y_1,$$

and

$$\int_{SO(q)} g\left((|X + i\pi(RY)|^2)\right) \nu(dR) =$$

$$\int_{[-1,1]^2} g\left(|X|^2 - \frac{1+u}{2}|Y|^2 + i\sqrt{2}|X||Y|\sqrt{1+u}s\right) \mu_{p,q-p}(dx)\mu_{p-1,p-1}(ds).$$

To finish the proof of the first formula (11), we observe that the two members of (11) coincide on $Y = 0$. On the other hand, the explicit formulation given in (12) shows that if g is analytic, then the right-hand side in (11) is also analytic in $(|X|^2, |Y|^2)$. Indeed, if we observe that the measure $\mu_{p-1,p-1}(ds)$ is symmetric, all odd powers of $|Y||X|$ in the polynomial extension of

$$\left(|X|^2 - \frac{1+u}{2}|Y|^2 + i\sqrt{2}|X||Y|\sqrt{1+us} \right)^n$$

will disappear through integration. We are therefore left with a series in $(|X|^2, |Y|^2)$.

It remains to see that two analytic harmonic bi-radial functions which coincide on $Y = 0$ coincide everywhere. Let $h(|X|^2, |Y|^2)$ an analytic bi-radial harmonic function on $\mathbb{R}^{p+q}$. The function h is a solution of

$$\left(x\partial_x^2 + \frac{p}{2}\partial_x + y\partial_y^2 + \frac{q}{2}\partial_y \right) h = 0.$$

Then, we see that if we write the expansion

$$h(x, y) = \sum_{n,m} a_{n,m} x^n y^m,$$

one has

$$a_{n,m+1} = -a_{n+1,m} \frac{(n+1)(n+p/2)}{(m+1)(m+q/2)}.$$

This shows that as soon as one knows $(a_{n,0})$, one knows h. This completes the proof of Lemma 5.3.   □

Proof. (Of Koornwinder's formula (5.2), for p and q integers.)

In the case where p and q are non-negative integer, it turns out that it once again relies on properties of the harmonic functions in the Euclidean space.

First, we lift both members on $\mathbb{R}^{p+q}$ and then we multiply them by R^k, where $R = |X|^2 + |Y|^2$.

As we have seen before, the function

$$(|X|^2 + |Y|^2)^k P_k \left(\frac{|X|^2 - |Y|^2}{|X|^2 + |Y|^2} \right)$$

is a bi-radial harmonic function, which is a polynomial in $|X|^2$ and $|Y|^2$.

It remains to apply Lemma 5.3 to conclude the proof.   □

If we want to extend the proof of Koornwinder's formula when p and q are no longer integers, then we just have to observe that, setting $S = |X|^2$ and $T = |Y|^2$, we used the fact that the fact that

$$H(S, T) = (S + T)^k P_k \left(\frac{S - T}{S + T} \right)$$

is a solution on $(0,\infty)^2$ of

$$\left(S\partial_S^2 + \frac{p}{2}\partial_S + T\partial_T^2 + \frac{q}{2}\partial_T\right)H = 0,$$

and that for any analytic function f, the function

$$K(S,T) = \int f\left(S - \frac{1+u}{2}T + i\sqrt{2}\sqrt{ST}\sqrt{1+us}\right)\,\mu_{p,q-p}(dx)\mu_{p-1,p-1}(ds)$$

is also a solution of the same equation (but this time, one has to compute that by brute force!).

Remarks. 1. The proof of Koornwinder's formula gives a representation, for analytic functions, of solutions $H(S,T)$ of $LH = 0$, where

$$L = S\partial_S^2 + \frac{p}{2}\partial_S + T\partial_T^2 + \frac{q}{2}\partial_T,$$

in terms of the boundary values $H(S,0)$. This is some kind of Poisson formula. In such a formula, one has (at least for bounded functions)

$$H(x,y) = \mathbb{E}_{x,y}(H(X_T,Y_T)),$$

where (X_s,Y_s) is the diffusion with generator L, and T the hitting time of the boundary.

Here, at least when $q \geq 2$, the boundary is polar and the set $\{y = 0\}$ is never attained. But the representation is given here through a complex variable (and of course our functions are unbounded). So it happens "as if" the process is willing to hit the boundary, provided one allows complex values (we do not know which is the meaning of that, of course). But it is certainly worth looking for more general integral representations of this type, with complex values on polar sets.

2. Of course, when q converges to p, the measure $\mu_{p,q-p}(ds)$ converges to the Dirac mass at the point 1, and Koornwinder's formula of Lemma 5.2 gives for the ultraspherical polynomials

$$\frac{P_k^{p,p}(x)}{P_k^{p,p}(1)} = \int_{[-1,1]^2}\left[x + i\frac{\sqrt{1-x^2}}{2}v\right]^k\,\mu_{p-1,p-1}(dv). \tag{13}$$

One may check directly this formula when $p \in \mathbb{N}$, through a much simpler argument, using harmonic functions in $\mathbb{R}^{p+1}$ instead of harmonic functions in $\mathbb{R}^{2p}$. This time, one has to extend the polynomial $P_k^{p,p}$ into

$$F_k(X) = |X|^k\,P_k^{p,p}\left(\frac{X}{|X|}\cdot e_1\right),$$

where e_1 is any point of the unit sphere.

References

1. A. Achour and K. Trimèche. Opérateurs de translation généralisée associés à un opérateur différentiel singulier sur un intervalle borné. *C. R. Acad. Sci. Paris Sér. A-B*, 288(7):A399–A402, 1979.

2. C. Ané, S. Blachère, D. Chafaï, P. Fougères, I. Gentil, F. Malrieu, C. Roberto, and G. Scheffer. *Sur les inégalités de Sobolev logarithmiques*, volume 10 of *Panoramas et Synthèses*. Société Mathématique de France, Paris, 2000. With a preface by D. Bakry and M. Ledoux.

3. R. Askey. Jacobi polynomials. I. New proofs of Koornwinder's Laplace type integral representation and Bateman's bilinear sum. *SIAM J. Math. Anal.*, 5:119–124, 1974.

4. R. Askey and J. Fitch. Integral representations for Jacobi polynomials and some applications. *J. Math. Anal. Appl.*, 26:411–437, 1969.

5. D. Bakry. Transformations de Riesz pour les semi-groupes symétriques. II. Étude sous la condition $\Gamma_2 \geq 0$. In *Séminaire de probabilités, XIX, 1983/84*, volume 1123 of *Lecture Notes in Math.*, pages 145–174. Springer, Berlin, 1985.

6. D. Bakry. Étude des transformations de Riesz dans les variétés riemanniennes à courbure de Ricci minorée. In *Séminaire de Probabilités, XXI*, volume 1247 of *Lecture Notes in Math.*, pages 137–172. Springer, Berlin, 1987.

7. D. Bakry. The Riesz transforms associated with second order differential operators. In *Seminar on Stochastic Processes, 1988 (Gainesville, FL, 1988)*, volume 17 of *Progr. Probab.*, pages 1–43. Birkhäuser Boston, Boston, MA, 1989.

8. D. Bakry and M. Echerbault. Sur les inégalités GKS. In *Séminaire de Probabilités, XXX*, volume 1626 of *Lecture Notes in Math.*, pages 178–206. Springer, Berlin, 1996.

9. D. Bakry and M. Emery. Diffusions hypercontractives. In *Séminaire de probabilités, XIX, 1983/84*, pages 177–206. Springer, Berlin, 1985.

10. W. Beckner. Inequalities in Fourier analysis. *Ann. of Math. (2)*, 102(1):159–182, 1975.

11. W. Beckner. Sobolev inequalities, the Poisson semigroup, and analysis on the sphere S^n. *Proc. Nat. Acad. Sci. U.S.A.*, 89(11):4816–4819, 1992.

12. N. H. Bingham. Random walk on spheres. *Z. Wahrscheinlichkeitstheorie und Verw. Gebiete*, 22:169–192, 1972.

13. H. Bloom, W.R. et Heyer. *Harmonic analysis of probability measures on hypergroups*. Walter de Gruyter, 1995.

14. B. L. J. Braaksma and B. Meulenbeld. Jacobi polynomials as spherical harmonics. *Nederl. Akad. Wetensch. Proc. Ser. A 71=Indag. Math.*, 30:384–389, 1968.

15. H. Brézis. *Analyse fonctionnelle*. Masson, Paris, 1983. Théorie et applications.

16. H. Chebli. Opérateurs de translation généralisée et semi-groupes de convolution. In *Théorie du potentiel et analyse harmonique (Journées Soc. Math. France, Inst. Recherche Math. Avancée, Strasbourg, 1973)*, pages 35–59. Lecture Notes in Math., Vol. 404. Springer, Berlin, 1974.

17. W. C. Connett, C. Markett, and A. L. Schwartz. Jacobi polynomials and related hypergroup structures. In *Probability measures on groups, X (Oberwolfach, 1990)*, pages 45–81. Plenum, New York, 1991.

18. W. C. Connett and A. L. Schwartz. Product formulas, hypergroups, and the Jacobi polynomials. *Bull. Amer. Math. Soc. (N.S.)*, 22(1):91–96, 1990.

19. W. C. Connett and A. L. Schwartz. Subsets of **R** which support hypergroups with polynomial characters. In *Proceedings of the International Conference on Orthogonality, Moment Problems and Continued Fractions (Delft, 1994)*, volume 65, pages 73–84, 1995.

20. E. B. Davies. *Heat kernels and spectral theory*. Cambridge University Press, Cambridge, 1990.

21. C. Dellacherie and P. A. Meyer. *Probabilités et potentiel*. Hermann, Paris, 1975. Chapitres I à IV, Édition entièrement refondue, Publications de l'Institut de Mathématique de l'Université de Strasbourg, No. XV, Actualités Scientifiques et Industrielles, No. 1372.

22. P. Diaconis. *Group representations in probability and statistics*. Institute of Mathematical Statistics, Hayward, CA, 1988.

23. A. Dijksma and T. H. Koornwinder. Spherical harmonics and the product of two Jacobi polynomials. *Nederl. Akad. Wetensch. Proc. Ser. A* **74**=*Indag. Math.*, 33:191–196, 1971.

24. G. Gasper. Linearization of the product of Jacobi polynomials. *Can. J. Math.*, 22:171–175,582–593, 1970.

25. G. Gasper. Positivity and the convolution structure for Jacobi series. *Ann. of Math.*, 2(93):112–118, 1971.

26. G. Gasper. Banach algebras for Jacobi series and positivity of a kernel. *Ann. of Math.*, 2(95):261–280, 1972.

27. R.B. Griffiths. Correlation in Ising ferromagnets. *J. Math. Phys.*, 8:478–489, 1967.

28. L. Gross. Logarithmic Sobolev inequalities. *Amer. J. Math.*, 97(4):1061–1083, 1975.

29. L. Gross. Logarithmic Sobolev inequalities and contractivity properties of semigroups. In *Dirichlet forms (Varenna, 1992)*, pages 54–88. Springer, Berlin, 1993.

30. J. Hadamard. Résolution d'une question relative aux déterminants. *Bulletin des Sciences Mathématiques*, 17:240–246, 1893.

31. I. Martin Isaacs. *Character theory of finite groups*. Dover Publications Inc., New York, 1994. Corrected reprint of the 1976 original [Academic Press, New York; MR0460423 (57 #417)].

32. D.G. Kelly and S. Sherman. General Griffiths' inequality on correlation in Ising ferromagnets. *J. Math.Phys.*, 9:466–484, 1968.

33. H. Kharaghani and B. Tayfeh-Rezaie. A Hadamard matrix of order 428. *J. Combin. Des.*, 13(6):435–440, 2005.

34. T. H. Koornwinder. Jacobi polynomials. II. An analytic proof of the product formula. *SIAM J. Math. Anal.*, 5:125–137, 1974.

35. T.H. Koornwinder. The addition formula for Jacobi polynomials and spherical harmonics. *SIAM J. Appl. Math.*, 25:236–246, 1973. Lie algebras: applications and computational methods (Conf., Drexel Univ., Philadelphia, Pa., 1972).

36. O. Mazet. *Semigroupes de Markov associés à une famille de polynômes orthogonaux*. PhD thesis, Université Paul Sabatier, Jan 1998.

37. R. E. A. C. Paley. On orthogonal matrices. *Journal of Mathematics and Physics*, 12:311–320, 1933.

38. G. Scheffer. Local Poincaré inequalities in non-negative curvature and finite dimension. *J. Funct. Anal.*, 198(1):197–228, 2003.

39. G. Szegö. *Orthogonal Polynomials*. American Mathematical Society, 4th edition, 1975.

40. A. Zettl. *Sturm-Liouville theory*, volume 121 of *Mathematical Surveys and Monographs*. American Mathematical Society, Providence, RI, 2005.

A new look at 'Markovian' Wiener-Hopf theory

David Williams

Department of Mathematics, Swansea University
Singleton Park, Swansea SA2 8PP, Wales, U.K.
e-mail: dw@reynoldston.com

Summary. This paper improves on the *theory* part of Williams [12][1] in that results *assumed* there are here *proved*. The Wiener-Hopf factorization is now formulated in terms of non-negative operators, so that problems concerning domains of infinitesimal generators and the like do not arise.

Duality results are proved for 'symmetric' cases by extending two important probabilistic ideas for the Markov-chain context from Kennedy [4] and finding a way to transfer the chain arguments to general Markov processes. The whole functional-analysis picture for 'symmetric' cases becomes much more illuminating under the assumption that a certain *strict-contraction property* holds. Then Hilbert-space spectral theory meshes extraordinarily well with Probability Theory. Even for a 2×2 Markov chain, the strict-contraction property may fail. On the other hand, Williams and Marles [14] showed that the property holds in a wide variety of important cases including some where it might be expected to fail. It is hoped to throw more light on the property in a subsequent paper.

1 Introduction and Summary

Let $X = \{X(r) : r \geq 0\}$ be a nice càdlàg (right-continuous, with left limits) strong Markov process on nice state-space E (locally compact with countable base) with transition semigroup $\{P(r) : r \geq 0\}$ and resolvent $\{R(\mu) : \mu > 0\}$. The idea behind the notation is that r will stand for 'real' time, the time-parameter of X. We shall use t to denote time for one of the Y-processes to be introduced. As usual, write X_r for $X(r)$, etc., when typographically neater.

We write $\mathbb{P}^x$ for the law of X when started at X and $\mathbb{E}^x$ for the associated expectation. Thus, for a non-negative Borel function f on E,

$$(P_r f)(x) := \mathbb{E}^x f(X_r), \qquad (R_\mu f)(x) := \mathbb{E}^x \int_0^\infty \mathrm{e}^{-\mu r} f(X_r) \, \mathrm{d}r.$$

---

[1] It does not seem to help with the explicit calculations given there.

We let E_+ be the closure of a connected open subset E_+° of E. We define E_-° to be the complement of E_+ and E_- to be the closure of E_-°. We write B for the 'boundary' $E_+ \cap E_-$, and make

Harmless Assumption[2] **(Bnull):** *we assume that, almost surely, X spends only a Lebesgue-null set of times in B.*

We let V be a function on E which is strictly positive on E_+°, strictly negative on E_-°. As usual, we introduce a coffin state ∂ isolated from E, and set $X(\infty) = \partial$ and $V(\partial) := 0$. We define

$$\Psi(r) := \int_0^r |V(X_q)| dq.$$

Harmless Assumption (PsiFin): *we assume that, almost surely, Ψ_r is finite for all $r > 0$.* This assumption is discussed later.

Killing according to a multiplicative functional. Let $c > 0$. Expand the sample space of X to contain a positive random variable ξ^c with

$$\mathbb{P}^{\cdot}[\xi^c > r \mid \{X(q) : q \geq 0\}] = \exp[-c\Psi(r)].$$

Then

$$X^c(r) := \begin{cases} X(r) & \text{if } r < \xi^c, \\ \partial & \text{otherwise,} \end{cases}$$

defines a strong Markov process. With $R^c(0)$ denoting the Green's function (0-resolvent) of X^c, $|V|$ denoting the operator of multiplication by the function $|V|$ and $\mathbf{1}$ the constant function $\mathbf{1}$ on E, we have

$$\big(R^c(0)|V|\mathbf{1}\big)(x) = \mathbb{E}^x \int_0^{\xi^c} |V(X_r)| dr = \mathbb{E}^x \int_0^\infty e^{-c\Psi(r)} |V(X_r)| dr$$

$$= \mathbb{E}^x \int_0^\infty e^{-c\Psi(r)} d\Psi(r) \leq c^{-1}. \tag{1}$$

Key definitions of τ and Y processes. Define, for $r \geq 0$ and $t \geq 0$,

$$\Phi^c(r) := \int_0^r V(X^c(q)) dq = \int_0^{r \wedge \xi^c} V(X(q)) dq,$$

$$\tau_\pm^c(t) := \inf\{r : \pm\Phi^c(r) > t\},$$

$$Y_\pm^c(t) := X^c(\tau_\pm^c(t)) = X(\tau_\pm^c(t)).$$

The above definitions are made with the usual conventions that the infimum of the empty set is ∞ and that $X^c(\infty) = \partial$. We note that $Y_\pm^c(t) \in E_\pm \cup \{\partial\}$ and that $Y_\pm^c$ has lifetime

$$\zeta_\pm^c = \sup\{\pm\Phi^c(r) : r > 0\}.$$

---

[2] A harmless assumption is one which holds in every case of interest.

The Strong Markov Theorem shows that $Y_{\pm}^c := \{Y_{\pm}^c(t) : t \geq 0\}$ is a strong Markov process on $E_{\pm} \cup \{\partial\}$. We let $\{Q_{\pm}^c(t) : t \geq 0\}$ be the transition function of $Y_{\pm}^c$, and $\{S_{\pm}^c(\mu) : \mu > 0\}$ its resolvent. Hence

$$\left(Q_{\pm}^c(t)f_{\pm}\right)(z) = \mathbb{E}^z f_{\pm}(Y_{\pm}^c(t))$$

with the convention that $f_{\pm}$ denotes a bounded, or else non-negative, function on $E_{\pm}$.

J and W operators. For $x \in E_-$, $y \in E_+$, define

$$\left(J_{-+}^c f_+\right)(x) := \mathbb{E}^x f_+(Y_+^c(0)), \qquad \left(J_{+-}^c f_-\right)(y) := \mathbb{E}^y f_-(Y_-^c(0)).$$

We introduce

$$\rho_+^c := \inf\{r > \tau_-^c(0) : \Phi^c(r) > 0\},$$

so that

$$\rho_+^c = \tau_-^c(0) + \tau_+^c(0) \circ \theta(\tau_-^c(0)),$$

with the usual time-shift $\theta(\cdot)$ operators. For $y \in E_+$ define

$$\left(W_+^c f_+\right)(y) := \mathbb{E}^y\left[f_+(X^c(\rho_+^c))\right], \quad \text{so} \quad W_+^c = J_{+-}^c J_{-+}^c.$$

We define ρ_-^c and W_-^c analogously.

When $c = 0$. The introduction of killing when $c > 0$ is important for boundedness results such as inequality (1). However, our primary interest is in the case when $c = 0$ and $\xi^c = \infty$. We drop the superscript c when $c = 0$, obtaining Φ, $\tau_{\pm}$, $\rho_{\pm}$, $Y_{\pm}$, $Q_{\pm}$, $S(\mu)$, J_{-+}, J_{+-}, $W_{\pm}$, etc.. We have, for example,

$$\left(Q_{\pm}^c(t)f_{\pm}\right)(z) = \mathbb{E}^z\left[f_{\pm}(Y_{\pm}(t)); t < \zeta_{\pm}^c\right].$$

[**Remark.** One of the motivations for our definition of X^c is that we could proceed from X to Y in two stages, first a standard (Volkonskii) time-transformation with V replaced by $|V|$, and then a second stage with V replaced by $\text{sgn}(V)$. But there are a number of advantages in proceeding with the 'direct' method adopted here.]

Suppressing the boundary B. The boundary $B = E_+ \cap E_-$ does not fit tidily into our picture. So, it is best to concentrate on $E \setminus B = E_+^\circ \cup E_-^\circ$.

For any 'sensible' situation, we shall have

$$J_{-+}(x, B) := \mathbb{P}^x\{X(\tau_+(0)) \in B\} = 0 \quad \text{for } x \in E_-^\circ. \tag{2}$$

(At the time of writing I haven't been able to produce any counterexample to (2) with X quasi-left-continuous. But I imagine that such counterexamples exist.)

The reason that (2) will hold for 'sensible' cases is as follows. Firstly, we have

$$\int_{s=0}^{t} f_+(Y_s)\,ds \;=\; \int_{s=0}^{t} f_+\big(X(\tau_+(s))\big)\,ds \;=\; \int_{q\in\tau_+[0,t]} f_+(X_q)V(X_q)\,dq,$$

whence

$$\left| \int_{s=0}^{t} f_+(Y_s)\,ds \right| \;\leq\; \int_{0}^{\tau_+(t)} |f_+(X_q)|\,|V(X_q)|\,dq;$$

and, on taking $f_+ = \chi_B$, we see that, almost surely, Y_+ spends zero time in B. Secondly, we have, for $x \in E_-^\circ$,

$$J_{-+}(x,B) \;=\; \mathbb{E}^x Q_+\big(\Psi(H_B); X(H_B), B\big),$$

where $H_B := \inf\{r : X_r \in B\}$. In all sensible examples, the law of $\Psi(H_B)$ conditional on $X(H_B)$ will have density with respect to Lebesgue measure, and so property (2) will follow.

Harmless Assumption (JBnull): *we proceed under the assumption that property (2), and the corresponding property with '+' and '−' interchanged, hold.*

Important Note. Of course, the set $\{t : Y_+(t) \in B\}$ may well be an *uncountable* set of measure 0. There is a sense in which, though this set plays a key rôle in the Probability, it cannot be seen by the Analysis.

Indeed, the somewhat cumbersome conventions now to be made are intended to keep in the reader's mind that the Analysis of operators cannot see the set B. When local times at B feature in Φ (see Stroock and Williams [9,10] and Williams and Andrews [13] for simple examples), we may well have to partition E as $E_+^\circ \cup E_-^\circ \cup B_+ \cup B_-$.

Conventions. For a nice function f on E, we write $f_+, f_-, f_+^\circ, f_-^\circ$ for the restrictions of f to $E_+, E_-, E_+^\circ, E_-^\circ$. But we also use f_+ [respectively, $f_-, f_+^\circ, f_-^\circ$] for a nice function on E_+ [respectively, $E_-, E_+^\circ, E_-^\circ$] when there is no prior 'f on E'. A function f_+° (for example) is extended if necessary to be 0 on E_-. We write $I_+, I_-, I_+^\circ, I_-^\circ$ for the identity operators acting on functions on $E_+, E_-, E_+^\circ, E_-^\circ$. We shall henceforth regard J_{-+}^c and W_+^c as operating on functions on E_+°, and J_{+-}^c and W_-^c as operating on functions on E_-°.

Towards the Wiener-Hopf (W-H) factorization. The first step towards thinking about the W-H factorization is to realize that (as is proved later)

$$R^c(0)|V|f \;=\; \begin{pmatrix} I_+^\circ & J_{+-}^c \\ J_{-+}^c & I_-^\circ \end{pmatrix} \begin{pmatrix} \Lambda_+^c & 0 \\ 0 & \Lambda_-^c \end{pmatrix} \begin{pmatrix} I_+^\circ & J_{+-}^c \\ J_{-+}^c & I_-^\circ \end{pmatrix} \begin{pmatrix} f_+^\circ \\ f_-^\circ \end{pmatrix} \quad \text{on } E \setminus B, \quad (3)$$

where, for $z \in E_{\pm}^{\circ}$,

$$(\Lambda_{\pm}^c f_{\pm}^{\circ})(z) := \mathbb{E}^z \int_0^{\xi^c} \chi_{\{\pm\Phi_r > 0\}} f_{\pm}^{\circ}(X_r) |V(X_r)| \, \mathrm{d}r. \tag{4}$$

The relation of (3) to the more traditional form of Wiener-Hopf factorization will be explained in a moment.

The great advantage of formulation (3) is that it involves only positive operators, so questions regarding domains do not arise. The Λ operators (though not the factorization (3)) were introduced in Joanne Kennedy's fine paper [4] on the Markov-chain case. In the context in which she worked, 'killing via Ψ' was unnecessary.

A formula for the Λ operators. For the chain case, Kennedy proved that

$$\Lambda_{\pm}^c = S_{\pm}^c(0)(I_{\pm}^{\circ} - W_{\pm}^c)^{-1} := S_{\pm}^c(0) \sum_{n=0}^{\infty} (W_{\pm}^c)^n, \tag{5}$$

where $I_{\pm}^{\circ}$ is the identity operator acting on functions on $E_{\pm}^{\circ}$. Note that, for now, we define the inverse as the Neumann sum. Kennedy also gave interesting results for the '(Markov-chain) W-H theory with noise' which had been introduced by Kennedy and Williams [5]. Chris Rogers [7], later but independently, used related ideas to obtain the (curious) time-reversal formula for the 'chain case with noise'. In this paper, we have to use somewhat different arguments to convince ourselves of the truth of (5); and we derive the result (3).

Formulae (3) and (5) constitute the Wiener-Hopf factorization.

The relation to the more usual formulation. Suppose for the moment that X is a Markov chain on a finite set E. Let $G^c, \tilde{G}_+^c$ and $\tilde{G}_-^c$ denote the generators of X^c, Y_+^c and Y_-^c. The usual form of the Wiener-Hopf factorization, going back to Barlow, Rogers and Williams [1], is that

$$V^{-1}G^c = \begin{pmatrix} I_+^{\circ} & J_{+-}^c \\ J_{-+}^c & I_-^{\circ} \end{pmatrix} \begin{pmatrix} \tilde{G}_+^c & 0 \\ 0 & -\tilde{G}_-^c \end{pmatrix} \begin{pmatrix} I_+^{\circ} & J_{+-}^c \\ J_{-+}^c & I_-^{\circ} \end{pmatrix}^{-1}. \tag{6}$$

Using the fact that

$$\begin{pmatrix} I_+^{\circ} & J_{+-}^c \\ J_{-+}^c & I_-^{\circ} \end{pmatrix} \begin{pmatrix} I_+^{\circ} & -J_{+-}^c \\ -J_{-+}^c & I_-^{\circ} \end{pmatrix} = \begin{pmatrix} (I_+^{\circ} - W_+^c) & 0 \\ 0 & (I_-^{\circ} - W_-^c) \end{pmatrix}$$
$$= \begin{pmatrix} I_+^{\circ} & -J_{+-}^c \\ -J_{-+}^c & I_-^{\circ} \end{pmatrix} \begin{pmatrix} I_+^{\circ} & J_{+-}^c \\ J_{-+}^c & I_-^{\circ} \end{pmatrix}, \tag{7}$$

we may derive (3) (with $\Lambda_{\pm}^c$ given by (5)) from (6) by algebraic manipulation.

But, of course, in general, (6) is plagued by difficulties of interpretation in regard to domains of generators, etc..

The 'symmetric' case

Assumption (Sym): *We now assume that the process X has symmetric positive transition density function on E:*

$$\{p(r; x, y) : r \geq 0;\ x, y \in E\}$$

with respect to some measure m on E with $m(B) = 0$. Thus, for $q, r > 0$, $\Gamma \in \mathcal{B}(E)$, $x, y \in E$,

$$\mathbb{P}^x(X_r \in \Gamma) = \int_\Gamma p(r; x, z)m(\mathrm{d}z),$$

$$p(r; x, y) = p(r; y, x) > 0;$$

$$p(q + r; x, y) = \int_E p(q; x, z)p(r; z, y)\, m(\mathrm{d}z).$$

The Analysis in which we are now interested cannot see the set B, and so cannot differentiate between $E_\pm$ and $E_\pm^\circ$, so there is no need to worry about the distinction. The process Y_+ may well spend an uncountable (but null) set of times in B, but its transition function is not aware of this.

We use $|V|m$ to denote the measure on E with

$$(|V|m)(\Gamma) = \int_\Gamma |V(z)|\, m(\mathrm{d}z).$$

We often state only the 'plus' versions of results, the 'minus' analogues then being obvious.

Harmless Assumption (VmK): *we assume that if K is a compact subset of E_+° or E_-°, then $(|V|m)(K) < \infty$.*

Harmless Assumption (Dens): *we shall assume that for $c \geq 0$, the kernel J_{-+}^c has a density which we shall write*

$$\{J_{-+}^c(x, y) : x \in E_-^\circ,\ y \in E_+^\circ\}$$

relative to the measure $(|V|m)_+^\circ$, the restriction of the measure $|V|m$ to E_+°. We make the analogous assumption in regard to J_{+-}. Thus,

$$J_{-+}^c(x, \Gamma) := \mathbb{P}^x\big(X(\tau_+^c(0)) \in \Gamma\big) = \int_\Gamma J_{-+}^c(x, z)|V(z)|\, m(\mathrm{d}z), \quad \Gamma \in \mathcal{B}(E_+^\circ).$$

Assumption (Dens) is highly plausible for most cases by the type of argument used to suggest Assumption (JBnull).

Here are the main results for the symmetric case. We concentrate on the case when $c = 0$. (Of course, the results for $c = 0$ actually imply the results for $c > 0$.) We define

$$L^2 = L^2(E, |V|m), \quad L^2_+ = L^2(E^\circ_+, (|V|m)^\circ_+), \quad L^2_- = L^2(E^\circ_-, (|V|m)^\circ_-).$$

THEOREM SymA. *The operator J_{-+} is bounded of norm at most 1 from L^2_+ to L^2_-. The operator J_{+-} is the Hilbert-adjoint of J_{-+}. Hence the operator $W_+ = J_{+-}J_{-+}$ is self-adjoint on L^2_+ of norm at most 1.*

To get a really tidy theory, we make the following strict-contraction assumption:

Assumption (SC): *we assume that $\|W_+\|_{L^2_+} < 1$.*

Notes explaining cases where this assumption is known to hold are given towards the end of this section.

Introduce a new scalar product $\langle \cdot, \ \cdot \rangle^\sim_+$ on L^2_+ via

$$\langle f^\circ_+, \ g^\circ_+ \rangle^\sim_+ := \langle f^\circ_+, \ (I_+ - W_+)g^\circ_+ \rangle_{L^2_+},$$

the new scalar product being equivalent to the old because $\|W_+\|_{L^2_+} < 1$. The new inner product makes L^2_+ into a new Hilbert space $(L^2)^\sim_+$. Make the analogous '$-$' definitions.

THEOREM SymB. *The transition semigroup $\{Q_+(t) : t \geq 0\}$ of Y_+ is a strongly continuous semigroup of self-adjoint operators on $(L^2)^\sim_+$, each of norm at most 1.*

Now, standard Functional Analysis shows that we can extend the map $t \mapsto Q_+(t)$ on $(0, \infty)$ to an analytic map $z \mapsto Q_+(z)$ on $\Re z > 0$ which is extendible to a strongly continuous semigroup $\{Q_+(z) : \Re z \geq 0\}$ of bounded operators on $(L^2)^\sim_+$. The family $\{Q_+(it) : t \in \mathbb{R}\}$ is a unitary group on $(L^2)^\sim_+$.

But this is not quite the way to think of things.

The $F^\pm$ spaces and quadratic forms. We are now guided by Williams [12]. Notice that (compare (7))

$$\begin{pmatrix} I^\circ_+ & J_{+-} \\ J_{-+} & I^\circ_- \end{pmatrix}^{-1} = \begin{pmatrix} (I^\circ_+ - W_+)^{-1} & 0 \\ 0 & (I^\circ_- - W_-)^{-1} \end{pmatrix} \begin{pmatrix} I^\circ_+ & -J_{+-} \\ -J_{-+} & I^\circ_- \end{pmatrix},$$

$(I^\circ_\pm - W_\pm)$ having bounded inverse on $L^2_\pm = L^2(E^\pm, (|V|m)^\circ_\pm)$ because, by Assumption (SC), each $W_\pm$ has norm strictly less than 1. We have a direct-sum decomposition

$$L^2 := L^2(E, |V|m) = F^+ \oplus F^-. \tag{8}$$

where F^+ and F^- are the important spaces:

$$F^+ := \{f \in L^2 : f^\circ_- = J_{-+}f^\circ_+\}, \qquad F^- := \{f \in L^2 : f^\circ_+ = J_{+-}f^\circ_-\}.$$

Indeed, for $f \in L^2$, we define

$$\tilde{f} := \begin{pmatrix} I_+^\circ & J_{+-} \\ J_{-+} & I_-^\circ \end{pmatrix}^{-1} f, \tag{9}$$

so that

$$f = \begin{pmatrix} I_+^\circ & J_{+-} \\ J_{-+} & I_-^\circ \end{pmatrix} \tilde{f} = \begin{pmatrix} I_+^\circ \\ J_{-+} \end{pmatrix} \tilde{f}_+^\circ + \begin{pmatrix} J_{+-} \\ I_-^\circ \end{pmatrix} \tilde{f}_-^\circ,$$

an expression for f as an element of $F^+ \oplus F^-$ which is easily shown to be unique. You are reminded that the space $L^2(E, |V|m)$ cannot see the boundary B.

It is now easily checked that we have the following decomposition of a signed inner product:

$$\langle f, \ Vg \rangle_{L^2(E,m)} = \langle \tilde{f}_+^\circ, \ \tilde{g}_+^\circ \rangle_{\tilde{+}} - \langle \tilde{f}_-^\circ, \ \tilde{g}_-^\circ \rangle_{\tilde{-}}. \tag{10}$$

Formally, if $G, \tilde{G}_+$ and $\tilde{G}_-$ are the generators of X, Y_+ and Y_-, then (compare (6))

$$(V^{-1}Gg)\tilde{} = \begin{pmatrix} \tilde{G}_+ & 0 \\ 0 & -\tilde{G}_- \end{pmatrix} \tilde{g}. \tag{11}$$

And, again formally, we have the Dirichlet-form result

$$\langle f, \ Gg \rangle_{L^2(E,m)} = \langle f, \ VV^{-1}Gg \rangle_{L^2(E,m)}$$
$$= \langle \tilde{f}_+^\circ, \ \tilde{G}_+ \tilde{g}_+^\circ \rangle_{\tilde{+}} + \langle \tilde{f}_-^\circ, \ \tilde{G}_- \tilde{g}_-^\circ \rangle_{\tilde{-}}. \tag{12}$$

Of course, the 'Dirichlet forms' (after a sign change!) on the right-hand side are not relative to the standard inner product; and this raises interesting questions.

The correct, namely Ray, picture. We should think of Y_+ as a Ray process on E, which jumps from E_-° to E_+° at time $0-$ according to the J_{-+} kernel. This is because for $x \in E_-^\circ$ and f_+° non-negative Borel on E_+°, we have (by definition of J_{-+})

$$\mathbb{E}^x f_+^\circ(Y_+(0)) = (J_{-+}f_+^\circ)(x).$$

See Rogers and Williams [8] for (as non-technical an account as can be of) the general theory of Ray processes. It is, of course, perfectly possible that if $x \in B$, then $Y_+(0) = x$. See Williams and Andrews [13] and Stroock and Williams [9,10] for Ray considerations in a simple Wiener-Hopf context, though a context to which our present theory does not apply because of the local-time terms in the definition of Φ in those papers. The transition semigroup of Y_+ should therefore be thought of as the semigroup $\{T_+(t)\}$ acting on F^+ via the usual Ray prescription

$$T_+(t) \begin{pmatrix} I_+^\circ \\ J_{-+} \end{pmatrix} f_+^\circ = \begin{pmatrix} I_+^\circ \\ J_{-+} \end{pmatrix} Q_+(t)f_+^\circ.$$

The operators $T_+(t)$ $(t > 0)$ are symmetric relative to the signed inner product $\langle f, \; Vg \rangle_{L^2(E,m)}$. We can transfer the action of $\{Q_+(z) : \Re z \geq 0\}$ to an action $\{T_+(z) : \; \Re z \; \geq \; 0\}$ on F^+. Note that $T_+(z) : F^+ \to F^+$, that $T_+(0)$ is the identity on F^+, and that $\{T_+(z) : \Re z \geq 0\}$ has the semigroup property.

We initially define for $t \in (-\infty, 0)$, $T_-(t) := Q_-(-t)$ on L_-^2 and then extend to F^- via

$$T_-(t) \begin{pmatrix} J_{+-} \\ I_-^{\circ} \end{pmatrix} f_-^{\circ} \; = \; \begin{pmatrix} J_{+-} \\ I_-^{\circ} \end{pmatrix} Q_-(-t) f_-^{\circ}.$$

We extend $\{T_-(t) : t < 0\}$ to $\{T_-(z) : \Re z \leq 0\}$.

The operator $H = V^{-1}G$ and the group generated by iH. Let $\tilde{G}_\pm$ be the infinitesimal generator, in the strict Hille-Yosida sense, of the self-adjoint semigroup on $(L^2)^{\widetilde{}}_\mp$ induced by the transition semigroup of $Y_\pm$. Keep in mind the tilde convention at (9). Define the operator H on $L^2 = L^2(E, |V|m)$ via

$$\mathcal{D}(H) \; := \; \{f \in L^2 : \tilde{f}_+^{\circ} \in \mathcal{D}(\tilde{G}_+), \; \tilde{f}_-^{\circ} \in \mathcal{D}(\tilde{G}_-)\},$$

and, for $f \in \mathcal{D}(H)$, let Hf be the unique element of L^2 with

$$(Hf)\widetilde{}_+ = \tilde{G}_+ \tilde{f}_+^{\circ}, \quad (Hf)\widetilde{}_- = -\tilde{G}_- \tilde{f}_-^{\circ}.$$

In the Hille-Yosida sense,

$$\text{for } \Re z \geq 0, \exp(zH) := T_+(z) \text{ on } F^+,$$
$$\text{for } \Re z \leq 0, \exp(zH) := T_-(z) \text{ on } F^-.$$

Since $\{T_+(it) : t \in \mathbb{R}\}$ on F^+ and $\{T_-(it) : t \in \mathbb{R}\}$ are unitary groups, the direct-sum decomposition (8) allows as to define a group $\{T(it) : t \in \mathbb{R}\}$ of uniformly bounded (but not unitary) operators on $L^2 = L^2(E, |V|m)$ restricting to $\{T_\pm(it) : t \in \mathbb{R}\}$ on $F^\pm$. This group has infinitesimal generator iH, and is the essential link between the left and right half-plane pictures. The operator H, with dense domain in L^2 should be regarded as the *definition* of $V^{-1}G$, where G is the infinitesimal generator of the self-adjoint semigroup on $L^2(E, m)$ induced by the transition function of X. The appropriateness of this definition is clear from equation (11). We are interpreting G and $H = V^{-1}G$ as operators on different spaces, so we are not thinking of the usual product of operators; but that's the way it has to be. Our probabilistic definitions avoid problems arising when V^{-1} is unbounded and/or when the generator G is horrible analytically (as may well occur for the general Markov process). Everywhere, the sensible strategy is to combine Probability with Analysis. Neither is satisfactory on its own.

Heuristics. We can get a better understanding in situations where there exist bounded eigenfunctions for H. Suppose that

$$Hf \; := \; V^{-1}Gf \; = \; \lambda f$$

where $\Re\lambda < 0$ and f is bounded on E. Then $Gf = \lambda V f$, so that if $M_r = \exp(-\lambda\Phi_r)f(X_r)$, then

$$\mathrm{d}M_r = \exp(-\lambda\Phi_r)\big[-\lambda V(X_r)f(X_r)\mathrm{d}r + (Gf)(X_r)\mathrm{d}r\big] + \mathrm{d}(\text{local martingale}).$$

Hence, M stopped at time $\tau_+(t)$ is a bounded martingale, and the Optional Stopping Theorem (OST) shows that

$$\text{for } y \in E_+,\; f(y) = \mathbb{E}^y e^{-\lambda t} f(Y_t), \text{ so that } Q_+(t)f_+^\circ = e^{\lambda t} f_+^\circ.$$

(Yes, we have assumed that $\Phi_r \to -\infty$ on $\{\tau_+(t) = \infty\}$.) Applying the OST at time $\tau_+(0)$ shows that $f_-^\circ = J_{-+}f_+^\circ$, so that, formally, $f \in F^+$.

Notes on the strict-contraction property. If X is a symmetric Markov chain on a finite set E, then Assumption (SC) holds if and only if $\sum V_i m_i \neq 0$.

The following results follow from those in Williams and Marles [14], Toland and Williams [11], and Williams [12].

If X is $\mathrm{BM}(\mathbb{R}^n)$, Brownian motion on $\mathbb{R}^n$, $m = \mathrm{Leb}(\mathbb{R}^n)$, Lebesgue measure on $\mathbb{R}^n$, E_+ is a half-space, $V = 1$ on E_+° and $V = -1$ on E_-°, then $\|W_+\|_{L_+^2} \leq \frac{1}{2}$. If X is $\mathrm{BM}(\mathbb{R}^3)$, $m = \mathrm{Leb}(\mathbb{R}^3)$, E_+ (or E_-) is a ball and $V = 1$ on E_+°, $V = -1$ on E_-°, then $\|W_+\|_{L_+^2} \leq \frac{1}{2}$.

If X is $\mathrm{BM}(\mathbb{R})$, $E = [0, \infty)$ and $V = 1$ on E_+° and V is any function on E_-° such that Assumption (PsiFin) holds, then $\|W_+\|_{L_+^2} \leq \frac{1}{2}$. I recently found that this follows from a remarkably simple new identity which I can prove but cannot explain in a sensible way. Since this identity seems to throw no light on the general case, I skip it here. A related identity may be found in Williams and Marles [14].

Williams and Andrews [13] illustrates for a simple diffusion case how it is sometimes possible to use quotient spaces when (SC) fails.

An interesting case is when $a > 0$, $\beta > -1$, $X = \mathrm{BM}(\mathbb{R})$, $E_+ = [0, \infty)$ and

$$V(x) = -a^{2+\beta}|x|^\beta \text{ if } x < 0, \quad V(y) = y^\beta \text{ if } y > 0.$$

Then

$$\|W_+\|_{L_+^2} = \frac{4a}{(1+a)^2}\sin^2\frac{\pi}{4+2\beta} < 1.$$

This suggests examples where $\|W_+\|_{L_+^2} = 1$, and where quotienting will not succeed in eliminating problems.

What next? The chief problem for the symmetric case is that of deciding when Assumption (SC) holds. The non-symmetric and 'noisy' cases certainly need further study.

Acknowledgement. I thank Dan Stroock, Brian Davies and Hubert Kalf for helpful comments. I also thank the referee who rightly asked that I clarify things in an earlier version, and I trust that he/she will find the paper now much improved.

But my main debt for the existence of this paper is owed to the skill of brain surgeons Robert Redfern and Uday Ghate and the supporting team at Morriston's Neurosurgery Unit.

2 Proofs of equations (5) and (3)

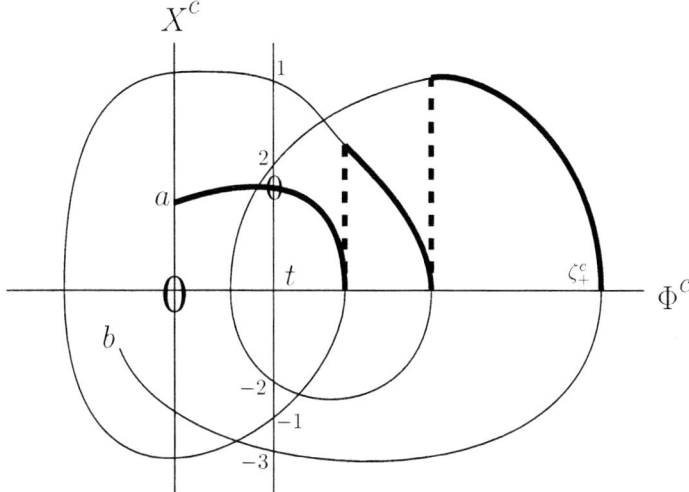

Here **we make no symmetry assumption, but we do assume (Bnull), (PsiFin) and (JBnull).** Everywhere in this section, we insist that

$$c > 0.$$

Our picture might be helpful even though it is a totally unrealistic representation of the track of the phase process (Φ^c, X^c) when X is BM($\mathbb{R}$), $E = [0, \infty)$, $V = 1$ on E_+°, $V = -1$ on E_-°. The (Φ^c, X^c) process starts at a and dies at b at time ξ^c. The 'bold' part is meant to illustrate the track of Y_+^c until it dies at its time ζ_+^c. In reality, on hitting 0, Y_+^c will make infinitely many jumps in a finite-interval: it is known that Y_+ will jump out from 0 with Lévy kernel density proportional to $y^{-3/2}$. The true path of (Φ^c, X^c) is hard to imagine, and any reasonably realistic simulation looks a mess.

We return to the general case. For almost every $t > 0$ (those positive Φ^c-values corresponding to times when $X^c \notin B$), the line $\Phi^c = t$ will cut the track in points corresponding to real times

$$\tau^{c,0}(t) := \tau_+^c(t), \quad \tau^{c,n+1}(t) := \tau^{c,n}(t) + \rho_+^c \circ \theta(\tau^{c,n}(t)) \quad (n \geq 0)$$

when X^c is in E_+° (see points $0, 1, 2$ in the picture) and at points corresponding to real times

$$\tau^{c,-n}(t) := \tau^{c,n-1}(t) + \tau_-^c(0) \circ \theta(\tau^{c,n-1}(t)) \quad (n \geq 1)$$

when X^c is in E_-° (see points $-1, -2, -3$ in the picture).

For almost all $t > 0$, it is true that, conditionally on $X^c(\tau^{c,0}(t))$ (which is $Y_+^c(t)$), the values $\{X^c(\tau^{c,n}(t)) : n \geq 0\}$ form a Markov chain on E_+° with one-step transition matrix W_+^c and hence with Green's function $(I_+^\circ - W_+^c)^{-1}$. Noting that $|\mathrm{d}\Phi_r^c| = |V(X_r^c)|\mathrm{d}r$, we therefore have

$$\mathbb{E}\left\{ \int_{r=0}^{\xi^c} f_+^\circ(X_r^c)\chi_{\{t \leq \Phi_r < t+\mathrm{d}t\}}|V(X_r^c)| \ \mathrm{d}r \ \middle| \ Y_+^c(t)\right\}$$
$$\geq \ \{(I_+^\circ - W_+^c)^{-1}f_+^\circ\}(Y_+^c(t)) \ \mathrm{d}t \quad (t < 0).$$

Hence, for $y \in E_+^\circ$, we have

$$(\Lambda_+^c f_+^\circ)(y) \ = \ \mathbb{E}^y \int_{r=0}^{\xi^c} f_+^\circ(X_r^c)\chi_{\{\Phi_r > 0\}}|V(X_r^c)| \ \mathrm{d}r$$
$$\geq \ \mathbb{E}^y \int_0^\infty \{(I_+^\circ - W_+^c)^{-1}f_+^\circ\}(Y_+^c(t)) \ \mathrm{d}t \ = \ \left\{S_+^c(0)(I_+^\circ - W_+^c)^{-1}f_+^\circ\right\}(y),$$
$$\tag{13}$$

one of two essential ideas on noiseless Wiener-Hopf factorization, proved with greater care for the case when X is a finite Markov chain, in Joanne Kennedy's paper.

Non-explosion. There is a subtle point here. Inequalities (13) and (1) tell us that $S_+^c(0)(I_+^\circ - W_+^c)^{-1}\mathbf{1}_+^\circ \leq c^{-1}\mathbf{1}_+^\circ$. This implies that for any $t > 0$, there will almost surely be only finitely many $\tau^{c,n}(t)$ $(n \geq 0)$ before time ξ^c. And now we see that

equality holds in (13).

Thus equation (5) is proved.

Clarification. For a moment, let us drop Assumption (PhiFin) that, almost surely, Ψ_t is finite for all t. But let us assume that, almost surely, Ψ_t is finite in some open t-interval. Let (using $c = 0$ in these definitions)

$$\tau^1(t) \ := \ \tau_+(t) + \rho_+ \circ \theta(\tau_+(t)), \qquad \tau^{n+1}(t) \ := \ \tau^n(t) + \rho_+ \circ \theta(\tau^n(t)).$$

Suppose that $\tau^n(t) \to \tau^\infty(t) < \infty$. We see that for any $c > 0$, we must have $\xi^c < \tau^\infty(t)$ almost surely; and the only way in which this can happen is if $\Psi(\tau^\infty(t)) = \infty$. We could replace Assumption (PhiFin) by the assumption that X dies at the explosion time of Ψ. But we shall stick to Assumption (PhiFin), any necessary modifications being obvious.

Back to the main argument. Kennedy could have carried her argument further, as follows. On thinking about $\tau^{c,-n}(t)$, we realize that

$$\mathbb{E}\left\{ f_-^\circ(X(\tau_t^{c,-n})) \ \middle| \ \mathcal{F}(\tau^{c,n-1}(t))\right\} \ = \ (J_{+-}^c f_+^\circ)\big(X(\tau^{c,n-1}(t))\big)$$

whence for $y \in E_+^\circ$,

$$\mathbb{E}^y \int_{r=0}^{\xi^c} f_-^{\circ}(X_r^c)\chi_{\{\Phi_r>0\}}|V(X_r^c)| \, \mathrm{d}r = \left\{ S_+^c(0)(I_+^{\circ} - W_+^c)^{-1}J_{+-}^c f_-^{\circ} \right\}(y).$$

(14)

Next, we realize that for $y \in E_+^{\circ}$,

$$\mathbb{E}^y \int_{r=0}^{\xi^c} f_-^{\circ}(X_r^c)\chi_{\{\Phi_r<0\}}|V(X_r^c)| \, \mathrm{d}r = \left\{ J_{+-}^c S_-^c(0)(I_-^{\circ} - W_-^c)^{-1}f_-^{\circ} \right\}(y),$$

(15)

and

$$\mathbb{E}^y \int_{r=0}^{\xi^c} f_+^{\circ}(X_r^c)\chi_{\{\Phi_r<0\}}|V(X_r^c)| \, \mathrm{d}r$$

$$= \left\{ J_{+-}^c S_-^c(0)(I_-^{\circ} - W_-^c)^{-1}J_{-+}^c f_+^{\circ} \right\}(y). \quad (16)$$

On summing equations (13), (14), (15) and (16), we obtain the 'E_+°' component of equation (3). Thus, equation (3) is proved.

3 Proofs of our symmetry results

We now also make the symmetry assumption, Assumption (Sym), and Assumptions (VmK) and (Dens). We continue to assume that

$$c > 0.$$

Proof that Λ_+^c is symmetric on L_+^2.   We now transliterate the second essential idea for noiseless Wiener-Hopf theory for chains from Joanne Kennedy's paper.

For $y, z \in E$, we define the law $\mathbb{P}_{r;y\to z}^{X\mathrm{BR}}$ of the X-bridge $X\mathrm{BR}_{r;y\to z}$ from y to z in time r via the usual formula that for $0 < r_1 < r_2 < \cdots < r_k < r$ and $y_1, y_2, \ldots, y_k \in E$,

$$\frac{\mathbb{P}_{r;y\to z}^{X\mathrm{BR}}\{X\mathrm{BR}(r_i) \in \mathrm{d}y_i : 1 \le i \le k\}}{m(\mathrm{d}y_1)m(\mathrm{d}y_2)\ldots m(\mathrm{d}y_k)}$$

$$= \frac{p(r_1; y, y_1)p(r_2 - r_1; y_1, y_2)\ldots p(r_k - r_{k-1}; y_{k-1}, y_k)p(r - r_k; y_k, z)}{p(r; y, z)}.$$

Then we have the identity in law:

$$\{X\mathrm{BR}_{r;y\to z}(r - q) : 0 \le q \le r\} \sim \{X\mathrm{BR}_{r;z\to y}(q) : 0 \le q \le r\}. \quad (17)$$

[**Remark.**   Of course, the process on the left is càglàd rather than càdlàg, but it has only countably many discontinuities.]

Now, for $y \in E_+^{\circ}$,

$$
\begin{aligned}
(\Lambda_+^c f_+^{\circ})(y) &= \mathbb{E}^y \int_0^{\infty} \mathrm{e}^{-c\Psi(r)} \chi_{\{\Phi_r > 0\}} f_+^{\circ}(X_r) |V(X_r)| \, \mathrm{d}r \\
&= \int_{r=0}^{\infty} \int_{z \in E_+^{\circ}} p(r; y, z) \mathbb{E}_{r; y \to z}^{XBR} \big\{ \mathrm{e}^{-c\Psi(r)} \chi_{\{\Phi_r > 0\}} \big\} f_+^{\circ}(z) |V(z)| \, \mathrm{d}z \mathrm{d}r.
\end{aligned}
$$

Thus, relative to the $(|V|m)_+^{\circ}$ measure, Λ_+^c has kernel density

$$
\int_{r=0}^{\infty} p(r; y, z) \mathbb{E}_{r; y \to z}^{XBR} \big\{ \mathrm{e}^{-c\Psi(r)} \chi_{\{\Phi_r > 0\}} \big\} \mathrm{d}r.
$$

However, $p(r; y, z) = p(r; z, y)$ and, by the time-reversal property (17),

$$
\mathbb{E}_{r; y \to z}^{XBR} \big\{ \mathrm{e}^{-c\Psi(r)} \chi_{\{\Phi_r > 0\}} \big\} = \mathbb{E}_{r; z \to y}^{XBR} \big\{ \mathrm{e}^{-c\Psi(r)} \chi_{\{\Phi_r > 0\}} \big\}. \tag{18}
$$

Hence

$$
\Lambda_+^c \text{ has symmetric kernel density relative to } (|V|m)_+^{\circ}. \tag{19}
$$

That's the transliterated Kennedy idea.

Application of Schur's test. Since

$$
\begin{aligned}
(\Lambda_+^c 1_+^{\circ})(y) &= \mathbb{E}^y \int_0^{\infty} \mathrm{e}^{-c\Psi(r)} \chi_{\{\Phi_r > 0\}} |V(X_r)| \, \mathrm{d}r \\
&\leq \mathbb{E}^y \int_0^{\infty} \mathrm{e}^{-c\Psi(r)} \mathrm{d}\Psi_r \leq c^{-1}, \tag{20}
\end{aligned}
$$

it follows from Schur's test (see, for example, Toland and Williams [11]) that

$$
\Lambda_+^c \text{ is self-adjoint on } L_+^2 = L^2(E_+^{\circ}, (|V|m)_+^{\circ}) \text{ with norm at most } c^{-1}. \tag{21}
$$

Proof of Theorem SymA. In her work on the chain case, Kennedy *assumed* that W_+^c is self-adjoint on L_+^2, something known for that chain case from London, McKean, Rogers and Williams [6]. (Williams [12] corrects some silly typing errors (mine!) there.) But we can *prove* the self-adjointness of W_+^c beginning with arguments similar to those above.

We have to be very careful about finiteness, and must work always with integrals of positive functions. We wish to use 'adjoints' for the moment without any Hilbert-space connotations. An operator A_+ acting on non-negative Borel functions on E_+° will be called $(|V|m)_+^{\circ}$ *symmetric* if

$$
(A_+ f_+^{\circ})(y) = \int_{E_+^{\circ}} A_+(y, z) f_+^{\circ}(z) |V(z)| \, m(\mathrm{d}z) \qquad (y \in E_+^{\circ}),
$$

where $A_+(\cdot, \cdot)$ is Borel and $A_+(y, z) = A_+(z, y) \geq 0$. If C_{-+} is a map from non-negative Borel functions on E_+° to non-negative Borel functions on E_-°

defined via

$$(C_{-+}f_+^\circ)(x) = \int_{E_+^\circ} C_{-+}(x,y)f_+^\circ(y)|V(y)|\, m(\mathrm{d}y),$$

with $C_{-+}(\cdot,\cdot)$ Borel and non-negative, we shall define for a non-negative function g_-° on E_-°,

$$((C_{-+})^* g_-^\circ)(y) := \int_{E_-^\circ} |V(x)|m(\mathrm{d}x)g_-^\circ(x)C_{-+}(x,y).$$

In an obvious sense, then, in terms of kernel densities relative to $|V|m$ measures, $(C_{-+})^*(y,x) = C_{-+}(x,y)$. Then Fubini's Theorem gives

$$\begin{aligned}
\langle f_+^\circ,\ (C_{-+})^* g_-^\circ \rangle_{(|V|m)_+^\circ} & \\
:= \int_{E_+^\circ} ((C_{-+})^* g_-^\circ)(y)f_+^\circ(y)|V(y)|\, m(\mathrm{d}y) & \\
= \int_{E^-} (C_{-+}f_+^\circ)(x)g_-^\circ(x)|V(x)|\, m(\mathrm{d}x) & \\
=: \langle C_{-+}f_+^\circ,\ g_-^\circ \rangle_{(|V|m)_-^\circ} & \\
\leq \infty. &
\end{aligned}$$

The symmetry of Λ_+^c means that

$$\Lambda_+^c \;=\; S_+^c(0)(I - W_+^c)^{-1} \;=\; (I - (W_+^c)^*)^{-1}S_+^c(0)^*. \qquad (22)$$

From equation (14) we see that, on taking kernel densities relative to $(|V|m)_-^\circ$,

$$\{\Lambda_+^c J_{+-}^c\}(y,x) = \int_0^\infty p(r;y,x)\mathbb{E}_{r;y\to x}^{\mathrm{XBR}}\{\mathrm{e}^{-c\Psi(r)}\chi_{\{\Phi_r > 0\}}\}\, \mathrm{d}r.$$

From the version of (15) with '+' and '−' interchanged, we see that on taking kernel densities relative to $(|V|m)_+^\circ$,

$$\{J_{-+}^c \Lambda_+^c\}(x,y) = \int_0^\infty \mathrm{e}^{-cr}p(r;x,y)\mathbb{E}_{r;x\to y}^{\mathrm{XBR}}\{\mathrm{e}^{-c\Psi(r)}\chi_{\{\Phi_r > 0\}}\}\, \mathrm{d}r.$$

Hence, by the time-reversal argument used before,

$$\Lambda_+^c J_{+-}^c \;=\; \Lambda_+^c (J_{-+}^c)^*.$$

From (22), we therefore have for a bounded non-negative function f_-° on E_-°,

$$\sum_{n=0}^\infty \left[(W_+^c)^*\right]^n S_+^c(0)^* J_{+-}^c f_-^\circ \;=\; \sum_{n=0}^\infty \left[(W_+^c)^*\right]^n S_+^c(0)^* (J_{-+}^c)^* f_-^\circ. \qquad (23)$$

each side of the above equation is $\Lambda_+^c J_{+-}^c f_-^\circ$ and is therefore bounded on E_+° because of (20) (and the fact that the J operators map bounded functions to bounded functions). By Fubini's Theorem, we may multiply both sides of (23) by $(W_+^c)^*$, and if we subtract this new equation from the original, we obtain

$$ S_+^c(0)^* J_{+-}^c \;=\; S_+^c(0)^*(J_{-+}^c)^*, \quad \text{so } (J_{+-}^c)^* S_+^c(0) \;=\; J_{-+}^c S_+^c(0). $$

Operating with the two sides of the last equation on $\mu S_+^c(\mu) f_+^\circ$, where f_+° is a non-negative element of $C_K(E_+^\circ)$, and subtracting the result from the original, we (rigorously) obtain from the resolvent equation that

$$ (J_{+-}^c)^* S_+^c(\mu) f_+^\circ \;=\; J_{-+}^c S_+^c(\mu) f_+^\circ. \tag{24} $$

But, since $f_+^\circ \in C_K(E_+^\circ)$ and Y is right-continuous (and spends a null set of times in B so that the value of $f_+^\circ(0)$ is irrelevant), we have as $\mu \to \infty$, for $x \in E_-^\circ$,

$$ \{J_{-+}^c \mu S_+^c(\mu) f_+^\circ\}(x) \;=\; \mu \mathbb{E}^x \int_0^\infty e^{-\mu t} f_+^\circ(Y_+^c(t))\,dt \tag{25} $$

$$ =\; \mathbb{E}^x \int_0^\infty e^{-t} f_+^\circ(Y_+^c(t/\mu))\,dt \;\longrightarrow\; \mathbb{E}^x f_+^\circ(Y_+^c(0)) \;=\; (J_{-+}^c f_+^\circ)(x). $$

On multiplying equation (24) by μ, letting $\mu \to \infty$, and using Fatou's Lemma, we therefore find from (24) that

$$ (J_{+-}^c)^* f_+^\circ \le J_{-+}^c f_+^\circ. $$

Thus, if $f_+^\circ \in C_K(E_+^\circ)$, $g_-^\circ \in C_K(E_-^\circ)$ and f_+° and g_-° are non-negative, we have with each 'inner product' relative to the appropriate one of $(|V|m)_-$ and $(|V|m)_+$,

$$ \langle J_{-+}^c f_+^\circ,\ g_-^\circ \rangle \ge \langle (J_{+-}^c)^* f_+^\circ,\ g_-^\circ \rangle = \langle f_+^\circ,\ J_{+-}^c g_-^\circ \rangle $$
$$ \ge \langle f_+^\circ,\ (J_{-+}^c)^* g_-^\circ \rangle = \langle J_{-+}^c f_+^\circ,\ g_-^\circ \rangle, $$

so we have equality all the way. Hence, relative to the $|V|m$ restrictions,

$$ (J_{-+}^c)^* = J_{-+}^c \tag{26} $$

and W_+^c is $(|V|m)_+^\circ$ symmetric. Since also, $W_+^c 1_+^\circ \le 1_+^\circ$, Schur's test shows that

W_+^c *is a self-adjoint map of norm at most 1 on* $L_+^2 = L^2(E_+^\circ, (|V|m)_+^\circ)$.
$$ \tag{27} $$

All that remains in order to prove Theorem SymA is to 'let $c \downarrow 0$'. Let f_+° be a non-negative element of $C_K(E_+^\circ)$ and g_-° a non-negative element of $C_K(E_-^\circ)$. Since

$$ \langle J_{-+}^c f_+^\circ,\ g_-^\circ \rangle_{L_-^2} \;=\; \langle f_+^\circ,\ J_{+-}^c g_-^\circ \rangle_{L_+^2}, $$

the Monotone-Convergence Theorem shows that

$$\left\langle J_{-+}f_+^\circ,\ g_-^\circ \right\rangle_{L_-^2}\ =\ \left\langle f_+^\circ,\ J_{+-}g_+^\circ \right\rangle_{L_+^2};$$

and we see that J_{-+} and J_{+-} are adjoints. Hence W_+ is symmetric on L_+^2, and Schur's test clinches that it is of norm at most 1. Each of J_{-+} and J_{+-} is therefore of norm at most 1. Since

$$\left\langle f_+^\circ,\ W_+f_+^\circ \right\rangle\ =\ \left\langle f_+^\circ,\ J_{-+}J_{+-}f_+^\circ \right\rangle\ =\ \left\langle J_{-+}f_+^\circ,\ J_{-+}f_+^\circ \right\rangle\ \geq\ 0,$$

W_+ is non-negative definite.

In the next section, 'letting $c \downarrow 0$' becomes more interesting.

4 The transition semigroup of Y_+

We now make the strict-contraction Assumption (SC) that

$$\|W_+\|_{L_+^2} < 1.$$

All the more is it true that $\|W_+^c\|_{L_+^2} < 1$ for $c > 0$.

We write $(L^2)_+^{\tilde{}c}$ for the space L_+^2 with the inner product

$$\left\langle f_+^\circ,\ g_+^\circ \right\rangle_+^{\tilde{}c}\ :=\ \left\langle f_+^\circ,\ (I_+^\circ - W_+^c)g_+^\circ \right\rangle_{L_+^2}.$$

Because of the norm equivalence which arises from the fact that both

$$(I_+^\circ - W_+^c) \text{ and } (I_+^\circ - W_+^c)^{-1}$$

are bounded even uniformly over c, there are absolute finite positive constants K_1 and K_2 such that for all $c \geq 0$,

$$K_1\|f_+^\circ\|_{L_+^2}\ \leq\ \|f_+^\circ\|_+^{\tilde{}c}\ \leq\ K_2\|f_+^\circ\|_{L_+^2}\qquad (f_+^\circ \in L_+^2)$$

and

$$K_1\|A_+\|_{L_+^2}\ \leq\ \|A_+\|_+^{\tilde{}c}\ \leq\ K_2\|A_+\|_{L_+^2}$$

where A_+ is an operator on L_+^2 and $\|A_+\|_+^{\tilde{}c}$ is its norm when interpreted as an operator on $(L^2)_+^{\tilde{}c}$.

We know that

$$\Lambda_+^c = S_+^c(0)(I_+^\circ - W_+^c)^{-1} \text{ is self-adjoint on } L_+^2 \text{ of norm at most } c^{-1}.$$

Hence $S_+^c(0) = \Lambda_+^c(I_+^\circ - W_+^c)$ and

$$\left\langle f_+^\circ,\ S_+^c(0)g_+^\circ \right\rangle_+^{\tilde{}c}\ =\ \left\langle f_+^\circ,\ (I_+^\circ - W_+^c)\Lambda_+^c(I_+^\circ - W_+^c)g_+^\circ \right\rangle_{L_+^2}$$

and it is clear that $S_+^c(0)$ is bounded self-adjoint on $(L^2)_+^{\tilde{}c}$.

Recall that $\{S_+^c(\mu) : \mu > 0\}$ is the resolvent of Y_+^c, and that $\{Q_+^c(t) : t \geq 0\}$ is its transition function on E_+°. Now, the Strong Markov Theorem applied at time $\tau_+^c(u)$ where $u > 0$ shows that, for $y \in E_+^\circ$,

$$\mathbb{E}^y \int_0^{\xi^c} \chi_{\{\Phi_r > u\}} f_+^\circ(X_r)|V(X_r)| \, dr = \{Q_+^c(u)\Lambda_+^c f_+^\circ\}(y),$$

so that, on multiplying by $\mu e^{-\mu u}$ and integrating over u from 0 to ∞, we have

$$\{\mu S_+^c(\mu)\Lambda_+^c f_+^\circ\}(y) = \mathbb{E}^y \int_0^{\xi^c} (1 - e^{-\mu \Phi_r})\chi_{\{\Phi_r > 0\}} f_+^\circ(X_r)|V(X_r)| \, dr.$$

Substituting $\Lambda_+^c = S_+^c(0)(I_+^\circ - W_+^c)^{-1}$ on the left-hand side, and using the resolvent equation, we find that

$$S_+^c(\mu)(I_+^\circ - W_+^c)^{-1} = \mathbb{E}^y \int_0^\infty e^{-c\Psi_r - \mu\Phi_r}\chi_{\{\Phi_r > 0\}} f_+^\circ(X_r)|V(X_r)| \, dr.$$

By exactly the same arguments as we used to prove results (19) and (21), we find that, for $c > 0$,

$$\text{for } \mu > 0, \ S_+^c(\mu) \text{ is bounded self-adjoint on } (L^2)_+^{\tilde{c}}. \tag{28}$$

And, of course, we know that

$$\{S_+^c(\mu) : \mu > 0\} \text{ satisfies the resolvent equation.} \tag{29}$$

Hilbert-space spectral theory now comes to our aid. We derive a number of results, assuming only (28) and (29). Really, we are using the Spectral Mapping Theorem, but we do not do so explicitly.

(a) *For $c > 0$*, spectrum$(S_+^c(\mu))_+^{\tilde{c}} \subseteq [0, \infty)$. Of course, spectrum$(\cdot)_+^{\tilde{c}}$ relates to the spectrum on $(L^2)_+^{\tilde{c}}$. This is in fact the same as the spectrum on L_+^2 because of the norm equivalence, but we wish to exploit self-adjointness.
Proof of (a). Suppose that $\mu > 0$ and that for some $\alpha > 0$, $-\alpha^{-1} \in$ spectrum$((S_+^c(\mu))_+^{\tilde{c}})$. Then there exists a sequence $(f_+^\circ(n))$ in $(L^2)_+^{\tilde{c}}$ with $\|f_+^\circ(n)\|_+^{\tilde{c}} = 1$ and $\{1 + \alpha S_+^c(\mu)\}f_+^\circ(n) \to 0$ in $(L^2)_+^{\tilde{c}}$. Since $S_+^c(\mu + \alpha)$ is bounded (we do not need to keep saying on which space),

$$S_+^c(\mu + \alpha)\{1 + \alpha S_+^c(\mu)\}f_+^\circ(n) \to 0$$

and hence by the resolvent equation $S_+^c(\mu)f_+^\circ(n) \to 0$. Hence,

$$f_+^\circ(n) = \{1 + \alpha S_+^c(\mu)\}f_+^\circ(n) - \alpha S_+^c(\mu)\}f_+^\circ(n) \longrightarrow 0,$$

a contradiction. Result (a) is proved.

(b) *For $c > 0$ and $\mu > 0$*, spectrum$(\mu S_+^c(\mu))_+^{\tilde{c}} \subseteq [0, 1]$.

Proof of (b). Suppose that $\sigma > 1$ and that $\sigma \in \mathrm{spectrum}((\mu S_+^c(\mu))_+^{\widetilde{c}})$. Then there exists a sequence $(f_+^\circ(n))$ in $(L^2)_+^{\widetilde{c}}$ with $\|f_+^\circ(n)\|_+^{\widetilde{c}} = 1$ and such that $\{1 - (\mu/\sigma)S_+^c(\mu)\}f_+^\circ(n) \to 0$ in $(L^2)_+^{\widetilde{c}}$. Now, $\mu - \mu/\sigma > 0$, and $S_+^c(\mu - \mu/\sigma)$ is bounded. So

$$S^c(\mu)f_+^\circ(n) \ = \ S_+^c\left(\mu - \frac{\mu}{\sigma}\right)\left[1 - \frac{\mu}{\sigma}S_+^c(\mu)f_+^\circ(n)\right] \ \longrightarrow \ 0;$$

and we find the same contradiction as before. Result (b) is proved.

(c) *For $c > 0$ and $\mu > 0$, $\|\mu S_+^c(\mu)\|_+^{\widetilde{c}} \leq 1$.*
Proof of (c). because norm equals spectral radius for self-adjoint operators, Result (c) is proved.

Results with $c = 0$.
(d) *For $\mu > 0$, $S_+(\mu)$ is bounded self-adjoint on $(L^2)_+^{\widetilde{\ }}$.*
Proof of (d). Fix $\mu > 0$. For a non-negative f_+° on E_+°, the Monotone-Convergence Theorem shows that

$$\|\mu S_+^c(\mu)(I_+^\circ - W_+^{c})^{-1}f_+^\circ\|_{L_+^2}^2 \ \uparrow \ \|\mu S_+(\mu)(I_+^\circ - W_+)^{-1}f_+^\circ\|_{L_+^2}^2 \ \leq \ \infty$$

as $c \downarrow 0$. Since $\|\mu S_+^c(\mu)\|_+^{\widetilde{c}} \leq 1$, uniform equivalence of norms clinches that $\mu S_+(\mu)(I_+^\circ - W_+)^{-1}$ is bounded on L_+^2, and hence $S_+(\mu)$ is bounded on $(L^2)_+^{\widetilde{\ }}$.

We have, for non-negative f_+°, g_+° on E_+°,

$$\left\langle S_+^c(\mu)(I - W_+^c)^{-1}f_+^\circ, \ g_+^\circ\right\rangle_{L_+^2} \ = \ \left\langle f_+^\circ, \ S_+^c(\mu)(I - W_+^c)^{-1}g_+^\circ\right\rangle_{L_+^2}.$$

On letting $c \downarrow 0$ and using the Monotone-Convergence Theorem, we find that $S_+(\mu)(I - W_+^c)^{-1}$ is self-adjoint on L_+^2; and we have seen before that this implies that $S_+(\mu)$ is self-adjoint on $(L^2)_+^{\widetilde{\ }}$.

(e) *For $\mu > 0$, $\|\mu S_+(\mu)\|_+^{\widetilde{\ }} \leq 1$.*
Proof of (e). Result (e) is true because we now know the analogues of (28) and (29) for $\{S_+(\mu) : \mu > 0\}$.

Completion of proof of Theorem SymB. We already know that $\{S_+(\mu) : \mu > 0\}$ is a family of non-negative-definite self-adjoint operators on $(L^2)_+^{\widetilde{\ }}$ with each $\mu S_+(\mu)$ of norm at most 1, and we know that the family satisfies the resolvent equation. To apply the Hille-Yosida Theorem, we need only prove that (see reminders below) as $\mu \to \infty$,

$$\left\langle \mu S_+(\mu)f_+^\circ, \ g_+^\circ\right\rangle \ \longrightarrow \ \left\langle f_+^\circ, \ g_+^\circ\right\rangle \text{ for } f_+^\circ, g_+^\circ \in C_K(E_+^\circ). \tag{30}$$

(Because of equivalence of norms, we can use the classical L_+^2 space.) Of course, result (30) follows from the right continuity of Y_+. Compare (25).

368   D. Williams

Some reminders. Because the $\mu S_+(\mu)$ are uniformly bounded and $C_K(E_+^\circ)$ is dense in E_+°, result (30) implies that

$$\langle \mu S_+(\mu) f_+^\circ, \ g_+^\circ \rangle \longrightarrow \langle f_+^\circ, \ g_+^\circ \rangle \text{ for } f_+^\circ, g_+^\circ \in L_+^2. \tag{31}$$

Since

$$S_+(\lambda) - S_+(\mu) + (\lambda - \mu)S_+(\mu)S_+(\lambda) = 0,$$

we have, for $f_+^\circ \in L_+^2$,

$$(\mu - \lambda)S_+(\mu)S_+(\lambda)f_+^\circ \to S_+(\lambda)f_+^\circ \quad \text{as } \mu \to \infty.$$

Hence, $G := \{g_+^\circ \in L_+^2 : \mu S_+(\mu)g_+^\circ \to g_+^\circ\}$ contains the common range $\mathcal{R} = S_+(\lambda)L_+^2$ (which is independent of λ). If $\mathcal{R}$ is not dense in L_+^2, then for some $h_+^\circ \neq 0$ in L_+^2, we shall have $\langle g_+^\circ, \ h_+^\circ \rangle = 0$ for all $h \in \mathcal{R}$. But then, by (31),

$$0 = \langle \mu S_+(\mu)h_+^\circ, \ h_+^\circ \rangle \longrightarrow \langle h_+^\circ, \ h_+^\circ \rangle,$$

a contradiction. So G is dense in L_+^2. Since the operators $\mu S_+(\mu)$ are uniformly bounded, G is closed, so that $G = L_+^2$. Hence,

$$\mu S_+(\mu)f_+^\circ \longrightarrow f_+^\circ \quad \text{for all } f_+^\circ \text{ in } (L^2)_+^{\widetilde{c}},$$

the more usual condition.

We now know that the transition semigroup $\{Q_+(t) : t \geq 0\}$ of Y_+ is a strongly continuous semigroup of self-adjoint operators each of norm at most 1. The remainder of the story in the Introduction then follows from such books as Hille and Phillips [3] and Davies [2].

References

1. Barlow, M.T., Rogers, L.C.G. and Williams, D., Wiener-Hopf factorization for matrices, *Séminaire de probabilités, XIV* (ed. J. Azéma and M. Yor), Lecture Notes in Math. **784** (Springer, Berlin, Heidelberg, New York, 1980), 324–331.
2. Davies, E. B., *One-Parameter Semigroups*, London Math. Soc. Monograph series, No. **15**, Academic Press, 1980.
3. Hille, E. and Phillips, R. S., *Functional Analysis and Semigroups, American Mathematical Society Colloquium Publications*, Volume **XXXI**, Providence, R. I., 1957.
4. Kennedy, J. E., A probabilistic view of some algebraic results in Wiener-Hopf theory for symmetrizable Markov chains. In *Stochastics and Quantum Mechanics* (edited by A. Truman and I. M. Davies) (World Scientific, 1992) 165–177.
5. Kennedy, J. E. and Williams, D., Probabilistic factorization of a quadratic matrix polynomial, *Math. Proc. Camb. Philos. Soc.*, **107**, 1990, 591–600.
6. London, R. R., McKean, H. P., Rogers, L. C. G. and Williams, D., A martingale approach to some Wiener-Hopf problems, II, *Séminaire de probabilités, XVI*, edited by J. Azéma and M. Yor, *Lecture Notes in Math.* **920**, (Springer, Berlin), 1982, 68–90.

7. Rogers, L. C. G., Time-reversal of the noisy Wiener-Hopf factorisation, *Proc. Symposia Pure Math.* **57**, (edited by M. C. Cranston and M. Pinsky) (Amer. Math. Soc., Providence RI, 1995), 129–135.

8. Rogers, L.C.G. and Williams, D., *Diffusions, Markov prcesses and martingales, Volume I: Foundations, Volume II: Itô calculus* (Cambridge University Press, 2000) (originally published by Wiley).

9. Stroock, D. W. and Williams, D., A simple PDE and Wiener-Hopf Riccati equations, *Comm. Pure Appl. Math.* **LVIII** (2005), 1116–1148.

10. Stroock, D. W. and Williams, D., Further study of a simple PDE, *Illinois J. Math.* **50** (2006), 961–989.

11. Toland, J. F. and Williams, D., On the Schur test for L^2-boundedness of positive integral operators with a Wiener-Hopf example, *J. Funct. Analysis* **160** (1998), 543–560.

12. Williams, D., Some aspects of Wiener-Hopf factorization, *Phil. Trans. Royal Soc. Lond. A* **335** (1991), 593–608.

13. Williams, D. and Andrews, S., Indefinite inner products: a simple illustrative example, *Math. Proc. Camb. Phil. Soc.* **141** (2006), 127–159.

14. Williams, D. and Marles, D. S., Surprising contraction properties of Wiener-Hopf operators, *Proc. Royal Soc. London, A* **455** (1999), 2151–2164.

Separability and completeness
for the Wasserstein distance

F. Bolley

Ecole Normale Supérieure de Lyon, Umpa (UMR 5669)
46 allée d'Italie, F-69364 Lyon cedex 7
Current address: Institut de Mathématiques, LSP (UMR C5583)
Université Paul Sabatier, Route de Narbonne, F-31062 Toulouse cedex 9
e-mail: bolley@cict.fr

Summary. We give an elementary proof that the Wasserstein distances, which play a basic role in optimal transportation issues, turn some spaces of probability measures into separable complete metric spaces.

Introduction

Let (X, d) be a separable complete metric space and $\mathcal{P}$ the set of Borel probability measures on X. Given $p > 0$, the $\mathbb{R} \cup \{+\infty\}$-valued map W_p defined on $\mathcal{P} \times \mathcal{P}$ by

$$W_p(\mu, \nu) = \inf_\pi \left(\iint_{X \times X} d(x, y)^p \, d\pi(x, y) \right)^{\frac{1}{p}} \qquad \text{if } 1 \le p$$

$$W_p(\mu, \nu) = \inf_\pi \iint_{X \times X} d(x, y)^p \, d\pi(x, y) \qquad \text{if } 0 < p < 1,$$

where π runs over the set of probability measures on $X \times X$ with marginals μ and ν, defines a metric on the subset $\mathcal{P}_p$ of measures μ in $\mathcal{P}$ such that $\int_X d(x_0, x)^p \, d\mu(x)$ be finite for some (and hence any) x_0 in X: it is called the Wasserstein distance of order p (see [1], [4], [5] or [6] for instance).

These distances are strongly linked to the theory of optimal transportation and have been widely used in various applications to partial differential equations, functional inequalities and probability theory. Some of them involve probability measures on infinite dimensional spaces such as the Wiener space of $\mathbb{R}^d$-valued continuous functions on the interval $[0, T]$ (as in [3] for instance) or some sets of probability measures on a phase space; this is a motivation to the general framework considered in this note, in which we shall prove:

Theorem. *If* (X, d) *is a separable complete metric space and* p *a positive number, then the metric space* $(\mathcal{P}_p, W_p)$ *is separable and complete.*

The completeness property is proven in [4] by comparing the W_p distances with the weaker Prohorov distance for which the property is known, and in [1] by means of a deep result by Kolmogorov. Here we shall give a more direct and elementary argument.

Let us actually note that these kind of properties can be studied as in [2] within the following broader scope of weighted spaces of probability measures.

Let (X, τ) be a topological space and ω be a real-valued continuous function on X, bounded by below by a positive constant, and let $\mathcal{P}_\omega$ denote the set of Borel probability measures μ on X such that $\int_X \omega(x) \, d\mu(x)$ be finite. We equip $\mathcal{P}_\omega$ with the natural weak topology defined by the set $\mathcal{C}_{b\omega}$ of real-valued continuous functions f on X such that $\omega^{-1} f$ be bounded on X: this topology, which will be denoted by $w\text{-}\mathcal{C}_{b\omega}$, is defined by the seminorms

$$\mu \mapsto \sup_{i=1,\ldots,n} \left| \int_X f_i(x) \, d\mu(x) \right|$$

for any finite family $f_1, \ldots, f_n$ of functions in $\mathcal{C}_{b\omega}$.

Then one can prove that if the topological space (X, τ) is separable (resp. separable and metrizable, resp. separable, metrizable and topologically complete), then so is $(\mathcal{P}_\omega, w\text{-}\mathcal{C}_{b\omega})$. Conversely if $(\mathcal{P}_\omega, w\text{-}\mathcal{C}_{b\omega})$ is separable (resp. separable, metrizable and topologically complete), then so is (X, τ) if (X, τ) is a priori metrizable.

When $\omega = 1$, that is, without weight, these properties are known. For a general ω, they can be proven either by building explicit distances on the considered spaces of probability measures, or by abstract functional methods as in [2].

In the case when (X, d) is a separable complete metric space and when $\omega = 1 + d(x_0, \cdot)^p$ for $p > 0$, then the $w\text{-}\mathcal{C}_{b\omega}$ topology on the set $\mathcal{P}_\omega = \mathcal{P}_p$ is metrized by the distance W_p: in particular $(\mathcal{P}_p, W_p)$ is separable and topologically complete.

The following two sections are devoted to a direct proof of the above theorem, which in particular ensures that $(\mathcal{P}_p, W_p)$ is complete.

1 Separability

In this section we prove that the metric space $(\mathcal{P}_p, W_p)$ is separable if (X, d) is a separable complete metric space and p is a positive number.

If $(x_n)_n$ is a sequence dense in (X, d) we actually prove that the countable set of all measures of the form $\sum_{n=1}^{N} b_n \delta_{x_n}$, where N is an integer number, the b_n's are N nonnegative rational numbers with unit sum and δ_x stands for the point mass at x, is dense in $(\mathcal{P}_p, W_p)$.

Let indeed μ be a given measure in $\mathcal{P}_p$ and ε be a given positive number.

1. We first approach μ by a measure $\mu_1 = \sum_{n=1}^{+\infty} a_n \delta_{x_n}$ where the a_n's are nonnegative real numbers with $\sum_{n=1}^{+\infty} a_n = 1$.

For this we note that X is covered by the balls $B(x_n, \varepsilon^{\max(1,1/p)})$ with centers x_n and radius $\varepsilon^{\max(1,1/p)}$, and is partitioned by the sets $\tilde{B}_n = B(x_n, \varepsilon^{\max(1,1/p)}) \setminus \bigcup_{k \le n-1} B(x_k, \varepsilon^{\max(1,1/p)})$, so that the $a_n = \mu[\tilde{B}_n]$ have unit sum. Moreover sending every point in $\tilde{B}_n$ onto x_n for each n defines a transport map between μ and $\mu_1 = \sum_{n=1}^{+\infty} a_n \delta_{x_n}$ with cost

$$\sum_{n=1}^{+\infty} \int_{\tilde{B}_n} |x - x_n|^p \, d\mu(x) \le \sum_{n=1}^{+\infty} a_n \, \varepsilon^{p \max(1,1/p)} = \varepsilon^{\max(p,1)};$$

consequently

$$W_p(\mu, \mu_1) \le \varepsilon.$$

2. Then we approach μ_1 by a measure $\mu_2 = \sum_{n=1}^{N} b_n \delta_{x_n}$ where the b_n's are nonnegative *rational* numbers with $\sum_{n=1}^{N} b_n = 1$.

First of all μ_1 belongs to $\mathcal{P}_p$ since it is at finite W_p distance from the measure μ in $\mathcal{P}_p$. Hence

$$\sum_{n=1}^{+\infty} a_n |x_n - x_1|^p = W_p(\mu_1, \delta_{x_1})^{\max(p,1)}$$

is finite since δ_{x_1} also belongs to $\mathcal{P}_p$. In particular there exists an integer N such that

$$\sum_{n=N+1}^{+\infty} a_n |x_n - x_1|^p \le \varepsilon^{\max(p,1)}.$$

For each $2 \le n \le N$ we now let b_n be a nonnegative rational number such that

$$0 \le a_n - b_n \le \varepsilon^{\max(p,1)} \left(\sum_{j=1}^{N} a_j |x_j - x_1|^p \right)^{-1} a_n,$$

and such that

$$b_1 = a_1 + \sum_{n=2}^{N}(a_n - b_n) + \sum_{n=N+1}^{+\infty} a_n$$

be rational: in particular the b_n's have unit sum. Moreover one can transport μ_1 onto $\mu_2 = \sum_{n=1}^{N} b_n \delta_{x_n}$ by keeping a b_n mass at x_n for each $n \le N$ and sending the remaining $a_n - b_n$ mass from x_n onto x_1, and sending the whole a_n mass from x_n onto x_1 for each $n \ge N+1$; the associated cost is

$$\sum_{n=1}^{N}(a_n - b_n)|x_n - x_1|^p + \sum_{n=N+1}^{+\infty} a_n|x_n - x_1|^p \le 2\,\varepsilon^{\max(p,1)},$$

so that

$$W_p(\mu_1, \mu_2) \le 2\,\varepsilon.$$

3. To sum up, we have approached μ by a measure μ_2 in $\mathcal{P}_p$, of the expected form and which is at most $3\,\varepsilon$ distant in W_p metric.

2 Completeness

In this section we prove that the metric space $(\mathcal{P}_p, W_p)$ is complete if (X, d) is a separable complete metric space and p is a positive number.

Let indeed $(\mu_n)_n$ be a Cauchy sequence in $(\mathcal{P}_p, W_p)$.

1. For $p \ge 1$ we first prove that $(\mu_n)_n$ is uniformly tight by adapting a classical proof of the Ulam lemma.

Let ε be a given positive number.

We note that $(\mu_n)_n$ is Cauchy in $(\mathcal{P}_1, W_1)$ since $W_1 \le W_p$. Hence there exists N such that $W_1(\mu_n, \mu_N) \le \varepsilon^2$ for any $n \ge N$ so that, for all n, there exists $j \le N$ such that

$$W_1(\mu_n, \mu_j) \le \varepsilon^2. \tag{1}$$

The finite family $(\mu_j)_{j \le N}$ is uniformly tight by Ulam's lemma, so there exists a compact set K such that

$$\mu_j(K) \ge 1 - \varepsilon$$

for any $j \le N$; hence one can choose q points $x_1, \ldots, x_q$ in X such that

$$\mu_j(U) \ge 1 - \varepsilon \tag{2}$$

for any $j \le N$, where $U = \bigcup_{k=1}^{q} B(x_k, \varepsilon)$.

Then let ϕ be the $\frac{1}{\varepsilon}$-Lipschitz function defined on X by

$$\phi(x) = \left(1 - \frac{d(x,U)}{\varepsilon}\right)^+.$$

Given j and n, if π is any joint measure on $X \times X$ with marginals μ_j and μ_n, then

$$\int_X \phi(x)\,d\mu_j(x) - \int_X \phi(y)\,d\mu_n(y) = \iint_{X \times X} (\phi(x) - \phi(y))\,d\pi(x,y)$$

$$\leq \frac{1}{\varepsilon} \iint_{X \times X} d(x,y)\,d\pi(x,y);$$

hence

$$\int_X \phi(x)\,d\mu_j(x) - \int_X \phi(y)\,d\mu_n(y) \leq \frac{1}{\varepsilon} W_1(\mu_j, \mu_n).$$

On the other hand $\mathbf{1}_U \leq \phi \leq \mathbf{1}_{U^\varepsilon}$ where $U^\varepsilon = \{x; d(x,U) < \varepsilon\}$, so

$$\int_X \phi(x)\,d\mu_j(x) \geq \mu_j(U) \qquad \text{and} \qquad \int_X \phi(y)\,d\mu_n(y) \leq \mu_n(U^\varepsilon).$$

Consequently

$$\mu_n(U^\varepsilon) \geq \mu_j(U) - \frac{1}{\varepsilon} W_1(\mu_j, \mu_n). \qquad (3)$$

Thus, by (1), (2) and (3), for any $\varepsilon > 0$ we have found q points $x_1, \ldots, x_q$ such that

$$\mu_n\left(X \setminus \bigcup_{k=1}^{q} B(x_k, 2\varepsilon)\right) \leq 2\varepsilon$$

for any n since $U^\varepsilon \subset \bigcup_{k=1}^{q} B(x_k, 2\varepsilon)$.

Therefore, replacing ε by $\varepsilon\,2^{-m-1}$ where m is any integer, there exist $q(m)$ points $x_1^m, \ldots, x_{q(m)}^m$ in X such that

$$\mu_n\left(X \setminus \bigcup_{k=1}^{q(m)} B(x_k^m, \varepsilon\,2^{-m})\right) \leq \varepsilon\,2^{-m}$$

for any n. In particular the set

$$S = \bigcap_{m=1}^{+\infty} \bigcup_{k=1}^{q(m)} B\left(x_k^m, \varepsilon\,2^{-m}\right)$$

is such that

$$\mu_n(X \setminus S) \le \sum_{m=1}^{+\infty} \mu_n\left(X \setminus \bigcup_{k=1}^{q(m)} B(x_k^m, \varepsilon\, 2^{-m})\right) \le \sum_{m=1}^{+\infty} \varepsilon\, 2^{-m} = \varepsilon$$

for any n.

On the other hand, for any ρ, and choosing m such that $\varepsilon\, 2^{-m} \le \rho$, the set S can be covered by the $q(m)$ balls $B(x_k^m, \varepsilon\, 2^{-m})$ with radius $\varepsilon\, 2^{-m} \le \rho$: in other words it is totally bounded, so that its closure $\overline{S}$ is compact since X is complete.

To sum up, the set $\overline{S}$ is compact and satisfies $\mu_n(X \setminus \overline{S}) \le \varepsilon$ for any n: this means that the sequence $(\mu_n)_n$ is indeed uniformly tight.

2. We deduce from step 1 that $(\mu_n)_n$ converges in $(\mathcal{P}_p, W_p)$ in the case when $p \ge 1$.

Indeed $(\mu_n)_n$ is uniformly tight by step 1, so by Prohorov's theorem there exists a subsequence $(\mu_{n'})_{n'}$ of $(\mu_n)_n$ converging to a probability measure μ on X for the narrow weak topology.

The distance $W_p(\mu, \mu_{n'})$ actually tends to 0 as n' goes to infinity. Let indeed $\pi_{n'm'}$ be a probability measure on $X \times X$ with marginals $\mu_{n'}$ and $\mu_{m'}$, optimal in the sense that

$$\iint_{X \times X} d(x, y)^p \, d\pi_{n'm'}(x, y) = W_p(\mu_{n'}, \mu_{m'})^p.$$

The sequence $(\mu_{n'})_{n'}$ is uniformly tight, hence so is $(\pi_{n'm'})_{n'}$ for given m'. Thus by Prohorov's theorem again there exists a subsequence $(\pi_{n''m'})_{n''}$ of $(\pi_{n'm'})_{n'}$ converging to a probability measure $\pi_{m'}$ on $X \times X$ for the narrow weak topology. Then by semicontinuity

$$\iint_{X \times X} d(x, y)^p \, d\pi_{m'}(x, y) \le \liminf_{n'' \to +\infty} \iint_{X \times X} d(x, y)^p \, d\pi_{n''m'}(x, y)$$
$$= \liminf_{n'' \to +\infty} W_p(\mu_{n''}, \mu_{m'})^p. \qquad (4)$$

But on one hand $\pi_{n''m'}$ has marginals $\mu_{n''}$ and $\mu_{m'}$, so at the limit (in n'') $\pi_{m'}$ has marginals μ and $\mu_{m'}$; hence

$$W_p(\mu, \mu_{m'})^p \le \iint_{X \times X} d(x, y)^p \, d\pi_{m'}(x, y) \qquad (5)$$

for any m'.

On the other hand the sequence $(\mu_{n'})_{n'}$ is Cauchy for the distance W_p, so for any $\varepsilon > 0$ and n'', m' large enough

$$W_p(\mu_{n''}, \mu_{m'}) \le \varepsilon. \qquad (6)$$

It finally follows from (4), (5) and (6) that

$$W_p(\mu, \mu_{m'}) \leq \varepsilon$$

for m' large enough, which means that μ belongs to $\mathcal{P}_p$ and that $W_p(\mu, \mu_{n'})$ indeed tends to 0 as n' goes to infinity.

Finally $W_p(\mu_n, \mu)$ tends to 0 as n goes to infinity since the whole sequence $(\mu_n)_n$ is Cauchy in $(\mathcal{P}_p, W_p)$.

3. We deduce from step 2 that $(\mu_n)_n$ converges in $(\mathcal{P}_p, W_p)$ in the case when $0 < p < 1$.

Indeed d^p is a distance on X which defines the same topology as d and (X, d^p) is complete if so is (X, d). Moreover $\mathcal{P}_p(X, d) = \mathcal{P}_1(X, d^p)$ and $W_p(X, d) = W_1(X, d^p)$ in obvious notation.

Thus, given an exponent $p \in]0, 1[$ and a metric d on X, the results associated with the exponent p and the metric d stem from the results proven in step 2 for the exponent 1 and the metric d^p.

Acknowledgements. The author thanks the referee for relevant comments which enabled to make this note more precise.

References

1. L. Ambrosio, N. Gigli, and G. Savaré. *Gradient flows in metric spaces and in the spaces of probability measures.* Birkhäuser, Basel, 2005.
2. F. Bolley. Applications du transport optimal à des problèmes de limites de champ moyen. Thèse de doctorat, Ecole Normale Supérieure de Lyon. Available at http://www.lsp.ups-tlse.fr/Fp/Bolley, 2005.
3. F. Bolley. Quantitative concentration inequalities on sample path space for mean field interaction. Preprint available at http://www.lsp.ups-tlse.fr/Fp/Bolley, 2005.
4. S. T. Rachev. *Probability metrics and the stability of stochastic models.* John Wiley and Sons, Chichester, 1991.
5. S. T. Rachev and L. Rüschendorf. *Mass transportation problems. Vol I and II.* Springer, New York, 1998.
6. C. Villani. *Topics in optimal transportation*, volume 58 of *Grad. Stud. Math.* AMS, Providence, 2003.

A probabilistic interpretation to the symmetries of a discrete heat equation

Nicolas Privault

Department of Mathematics, City University of Hong Kong
83 Tat Chee Avenue, Kowloon Tong, Hong Kong
e-mail: nprivaul@cityu.edu.hk

Summary. A probabilistic interpretation is constructed for the symmetry group G of the finite difference-differential equation $\partial_t \eta(x,t) = \eta(x,t) - \eta(x+1,t)$ using the Doob transform for Markov (jump) processes. While the first three generators of G correspond to the identity and to space and time shifts, we show that in this interpretation the fourth generator of G is associated to time dilations and is linked to a creation operator on the Poisson space.

Key words: Key words: Finite difference equations, symmetries, jump processes, Doob transform.

Classification: 39A12, 34C14, 60J25, 81S25.

1 Introduction

Symmetry groups of partial differential equations have been extensively studied, see e.g. [9], [4] for the heat equation, and [8] for finite difference equations. Recently, probabilistic interpretations of the symmetries of the classical heat equation

$$\frac{\partial u}{\partial t}(x,t) = \frac{1}{2}\Delta u(x,t), \qquad x \in \mathbf{R}^d, \quad t \in \mathbf{R}_+, \tag{1}$$

have been provided in [6], [7], using a family of reversible diffusion processes.

In this paper we study in a similar way the symmetry group G of the simple finite difference-differential equation

$$L\eta(k,t) := \frac{\partial \eta}{\partial t}(k,t) + \eta(k+1,t) - \eta(k,t) = 0, \qquad k \in \mathbf{R}, \quad t \in \mathbf{R}_+. \tag{2}$$

Let P and S denote the creation on Poisson space and the right shift operators defined as

$$P\eta(k,t) = k\eta(k-1,t) - t\eta(k,t) \quad \text{and} \quad S\eta(k,t) = \eta(k+1,t).$$

The operator P is a finite dimensional creation operator, due to its action on multiple Poisson stochastic integrals.

Our main results can be summarized as follows.

a) We show that the symmetry group of (2) has 4 generators denoted by $(N_i)_{i=1,...,4}$, where

- $N_1 = I$ the identity,
- $N_2 = \dfrac{\partial}{\partial t}$ which generates the group of time shifts,
- $N_3 = \dfrac{\partial}{\partial k}$ which generates the group of space shifts ,

and N_4 is written as $N_4 = PS$, i.e. any element N of the Lie algebra associated to (2) can be written as

$$ N = \alpha_1 I + \alpha_2 \frac{\partial}{\partial t} + \alpha_3 \frac{\partial}{\partial k} + \alpha_4 PS, \qquad \alpha_1,\ \alpha_2,\ \alpha_3,\ \alpha_4 \in \mathbf{R}. $$

b) We provide a probabilistic interpretation for G using Markovian Bernstein processes, with a particular attention given to N_4 which is shown to generate time dilations in the following sense. Given η a strictly positive solution of $L\eta = 0$, let $(Z_\eta(t))_{t \in \mathbf{R}_+}$ denote the Markov process with generator $\mathcal{L}_\eta$ defined by the Doob transformation

$$ \mathcal{L}_\eta f(k,t) := \frac{1}{\eta(k,t)} L(\eta f)(k,t). $$

Then $(Z_\eta(t))_{t \in \mathbf{R}_+}$ and $(Z_{e^{\beta N_4}\eta}(t))_{t \in \mathbf{R}_+}$ are linked by the relation in distribution

$$ Z_{e^{\beta N_4}\eta}(t) \simeq Z_\eta(e^\beta t), \qquad t \in \mathbf{R}_+. $$

Dual versions of a) and b) are obtained by time reversal.

We proceed as follows. The Doob transformation, which defines the time reversible Markov processes on which our probabilistic interpretation is based, is recalled in Section 2. In Section 3 we consider the symmetries of the discrete heat equation (2), and in Section 4 their probabilistic interpretation is constructed. In Section 5 we state the corresponding Girsanov theorem. In the appendix (Section 6) we recall some elements of normal martingale theory and quantum stochastic calculus to prove, in a general framework, the commutation relations satisfied by P and L.

2 Doob transform and reversible Markov processes

Consider $(X_t)_{t \in \mathbf{R}_+}$ a Markov process whose forward and backward generators H and H^* are assumed to be mutually adjoint, i.e. $H^\dagger = H^*$, with respect to a given reference measure λ. Consider $\eta(k,t)$, resp. $\eta^*(k,t)$, a strictly positive solution of the partial integro-differential equation

$$-\frac{\partial \eta}{\partial t}(k,t) = H\eta(k,t), \tag{3}$$

resp.

$$\frac{\partial \eta^*}{\partial t}(k,t) = H^*\eta^*(k,t). \tag{4}$$

To η and η^* we associate the (non homogeneous) Markov jump processes $(Z_\eta(t))_{t\in R_+}$ and $(Z^*_{\eta^*}(t))_{t\in R_+}$ whose respective forward and backward generators $\mathcal{L}_\eta$ and $\mathcal{L}^*_{\eta^*}$ are given by the Doob transforms

$$\mathcal{L}_\eta f(k,t) := \frac{1}{\eta(k,t)}\left(H + \frac{\partial}{\partial t}\right)(\eta f)(k,t) \tag{5}$$

and

$$\mathcal{L}^*_{\eta^*} f(k,t) := \frac{1}{\eta^*(k,t)}\left(H^* - \frac{\partial}{\partial t}\right)(\eta^* f)(k,t). \tag{6}$$

The processes $(Z_\eta(t))_{t\in R_+}$ and $(Z^*_{\eta^*}(t))_{t\in R_+}$ are called Bernstein processes [13], and by construction $\mathcal{L}^*_{\eta^*}$ is adjoint of $\mathcal{L}_\eta$ with respect to the measure of density $\eta(k,t)\eta^*(k,t)$ with respect to $\lambda(dk)$. Moreover we have the following proposition.

Proposition 1. *For any $0 \le u < v$, both $(Z_\eta(t))_{t\in[u,v]}$ and $(Z^*_{\eta^*}(t))_{t\in[u,v]}$ have distribution*

$$\eta(k,t)\eta^*(k,t)\lambda(dk), \qquad t \in [u,v],$$

provided they are respectively started with the initial and terminal distributions

$$\eta(k,u)\eta^*(k,u)\lambda(dk) \qquad and \qquad \eta(k,v)\eta^*(k,v)\lambda(dk).$$

Proof. Let us show that $Z_\eta(t)$ has density $\eta(\cdot,t)\eta^*(\cdot,t)$ at time $t \in [u,v]$. For all $f \in \mathcal{S}(\mathbf{R})$ we have

$$\frac{d}{dt}\langle \eta\eta^*(\cdot,t), f(\cdot)\rangle_{L^2(d\lambda)}$$

$$= \left\langle f(\cdot), \eta^*(\cdot,t)\frac{\partial \eta}{\partial t}(\cdot,t)\right\rangle_{L^2(d\lambda)} + \left\langle f(\cdot), \eta(\cdot,t)\frac{\partial \eta^*}{\partial t}(\cdot,t)\right\rangle_{L^2(d\lambda)}$$

$$= \left\langle f(\cdot), \eta^*(\cdot,t)\frac{\partial \eta}{\partial t}(\cdot,t)\right\rangle_{L^2(d\lambda)} + \langle f(\cdot), \eta(\cdot,t)H^*\eta^*(\cdot,t)\rangle_{L^2(d\lambda)}$$

$$= \left\langle \eta^*(\cdot,t), f(\cdot)\frac{\partial \eta}{\partial t}(\cdot,t)\right\rangle_{L^2(d\lambda)} + \langle \eta^*(\cdot,t), H(f\eta)(\cdot,t)\rangle_{L^2(d\lambda)}$$

$$= \langle \eta\eta^*(\cdot,t), \mathcal{L}_\eta f(\cdot)\rangle_{L^2(d\lambda)}.$$

Hence by standard arguments,

$$\langle \eta\eta^*(\cdot,t), f(\cdot)\rangle_{L^2(d\lambda)} = \sum_{n=0}^{\infty} \frac{(t-u)^n}{n!} \frac{d^n}{dt^n} \langle \eta\eta^*(\cdot,t), f(\cdot)\rangle_{L^2(d\lambda)} \Big|_{t=u}$$

$$= \sum_{n=0}^{\infty} \frac{(t-u)^n}{n!} \langle \eta\eta^*(\cdot,u), (\mathcal{L}_\eta)^n f(\cdot)\rangle_{L^2(d\lambda)}$$

$$= \langle \eta\eta^*(\cdot,u), e^{(t-u)\mathcal{L}_\eta} f(\cdot)\rangle_{L^2(d\lambda)}$$

$$= \int e^{(t-u)\mathcal{L}_\eta} f(k)\eta\eta^*(\cdot,u)\lambda(dk), \qquad t \in [u,v].$$

Similarly we have

$$\frac{d}{dt}\langle \eta\eta^*(\cdot,t), f(\cdot)\rangle_{L^2(d\lambda)}$$

$$= \left\langle f(\cdot), \eta(\cdot,t)\frac{\partial\eta^*}{\partial t}(\cdot,t)\right\rangle_{L^2(d\lambda)} + \left\langle f(\cdot), \eta^*(\cdot,t)\frac{\partial\eta}{\partial t}(\cdot,t)\right\rangle_{L^2(d\lambda)}$$

$$= \left\langle f(\cdot), \eta(\cdot,t)\frac{\partial\eta^*}{\partial t}(\cdot,t)\right\rangle_{L^2(d\lambda)} - \langle f(\cdot), \eta^*(\cdot,t)H\eta(\cdot,t)\rangle_{L^2(d\lambda)}$$

$$= \left\langle \eta(\cdot,t), f(\cdot)\frac{\partial\eta^*}{\partial t}(\cdot,t)\right\rangle_{L^2(d\lambda)} - \langle \eta(\cdot,t), H^*(f\eta^*)(\cdot,t)\rangle_{L^2(d\lambda)}$$

$$= -\langle \eta\eta^*(\cdot,t), \mathcal{L}_{\eta^*}^* f(\cdot)\rangle_{L^2(d\lambda)},$$

hence

$$\langle \eta\eta^*(\cdot,t), f(\cdot)\rangle_{L^2(d\lambda)} = \sum_{n=0}^{\infty} \frac{(t-v)^n}{n!} \frac{d^n}{dt^n} \langle \eta\eta^*(\cdot,t), f(\cdot)\rangle_{L^2(d\lambda)} \Big|_{t=v}$$

$$= \sum_{n=0}^{\infty} \frac{(v-t)^n}{n!} \langle \eta\eta^*(\cdot,v), (\mathcal{L}_{\eta^*}^*)^n f(\cdot)\rangle_{L^2(d\lambda)}$$

$$= \langle \eta\eta^*(\cdot,v), e^{(v-t)\mathcal{L}_{\eta^*}^*} f(\cdot)\rangle_{L^2(d\lambda)}$$

$$= \int e^{(v-t)\mathcal{L}_{\eta^*}^*} f(k)\eta\eta^*(k,v)\lambda(dk), \qquad t \in [u,v].$$

$\square$

Recall that a Theorem of Beurling [1] states that any given initial and final probability densities π_u and π_v on $\mathbf{R}^d$ can be written in the product forms $\pi_u(k) = \eta(k,u)\eta^*(k,u)$ and $\pi_v(k) = \eta(k,v)\eta^*(k,v)$ where η, η^* are suitably chosen solutions of (3) and (4). Hence the processes $(Z_\eta(t))_{t\in[u,v]}$ and $(Z_{\eta^*}^*(t))_{t\in[u,v]}$ can be used to construct Markovian bridges having arbitrary prescribed absolutely continuous initial and final distributions. Thus they provide a Euclidean probabilistic interpretation of a given quantum system with Hamiltonian H, in which complex conjugation is replaced with time reversal, cf. [13], [11], [12].

Let now

$$L = \frac{\partial}{\partial t} + H, \quad \text{and} \quad L^* = -\frac{\partial}{\partial t} + H^*.$$

Given $A \in G$ and $B^* \in G^*$ in the symmetry groups G, G^* of L and L^*, a natural question is to consider the application

$$\{\eta, \eta^*\} \mapsto \{A\eta, B^*\eta^*\}$$

which acts on respective solutions of $L\eta = 0$ and $L^*\eta^* = 0$, and to determine the associated pathwise transformations which map

$$(Z_\eta(t))_{t \in [u,v]} \quad \text{to} \quad (Z_{A\eta}(t))_{t \in [u,v]}$$

and

$$(Z^*_{\eta^*}(t))_{t \in [u,v]} \quad \text{to} \quad (Z_{B^*\eta^*}(t))_{t \in [u,v]}.$$

In this way we construct a probabilistic representation of the symmetry groups of (3) and (4) using Bernstein jump processes.

In other terms we would like to study the relationship between $\mathcal{L}_\eta$ and $\mathcal{L}_{A\eta}$, resp. $\mathcal{L}^*_{\eta^*}$ and $\mathcal{L}^*_{A\eta^*}$, and to determine of a mapping $\phi : \mathbf{R} \times \mathbf{R}_+ \to \mathbf{R} \times \mathbf{R}_+$ such that

$$(A\eta \cdot B^*\eta^*) \circ \phi(k,t) = (\eta \cdot \eta^*)(k,t), \quad (k,t) \in \mathbf{R} \times \mathbf{R}_+.$$

This procedure is trivial for the pairs $(e^{\beta N_k}, e^{-\beta N_k^*})$, $\beta \in \mathbf{R}$, $k = 1, 2, 3$, described in the introduction, for which the associated transformations are the identity and time and space shifts, respectively. In this paper we investigate the role played by N_4, and we show that it is associated to time dilations and a creation operator.

3 Lie algebra generators

Consider $(N_t)_{t \in \mathbf{R}_+}$ a standard Poisson process with forward/backward generators

$$L\eta(k,t) := \frac{\partial \eta}{\partial t}(k,t) + \eta(k+1,t) - \eta(k,t), \tag{7}$$

and

$$L^*\eta^*(k,t) := -\frac{\partial \eta^*}{\partial t}(k,t) + \eta^*(k-1,t) - \eta^*(k,t). \tag{8}$$

In other terms we have

$$Hf(k,t) = f(k+1,t) - f(k,t) \quad \text{and} \quad H^*g(k,t) = g(k-1,t) - g(k,t),$$

which are mutually adjoint with respect to the counting measure

$$\lambda(dl) = \sum_{k \in \mathbb{Z}} \delta_k(dl),$$

where δ_x denotes the Dirac measure at $x \in \mathbf{R}$. For example the standard Poisson process is obtained with

$$\eta(k,t) = 1 \quad \text{and} \quad \eta^*(k,t) = 1_{\{k \geq 0\}} \frac{t^k}{k!} e^{-t}.$$

Recall that the Charlier polynomials $(C_n(k,t))_{n \in \mathbf{N}}$ of parameter $t \in \mathbf{R}$, defined through their generating function

$$\psi_\lambda(k) = \sum_{n=0}^{\infty} \frac{\lambda^n}{n!} C_n(k,t) = e^{-\lambda t}(1+\lambda)^k, \quad \lambda \in (-1,1), \quad k, t \in \mathbf{R},$$

or by

$$C_0(k,t) = 1, \qquad C_1(k,t) = k - t,$$

and the recurrence relation

$$C_{n+1}(k,t) = (k - n - t)C_n(k,t) - ntC_{n-1}(k,t), \tag{9}$$

are solutions of $L\eta(k,t) = 0$, $n \in \mathbf{N}$, with the initial condition $\eta(k,0) = \prod_{i=1}^{n}(k-i)$.

Definition 1. *Let the creation operator P be defined as*

$$P\eta(k,t) = k\eta(k-1,t) - t\eta(k,t).$$

The Charlier polynomials satisfy the relation

$$PC_n(k,t) = kC_n(k-1,t) - tC_n(k,t) = C_{n+1}(k,t), \qquad n \in \mathbf{N}.$$

We have the commutation relation

$$[P,L] = 0, \tag{10}$$

which is easily proved by direct calculation or as a consequence of the more general result (Proposition 10) stated in the appendix.

Definition 2. *Let $\mathcal{G}$ be the Lie algebra with commutator $[\cdot, \cdot]$ spanned by all first-order differential operators with smooth coefficients, of the form*

$$N = \alpha(k,t)I + \beta(k,t)\frac{\partial}{\partial t} + \gamma(k,t)\frac{\partial}{\partial k}, \tag{11}$$

where I denotes the identity, and verifying the stability property

$$L\eta = 0 \quad \Rightarrow \quad LN\eta = 0.$$

Similarly, let $\mathcal{G}^*$ denote the Lie algebra spanned by all first order differential operators N^* of the form (11) and satisfying

$$L^*\eta^* = 0 \quad \Rightarrow \quad L^*N^*\eta^* = 0.$$

Definition 3. *Let R be defined by $R\eta(k,t) = \eta(-k,-t)$, $k \in \mathbf{R}$, $t \in [u,v]$.*

Note that we have
$$H^* = RHR, \qquad L^* = RLR.$$

In the same way, any element A of the symmetry group G can be associated to an element $A^* := RAR$ of G^*, and the mapping $N \mapsto N^*$ is an isomorphism from $\mathcal{G}$ onto $\mathcal{G}^*$, and from G onto G^*.

Next, we state representation results for the elements of $\mathcal{G}$ and $\mathcal{G}^*$.

Proposition 2. *On the solution space*
$$\mathrm{Ker}\,(L) = \left\{ \eta \;:\; \frac{\partial \eta}{\partial t}(k,t) + \eta(k+1,t) - \eta(k,t) = 0 \right\}$$

of L, any element of $\mathcal{G}$ can be written as
$$N = \alpha_1 I + \alpha_2 \frac{\partial}{\partial t} + \alpha_3 \frac{\partial}{\partial k} + \alpha_4 PS, \qquad (12)$$

for some $\alpha_1, \alpha_2, \alpha_3, \alpha_4 \in \mathbf{R}$.

Proof. We start by showing that any element N of $\mathcal{G}$ can be represented as
$$N = (\alpha + (k-t)\lambda)I + (\xi + \lambda t)\frac{\partial}{\partial t} + \gamma \frac{\partial}{\partial k}, \qquad (13)$$

for some $a, \xi, \lambda, \gamma \in \mathbf{R}$. Indeed, a necessary and sufficient condition for an element N of the form (11) to belong to $\mathcal{G}$ is the existence of a function $\lambda(k,t)$ such that
$$[N, L] = \lambda(k,t)L, \qquad (14)$$

on $\{\eta \;:\; L\eta = 0\}$. Now, Relation (14) reads

$NL\eta(k,t) - LN\eta(k,t)$

$\displaystyle = \frac{\partial \gamma}{\partial t}(k,t)\frac{\partial \eta}{\partial k}(k,t) + \gamma(k,t)\frac{\partial^2 \eta}{\partial t \partial k}(k,t) + \eta(k,t)\frac{\partial \alpha}{\partial t}(k,t) + \alpha(k,t)\frac{\partial \eta}{\partial t}(k,t)$

$\displaystyle + \frac{\partial \beta}{\partial t}(k,t)\frac{\partial \eta}{\partial t}(k,t) + \beta(k,t)\frac{\partial^2 \eta}{\partial t^2}(k,t)$

$\displaystyle + \gamma(k+1,t)\frac{\partial \eta}{\partial k}(k+1,t) + \alpha(k+1,t)\eta(k+1,t) + \beta(k+1,t)\frac{\partial \eta}{\partial t}(k+1,t)$

$\displaystyle - \gamma(k,t)\frac{\partial \eta}{\partial k}(k,t) - \alpha(k,t)\eta(k,t) - \beta(k,t)\frac{\partial \eta}{\partial t}(k,t)$

$\displaystyle - \left(\gamma(k,t)\frac{\partial^2 \eta}{\partial k \partial t}(k,t) + \gamma(k,t)\frac{\partial \eta}{\partial k}(k+1,t) - \gamma(k,t)\frac{\partial \eta}{\partial k}(k,t) \right.$

$\displaystyle + \alpha(k,t)\frac{\partial \eta}{\partial t}(k,t) + \alpha(k,t)\eta(k+1,t) - \alpha(k,t)\eta(k,t)$

$\displaystyle \left. + \beta(k,t)\frac{\partial^2 \eta}{\partial t^2}(k,t) + \beta(k,t)\frac{\partial \eta}{\partial t}(k+1,t) - \beta(k,t)\frac{\partial \eta}{\partial t}(k,t) \right)$

$\displaystyle = \lambda(k,t)\left(\frac{\partial \eta}{\partial t}(k,t) + \eta(k+1,t) - \eta(k,t) \right)$

$\displaystyle = \lambda(k,t)L\eta(k,t),$

from which we deduce

$$\frac{\partial \gamma}{\partial t}(k,t) = 0,$$

$$\gamma(k+1,t) - \gamma(k,t) = 0,$$

$$\frac{\partial \alpha}{\partial t}(k,t) = -\lambda(k,t),$$

$$\alpha(k+1,t) - \alpha(k,t) = \lambda(k,t),$$

$$\frac{\partial \beta}{\partial t}(k,t) = \lambda(k,t),$$

$$\beta(k+1,t) - \beta(k,t) = 0.$$

It follows that $\lambda(k+1,t) = \lambda(k,t)$ and $\alpha(k,t) = a(t) - k\lambda(t)$, hence

$$a'(t) - k\lambda'(t) = -\lambda(t),$$

which implies $\lambda'(t) = 0$ and $a'(t) = -\lambda(t)$, i.e. $a(t) = a - \lambda t$, for some $a, \lambda \in \mathbf{R}$. On the other hand we have

$$\alpha(k,t) = a + (k-t)\lambda, \qquad \beta(k,t) = \xi + \lambda t, \qquad \gamma(k,t) = \gamma,$$

for some $\xi, \gamma \in \mathbf{R}$, which proves (13). Finally we have

$$N\eta(k,t) = \alpha_3 \frac{\partial \eta}{\partial k}(k,t) + (\alpha_1 + (k-t)\alpha_4)\eta(k,t) + (\alpha_2 + \alpha_4 t)\frac{\partial \eta}{\partial t}(k,t)$$

$$= \alpha_3 \frac{\partial \eta}{\partial k}(k,t) + (\alpha_1 + (k-t)\alpha_4)\eta(k,t) + \alpha_4 t(\eta(k,t) - \eta(k+1,t)) + \alpha_2 \frac{\partial \eta}{\partial t}(k,t)$$

$$= \alpha_3 \frac{\partial \eta}{\partial k}(k,t) + \alpha_1 \eta(k,t) + \alpha_4(k\eta(k,t) - t\eta(k+1,t)) + \alpha_2 \frac{\partial \eta}{\partial t}(k,t)$$

$$= \alpha_3 \frac{\partial \eta}{\partial k}(k,t) + \alpha_1 \eta(k,t) + \alpha_4 PS\eta(k,t) + \alpha_2 \frac{\partial \eta}{\partial t}(k,t).$$

$\square$

The first three generators N_1, N_2, N_3, resp. N_1^*, N_2^*, N_3^*, of $\mathcal{G}$, resp. $\mathcal{G}^*$ are given by

$$N_1 = N_1^* = I, \quad N_2 = -N_2^* = \frac{\partial}{\partial t}, \quad \text{and} \quad N_3 = -N_3^* = \frac{\partial}{\partial k},$$

and we let $N_4 = PS$, i.e. (12) is written as

$$N = \alpha_1 N_1 + \alpha_2 N_2 + \alpha_3 N_3 + \alpha_4 N_4.$$

We have the commutation table

$[\cdot,\cdot]$	N_1	N_2	N_3	N_4
N_1	0	0	0	0
N_2	0	0	0	$-N_1$
N_3	0	0	0	N_1
N_4	0	$-N_1$	N_1	0

The operators P^* and S^* satisfy

$$P^*\eta^*(k,t) = -k\eta^*(k+1,t) + t\eta^*(k,t), \quad \text{and} \quad S^*\eta^*(k,t) = \eta^*(k-1,t),$$

with $S^* = e^{N_3^*}$, and the relations

$$[P^*, S^*] = -I, \qquad [P^*, L^*] = 0,$$

We may also let

$$C_n^*(k,t) := RC_n(k,t) = C_n(-k,-t), \qquad n \in \mathbf{N}, \quad k,t \in \mathbf{R},$$

and in this case

$$P^*C_n^*(k,t) = kC_n^*(k+1,t) - tC_n^*(k,t) = C_{n+1}^*(k,t), \qquad n \in \mathbf{N}.$$

Similarly to the above, we have the following proposition, with $N_4 = P^*S^*$.

Proposition 3. *On the solution space*

$$\mathrm{Ker}\,(L^*) = \left\{ \eta^* \; : \; -\frac{\partial \eta^*}{\partial t}(k,t) + \eta^*(k-1,t) - \eta^*(k,t) = 0 \right\}$$

of L^*, any element of $\mathcal{G}^*$ can be written as

$$N^* = \alpha_1 N_1^* + \alpha_2 N_2^* + \alpha_3 N_3^* + \alpha_4 N_4^*,$$

for some $\alpha_1, \alpha_2, \alpha_3, \alpha_4 \in \mathbf{R}$.

4 Probabilistic interpretation of G

Consider again two respective (λ-a.e. strictly positive) solutions $\eta(k,t)$, $\eta^*(k,t)$ of the finite difference partial differential equations

$$L\eta(k,t) = 0 \quad \text{and} \quad L^*\eta^*(k,t) = 0.$$

Using (5) and (6), the forward and backward generators of $(Z_\eta(t))_{t \in \mathbf{R}_+}$ and $(Z_{\eta^*}^*(t))_{t \in \mathbf{R}_+}$ can be computed as follows:

$$\mathcal{L}_\eta f(k) = \frac{\eta(k+1,t)}{\eta(k,t)}(f(k+1) - f(k)),$$

and

$$\mathcal{L}_{\eta^*}^* f(k) = \frac{\eta^*(k-1,t)}{\eta^*(k,t)}(f(k-1) - f(k)).$$

In particular $(Z_\eta(t))_{t\in R_+}$ and $(Z_{\eta^*}^*(t))_{t\in R_+}$ are point processes with respective forward and backward intensities

$$\frac{\eta(Z_\eta(t^-)+1,t)}{\eta(Z_\eta(t^-),t)} \quad \text{and} \quad \frac{\eta^*(Z_{\eta^*}^*(t^+)-1,t)}{\eta^*(Z_{\eta^*}^*(t^+),t)}.$$

We will denote by $(\mathcal{F}_t)_{t\in[u,v]}$, resp. $(\mathcal{F}_t^*)_{t\in[u,v]}$ the forward, resp. backward filtration generated by the transformations $(Z_\eta(t))_{t\in[u,v]}$, resp. $(Z_{\eta^*}^*(t))_{t\in[u,v]}$. The generators $(N_1, -N_1^*) = (I, -I)$, $(N_2, -N_2^*) = \left(\dfrac{\partial}{\partial t}, \dfrac{\partial}{\partial t}\right)$, and $(N_3, -N_3^*) = \left(\dfrac{\partial}{\partial k}, \dfrac{\partial}{\partial k}\right)$, are respectively associated to:

- the identity:

$$e^{\alpha N_1}\eta(k,s)e^{-\alpha N_1^*}\eta^*(k,s) = \eta(k,s)\eta^*(k,s),$$

- time translations:

$$e^{\alpha N_2}\eta(k,s)e^{-\alpha N_2^*}\eta^*(k,s) = \eta(k,s+\alpha)\eta^*(k,s+\alpha),$$

corresponding to $(Z_\eta(s))_{s\in[u,v]} \mapsto (Z_\eta(s+t))_{s\in[u+t,v+t]}$,
- space translations:

$$e^{\alpha N_3}\eta(k,s)e^{-\alpha N_3^*}\eta^*(k,s) = \eta(k+\alpha,s)\eta^*(k+\alpha,s),$$

corresponding to $(Z_\eta(s))_{s\in[u,v]} \mapsto (k+Z_\eta(s))_{s\in[u,v]}$,

and similarly for $(Z_{\eta^*}^*(s))_{s\in[u,v]}$.

We now focus on the role of the operators $N_4 = PS$ and $N_4^* = P^*S^*$, and show that they are linked to pathwise transformations of $(Z_\eta(t))_{t\in R_+}$ and $(Z_{\eta^*}^*(t))_{t\in R_+}$ by time dilations. We identify the one-parameter semigroup of solutions of $L\eta = 0$ generated by $N_4 = PS$ and we show that it corresponds to time dilations.

Proposition 4. *For all $\eta \in \mathcal{S}$ and all $k \in R$, $t \in R_+$, $\beta \in R$ we have*

$$e^{\beta N_4}\eta(k,t) = \exp(k\beta - (e^\beta - 1)t)\eta(k, e^\beta t), \tag{15}$$

and

$$e^{\beta N_4^*}\eta^*(k,t) = \exp(-k\beta + (e^\beta - 1)t)\eta^*(k, e^\beta t). \tag{16}$$

Proof. For convenience of notation we make the change of variable $\beta = \log(1+\alpha)$, $\alpha > -1$. Let

$$\eta_\alpha(k,t) = (1+\alpha)^k e^{-\alpha t}\eta(k,(1+\alpha)t), \qquad k \in R, \quad t \in R_+, \quad \eta \in \mathcal{S}.$$

We have

$$\frac{\partial \eta_\alpha}{\partial \alpha}(k,t) = k(1+\alpha)^{k-1}e^{-\alpha t}\eta(k,(1+\alpha)t)$$

$$-t(1+\alpha)^k e^{-\alpha t}\eta(k,(1+\alpha)t) + t(1+\alpha)^k e^{-\alpha t}\frac{\partial \eta}{\partial t}(k,(1+\alpha)t)$$

$$= k(1+\alpha)^{k-1}e^{-\alpha t}\eta(k,(1+\alpha)t) - t(1+\alpha)^k e^{-\alpha t}\eta(k,(1+\alpha)t)$$

$$+t(1+\alpha)^k e^{-\alpha t}(\eta(k,(1+\alpha)t) - \eta(k+1,(1+\alpha)t))$$

$$= k(1+\alpha)^{k-1}e^{-\alpha t}\eta(k,(1+\alpha)t) - t(1+\alpha)^k e^{-\alpha t}\eta(k+1,(1+\alpha)t)$$

$$= (1+\alpha)^{-1}\left(kS\eta_\alpha(k-1,t) - tS\eta_\alpha(k,t)\right)$$

$$= (1+\alpha)^{-1}PS\eta_\alpha(k,t),$$

which shows that η_α is solution in $\mathcal{S}$ of

$$\begin{cases} \dfrac{\partial \eta_\alpha}{\partial \alpha} = \dfrac{1}{1+\alpha}N_4\eta_\alpha, & \alpha > -1, \\[2mm] \eta_0 = \eta, \end{cases}$$

hence $\eta_\alpha = \exp(\log(1+\alpha)N_4)\eta$, $\alpha > -1$. Relation (16) can be obtained by direct transfer using the mapping R, or from the solution of the equation

$$\begin{cases} \dfrac{\partial \eta_\alpha^*}{\partial \alpha} = \dfrac{1}{1+\alpha}N_4^*\eta_\alpha^*, & \alpha > -1, \\[2mm] \eta_0^* = \eta^*, \end{cases} \tag{17}$$

which is given by

$$\eta_\alpha^*(k,t) = \exp(\log(1+\alpha)N_4^*)\eta^*(k,t) = (1+\alpha)^{-k}e^{\alpha t}\eta^*(k,(1+\alpha)t),$$

since we have

$$\frac{\partial \eta_\alpha^*}{\partial \alpha}(k,t) = -k(1+\alpha)^{-k-1}e^{\alpha t}\eta^*(k,(1+\alpha)t)$$

$$+t(1+\alpha)^{-k}e^{\alpha t}\eta^*(k,(1+\alpha)t) + t(1+\alpha)^{-k}e^{-\alpha t}\frac{\partial \eta^*}{\partial t}(k,(1+\alpha)t)$$

$$= -k(1+\alpha)^{-k-1}e^{-\alpha t}\eta^*(k,(1+\alpha)t) + t(1+\alpha)^{-k}e^{\alpha t}\eta^*(k,(1+\alpha)t)$$

$$+t(1+\alpha)^{-k}e^{\alpha t}(\eta^*(k-1,(1+\alpha)t) - \eta^*(k,(1+\alpha)t))$$

$$= -(1+\alpha)^{-1}\Big(k(1+\alpha)^{-k}e^{-\alpha t}\eta(k,(1+\alpha)t)$$

$$-t(1+\alpha)^{-(k-1)}e^{-\alpha t}\eta(k-1,(1+\alpha)t)\Big)$$

$$= (1+\alpha)^{-1}P^*S^*\eta_\alpha^*(k,t). \qquad \square$$

The following proposition is now obvious and shows that $(Z_{\eta_\alpha}(s))_{s\in[u,v]}$, $(Z_{\eta_\alpha^*}^*(s))_{s\in[u,v]}$ have same laws as $(Z_\eta((1+\alpha)s))_{s\in[u,v]}$ and $(Z_{\eta^*}^*((1+\alpha)s))_{s\in[u,v]}$.

Proposition 5. *We have*

$$e^{\beta N_4}\eta(k,t)e^{-\beta N_4^*}\eta^*(k,t) = (\eta^*\eta)(k,e^{\beta}t), \qquad k \in \mathbf{R}, \quad t \in \mathbf{R}_+.$$

In other terms, $(Z_{\eta_\alpha}(s))_{s\in[u,v]}$ and $(Z_{\eta_\alpha^*}^*(s))_{s\in[u,v]}$ have the forward and backward intensities

$$\frac{\eta_\alpha(Z_{\eta_\alpha}(t^-)+1,t)}{\eta_\alpha(Z_{\eta_\alpha}(t^-),t)} = (1+\alpha)\frac{\eta(Z_{\eta_\alpha}(t^-)+1,t)}{\eta(Z_{\eta_\alpha}(t^-),t)},$$

and

$$\frac{\eta_\alpha^*(Z_{\eta_\alpha^*}^*(t^+)-1,t)}{\eta_\alpha^*(Z_{\eta_\alpha^*}^*(t^+),t)} = (1+\alpha)\frac{\eta(Z_{\eta_\alpha^*}^*(t^+)-1,t)}{\eta(Z_{\eta_\alpha^*}^*(t^+),t)},$$

i.e. $(Z_{\eta_\alpha}(\tau^{-1}(t)))_{t\in\mathbf{R}_+}$ is a standard Poisson process, where the time change $\tau(\cdot)$ is defined by

$$\tau(t) = \int_0^t \frac{\eta_\alpha(Z_{\eta_\alpha}(s)+1,s)}{\eta_\alpha(Z_{\eta_\alpha}(s),s)}ds = (1+\alpha)\int_0^t \frac{\eta(Z_{\eta_\alpha}(s)+1,s)}{\eta(Z_{\eta_\alpha}(s),s)}ds, \qquad t \in \mathbf{R}_+.$$

5 Absolute continuity

The aim of this section is to study the change of measure generated by the transformations

$$(Z_\eta(t))_{t\in[u,v]} \mapsto (Z_\zeta(t))_{t\in[u,v]}$$

and

$$(Z_{\eta^*}^*(t))_{t\in[u,v]} \mapsto (Z_{\zeta^*}^*(t))_{t\in[u,v]},$$

given η, ζ, resp. η^*, ζ^*, two a.e. strictly positive solutions of $\{L\eta = 0\}$, resp. $\{L^*\eta^* = 0\}$.

Clearly, the (unconditional) density of $Z_\zeta(t)$ with respect to $Z_\eta(t)$ is

$$\Psi(Z_\eta(t),t) = \frac{\zeta(Z_\eta(t),t)}{\eta(Z_\eta(t),t)}\frac{\zeta^*(Z_\eta(t),t)}{\eta^*(Z_\eta(t),t)}, \qquad t \in [u,v],$$

as follows from

$$E[f(Z_\zeta(t))] = \int f(k)\zeta(k,t)\zeta^*(k,t)\lambda(dk)$$

$$= \int f(k)\Psi(k,t)\eta(k,t)\eta^*(k,t)\lambda(dk)$$

$$= E[f(Z_\eta(t))\Psi(Z_\eta(t),t)].$$

Similarly the density of $Z_{\zeta^*}^*(t)$ with respect to $Z_{\eta^*}^*(t)$ is

$$\Psi(Z_{\eta^*}^*(t),t) = \frac{\zeta(Z_{\eta^*}^*(t),t)}{\eta(Z_{\eta^*}^*(t),t)}\frac{\zeta^*(Z_{\eta^*}^*(t),t)}{\eta^*(Z_{\eta^*}^*(t),t)}, \qquad t \in [u,v].$$

We now turn to conditional densities.

Proposition 6. *The density of* $(Z_\zeta(u))_{s \leq u \leq t}$ *with respect to* $(Z_\eta(u))_{s \leq u \leq t}$ *given* $\mathcal{F}_s$ *is*

$$\Lambda_{s,t}(Z_\zeta(s), Z_\eta(t)) = \frac{\zeta(Z_\eta(t), t)}{\eta(Z_\eta(t), t)} \frac{\eta(Z_\zeta(s), s)}{\zeta(Z_\zeta(s), s)}.$$

Proof. By (5), the process $(Z_\eta(s))_{s \in [u,v]}$ has the forward transition semigroup

$$p_\eta(t, k, u, dl) = \frac{\eta(l, u)}{\eta(k, t)} h(t, k, u, dl),$$

where $h(s, l, t, dk)$ denotes the kernel of $e^{(t-s)H}$, i.e.

$$e^{(t-s)H} f(l) = \int f(k) h(s, l, t, dk), \qquad l \in \mathbf{R}, \quad 0 \leq s < t.$$

By the Markov property of $(Z_\eta(t))_{t \in [u,v]}$, it suffices to check that the finite-dimensional distributions satisfy:

$$E[f(Z_\eta(t_1)), \ldots, Z_\eta(t_n)) \Lambda_{s,t}(Z_\zeta(s), Z_\eta(t)) | Z_\eta(s) = k]$$
$$= \int \cdots \int f(k_1, \ldots, k_n) h(s, k, t_1, dk_1) \cdots h(t_n, k_n, t, dl) \Lambda_{s,t}(k, l) \frac{\eta(l, t)}{\eta(k, s)}$$
$$= \int \cdots \int f(k_1, \ldots, k_n) h(s, k, t_1, dk_1) \cdots h(t_n, k_n, t, dl) \frac{\zeta(l, t)}{\zeta(k, s)}$$
$$= E[f(Z_\zeta(t_1), \ldots, Z_\zeta(t_n)) | Z_\zeta(s) = k],$$

for all $s \leq t_1 < \cdots < t_n \leq t$, $f \in \mathcal{C}_b(\mathbf{R})$. $\qquad\square$

Note that the unconditional density $\Psi(Z_\eta(t), t)$ can be recovered from the conditional density $\Lambda_{s,t}(Z_\zeta(s), Z_\eta(t))$, as follows:

$$E[f(Z_\zeta(t))] = E[E[f(Z_\zeta(t)) \mid Z_\zeta(s)]]$$
$$= E[E[f(Z_\eta(t)) \Lambda_{s,t}(Z_\zeta(s), Z_\eta(t)) \mid Z_\zeta(s)]]$$
$$= E[f(Z_\eta(t)) \Lambda_{s,t}(Z_\zeta(s), Z_\eta(t))]$$
$$= E\left[f(Z_\eta(t)) \int \Lambda_{s,t}(l, Z_\eta(t)) dP(Z_\zeta(s) = l \mid Z_\eta(t)) \right], \quad f \in \mathcal{C}_b^\infty(\mathbf{R}),$$

which implies

$$\Psi(k, t) = \int \Lambda_{s,t}(l, k) dP(Z_\zeta(s) = l \mid Z_\eta(t) = k),$$

$k \in \mathbf{R}$, $0 \leq s \leq t$. More explicitly, using the duality

$$h(s, l, t, dk)\lambda(dl) = h^*(s, dl, t, k)\lambda(dk) \tag{18}$$

between H and H^* we have

$$\Psi(k,t) = \frac{\zeta(k,t)}{\eta(k,t)} \frac{\zeta^*(k,t)}{\eta^*(k,t)}$$

$$= \int \zeta(k,t)\zeta^*(l,s) \frac{h^*(s,dl,t,k)}{\eta^*(k,t)\eta(k,t)}$$

$$= \int \frac{\zeta(k,t)}{\zeta(l,s)} \frac{h(s,l,t,dk)}{\eta^*(k,t)\eta(k,t)\lambda(dk)} \zeta(l,s)\zeta^*(l,s)\lambda(dl)$$

$$= \int \frac{\zeta(k,t)}{\zeta(l,s)} \frac{\eta(l,s)}{\eta(k,t)} \frac{h(s,l,t,dk)}{\eta^*(k,t)\eta(k,t)\lambda(dk)} \frac{\eta(k,t)}{\eta(l,s)} \zeta(l,s)\zeta^*(l,s)\lambda(dl)$$

$$= \int \Lambda_{s,t}(l,k) \frac{p_\eta(s,l,t,dk)}{\eta^*(k,t)\eta(k,t)\lambda(dk)} dP(Z_\zeta(s) = l)$$

$$= \int \Lambda_{s,t}(l,k) dP(Z_\zeta(s) = l \mid Z_\eta(t) = k).$$

The analogous time reversed statement on conditional densities is as follows.

Proposition 7. *The density of* $(Z^*_{\zeta^*}(u))_{s \leq u \leq t}$ *with respect to* $(Z^*_{\eta^*}(u))_{s \leq u \leq t}$ *given* $\mathcal{F}^*_t$ *is*

$$\Lambda^*_{s,t}(k,l) = \frac{\zeta^*(Z^*_{\zeta^*}(s),s)}{\eta^*(Z^*_{\eta^*}(t),t)} \frac{\eta^*(Z^*_{\zeta^*}(s),s)}{\zeta^*(Z^*_{\eta^*}(t),t)}.$$

Proof. By (6), the process $(Z^*_\eta(s))_{s \in [u,v]}$ has the backward transition semi-group

$$p^*_{\eta^*}(s,dj,t,k) = \frac{\eta^*(j,s)}{\eta^*(k,t)} h(s,dj,t,k),$$

where $h^*(s,dl,t,k)$ denotes the kernels of $e^{(t-s)H^*}$, i.e.

$$e^{(t-s)H^*} f(k) = \int f(k)h^*(s,dl,t,k), \qquad k \in \mathbf{R}, \quad 0 \leq s < t.$$

It suffices to check that the finite-dimensional distributions satisfy:

$$E[f(Z^*_{\eta^*}(t_1),\ldots,Z^*_{\eta^*}(t_n))\Lambda^*_{s,t}(Z^*_{\zeta^*}(s),Z^*_{\eta^*}(t))|Z^*_{\eta^*}(t) = l]$$

$$= \int \cdots \int f(k_1,\ldots,k_n)h^*(s,dk,t_1,k_1)\cdots h^*(t_n,dk_n,t,l)\Lambda^*_{s,t}(k,l)\frac{\eta^*(k,s)}{\eta^*(l,t)}$$

$$= \int \cdots \int f(k_1,\ldots,k_n)h^*(s,dk,t_1,k_1)\cdots h^*(t_n,dk_n,t,k)\frac{\zeta^*(k,s)}{\zeta^*(l,t)}$$

$$= E[f(Z^*_{\zeta^*}(t_1),\ldots,Z^*_{\zeta^*}(t_n))|Z^*_{\zeta^*}(t) = l],$$

for all $s \leq t_1 < \cdots < t_n \leq t$. $\qquad \square$

Similarly to the above, the unconditional density $\Psi(Z^*_{\eta^*}(s), s)$ can be recovered from the conditional density $\Lambda^*_{s,t}(Z^*_{\eta^*}(s), Z^*_{\eta^*}(t))$:

$$
\begin{aligned}
E[f(Z^*_{\zeta^*}(s))] &= E[E[f(Z^*_{\zeta^*}(s)) \mid Z^*_{\zeta^*}(t)]] \\
&= E[E[f(Z^*_{\eta^*}(s)) \Lambda^*_{s,t}(Z^*_{\zeta^*}(s), Z^*_{\eta^*}(t)) \mid Z^*_{\eta^*}(t)]] \\
&= E[f(Z^*_{\eta^*}(s)) \Lambda^*_{s,t}(Z^*_{\zeta^*}(t), Z^*_{\eta^*}(s))] \\
&= E[E[f(Z^*_{\eta^*}(s)) \Lambda^*_{s,t}(Z^*_{\zeta^*}(t), Z^*_{\eta^*}(s)) \mid Z^*_{\eta^*}(t)]] \\
&= E\left[f(Z^*_{\eta^*}(s)) \int \Lambda^*_{s,t}(Z^*_{\eta^*}(s), k) dP(Z^*_{\zeta^*}(t) = k \mid Z^*_{\eta^*}(s)) \right],
\end{aligned}
$$

hence

$$
\Psi(l, s) = \int \Lambda^*_{s,t}(l, k) dP(Z^*_{\zeta^*}(t) = k \mid Z^*_{\eta^*}(s) = l),
$$

$k \in \mathbf{R}$, $0 \le s \le t$. Again, the above calculation can be confirmed as follows using (18):

$$
\begin{aligned}
\Psi(l, s) &= \frac{\zeta(l, s)}{\eta(l, s)} \frac{\zeta^*(l, s)}{\eta^*(l, s)} \\
&= \int \zeta(k, t) \zeta^*(l, s) \frac{h(s, l, t, dk)}{\eta^*(l, s) \eta(l, s)} \\
&= \int \frac{\zeta^*(l, s)}{\zeta^*(k, t)} \frac{h^*(s, dl, t, k)}{\eta^*(l, s) \eta(l, s) \lambda(dl)} \zeta(k, t) \zeta^*(k, t) \lambda(dk) \\
&= \int \frac{\zeta^*(l, s)}{\zeta^*(k, t)} \frac{\eta^*(k, t)}{\eta^*(l, s)} \frac{h^*(s, dl, t, k)}{\eta^*(l, s) \eta(l, s) \lambda(dl)} \frac{\eta^*(l, s)}{\eta^*(k, t)} \zeta(l, s) \zeta^*(l, s) \lambda(dk) \\
&= \int \Lambda^*_{s,t}(l, k) \frac{p^*_{\eta^*}(s, dl, t, k)}{\eta^*(l, s) \eta(l, s) \lambda(dl)} dP(Z^*_{\zeta^*}(s) = l) \\
&= \int \Lambda^*_{s,t}(l, k) dP(Z^*_{\zeta^*}(t) = k \mid Z^*_{\eta^*}(s) = l).
\end{aligned}
$$

The density processes $\dfrac{\zeta(Z_\eta(t), t)}{\eta(Z_\eta(t), t)}$ and $\dfrac{\zeta^*(Z^*_\eta(t), t)}{\eta^*(Z^*_\eta(t), t)}$ are forward/backward martingales since by construction,

$$
\mathcal{L}_\eta \left(\frac{\zeta}{\eta} \right)(k, t) = 0,
$$

and

$$
\mathcal{L}^*_{\eta^*} \left(\frac{\zeta^*}{\eta^*} \right)(k, t) = 0.
$$

This is the case in particular for

$$
\frac{e^{\beta N_4} \eta}{\eta}(Z_\eta(t), t) = \exp(\beta Z_\eta(t) - (e^\beta - 1)t) \frac{\eta(Z_\eta(t), e^\beta t)}{\eta(Z_\eta(t), t)},
$$

and

$$\frac{e^{\beta N_4^* } \eta^*}{\eta^*}(Z_{\eta^*}^*(t),t) = \exp(-\beta Z_{\eta^*}^*(t) + (e^\beta - 1)t)\frac{\eta^*(Z_{\eta^*}^*(t),e^\beta t)}{\eta^*(Z_{\eta^*}^*(t),t)}.$$

In the framework of the previous section the Girsanov densities are given by the classical expression

$$\Psi(k,t) = \frac{\eta_\alpha(k,t)\eta_\alpha^*(k,t)}{\eta(k,t)\eta^*(k,t)} = \frac{\eta(k,(1+\alpha)t)\eta^*(k,(1+\alpha)t)}{\eta(k,t)\eta^*(k,t)},$$

and

$$\begin{aligned}
\Lambda_{s,t}(Z_{\eta_\alpha}(s),Z_\eta(t)) &= \frac{\eta_\alpha(Z_\eta(t),t)}{\eta(Z_\eta(t),t)}\frac{\eta(Z_{\eta_\alpha}(s),s)}{\eta_\alpha(Z_{\eta_\alpha}(s),s)} \\
&= (1+\alpha)^{Z_\eta(t)-Z_{\eta_\alpha}(s)}e^{-\alpha(t-s)} \\
&\quad \cdot\frac{\eta(Z_\eta(t),(1+\alpha)t)\eta(Z_{\eta_\alpha}(s),s)}{\eta(Z_\eta(t),t)\eta(Z_{\eta_\alpha}(s),(1+\alpha)s)},
\end{aligned}$$

$$\begin{aligned}
\Lambda_{s,t}^*(Z_{\eta_\alpha}^*(s),Z_\eta^*(t)) &= \frac{\eta_\alpha^*(Z_\eta^*(t),t)}{\eta^*(Z_\eta^*(t),t)}\frac{\eta^*(Z_{\eta_\alpha^*}^*(s),s)}{\eta_\alpha^*(Z_{\eta_\alpha^*}^*(s),s)} \\
&= (1+\alpha)^{-(Z_{\eta^*}^*(t)-Z_{\eta_\alpha^*}^*(s))}e^{\alpha(t-s)} \\
&\quad \cdot\frac{\eta^*(Z_\eta^*(t),(1+\alpha)t)\eta^*(Z_{\eta_\alpha}^*(s),s)}{\eta^*(Z_{\eta^*}^*(t),t)\eta^*(Z_{\eta_\alpha^*}^*(s),(1+\alpha)s)}.
\end{aligned}$$

For example in the case of the forward Poisson process we can take

$$\eta(k,t) = 1 \quad \text{and} \quad \eta^*(k,t) = 1_{\{k\geq0\}}\frac{t^k}{k!}e^{-t},$$

in this case the Girsanov densities are given by

$$\Psi(Z_\eta(t),t) = (1+\alpha)^{Z_\eta(t)}e^{-\alpha t}, \qquad \Psi(Z_\eta^*(t),t) = (1+\alpha)^{Z_\eta^*(t)}e^{-\alpha t},$$

and

$$\Lambda_{s,t}(Z_{\eta_\alpha}(s),Z_\eta(t)) = 1, \qquad \Lambda_{s,t}^*(Z_{\eta_\alpha^*}^*(s),Z_\eta^*(t)) = (1+\alpha)^{Z_{\eta^*}^*(t)-Z_{\eta_\alpha^*}^*(s)}e^{-\alpha(t-s)}.$$

6 Appendix

In this section we note (using quantum stochastic calculus and normal martingales, see Proposition 10 below) that the commutation property (10) of the creation operator P with the generator L actually holds in both the Poisson and Wiener cases. Let $(M_t)_{t \in \mathbf{R}_+}$ be a martingale with deterministic angle bracket $d\langle M_t, M_t \rangle = dt$. The multiple stochastic integral $I_n(f_n)$ is defined as

$$I_n(f_n) = n! \int_0^\infty \int_0^{t_n} \cdots \int_0^{t_2} f_n(t_1, \ldots, t_n) dM_{t_1} \cdots dM_{t_n},$$

$f_n \in L^2(\mathbf{R}_+)^{\circ n}$, $n \geq 1$, where $L^2(\mathbf{R}_+)^{\circ n}$ is the space of symmetric square integrable functions on $\mathbf{R}_+^n$, with the isometry property

$$E[I_n(f_n) I_m(g_m)] = n! \mathbf{1}_{\{n=m\}} \langle f_n, g_m \rangle_{L^2(\mathbf{R}_+)^{\circ n}}. \tag{19}$$

We assume that $(M_t)_{t \in \mathbf{R}_+}$ has the chaos representation property (CRP), i.e. every $F \in L^2(\Omega, \mathcal{F}, P)$ has a decomposition

$$F = \sum_{n=0}^\infty I_n(f_n).$$

If $(M_t)_{t \in \mathbf{R}_+}$ is in $L^4(\Omega, \mathcal{F}, P)$ then the CRP implies the existence of a square-integrable predictable process $(\phi_t)_{t \in \mathbf{R}_+}$ such that the structure equation

$$d[M_t, M_t] = dt + \phi_t dM_t, \quad t \in \mathbf{R}_+, \tag{20}$$

is satisfied, cf. Proposition 2 of [3]. Recall that $(M_t)_{t \in \mathbf{R}_+}$ is a compensated Poisson process if $\phi_t = 1$, $t \in \mathbf{R}_+$, and is a Brownian motion for $\phi_t = 0$, $t \in \mathbf{R}_+$. The class of normal martingales also includes the Azéma martingales and we have the following change of variable formula, cf. [3].

Proposition 8. *For $\eta \in \mathcal{C}^2(\mathbf{R} \times \mathbf{R}_+)$, let*

$$\nabla_\phi \eta(k,t) := \begin{cases} \dfrac{\eta(k + \phi_t, t) - \eta(k,t)}{\phi_t}, & \phi_t \neq 0, \\[2ex] \dfrac{\partial \eta}{\partial k}(k,t), & \phi_t = 0, \end{cases}$$

and

$$L_\phi \eta(k,t) := \begin{cases} \dfrac{1}{\phi_t^2}\left(\eta(k+\phi_t, t) - \eta(k,t) - \phi_t \dfrac{\partial \eta}{\partial k}(k,t)\right) + \dfrac{\partial \eta}{\partial t}(k,t), & \phi_t \neq 0, \\[2ex] \dfrac{1}{2}\dfrac{\partial^2 \eta}{\partial k^2}(k,t) + \dfrac{\partial \eta}{\partial t}(k,t), & \phi_t = 0, \end{cases} \tag{21}$$

We have

$$\eta(M_t, t) = \eta(M_0, 0) + \int_0^t \nabla_\phi \eta(M_{s^-}, s) dM_s + \int_0^t L_\phi \eta(M_s, s) ds, \tag{22}$$

$\eta \in \mathcal{C}^2(\mathbf{R} \times \mathbf{R}_+)$.

Let

$$D : L^2(\Omega, d\mathbb{P}) \longrightarrow L^2(\Omega \times \mathbf{R}_+, d\mathbb{P} \times dt)$$

denote the (unbounded) closable gradient operator defined as

$$D_t F = \sum_{n=1}^{\infty} n I_{n-1}(f_n(*, t)), \quad d\mathbb{P} \times dt - a.e.,$$

with $F = \sum_{n=0}^{\infty} I_n(f_n)$. Let also the divergence operator $\delta : L^2(\Omega \times \mathbf{R}_+, d\mathbb{P} \times dt) \longrightarrow L^2(\Omega, d\mathbb{P})$ be defined as

$$\delta(u) = \sum_{n=0}^{\infty} I_{n+1}(\tilde{f}_{n+1}), \quad d\mathbb{P} - a.e.,$$

with $u_t = \sum_{n=0}^{\infty} I_n(f_{n+1}(*, t))$, $t \in \mathbf{R}_+$, and $\tilde{f}_{n+1}$ denotes the symmetrization of f_{n+1} in $n+1$ variables. The stochastic differentials da_t^-, da_t^+ and da_t° can be defined through the identities

$$\int_0^\infty U_t da_t^- F = \int_0^\infty U_t D_t F dt, \quad \int_0^\infty U_t da_t^+ F = \delta(U.F), \quad \int_0^\infty U_t da_t^\circ F = \delta(U.D.F),$$

$F \in \mathcal{S}$, for $(U_t)_{t \in \mathbf{R}_+}$ an adapted operator-valued process satisfying suitable domain conditions. They satisfy the Hudson-Parthasarathy Itô table [5]:

$$
\begin{array}{|c|c|c|c|}
\hline
\cdot & da_t^- & da_t^\circ & da_t^+ \\
\hline
da_t^+ & 0 & 0 & 0 \\
\hline
da_t^\circ & 0 & da_t^\circ & da_t^+ \\
\hline
da_t^- & 0 & da_t^- & dt \\
\hline
\end{array}
\tag{23}
$$

On the other hand, Proposition 18 of [2], which can be interpreted as $dM_t = da_t^- + da_t^+ + \phi_t da_t^\circ$, yields the following representation of the multiplication operator by $\eta(M_t, t)$.

Proposition 9. *The multiplication operator ζ_t^η by $\eta(M_t, t)$ has the decomposition*

$$\zeta_t^\eta = \zeta_0^\eta + \int_0^t \nabla_\phi \eta(M_s, s) da_s^- + \int_0^t \nabla_\phi \eta(M_s, s) da_s^+ + \int_0^t \phi_s \nabla_\phi \eta(M_s, s) da_s^\circ$$

$$+ \int_0^t L_\phi \eta(M_s, s) ds. \tag{24}$$

In other terms we have

$$\eta(M_t, t) F = \eta(M_0, 0) F + \int_0^t \nabla_\phi \eta(M_s, s) D_s F ds + \delta(1_{[0,t]}(\cdot) F \nabla_\phi \eta(M., \cdot))$$

$$+ \delta(1_{[0,t]}(\cdot) F \phi. \nabla_\phi \eta(M., \cdot) D.F) + F \int_0^t L_\phi \eta(M_s, s) ds,$$

for sufficiently regular F.

Wiener case

We have

$$Pf(x,t) = xf(x,t) - tf'(x,t),$$

and $(P^n \mathbf{1}(\cdot,t))_{n\in\mathbf{N}} = (H_n(\cdot,t))_{n\in\mathbf{N}}$ is the family of Hermite polynomials with parameter $t > 0$.

Poisson case

We have

$$Pf(k,t) = kf(k-1,t) - tf(k,t),$$

and $(P^n \mathbf{1}(\cdot,t))_{n\in\mathbf{N}} = (C_n(\cdot,t))_{n\in\mathbf{N}}$ is the family of Charlier polynomials with parameter $t > 0$.

It is known that $I_n(1^{\circ n}_{[0,t]})$ is function of M_t only in the Wiener and Poisson cases, i.e. for constant deterministic ϕ, cf. [10]. More precisely we have the following.

Remark 1. The operator $P : \mathcal{C}(\mathbf{R} \times \mathbf{R}_+) \to \mathcal{C}(\mathbf{R} \times \mathbf{R}_+)$ satisfies the relation

$$[Pf](X_t,t) = a_t^+[f(X_t,t)], \qquad f \in \mathcal{C}(\mathbf{R} \times \mathbf{R}_+), \tag{25}$$

with $X_t = M_t$ in the Wiener case and $X_t = M_t + t$ in the Poisson case, i.e. for constant deterministic ϕ, cf. [10].

For $f = \mathbf{1}$, (25) reads:

$$[P^n \mathbf{1}](X_t,t) = I_n(1^{\circ n}_{[0,t]}).$$

The mapping P can be seen as a finite-dimensional projection of the creation operator a_t^+. Finally we give a proof of the commutation relation (10) in the general context of normal martingales.

Proposition 10. *We have*

$$[L_\phi, P] = 0 \quad and \quad [\nabla_\phi, P] = I \quad .$$

Proof. From (24) and the Itô table (23) we get

$$
\begin{aligned}
d(a_t^+ \zeta_t^\eta) &= a_t^+ d\zeta_t^\eta + \zeta_t^\eta da_t^+ + da_t^+ \cdot d\zeta_t^\eta \\
&= a_t^+ d\zeta_t^\eta + \zeta_t^\eta da_t^+ + \eta(X_t,t)da_t^+ \cdot (da_t^- + da_t^+ + \phi_t da_t^\circ) \\
&\quad + L_\phi \eta(X_t,t)da_t^+ \cdot dt \\
&= a_t^+ d\zeta_t^\eta + \zeta_t^\eta da_t^+ \\
&= a_t^+ \nabla_\phi \eta(X_t,t)da_t^+ + a_t^+ (L_\phi \eta(X_t,t))dt + \eta(X_t,t)da_t^+ \\
&= (P\nabla_\phi \eta(X_t,t) + \eta(X_t,t))da_t^+ + PL_\phi \eta(X_t,t)dt,
\end{aligned}
$$

hence

$$d(a_t^+ \zeta_t^\eta)\mathbf{1} = (P\nabla_\phi \eta(X_t,t) + \eta(X_t,t))dX_t + PL_\phi \eta(X_t,t)dt.$$

On the other hand from the classical Itô formula (22) we have

$$d(P\eta(X_t,t)) = \nabla_\phi P\eta(X_t,t)dX_t + L_\phi P\eta(X_t,t)dt.$$

The identification $d(a_t^+ \zeta_t^f)\mathbf{1} = d(P\eta(X_t,t))$ due to Definition 1 we obtain

$$\nabla_\phi P\eta = \eta + P\nabla_\phi \eta, \quad \text{and} \quad PL_\phi \eta = L_\phi P\eta.$$

$\square$

The relation $[\nabla_\phi, P] = I$ can be viewed as a one-dimensional projection of the canonical commutation relation

$$D_t \delta(u) = \delta(D_t u) + u_t$$

between D and δ, or $[a_t^-, a_t^+] = tI$.

References

1. A. Beurling. An automorphism of product measures. *Ann. of Math. (2)*, 72:189–200, 1960.
2. Ph. Biane. Calcul stochastique non-commutatif. In *Ecole d'Eté de Probabilités de Saint-Flour*, volume 1608 of *Lecture Notes in Mathematics*. Springer-Verlag, 1993.
3. M. Émery. On the Azéma martingales. In *Séminaire de Probabilités XXIII*, volume 1372 of *Lecture Notes in Mathematics*, pages 66–87. Springer Verlag, 1990.
4. T. Hida. *Brownian Motion*. Springer Verlag, 1981.
5. R.L. Hudson and K. R. Parthasarathy. Quantum Ito's formula and stochastic evolutions. *Comm. Math. Phys.*, 93(3):301–323, 1984.
6. P. Lescot and J.-C. Zambrini. Isovectors for the Hamilton-Jacobi-Bellman equation, formal stochastic differentials and first integrals in Euclidean Quantum Mechanics. In R. Dalang, M. Dozzi, and F. Russo, editors, *Seminar on Stochastic Analysis, Random Fields and Applications (Ascona, 2002)*, volume 58 of *Progress in Probability*, pages 187–202. Birkhäuser, Basel, 2004. See also C. R. Math. Acad. Sci. Paris 335 (2002) 263–266.
7. P. Lescot and J.-C. Zambrini. Probabilistic deformation of contact geometry, diffusion processes and their quadratures. In R. Dalang, M. Dozzi, and F. Russo, editors, *Seminar on Stochastic Analysis, Random Fields and Applications (Ascona, 2005)*, volume 59 of *Progress in Probability*, pages 205–227, Basel, 2008. Birkhäuser.
8. D. Levi, L. Vinet, and P. Winternitz, editors. *Symmetries and integrability of difference equations*, volume 9 of *CRM Proceedings & Lecture Notes*, Providence, RI, 1996. American Mathematical Society.
9. P.J. Olver. *Applications of Lie groups to differential equations*. Springer-Verlag, New York, second edition, 1993.
10. N. Privault, J.L. Solé, and J. Vives. Chaotic Kabanov formula for the Azéma martingales. *Bernoulli*, 6(4):633–651, 2000.

11. N. Privault and J.-C. Zambrini. Markovian bridges and reversible diffusion processes with jumps. *Ann. Inst. H. Poincaré Probab. Statist.*, 40(5):599–633, 2004.
12. N. Privault and J.-C. Zambrini. Euclidean quantum mechanics in the momentum representation. *Journal of Mathematical Physics*, 46, 2005.
13. J.-C. Zambrini. Variational processes and stochastic versions of mechanics. *J. Math. Phys.*, 27(9):2307–2330, 1986.

On the tail distributions of the supremum and the quadratic variation of a càdlàg local martingale

Shunsuke Kaji

Department of Mathematics, Graduate School of Science, Osaka University
Machikaneyamachou 1-1, Toyonaka, Osaka Japan 560-0043
e-mail: kaji@math.sci.osaka-u.ac.jp

1 Introduction

We study a tail property of the distribution of the supremum and the quadratic variation of a local martingale. In the case when the local martingale is continuous, there are works by Azéma, Gundy, and Yor [1], Novikov [9], Elworthy, Li, and Yor [2], Madan and Yor [8], Takaoka [10] etc. Recently, Liptser and Novikov [7] extended these studies to the case of a local martingale with uniformly bounded jumps; here is their main result:

Theorem 1.1 *Let $M = \{M_t\}_{t \in \mathbf{R}_+}$ be a locally square integrable càdlàg martingale defined on a filtered probability space $(\Omega, \mathcal{F}, \{\mathcal{F}_t\}_{t \in \mathbf{R}_+}, P)$ with standard general conditions. Assume that $\langle M \rangle_\infty = \lim_{t \to \infty} \langle M \rangle_t < \infty$ a.s and $\{M_\tau^+\}_{\tau \in \mathcal{T}}$ is uniformly integrable, where $\mathcal{T}$ is the set of stopping times τ. Then*

(i) $0 \leq E[M_\infty] \leq E[M_\infty^+] < \infty.$

Furthermore,

(ii) if $\{\Delta M_\tau\}_{\tau \in \mathcal{T}}$ is uniformly integrable, then

$$\lim_{\lambda \to \infty} \lambda P\big(\sup_{t \in \mathbf{R}_+} (M_t^-) > \lambda \big) = E[M_\infty];$$

(iii) if $|\Delta M| \leq K$ and $E[e^{\epsilon M_\infty}] < \infty$ for some $K > 0$ and $\epsilon > 0$, then

$$\lim_{\lambda \to \infty} \lambda P\big(\sqrt{\langle M \rangle_\infty} > \lambda \big) = \lim_{\lambda \to \infty} \lambda P\big(\sqrt{[M]_\infty} > \lambda \big) = \sqrt{\frac{2}{\pi}} E[M_\infty].$$

We shall extend this theorem to the case of a local martingale whose jumps are no longer uniformly bounded; we shall only assume the existence of some exponential moment involving the compensator of the counting measure of ΔM. In Remark 3.1, we will explain how our main result includes Theorem 1.1.

2 Preliminaries

Let $(\Omega, \mathcal{F}, P)$ be a complete probability space with a right-continuous filtration $\{\mathcal{F}_t\}_{t \in \mathbf{R}_+}$ verifying $\mathcal{F}_0 \supset \mathcal{N}$, where $\mathbf{R}_+ = [0, \infty)$ and $\mathcal{N}$ is the class of P-null sets; and let $M = \{M_t\}_{t \in \mathbf{R}_+}$ be a càdlàg local martingale with $M_0 = 0$ defined on $(\Omega, \mathcal{F}, \{\mathcal{F}_t\}_{t \in \mathbf{R}_+}, P)$.

2.1 A random measure and an integral process

Definition 2.1 *A random measure ν on $\mathbf{R}_+ \times \mathbf{R}$ is a family $\{\nu(\omega, dt dx)\}_{\omega \in \Omega}$ of measures on $(\mathbf{R}_+ \times \mathbf{R}, \mathcal{B}(\mathbf{R}_+) \times \mathcal{B}(\mathbf{R}))$, satisfying $\nu(\cdot, \{0\} \times \mathbf{R}) = 0$ identically.*

We introduce the σ-field $\tilde{\mathcal{P}} = \mathcal{P} \times \mathcal{B}(\mathbf{R})$ defined on $\Omega \times \mathbf{R}_+ \times \mathbf{R}$, where $\mathcal{P}$ is the predictable σ-field on $\Omega \times \mathbf{R}_+$.

Definition 2.2 *For any $\tilde{\mathcal{P}}$-measurable function $\alpha(\omega, t, x)$ we define the integral process $\alpha * \nu$ as*

$$(\alpha * \nu)_t = \begin{cases} \int_{(0,t] \times \mathbf{R}} \alpha(\cdot, s, x) \nu(\cdot, ds\, dx) & \text{if } \int_{(0,t] \times \mathbf{R}} |\alpha(\cdot, s, x)| \nu(\cdot, ds\, dx) < \infty \\ \infty & \text{otherwise.} \end{cases}$$

Analogously, a random measure ν is predictable if the process $\alpha * \nu$ is predictable for every $\tilde{\mathcal{P}}$-measurable function α.

2.2 A counting measure and its compensator

Introduce the random measure μ such that $\mu(\cdot, (0, t] \times U) = \sum_{0 < s \leq t} 1_U(\Delta M_s)$ for all $t \in (0, \infty)$ and $U \in \mathcal{B}(\mathbf{X})$, where $\mathbf{X} = \mathbf{R} - \{0\}$; then there exist $U_n \in \mathcal{B}(\mathbf{X}), n = 1, 2, \ldots$ such that $U_n \nearrow \mathbf{X}$ and $E[\mu(\cdot, (0, t] \times U_n)] < \infty$ for all $t > 0, n = 1, 2, \ldots$ (see Proposition 2.1.16 of [5]). So, given μ, there exists a unique random measure $\hat{\mu}$ with the following properties (see Theorem 2.1.8 of [5]):

(i) $\hat{\mu}$ is predictable, and
(ii) for every $\tilde{\mathcal{P}}$-measurable function α such that $|\alpha| * \mu \in \mathcal{A}_{loc}^+$, $\alpha * \mu - \alpha * \hat{\mu}$ is a local martingale,

where $\mathcal{A}_{loc}^+$ is the class of all increasing and locally integrable processes.

Definition 2.3 *$\hat{\mu}$ is called the* compensator *of μ. Moreover, we define a random measure $\hat{\mu}^c$ by*

$$\hat{\mu}^c(\cdot, ds\, dx) = 1_{\{\mu(\cdot, \{s\} \times \mathbf{X}) = 0\}} \mu(\cdot, ds\, dx).$$

2.3 A canonical representation for a local martingale

First, assuming M is locally square integrable, we need the definition of a stochastic integral with respect to $\mu - \hat{\mu}$.

Definition 2.4 *Given a $\tilde{\mathcal{P}}$-measurable function α such that $(\tilde{\alpha}^2 * \mu)^{\frac{1}{2}} \in \mathcal{A}_{loc}^+$, where $\tilde{\alpha}(\cdot, t, x) = \alpha(\cdot, t, x) - \int_{\mathbf{X}} \alpha(\cdot, t, x) \hat{\mu}(\cdot, \{t\}, dx)$, what we call the stochastic integral $\alpha * (\mu - \hat{\mu})$ with respect to $\mu - \hat{\mu}$ is any purely discontinuous[1] local martingale X such that $\tilde{\alpha}$ and ΔX are indistinguishable. (To justify this definition, see Definition 2.1.27 of [5].)*

According to Corollary 2.2.38 of [5], one has the canonical decomposition

$$M = M^c + x * (\mu - \hat{\mu}),$$

where M^c is a continuous local martingale with $M_0^c = 0$ and $\sum_{0 < s \le \cdot} (\Delta M_s)^2 \in \mathcal{A}_{loc}^+$. Moreover, recall that $\int_{\mathbf{X}} x \hat{\mu}(\cdot, \{t\}, dx) = 0$ a.s. for $t > 0$ and that $\langle x * (\mu - \hat{\mu}) \rangle_t = (x^2 * \hat{\mu})_t < \infty$ a.s. for $t > 0$.

2.4 A stochastic exponential

Definition 2.5 *Let $X = \{X_t\}_{t \in \mathbf{R}_+}$ be a càdlàg semimartingale with $X_0 = 0$ defined on $(\Omega, \mathcal{F}, \{\mathcal{F}_t\}_{t \in \mathbf{R}_+}, P)$. The stochastic exponential $\mathcal{E}(X) = \{\mathcal{E}(X)_t\}_{t \in \mathbf{R}_+}$ is defined as the (up to indistinguishability unique) solution $Y = \{Y_t\}_{t \in \mathbf{R}_+}$ to the SDE*

$$Y_t = 1 + \int_0^t Y_{s-} dX_s, \quad t \in \mathbf{R}_+.$$

According to Theorem 2.19 in Kallsen and Shiryaev [6], the stochastic exponential $\mathcal{E}\left(-\lambda M^c + \frac{e^{-\lambda x} - 1}{1 + \Delta(\psi_\lambda * \hat{\mu})} * (\mu - \hat{\mu})\right)$ is a nonnegative local martingale and it follows that

$$\mathcal{E}\left(-\lambda M^c + \frac{e^{-\lambda x} - 1}{1 + \Delta(\psi_\lambda * \hat{\mu})} * (\mu - \hat{\mu})\right)$$
$$= \exp\left\{-\lambda M - \frac{\lambda^2}{2}\langle M^c \rangle - (\psi_\lambda * \hat{\mu}^c) - \sum_{0 < s \le \cdot} \log(1 + \Delta(\psi_\lambda * \hat{\mu})_s)\right\},$$

where $\lambda > 0$ and $\psi_\lambda(x) = e^{-\lambda x} - 1 + \lambda x$. We shall write the expression above as $\mathcal{E}(\lambda)$.

---

[1] A local martingale X is called purely discontinuous if $X_0 = 0$ and if XY is a local martingale for any continuous local martingale Y.

3 Main result

Assume that the limit $M_\infty = \lim_{t \to \infty} M_t$ exists a.s. in $\mathbf{R}$ and that $\{M_\tau^-\}_{\tau \in \mathcal{T}}$ is uniformly integrable, where $\mathcal{T}$ is the set of all stopping times. First, here is our result concerning the tail distribution of the supremum of M:

Theorem 3.1 *One has*

(i) $-\infty < -E[M_\infty^-] \le E[M_\infty] \le 0.$

Moreover, if $\{\Delta M_\tau\}_{\tau \in \mathcal{T}}$ *is uniformly integrable, then*

(ii) $\displaystyle \lim_{\lambda \to \infty} \lambda P\big(\sup_{t \in \mathbf{R}_+} M_t > \lambda \big) = -E[M_\infty].$

Second, here is a statement concerning the tail distribution of the quadratic variation of M:

Theorem 3.2 *Assume that M is locally square integrable, with $\lim_{t \to \infty} \langle M \rangle_t = \langle M \rangle_\infty < \infty$ a.s., and that there exists $\lambda_0 > 0$ such that*

(∗) $E[\exp\{\lambda_0 M_\infty^- + (|\phi_{\lambda_0}| 1_{\{|x|>K\}} * \hat{\mu})_\infty\}] < \infty$

for some $K > 0$, where $\phi_\lambda(x) = e^{-\lambda x} - 1 + \lambda x - \frac{\lambda^2}{2} x^2$. Then

(i) $\lim_{\lambda \to \infty} \lambda P\big(\sqrt{\langle M \rangle_\infty} > \lambda \big) = -\sqrt{\frac{2}{\pi}}\, E[M_\infty],$

(ii) $\lim_{\lambda \to \infty} \lambda P\big(\sqrt{[M]_\infty} > \lambda \big) = -\sqrt{\frac{2}{\pi}}\, E[M_\infty].$

Corollary 3.1 *Under the assumptions of Theorems 3.1 and 3.2, $\{M_\tau\}_{\tau \in \mathcal{T}}$ is uniformly integrable if and only if*

$$\lim_{\lambda \to \infty} \lambda P\big(\sup_{t \in \mathbf{R}_+} M_t > \lambda \big) = \lim_{\lambda \to \infty} \lambda P\big(\sqrt{\langle M \rangle_\infty} > \lambda \big)$$

$$= \lim_{\lambda \to \infty} \lambda P\big(\sqrt{[M]_\infty} > \lambda \big) = 0.$$

Remark 3.1 *If for some $\lambda_0 > 0$ and $K > 0$*

$$E[\exp(\lambda_0 M_\infty^-)] < \infty \quad and \quad |\Delta M| \le K,$$

then (∗) *is satisfied, because $|\Delta M| \le K$ implies $\mu(\cdot, \mathbf{R}_+ \times \{|x| > K\}) = 0$ a.s. which gives the result $\hat{\mu}(\cdot, \mathbf{R}_+ \times \{|x| > K\}) = 0$ a.s.*

4 Two elementary lemmas

This section prepares the proofs of Theorems 3.1 and 3.2 with two elementary facts; their proofs are left to the reader.

Lemma 4.1 *Put* $\rho(\lambda, x) = \frac{e^{-\lambda x} - 1}{\lambda} + x - \frac{\lambda}{2}x^2$. *For each fixed* $x \neq 0$, *the function* $\lambda \to |\rho(\lambda, x)|$ *is increasing.*

Lemma 4.2 *Fix* $K > 0$ *and put* $\phi(\lambda, x) = e^{-\lambda x} - 1 + \lambda x - \frac{\lambda^2}{2}x^2$. *For each fixed* $\lambda > 0$ *there exist* $c_1, c_2 > 0$ *such that*

$$\forall x \neq 0 \qquad x^2 \leq c_1 |\phi(\lambda, x)| 1_{\{|x|>K\}} + c_2 \sqrt{|x|} 1_{\{0<|x|\leq K\}} \, .$$

5 Proof of Theorem 3.1

5.1 Proof of (i)

Let $\{\tau_n\}_{n=1}^{\infty}$ be a sequence of stopping times such that $\tau_n \nearrow \infty$ and M^{τ_n} is a martingale for each n. Then $E[M_{\tau_n}^+] = E[M_{\tau_n}^-]$, $n \geq 1$. Since $\{M_\tau^-\}_{\tau \in \mathcal{T}}$ is uniformly integrable, we have $\lim_{t \to \infty} E[M_t^-] = E[M_\infty^-] < \infty$. By Fatou's lemma the last two results imply

$$E[M_\infty^+] \leq \liminf_{n \to \infty} E[M_{\tau_n}^+] = \liminf_{n \to \infty} E[M_{\tau_n}^-] = E[M_\infty^-] < \infty.$$

Hence, $-\infty < -E[M_\infty^-] \leq E[M_\infty](= E[M_\infty^+] - E[M_\infty^-]) \leq 0$ holds.

5.2 Proof of (ii)

Set $\tau_\lambda = \inf\{t > 0 | M_t > \lambda\}$ where $\inf \emptyset = \infty$; then $M_t \leq \lambda$ for all $0 < t < \tau_\lambda$. Therefore $M_{\tau_\lambda-} \leq \lambda$. The last two facts imply $M_t^{\tau_\lambda} \leq \lambda + |\Delta M_{\tau_\lambda}|$ for all $t \in \mathbf{R}_+$, and therefore $|M_t^{\tau_\lambda}| = M_t^{\tau_\lambda} + 2M_{\tau_\lambda \wedge t}^- \leq \lambda + |\Delta M_{\tau_\lambda \wedge t}| + 2M_{\tau_\lambda \wedge t}^-$ for all $t \in \mathbf{R}_+$. By uniform integrability of $\{M_\tau^-\}_{\tau \in \mathcal{T}}$ and uniform integrability of $|\Delta M_\tau|_{\tau \in \mathcal{T}}$, the last result implies uniform integrability of $\{M_t^{\tau_\lambda}\}_{t \in \mathbf{R}_+}$, and so $E[M_{\tau_\lambda}] = 0$.

Now, observe that

$$0 = E[M_{\tau_\lambda}] = E[M_\infty; \sup_{t \in \mathbf{R}_+} M_t \leq \lambda] + E[M_{\tau_\lambda}; \sup_{t \in \mathbf{R}_+} M_t > \lambda].$$

Since $\{\tau_\lambda < \infty\} = \{\sup_{t \in \mathbf{R}_+} M_t > \lambda\}$, there exists a sequence of stopping times $\{\tau_\lambda^n\}_{n=1}^{\infty}$ such that $M_{\tau_\lambda^n} > \lambda$ for all n and $\tau_\lambda^n \searrow \tau_\lambda$ on $\{\sup_{t \in \mathbf{R}_+} M_t > \lambda\}$, and therefore $M_{\tau_\lambda} \geq \lambda$ on that set. This and the last result imply

$$0 \geq E[M_\infty; \sup_{t \in \mathbf{R}_+} M_t \leq \lambda] + \lambda P\big(\sup_{t \in \mathbf{R}_+} M_t > \lambda\big),$$

and so

$$0 \geq E[M_\infty] + \limsup_{\lambda \to \infty} \lambda P\left(\sup_{t \in \mathbf{R}_+} M_t > \lambda \right). \qquad (1)$$

On the other hand, we have

$$0 = E[M_\infty; \sup_{t \in \mathbf{R}_+} M_t \leq \lambda] + E[M_{\tau_\lambda}; \sup_{t \in \mathbf{R}_+} M_t > \lambda]$$

$$\leq E[M_\infty; \sup_{t \in \mathbf{R}_+} M_t \leq \lambda] + E[\lambda + |\Delta M_{\tau_\lambda}|; \sup_{t \in \mathbf{R}_+} M_t > \lambda]$$

$$\leq E[M_\infty; \sup_{t \in \mathbf{R}_+} M_t \leq \lambda] + \lambda P(\sup_{t \in \mathbf{R}_+} M_t > \lambda) + E[|\Delta M_{\tau_\lambda}|].$$

Now, since M_t tends to M_∞ when $t \to \infty$, M_{t-} also tends to M_∞, and $\Delta M_t \to 0$. So, by uniform integrability of $\{\Delta M_\tau\}_{\tau \in \mathcal{T}}$, we have $\lim_{\lambda \to \infty} E[|\Delta M_{\tau_\lambda}|] = 0$ and letting $\lambda \to \infty$ in the above inequality gives

$$0 \leq E[M_\infty] + \liminf_{\lambda \to \infty} \lambda P\left(\sup_{t \in \mathbf{R}_+} M_t > \lambda \right). \qquad (2)$$

From the inequalities (1) and (2) we obtain the desired conclusion.

6 Proof of Theorem 3.2

First, notice that (i) of the theorem 3.1 holds and $\langle M \rangle_\infty < \infty$ a.s. implies the existence of $[M]_\infty < \infty$ a.s.(see [5]).

6.1 Five lemmas

Before we present five lemmas, we point out that for all $0 < \lambda \leq \lambda_0$

$$(|\phi_\lambda| * \hat{\mu})_\infty = \lim_{t \to \infty} (|\phi_\lambda| * \hat{\mu})_t < \infty \quad \text{a.s.} \qquad (3)$$

$$(\psi_\lambda * \hat{\mu})_\infty = \lim_{t \to \infty} (\psi_\lambda * \hat{\mu})_t < \infty \quad \text{a.s., and} \qquad (4)$$

$$\mathcal{E}(\lambda)_\infty = \lim_{t \to \infty} \mathcal{E}(\lambda)_t < \infty \quad \text{a.s.} \qquad (5)$$

To see (3), observe that

$$(|\phi_\lambda| * \hat{\mu})_\infty = (|\phi_\lambda| 1_{\{|x| \leq K\}} * \hat{\mu})_\infty + (|\phi_\lambda| 1_{\{|x| > K\}} * \hat{\mu})_\infty. \qquad (6)$$

By Lemma 4.1 the second term of the right-hand side of (6) is

$$\leq (|\phi_{\lambda_0}| 1_{\{|x| > K\}} * \hat{\mu})_\infty,$$

which is $< \infty$ a.s. by assumption $(*)$, and by choosing the constant $c_0 > 0$ such that

$$\left| e^{-x} - 1 + x - \frac{x^2}{2} \right| \le c_0 |x|^3 \quad \text{for all } |x| \le \lambda_0 K, \tag{7}$$

the first term of the right-hand side of (6) is

$$\le c_0 \lambda^3 (|x|^3 1_{\{|x| \le K\}} * \hat{\mu})_\infty \le c_0 \lambda^3 K (x^2 1_{\{|x| \le K\}} * \hat{\mu})_\infty,$$

which is $\le c_0 \lambda^3 K \langle M \rangle_\infty < \infty$ a.s. by $(x^2 * \hat{\mu})_\infty \le \langle M \rangle_\infty < \infty$ a.s. Hence,(3) holds. On the other hand, to see (4), observe

$$(\psi_\lambda * \hat{\mu})_\infty = (\phi_\lambda * \hat{\mu})_\infty + \frac{\lambda^2}{2} (x^2 * \hat{\mu})_\infty. \tag{8}$$

By Lemma 4.1 the first term on the right-hand side of (8) is

$$\le (|\phi_{\lambda_0}| * \hat{\mu})_\infty,$$

which is $< \infty$ a.s. by (3), and by $(x^2 * \hat{\mu})_\infty \le \langle M \rangle_\infty < \infty$ a.s. the second term of the right-hand side of (8) is

$$\le \frac{\lambda_0{}^2}{2} \langle M \rangle_\infty < \infty \quad \text{a.s.}$$

Hence, (4) holds. Finally, it follows that $(\psi_\lambda * \hat{\mu}^c)_t \le (\psi_\lambda * \hat{\mu})_\infty$ and

$$\sum_{0 < s \le t} \log(1 + \Delta(\psi_\lambda * \hat{\mu})_s) \le \sum_{0 < s \le t} \Delta(\psi_\lambda * \hat{\mu})_s \le (\psi_\lambda * \hat{\mu})_\infty,$$

where the right-hand sides of the above two inequalities are $< \infty$ a.s. by (4). Hence, (5) holds.

Lemma 6.1

$$\lim_{\lambda \to 0} \frac{1}{\lambda} \left(E[e^{-\lambda M_\infty - \frac{\lambda^2}{2} \langle M \rangle_\infty}] - E[e^{-\frac{\lambda^2}{2} \langle M \rangle_\infty}] \right) = -E[M_\infty]$$

Proof: First, it is easy to see that

$$\frac{1}{\lambda} \left(e^{-\lambda M_\infty - \frac{\lambda^2}{2} \langle M \rangle_\infty} - e^{-\frac{\lambda^2}{2} \langle M \rangle_\infty} \right) = e^{-\frac{\lambda^2}{2} \langle M \rangle_\infty} \left(\frac{e^{-\lambda M_\infty} - 1}{\lambda} \right)$$

$$\to -M_\infty \text{ a.s. as } \lambda \to 0 \tag{9}$$

Moreover, by using the fact:

$$\left| \frac{e^{\nu x} - 1}{\nu} \right| \le e^{\nu x} 1_{\{x \ge 0\}} + x^- 1_{\{x \le 0\}}, \ \nu > 0,$$

we see that for all $0 < \lambda \le \lambda_0$

$$\left|\frac{1}{\lambda}(e^{-\lambda M_\infty - \frac{\lambda^2}{2}\langle M\rangle_\infty} - e^{-\frac{\lambda^2}{2}\langle M\rangle_\infty})\right| = \left|e^{-\frac{\lambda^2}{2}\langle M\rangle_\infty}\left(\frac{e^{-\lambda M_\infty}-1}{\lambda}\right)\right|$$
$$\leq \left|\frac{e^{-\lambda M_\infty}-1}{\lambda}\right|$$
$$\leq e^{-\lambda M_\infty}1_{\{M_\infty \leq 0\}} + M_\infty^+ 1_{\{M_\infty \geq 0\}}$$
$$\leq e^{\lambda_0 M_\infty^-} + M_\infty^+ . \tag{10}$$

Then, notice that $E[e^{\lambda_0 M_\infty^-} + M_\infty^+] < \infty$ holds (by assumption $(*)$ and (i) of Theorem 3.1). According to the dominated convergence theorem, $(9),(10)$, and the above notation imply the desired conclusion.

Lemma 6.2

$$E[\mathcal{E}(\lambda)_\infty] = 1 \qquad 0 < \forall \lambda < \lambda_0$$

Proof: Fix $0 < \lambda < \lambda_0$. We know that $\{\mathcal{E}(\lambda)_t\}_{t\in\mathbf{R}_+}$ is a nonnegative local martingale and has the initial value 1. So, to be complete, it suffices to show that $\{\mathcal{E}(\lambda)_\tau\}_{\tau\in\mathcal{T}}$ is uniformly integrable. To do so, it is enough to show the following two conditions:

(a) $\mathcal{E}(\lambda)_\tau \leq E[e^{\lambda M_\infty^-}|\mathcal{F}_\tau]$ for all $\tau \in \mathcal{T}$;
(b) $\{E[e^{\lambda M_\infty^-}|\mathcal{F}_\tau]\}_{\tau\in\mathcal{T}}$ is uniformly integrable.

Before we show these conditions, we notice that

$$E[e^{\lambda M_\infty^-}] \leq E[e^{\lambda_0 M_\infty^-}] < \infty$$

holds. Now, to show (a), observe that

$$E[M_\infty^-|\mathcal{F}_\tau] \geq M_\tau^-, \ \tau \in \mathcal{T},$$

which holds by Jensen's inequality and by uniform integrability of $\{M_\tau^-\}_{\tau\in\mathcal{T}}$. By this observation and Jensen's inequality

$$e^{\lambda M_\tau^-} \leq e^{E[\lambda M_\infty^-|\mathcal{F}_\tau]} \leq E[e^{\lambda M_\infty^-}|\mathcal{F}_\tau], \ \tau \in \mathcal{T},$$

and then, since it is trivial that $\mathcal{E}(\lambda)_\tau \leq e^{\lambda M_\tau^-}, \tau \in \mathcal{T}$, (a) holds. Finally, we will show (b). Pick $1 < p \leq \frac{\lambda_0}{\lambda}$. So we obtain $p\lambda M_\infty^- \leq \lambda_0 M_\infty^-$, and then, by Jensen's inequality and assumption $(*)$ it follows that

$$E[(E[e^{\lambda M_\infty^-}|\mathcal{F}_\tau])^p] \leq E[E[e^{p\lambda M_\infty^-}|\mathcal{F}_\tau]] = E[e^{p\lambda M_\infty^-}] \leq E[e^{\lambda_0 M_\infty^-}] < \infty, \ \tau \in \mathcal{T},$$

and so $\sup_{\tau\in\mathcal{T}} E[(E[e^{\lambda M_\infty^-}|\mathcal{F}_\tau])^p] < \infty$. Hence, (b) holds.

Lemma 6.3

$$\lim_{\lambda\to 0} E[e^{\lambda_0 M_\infty^- - \frac{\lambda^2}{2}(x^2*\hat{\mu}^c)_\infty} \cdot \frac{1}{\lambda}|e^{-(\phi_\lambda*\hat{\mu}^c)_\infty} - 1|] = 0$$

Proof: According to the dominated convergence theorem, to prove this lemma, it suffices to show the following facts: For all $0 < \lambda < \lambda_0 \wedge \frac{1}{4c_0 K}$,

(a) $e^{-\frac{\lambda^2}{2}(x^2 * \hat{\mu}^c)_\infty} \cdot \frac{1}{\lambda}|1 - e^{-(\phi_\lambda * \hat{\mu}^c)_\infty}| \leq 4c_0 K e^{-1} e^{\lambda_0 \Phi_\infty^c} + \Phi_\infty^c e^{\lambda \Phi_\infty^c}$,

where $\Phi^c = \frac{1}{\lambda_0}(|\phi_{\lambda_0}| 1_{\{|x| > K\}} * \hat{\mu}^c)$. For all $0 < \lambda < \lambda_0$,

(b) $E[e^{\lambda_0 M_\infty^-} \Phi_\infty^c e^{\lambda \Phi_\infty^c}] < \infty$.

(c) $\lim_{\lambda \to 0} \frac{1}{\lambda}|1 - e^{-(\phi_\lambda * \hat{\mu}^c)_\infty}| = 0$ a.s.

First, we will show (a). Fix $0 < \lambda < \lambda_0 \wedge \frac{1}{4c_0 K}$. By the fact $|\frac{1 - e^{\epsilon x}}{\epsilon}| \leq |x| e^{\epsilon |x|}$, valid for $\epsilon > 0$ (recall $e^x = \sum_{n=0}^\infty \frac{x^n}{n!}$), we have

$$\frac{1}{\lambda}|1 - e^{-(\phi_\lambda * \hat{\mu}^c)_\infty}| \leq \frac{1}{\lambda}|(\phi_\lambda * \hat{\mu}^c)_\infty| \cdot e^{|(\phi_\lambda * \hat{\mu}^c)_\infty|}$$
$$\leq \frac{1}{\lambda}(|\phi_\lambda| * \hat{\mu}^c)_\infty \cdot e^{(|\phi_\lambda| * \hat{\mu}^c)_\infty},$$

and so

$$e^{-\frac{\lambda^2}{2}(x^2 * \hat{\mu}^c)_\infty} \cdot \frac{1}{\lambda}|1 - e^{-(\phi_\lambda * \hat{\mu}^c)_\infty}|$$
$$\leq \frac{1}{\lambda}(|\phi_\lambda| * \hat{\mu}^c)_\infty e^{-\frac{\lambda^2}{4}(x^2 * \hat{\mu}^c)_\infty} \cdot e^{((|\phi_\lambda| - \frac{\lambda^2}{4} x^2) * \hat{\mu}^c)_\infty} = I_1 \cdot I_2.$$

We will compute I_1. By (7) we have

$$I_1 = \{\frac{1}{\lambda}(|\phi_\lambda| 1_{\{|x| \leq K\}} * \hat{\mu}^c)_\infty + \frac{1}{\lambda}(|\phi_\lambda| 1_{\{|x| > K\}} * \hat{\mu}^c)_\infty\}$$
$$\times e^{-\frac{\lambda^2}{4}(x^2 1_{\{|x| \leq K\}} * \hat{\mu}^c)_\infty} e^{-\frac{\lambda^2}{4}(x^2 1_{\{|x| > K\}} * \hat{\mu}^c)_\infty}$$
$$\leq \frac{1}{\lambda}(|\phi_\lambda| 1_{\{|x| \leq K\}} * \hat{\mu}^c)_\infty e^{-\frac{\lambda^2}{4}(x^2 1_{\{|x| \leq K\}} * \hat{\mu}^c)_\infty} + \frac{1}{\lambda}(|\phi_\lambda| 1_{\{|x| > K\}} * \hat{\mu}^c)_\infty$$
$$\leq 4c_0 K \cdot \frac{\lambda^2}{4}(x^2 1_{\{|x| \leq K\}} * \hat{\mu}^c)_\infty e^{-\frac{\lambda^2}{4}(x^2 1_{\{|x| \leq K\}} * \hat{\mu}^c)_\infty} + \frac{1}{\lambda}(|\phi_\lambda| 1_{\{|x| > K\}} * \hat{\mu}^c)_\infty.$$

From $xe^{-x} \leq e^{-1}$, the right-hand side of the last inequality is

$$\leq 4c_0 K e^{-1} + \frac{1}{\lambda}(|\phi_\lambda| 1_{\{|x| > K\}} * \hat{\mu}^c)_\infty.$$

We now compute I_2. By (7) we have

$$I_2 \leq e^{((|\phi_\lambda| - \frac{\lambda^2}{4} x^2) 1_{\{|x| \leq K\}} * \hat{\mu}^c)_\infty + ((|\phi_\lambda| - \frac{\lambda^2}{4} x^2) 1_{\{|x| > K\}} * \hat{\mu}^c)_\infty}$$
$$\leq e^{((c_0 \lambda^3 K - \frac{\lambda^2}{4}) x^2 1_{\{|x| \leq K\}} * \hat{\mu}^c)_\infty + (|\phi_\lambda| 1_{\{|x| > K\}} * \hat{\mu}^c)_\infty},$$

and then since $0 < \lambda < \lambda_0 \wedge \frac{1}{4c_0 K}$ the right-hand side of the last inequality is

$$\leq e^{0 + (|\phi_\lambda| 1_{\{|x| > K\}} * \hat{\mu}^c)_\infty}.$$

Therefore, it follows that

$$I_1 \cdot I_2 \leq \left\{ 4c_0 K e^{-1} + \frac{1}{\lambda}(|\phi_\lambda|1_{\{|x|>K\}}*\hat{\mu}^c)_\infty \right\} e^{(|\phi_\lambda|1_{\{|x|>K\}}*\hat{\mu}^c)_\infty}$$

$$\leq 4c_0 K e^{-1} e^{(|\phi_\lambda|1_{\{|x|>K\}}*\hat{\mu}^c)_\infty}$$
$$+ \frac{1}{\lambda}(|\phi_\lambda|1_{\{|x|>K\}}*\hat{\mu}^c)_\infty e^{(|\phi_\lambda|1_{\{|x|>K\}}*\hat{\mu}^c)_\infty},$$

and then by Lemma 4.1 the right-hand side of the last inequality is

$$\leq 4c_0 K e^{-1} e^{\lambda_0 \Phi_\infty^c} + \Phi_\infty^c e^{\lambda \Phi_\infty^c}.$$

Hence, (a) holds.

Second, by using the idea of the proof of Lemma 1 in Galtchouk and Novikov [4] we will show (b). Fix $0 < \lambda < \lambda_0$. Choose $\epsilon > 0$ such that $\lambda + \epsilon < \lambda_0$. Given this ϵ there exists a positive constant ρ_ϵ such that

$$e^{(\lambda+\epsilon)x} > xe^{\lambda x} \qquad \text{for } x > \rho_\epsilon.$$

Then we have

$$E[e^{\lambda_0 M_\infty^-} \Phi_\infty^c e^{\lambda \Phi_\infty^c}] = E[e^{\lambda_0 M_\infty^-} \Phi_\infty^c e^{\lambda \Phi_\infty^c}; \Phi_\infty^c \leq \rho_\epsilon] + E[e^{\lambda_0 M_\infty^-} \Phi_\infty^c e^{\lambda \Phi_\infty^c}; \Phi_\infty^c > \rho_\epsilon]$$
$$\leq \rho_\epsilon e^{\lambda \rho_\epsilon} E[e^{\lambda_0 M_\infty^-}] + E[e^{\lambda_0 M_\infty^-} e^{(\lambda+\epsilon)\Phi_\infty^c}]$$
$$\leq \rho_\epsilon e^{\lambda_0 \rho_\epsilon} E[e^{\lambda_0 M_\infty^-}] + E[e^{\lambda_0 M_\infty^- + \lambda_0 \Phi_\infty^c}].$$

The last inequality and assumption $(*)$ imply that (b) holds.

Finally, we show (c). From the fact $|\frac{1-e^{\epsilon x}}{\epsilon}| \leq |x|e^{\epsilon|x|}$ for $\epsilon > 0$, we deduce that for any $0 < \lambda < \lambda_0$

$$\frac{1}{\lambda}|1 - e^{-(\phi_\lambda * \hat{\mu}^c)_\infty}| \leq \frac{1}{\lambda}|(\phi_\lambda * \hat{\mu}^c)_\infty| e^{|(\phi_\lambda * \hat{\mu}^c)_\infty|}$$
$$\leq \frac{1}{\lambda}(|\phi_\lambda| * \hat{\mu}^c)_\infty e^{(|\phi_\lambda| * \hat{\mu}^c)_\infty},$$

and moreover by Lemma 4.1 the right-hand side of the last inequality is

$$\leq \frac{1}{\lambda}(|\phi_\lambda| * \hat{\mu}^c)_\infty e^{(|\phi_{\lambda_0}| * \hat{\mu}^c)_\infty},$$

where $e^{(|\phi_{\lambda_0}| * \hat{\mu}^c)_\infty} < \infty$ a.s. by $(*)$. Therefore, to get (c), it suffices to show

$$\lim_{\lambda \to 0} \frac{1}{\lambda}(|\phi_\lambda| * \hat{\mu}^c)_\infty = 0 \quad a.s.$$

According to the dominated convergence theorem with respect to the random measure $\hat{\mu}^c(\cdot, ds\, dx)$, the two facts $\lim_{\lambda \to 0} \frac{\phi_\lambda}{\lambda} = 0$ and $|\frac{\phi_\lambda}{\lambda}| \leq |\frac{\phi_{\lambda_0}}{\lambda_0}|$ for all $0 < \lambda < \lambda_0$ (see Lemma 4.1) and $\frac{1}{\lambda_0}(|\phi_{\lambda_0}| * \hat{\mu}^c)_\infty < \infty$ a.s., which holds by assumption $(*)$, imply the desired convergence.

Lemma 6.4 *Define a process by*

$$\eta^\lambda = \sum_{0 < s \le \cdot} \{\frac{\lambda^2}{2} \Delta(x^2 * \hat{\mu})_s - \log(1 + \Delta(\psi_\lambda * \hat{\mu})_s)\}$$

for each $0 < \lambda \le \lambda_0$ and choose the positive constant c_1 such that

$$|e^{-x} - 1 + x| \le c_1 x^2 \quad \text{for all } |x| \le \lambda_0 K. \tag{11}$$

Then, for each λ,

(i)
$$|\eta^\lambda| \le \sum_{0 < s \le \cdot} \{\Delta(|\phi_\lambda| 1_{\{|x| > K\}} * \hat{\mu})_s + \frac{\lambda^2}{2} \Delta(x^2 1_{\{|x| > K\}} * \hat{\mu})_s$$
$$+ C\lambda^3 \Delta(x^2 1_{\{|x| \le K\}} * \hat{\mu})_s\},$$

where $C = c_0 K + \frac{c_1^2 \lambda_0 K^2}{2}$ and c_0 is the constant in (7); and

(ii) $\eta_\infty^\lambda = \lim_{t \to \infty} \eta_t^\lambda < \infty$ *a.s..*

Moreover,

(iii) $\lim_{\lambda \to 0} E[e^{\lambda_0 M_\infty^- - \frac{\lambda^2}{2} \sum_{0 < s < \infty} \Delta(x^2 * \hat{\mu})_s} \cdot \frac{1}{\lambda} |e^{\eta_\infty^\lambda} - 1|] = 0.$

Proof: First, we will show (i). Fix $0 < \lambda \le \lambda_0$. From the fact $x \ge \log(1 + x)$ for all $x \ge 0$, we have that for all $s \in (0, \infty)$

$$\log(1 + \Delta(\psi_\lambda * \hat{\mu})_s) - \frac{\lambda^2}{2} \Delta(x^2 * \hat{\mu})_s \le \Delta(\psi_\lambda * \hat{\mu})_s - \frac{\lambda^2}{2} \Delta(x^2 * \hat{\mu})_s \tag{12}$$
$$= \Delta(\phi_\lambda * \hat{\mu})_s.$$

On the other hand, it follows that for all $s \in (0, \infty)$

$$\log(1 + \Delta(\psi_\lambda * \hat{\mu})_s) - \frac{\lambda^2}{2} \Delta(x^2 * \hat{\mu})_s$$
$$\ge \log(1 + \Delta(\psi_\lambda 1_{\{|x| \le K\}} * \hat{\mu})_s) - \frac{\lambda^2}{2} \Delta(x^2 * \hat{\mu})_s$$
$$\ge \Delta(\psi_\lambda 1_{\{|x| \le K\}} * \hat{\mu})_s - \frac{1}{2} \{\Delta(\psi_\lambda 1_{\{|x| \le K\}} * \hat{\mu})_s\}^2 - \frac{\lambda^2}{2} \Delta(x^2 * \hat{\mu})_s$$
$$\ge \Delta(\phi_\lambda 1_{\{|x| \le K\}} * \hat{\mu})_s - \frac{1}{2} \{\Delta(\psi_\lambda 1_{\{|x| \le K\}} * \hat{\mu})_s\}^2 - \frac{\lambda^2}{2} \Delta(x^2 1_{\{|x| > K\}} * \hat{\mu})_s.$$

The first line of the last inequality comes from $\psi_\lambda \ge \psi_\lambda 1_{\{|x| \le K\}}$, the second line from $\log(1 + x) \ge x - \frac{1}{2}x^2$ for all $x \ge 0$, and the third line from $(x^2 * \hat{\mu})_s = (x^2 1_{\{|x| > K\}} * \hat{\mu})_s + (x^2 1_{\{|x| \le K\}} * \hat{\mu})_s$. Therefore, by (12) and the last inequality we obtain for all $s \in (0, \infty)$

$$\left|\log(1 + \Delta(\psi_\lambda * \hat\mu)_s) - \frac{\lambda^2}{2}\Delta(x^2 * \hat\mu)_s\right|$$

$$\leq \Delta(|\phi_\lambda| * \hat\mu)_s + \frac{1}{2}\{\Delta(\psi_\lambda 1_{\{|x|\leq K\}} * \hat\mu)_s\}^2 + \frac{\lambda^2}{2}\Delta(x^2 1_{\{|x|>K\}} * \hat\mu)_s$$

$$\leq \Delta(|\phi_\lambda| 1_{\{|x|>K\}} * \hat\mu)_s + \Delta(|\phi_\lambda| 1_{\{|x|\leq K\}} * \hat\mu)_s$$

$$+ \frac{1}{2}\{\Delta(\psi_\lambda 1_{\{|x|\leq K\}} * \hat\mu)_s\}^2 + \frac{\lambda^2}{2}\Delta(x^2 1_{\{|x|>K\}} * \hat\mu)_s.$$

Then, by (7) and (11) the right-hand side of the last inequality is

$$\leq \Delta(|\phi_\lambda| 1_{\{|x|>K\}} * \hat\mu)_s + c_0 K \lambda^3 \Delta(x^2 1_{\{|x|\leq K\}} * \hat\mu)_s$$

$$+ \frac{1}{2}\{c_1 \lambda^2 \Delta(x^2 1_{\{|x|\leq K\}} * \hat\mu)_s\}^2 + \frac{\lambda^2}{2}\Delta(x^2 1_{\{|x|>K\}} * \hat\mu)_s.$$

By Jensen's inequality the third term above is $\leq \frac{c_1^2 \lambda^4}{2}\Delta(x^4 1_{\{|x|\leq K\}} * \hat\mu)_s$, and so the estimate becomes

$$\leq \Delta(|\phi_\lambda| 1_{\{|x|>K\}} * \hat\mu)_s + c_0 K \lambda^3 \Delta(x^2 1_{\{|x|\leq K\}} * \hat\mu)_s$$

$$+ \frac{c_1^2 \lambda^4}{2}\Delta(x^4 1_{\{|x|\leq K\}} * \hat\mu)_s + \frac{\lambda^2}{2}\Delta(x^2 1_{\{|x|>K\}} * \hat\mu)_s$$

$$\leq \Delta(|\phi_\lambda| 1_{\{|x|>K\}} * \hat\mu)_s + c_0 K \lambda^3 \Delta(x^2 1_{\{|x|\leq K\}} * \hat\mu)_s$$

$$+ \frac{c_1^2 K^2 \lambda^4}{2}\Delta(x^2 1_{\{|x|\leq K\}} * \hat\mu)_s + \frac{\lambda^2}{2}\Delta(x^2 1_{\{|x|>K\}} * \hat\mu)_s$$

$$\leq \Delta(|\phi_\lambda| 1_{\{|x|>K\}} * \hat\mu)_s + C\lambda^3 \Delta(x^2 1_{\{|x|\leq K\}} * \hat\mu)_s + \frac{\lambda^2}{2}\Delta(x^2 1_{\{|x|>K\}} * \hat\mu)_s.$$

Hence, (i) holds.

Moreover, (ii) holds by the assumptions (*) and $\langle M \rangle_\infty < \infty$ a.s., because the right-hand side of the inequality of (i) is

$$\leq (|\phi_{\lambda_0}| 1_{\{|x|>K\}} * \hat\mu)_\infty + \{C\lambda_0^3 + \frac{\lambda_0^2}{2}\}\langle M \rangle_\infty.$$

Next, we choose the positive constant c_2 such that $x^2 \leq c_2 |\phi_{\lambda_0}(x)|$ for all $|x| > K$ (see Lemma 4.2). According to the dominated convergence theorem, to show (iii), it suffices to show the following three statements:

For all $0 < \lambda < \lambda_0 \wedge \frac{1}{4C}$,

(a) $e^{-\frac{\lambda^2}{2}\sum_{0<s<\infty}\Delta(x^2 * \hat\mu)_s} \cdot \frac{1}{\lambda}|1 - e^{\eta_\infty^\lambda}| \leq 4Ce^{-1}e^{\lambda_0 \Phi_\infty^d} + \{1 + \frac{c_2\lambda_0^2}{2}\}\Phi_\infty^d e^{\lambda \Phi_\infty^d}$,
where $\Phi^d = \frac{1}{\lambda_0}\sum_{0<s\leq}\{\Delta(|\phi_{\lambda_0}| 1_{\{|x|>K\}} * \hat\mu)_s$.

For all $0 < \lambda < \lambda_0$,

(b) $E[e^{\lambda_0 M_\infty^-}\Phi_\infty^d e^{\lambda \Phi_\infty^d}] < \infty$.

(c) $\lim_{\lambda \to 0}\frac{1}{\lambda}|1 - e^{\eta_\infty^\lambda}| = 0$ a.s.

We will show (a). Fix $0 < \lambda < \lambda_0 \wedge \frac{1}{4C}$. By the fact $\left|\frac{1 - e^{\epsilon x}}{\epsilon}\right| \le |x| e^{\epsilon|x|}, \epsilon > 0$, we have

$$e^{-\frac{\lambda^2}{2} \sum_{0 < s < \infty} \Delta(x^2 * \hat{\mu})_s} \cdot \frac{1}{\lambda}\left|1 - e^{\eta_\infty^\lambda}\right| \le e^{-\frac{\lambda^2}{2} \sum_{0 < s < \infty} \Delta(x^2 * \hat{\mu})_s} \cdot \left|\frac{\eta_\infty^\lambda}{\lambda}\right| e^{|\eta_\infty^\lambda|},$$

and then by (i) the right-hand side of the last inequality is

$$= e^{-\frac{\lambda^2}{2} \sum_{0 < s < \infty} \Delta(x^2 * \hat{\mu})_s}$$
$$\times \frac{1}{\lambda} \sum_{0 < s \le \infty} \{\Delta(|\phi_\lambda| 1_{\{|x| > K\}} * \hat{\mu})_s + \frac{\lambda^2}{2}\Delta(x^2 1_{\{|x| > K\}} * \hat{\mu})_s$$
$$+ C\lambda^3 \Delta(x^2 1_{\{|x| \le K\}} * \hat{\mu})_s\} \times e^{|\eta_\infty^\lambda|}$$
$$\le e^{|\eta_\infty^\lambda| - \frac{\lambda^2}{2} \sum_{0 < s < \infty} \Delta(x^2 * \hat{\mu})_s} \sum_{0 < s \le \infty} \Delta\left(\left|\frac{\phi_\lambda}{\lambda}\right| 1_{\{|x| > K\}} * \hat{\mu}\right)_s$$
$$+ e^{|\eta_\infty^\lambda| - \frac{\lambda^2}{2} \sum_{0 < s < \infty} \Delta(x^2 * \hat{\mu})_s} \sum_{0 < s < \infty} \frac{\lambda}{2}\Delta(x^2 1_{\{|x| > K\}} * \hat{\mu})_s$$
$$+ e^{|\eta_\infty^\lambda| - \frac{\lambda^2}{2} \sum_{0 < s < \infty} \Delta(x^2 * \hat{\mu})_s} \sum_{0 < s < \infty} C\lambda^2 \Delta(x^2 1_{\{|x| \le K\}} * \hat{\mu})_s$$
$$= J_1 + J_2 + J_3.$$

First, we compute J_1. By (i) we have

$$J_1 \le e^{\sum_{0 < s < \infty} \{\Delta(|\phi_\lambda| 1_{\{|x| > K\}} * \hat{\mu})_s + (C\lambda^3 - \frac{\lambda^2}{2})\Delta(x^2 1_{\{|x| \le K\}} * \hat{\mu})_s\}}$$
$$\times \sum_{0 < s \le \infty} \Delta\left(\left|\frac{\phi_\lambda}{\lambda}\right| 1_{\{|x| > K\}} * \hat{\mu}\right)_s.$$

Here, recall $0 < \lambda < \lambda_0 \wedge \frac{1}{4C}$. So the right-hand side of the last inequality is

$$\le e^{\sum_{0 < s < \infty} \{\Delta(|\phi_\lambda| 1_{\{|x| > K\}} * \hat{\mu})_s + 0\}} \sum_{0 < s \le \infty} \Delta\left(\left|\frac{\phi_\lambda}{\lambda}\right| 1_{\{|x| > K\}} * \hat{\mu}\right)_s,$$

and by Lemma 4.1 the above is

$$\le \Phi_\infty^d e^{\lambda \Phi_\infty^d}.$$

Second, we compute J_2. By (i) we have

$$J_2 \le e^{\sum_{0 < s < \infty} \{\Delta(|\phi_\lambda| 1_{\{|x| > K\}} * \hat{\mu})_s + (C\lambda^3 - \frac{\lambda^2}{2})\Delta(x^2 1_{\{|x| \le K\}} * \hat{\mu})_s\}}$$
$$\times \frac{\lambda}{2} \sum_{0 < s \le \infty} \Delta(x^2 1_{\{|x| > K\}} * \hat{\mu})_s.$$

By $0 < \lambda < \lambda_0 \wedge \frac{1}{4C}$ the right-hand side of the last inequality is

$$\leq e^{\sum_{0<s<\infty}\{\Delta(|\phi_\lambda|1_{\{|x|>K\}}*\hat{\mu})_s+0\}} \cdot \frac{\lambda}{2} \sum_{0<s\leq\infty} \Delta(x^2 1_{\{|x|>K\}}*\hat{\mu})_s,$$

and moreover by Lemma 4.1 and 4.2 the above is

$$\leq \frac{c_2\lambda_0^2}{2} \Phi_\infty^d e^{\lambda\Phi_\infty^d}.$$

Finally, we compute J_3. By (i) we have

$$J_3 \leq e^{\sum_{0<s<\infty}\Delta(|\phi_\lambda|1_{\{|x|>K\}}*\hat{\mu})_s} \cdot 4C \cdot \frac{\lambda^2}{4} \sum_{0<s<\infty} \Delta(x^2 1_{\{|x|\leq K\}}*\hat{\mu})_s$$

$$\times e^{-\frac{\lambda^2}{4}\sum_{0<s<\infty}\Delta(x^2 1_{\{|x|\leq K\}}*\hat{\mu})_s}$$

$$\times e^{(C\lambda^3-\frac{\lambda^2}{4})\sum_{0<s<\infty}\Delta(x^2 1_{\{|x|\leq K\}}*\hat{\mu})_s}.$$

By $0 < \lambda < \lambda_0 \wedge \frac{1}{4C}$ the right-hand side of the last inequality is

$$\leq e^{\sum_{0<s<\infty}\Delta(|\phi_\lambda|1_{\{|x|>K\}}*\hat{\mu})_s} \cdot 4C \cdot \frac{\lambda^2}{4} \sum_{0<s<\infty} \Delta(x^2 1_{\{|x|\leq K\}}*\hat{\mu})_s$$

$$\times e^{-\frac{\lambda^2}{4}\sum_{0<s<\infty}\Delta(x^2 1_{\{|x|\leq K\}}*\hat{\mu})_s}$$

$$\times \exp(0),$$

and moreover by Lemma 4.1 and the fact $xe^{-x} \leq e^{-1}$ the above is

$$\leq e^{\lambda_0\Phi_\infty^d} \cdot \frac{4C}{e} \cdot 1.$$

Therefore, the three inequalities with respect to J_1, J_2, and J_3 imply

$$J_1 + J_2 + J_3 \leq (1 + \frac{c_2\lambda_0^2}{2})\Phi_\infty^d e^{\lambda\Phi_\infty^d} + \frac{4C}{e}e^{\lambda_0\Phi_\infty^d},$$

and hence (a) holds.

Second, (b) can be shown by the same method as for (b) in lemma 6.3. Finally, we will show (c). By the fact $|\frac{1-e^{\epsilon x}}{\epsilon}| \leq |x|e^{\epsilon|x|}$ for $\epsilon > 0$, we have that for any $0 < \lambda < \lambda_0$

$$\frac{1}{\lambda}|1 - e^{\eta_\infty^\lambda}| \leq \frac{1}{\lambda}|\eta_\infty^\lambda|e^{|\eta_\infty^\lambda|}.$$

Then, the last inequality and (i) imply that for any $0 < \lambda < \lambda_0$

$$\frac{1}{\lambda}|1 - e^{\eta_\infty^\lambda}| \leq \frac{1}{\lambda}Le^L,$$

where

$$L = \sum_{0<s<\infty} \{\Delta(|\phi_\lambda|1_{\{|x|>K\}}*\hat{\mu})_s + \frac{\lambda^2}{2}\Delta(x^21_{\{|x|>K\}}*\hat{\mu})_s$$
$$+C\lambda^3\Delta(x^21_{\{|x|\leq K\}}*\hat{\mu})_s\}.$$

Since $\sum_{0<s<\infty} \Delta(x^2*\hat{\mu})_s \leq \langle M\rangle_\infty$, the right-hand side of the last inequality is

$$\leq \sum_{0<s<\infty} \Delta(\{|\frac{\phi_\lambda}{\lambda}|1_{\{|x|>K\}} + \frac{\lambda}{2}x^21_{\{|x|>K\}} + C\lambda^2x^21_{\{|x|\leq K\}}\}*\hat{\mu})_s$$
$$\times e^{(|\phi_\lambda|1_{\{|x|>K\}}*\hat{\mu})_\infty + (\frac{\lambda^2}{2}+C\lambda^3)\langle M\rangle_\infty}$$

$$= K_1 \times K_2.$$

By Lemma 4.1 and the assumptions $(*)$ and $\langle M\rangle_\infty < \infty$ a.s.

$$K_2 \leq e^{(|\phi_{\lambda_0}|1_{\{|x|>K\}}*\hat{\mu})_\infty + (\frac{\lambda_0{}^2}{2}+C\lambda_0{}^3)\langle M\rangle_\infty} < \infty \text{ a.s.}$$

holds, and furthermore

$$K_1 \leq (\{|\frac{\phi_\lambda}{\lambda}|1_{\{|x|>K\}} + \frac{\lambda}{2}x^21_{\{|x|>K\}} + C\lambda^2x^21_{\{|x|\leq K\}}\}*\hat{\mu})_\infty$$

holds. Therefore, to get (c), it suffices to show that the right-hand side of the last inequality converges to 0 a.s. as $\lambda \to 0$. By using the fact $\lim_{\lambda\to 0}\frac{\phi_\lambda}{\lambda} = 0$ we have that for each x

$$\lim_{\lambda\to 0}\{\frac{\phi_\lambda}{\lambda}|1_{\{|x|>K\}} + \frac{\lambda}{2}x^21_{\{|x|>K\}} + C\lambda^2x^21_{\{|x|\leq K\}}\} = 0$$

and by Lemma 4.1 we have that for each x

$$|\frac{\phi_\lambda}{\lambda}|1_{\{|x|>K\}} + \frac{\lambda}{2}x^21_{\{|x|>K\}} + C\lambda^2x^21_{\{|x|\leq K\}} \leq |\frac{\phi_{\lambda_0}}{\lambda_0}|1_{\{|x|>K\}} + (\frac{\lambda_0}{2}+C\lambda_0{}^2)x^2.$$

Then, according to the dominated convergence theorem with respect to the random measure $\hat{\mu}(\cdot, ds\,dx)$, the last two results and the assumptions $(*)$ and $\langle M\rangle_\infty < \infty$ a.s. imply the desired convergence. Hence, (c) holds.

Lemma 6.5

$$\lim_{\lambda\to 0}\frac{1}{\lambda}\{E[e^{-\lambda M_\infty - \frac{\lambda^2}{2}\langle M\rangle_\infty}] - E[\mathcal{E}(\lambda)_\infty]\} = 0$$

Proof: Fix $0 < \lambda < \lambda_0$. First, since we know $\langle M\rangle_\infty = \langle M^c\rangle_\infty + (x^2*\hat{\mu})_\infty$, we have

$$\left|\frac{1}{\lambda}\{e^{-\lambda M_\infty - \frac{\lambda^2}{2}\langle M\rangle_\infty} - \mathcal{E}(\lambda)_\infty\}\right|$$

$$= \left|\frac{1}{\lambda}\{e^{-\lambda M_\infty - \frac{\lambda^2}{2}\langle M\rangle_\infty}\right.$$

$$\left. -e^{-\lambda M_\infty - \frac{\lambda^2}{2}\langle M\rangle_\infty + \frac{\lambda^2}{2}(x^2*\hat\mu)_\infty - (\psi_\lambda*\hat\mu^c)_\infty - \sum_{0<s<\infty}\log(1+\Delta(\psi_\lambda*\hat\mu)_s)}\}\right|,$$

and then, since we know $(x^2*\hat\mu)_\infty = (x^2*\hat\mu^c)_\infty + \sum_{0<s<\infty}\Delta(x^2*\hat\mu)_s$, the right-hand side of the last equality is

$$= \left|\frac{1}{\lambda}\{e^{-\lambda M_\infty - \frac{\lambda^2}{2}\langle M\rangle_\infty}\right.$$

$$-e^{-\lambda M_\infty - \frac{\lambda^2}{2}\langle M\rangle_\infty + \frac{\lambda^2}{2}\sum_{0<s<\infty}\Delta(x^2*\hat\mu)_s + \frac{\lambda^2}{2}(x^2*\hat\mu^c)_\infty}$$

$$\left. \times e^{-(\psi_\lambda*\hat\mu^c)_\infty - \sum_{0<s<\infty}\log(1+\Delta(\psi_\lambda*\hat\mu)_s)}\}\right|$$

$$= \left|e^{-\lambda M_\infty - \frac{\lambda^2}{2}\langle M\rangle_\infty + \sum_{0<s<\infty}\{\frac{\lambda^2}{2}\Delta(x^2*\hat\mu)_s - \log(1+\Delta(\psi_\lambda*\hat\mu)_s)\}}\right.$$

$$\left. \times \frac{1}{\lambda}\{e^{\sum_{0<s<\infty}\{\log(1+\Delta(\psi_\lambda*\hat\mu)_s) - \frac{\lambda^2}{2}\Delta(x^2*\hat\mu)_s\}} - e^{-(\phi_\lambda*\hat\mu^c)_\infty}\}\right|$$

$$\le e^{\lambda_0 M_\infty^- - \frac{\lambda^2}{2}(x^2*\hat\mu^c)_\infty - \sum_{0<s<\infty}\log(1+\Delta(\psi_\lambda*\hat\mu)_s)}$$

$$\times \left\{\frac{1}{\lambda}|e^{\sum_{0<s<\infty}\{\log(1+\Delta(\psi_\lambda*\hat\mu)_s) - \frac{\lambda^2}{2}\Delta(x^2*\hat\mu)_s\}} - 1| + \frac{1}{\lambda}|e^{-(\phi_\lambda*\hat\mu^c)_\infty} - 1|\right\}$$

$$\le e^{\lambda_0 M_\infty^- - \frac{\lambda^2}{2}(x^2*\hat\mu^c)_\infty} \cdot \frac{1}{\lambda}|e^{-(\phi_\lambda*\hat\mu^c)_\infty} - 1|$$

$$+ e^{\lambda_0 M_\infty^- - \frac{\lambda^2}{2}\sum_{0<s<\infty}\Delta(x^2*\hat\mu)_s} \cdot \frac{1}{\lambda}\left|e^{\sum_{0<s<\infty}\{\frac{\lambda^2}{2}\Delta(x^2*\hat\mu)_s - \log(1+\Delta(\psi_\lambda*\hat\mu)_s)\}} - 1\right|$$

$$= e^{\lambda_0 M_\infty^- - \frac{\lambda^2}{2}(x^2*\hat\mu^c)_\infty} \cdot \frac{1}{\lambda}|e^{-(\phi_\lambda*\hat\mu^c)_\infty} - 1|$$

$$+ e^{\lambda_0 M_\infty^- - \frac{\lambda^2}{2}\sum_{0<s<\infty}\Delta(x^2*\hat\mu)_s} \cdot \frac{1}{\lambda}|e^{\eta_\infty^\lambda} - 1|.$$

Therefore we have

$$\frac{1}{\lambda}\left|E[e^{-\lambda M_\infty - \frac{\lambda^2}{2}\langle M\rangle_\infty}] - E[\mathcal{E}(\lambda)_\infty]\right|$$

$$\le E[e^{\lambda_0 M_\infty^- - \frac{\lambda^2}{2}(x^2*\hat\mu^c)_\infty} \cdot \frac{1}{\lambda}|e^{-(\phi_\lambda*\hat\mu^c)_\infty} - 1|]$$

$$+ E[e^{\lambda_0 M_\infty^- - \frac{\lambda^2}{2}\sum_{0<s<\infty}\Delta(x^2*\hat\mu)_s} \cdot \frac{1}{\lambda}|e^{\eta_\infty^\lambda} - 1|].$$

Lemmas 6.3, 6.4, and the last inequality imply the desired conclusion.

6.2 A Tauberian theorem

Theorem 6.1 ([3]) *Let X be an $\mathbf{R}_+$-valued random variable such that* $\lim_{\lambda\to 0}\frac{1}{\lambda}(1 - E[e^{-\frac{\lambda^2}{2}X}])$ *exists in $\mathbf{R}$, then*

$$\sqrt{\frac{2}{\pi}}\lim_{\lambda\to 0}\frac{1}{\lambda}(1 - E[e^{-\frac{\lambda^2}{2}X}]) = \lim_{\lambda\to\infty}\lambda P(\sqrt{X} > \lambda).$$

6.3 Proof of (i)

According to lemmas 6.2 and 6.5, we have

$$\lim_{\lambda\to 0}\frac{1}{\lambda}(E[e^{-\lambda M_\infty - \frac{\lambda^2}{2}\langle M\rangle_\infty}] - 1) = 0.$$

Moreover, by lemma 6.1 and this result,

$$\lim_{\lambda\to 0}\frac{1}{\lambda}(1 - E[e^{-\frac{\lambda^2}{2}\langle M\rangle_\infty}]) = -E[M_\infty]$$

holds. Then, by using the Tauberian theorem 6.1, the last result implies

$$\lim_{\lambda\to\infty}\lambda P(\sqrt{\langle M\rangle_\infty} > \lambda) = -\sqrt{\frac{2}{\pi}}E[M_\infty].$$

6.4 Proof of (ii)

Set $L_t = [M]_t - \langle M\rangle_t$ for $t > 0$. We divide the proof into two steps.

Step 1. Since the definition of L gives $L = (x^2 * \mu) - (x^2 * \hat{\mu})$, it follows that for every stopping time $\tau \in \mathcal{T}$

$$E\big[|L_\tau|\big] \le E\big[(x^2 * \mu)_\tau\big] + E\big[(x^2 * \hat{\mu})_\tau\big] = 2E\big[(x^2 * \hat{\mu})_\tau\big],$$

and moreover, by Lemma 4.2 the right hand side of the last inequality is

$$\le E\big[2c_1(|\phi_{\lambda_0}|1_{\{|x|>K\}} * \hat{\mu})_\tau + 2c_2(\sqrt{|x|} * \hat{\mu})_\tau\big]$$

for some $c_1, c_2 > 0$. Therefore, we have

$$E\big[|L_\tau|\big] \le E\big[2c_1(|\phi_{\lambda_0}|1_{\{|x|>K\}} * \hat{\mu})_\tau + 2c_2(\sqrt{|x|} * \hat{\mu})_\tau\big] \quad \text{for all } \tau \in \mathcal{T}.$$

Then, according to the Lenglart domination property (see lemma 3.30 in [5]), the last inequality implies that for any $\epsilon > 0$

$$P(\sup_{t\in\mathbf{R}_+}|L_t| \ge \lambda^2)$$

$$\le \frac{\epsilon\sqrt{\lambda}}{\lambda^2} + P\big(2c_1(|\phi_{\lambda_0}|1_{\{|x|>K\}} * \hat{\mu})_\infty + 2c_2(\sqrt{|x|} * \hat{\mu})_\infty \ge \epsilon\sqrt{\lambda}\big). \quad (13)$$

Now, fix $\epsilon > 0$. It is easy to see that

$$P\big(2c_1(|\phi_{\lambda_0}|1_{\{|x|>K\}}*\hat{\mu})_\infty + 2c_2(\sqrt{|x|}*\hat{\mu})_\infty \ge \epsilon\sqrt{\lambda}\big)$$
$$\le P\Big((|\phi_{\lambda_0}|1_{\{|x|>K\}}*\hat{\mu})_\infty \ge \frac{\epsilon\sqrt{\lambda}}{4c_1}\Big) + P\Big((\sqrt{|x|}*\hat{\mu})_\infty \ge \frac{\epsilon\sqrt{\lambda}}{4c_2}\Big).$$

By Chebyshev's inequality and assumption $(*)$, the first term in the right-hand side is

$$\le e^{-\frac{\epsilon\sqrt{\lambda}}{4c_1}} E[e^{(|\phi_{\lambda_0}|1_{\{|x|>K\}}*\hat{\mu})_\infty}];$$

and since $(x^2*\hat{\mu})_\infty \le \langle M\rangle_\infty$ and $\{(\sqrt{|x|}*\hat{\mu})_\infty\}^4 \le (x^2*\hat{\mu})_\infty$ by Jensen's inequality, the second term is

$$\le P\Big((x^2*\hat{\mu})_\infty \ge \frac{\epsilon^4\lambda^2}{16c_2{}^4}\Big) \le P\Big(\langle M\rangle_\infty \ge \frac{\epsilon^4\lambda^2}{16c_2{}^4}\Big),$$

and so we have

$$P\big(2c_1(|\phi_{\lambda_0}|1_{\{|x|>K\}}*\hat{\mu})_\infty + 2c_2(\sqrt{|x|}*\hat{\mu})_\infty \ge \epsilon\sqrt{\lambda}\big) \qquad (14)$$
$$\le e^{-\frac{\epsilon\sqrt{\lambda}}{4c_1}} E[e^{(|\phi_{\lambda_0}|1_{\{|x|>K\}}*\hat{\mu})_\infty}] + P\Big(\langle M\rangle_\infty \ge \frac{\epsilon^4\lambda^2}{16c_2{}^4}\Big).$$

By (13) and (14),

$$\lambda P\big(\sup_{t\in\mathbf{R}_+}|L_t| \ge \lambda^2\big) \le \frac{\epsilon\sqrt{\lambda}}{\lambda} + \lambda e^{-\frac{\epsilon\sqrt{\lambda}}{4c_1}} E[e^{(|\phi_{\lambda_0}|1_{\{|x|>K\}}*\hat{\mu})_\infty}]$$
$$+ \lambda P\Big(\langle M\rangle_\infty \ge \frac{\epsilon^4\lambda^2}{16c_2{}^4}\Big),$$

and therefore, by (i) of theorem 3.2,

$$\limsup_{\lambda\to\infty}\lambda P\big(\sup_{t\in\mathbf{R}_+}|L_t| \ge \lambda^2\big) \le 0 + 0 + \frac{4c_2{}^2}{\epsilon^2}\Big(-\sqrt{\frac{2}{\pi}}E[M_\infty]\Big).$$

Letting $\epsilon \to \infty$ gives $\limsup_{\lambda\to\infty}\lambda P\big(\sup_{t\in\mathbf{R}_+}|L_t| \ge \lambda^2\big) \le 0$, that is,

$$\lim_{\lambda\to\infty}\lambda P\big(\sup_{t\in\mathbf{R}_+}|L_t| \ge \lambda^2\big) = 0.$$

Step 2. By the definition of L we have $[M]_\infty \le \langle M\rangle_\infty + \sup_{t\in\mathbf{R}_+} L_t$, and therefore, for every $0 < \epsilon < 1$ and $\lambda > 0$,

$$\{[M]_\infty \le \lambda^2\} \supset \{\sup_{t\in\mathbf{R}_+} L_t \le \epsilon\lambda^2\} \cap \{\langle M\rangle_\infty \le (1-\epsilon)\lambda^2\},$$

and so,

$$\lambda P([M]_\infty > \lambda^2) \leq \lambda P(\sup_{t \in \mathbf{R}_+} L_t > \epsilon \lambda^2) + \lambda P(\langle M \rangle_\infty > (1 - \epsilon)\lambda^2)$$

$$\leq \frac{1}{\sqrt{\epsilon}} \cdot \sqrt{\epsilon} \, \lambda P\Big(\sup_{t \in \mathbf{R}_+} |L_t| > \epsilon \lambda^2\Big)$$

$$+ \frac{1}{\sqrt{1 - \epsilon}} \cdot \sqrt{1 - \epsilon} \, \lambda P\Big(\sqrt{\langle M \rangle_\infty} > \sqrt{1 - \epsilon}\lambda\Big).$$

By Step 1 and (i) of theorem 3.2 the last inequality implies

$$\limsup_{\lambda \to \infty} \lambda P([M]_\infty > \lambda^2) \leq 0 + \frac{1}{\sqrt{1 - \epsilon}}\Big(-\sqrt{\frac{2}{\pi}} \, E[M_\infty]\Big),$$

and hence, letting ϵ tend to 0,

$$\limsup_{\lambda \to \infty} \lambda P([M]_\infty > \lambda^2) \leq -\sqrt{\frac{2}{\pi}} \, E[M_\infty]. \qquad (15)$$

On the other hand, $\langle M \rangle_\infty \leq [M]_\infty + \sup_{t \in \mathbf{R}_+} |L_t|$ by the definition of L, and therefore, for every $\epsilon > 0$ and $\lambda > 0$,

$$\{\langle M \rangle_\infty \leq (1 + \epsilon)\lambda^2\} \supset \{\sup_{t \in \mathbf{R}_+} |L_t| \leq \epsilon \lambda^2\} \cap \{[M]_\infty \leq \lambda^2\};$$

so,

$$\frac{1}{\sqrt{1 + \epsilon}} \cdot \sqrt{1 + \epsilon} \, \lambda P(\langle M \rangle_\infty > (1 + \epsilon)\lambda^2) \leq \frac{1}{\sqrt{\epsilon}} \cdot \sqrt{\epsilon} \, \lambda P\Big(\sup_{t \in \mathbf{R}_+} |L_t| > \epsilon \lambda^2\Big)$$

$$+ \lambda P\Big(\sqrt{[M]_\infty} > \lambda\Big).$$

By Step 1 and (i) of theorem 3.2, this implies

$$\frac{1}{\sqrt{1 + \epsilon}}\Big(-\sqrt{\frac{2}{\pi}} E[M_\infty]\Big) \leq 0 + \liminf_{\lambda \to \infty} \lambda P([M]_\infty > \lambda^2),$$

and hence, since ϵ is arbitrary,

$$-\sqrt{\frac{2}{\pi}} E[M_\infty] \leq \liminf_{\lambda \to \infty} \lambda P([M]_\infty > \lambda^2). \qquad (16)$$

The desired conclusion follows from (15) and (16).

References

1. Azéma, J., Gundy, R. and Yor, M. (1980). Sur l'intégrabilité uniforme des martingales continues. Séminaire de Probabilités XIV, LNM 784, Springer, pp. 249-304.
2. Elworthy, K.D., Li, X.M. and Yor, M. (1997). On the tails of the supremum and the quadratic variation of strictly local martingales. Séminaire de Probabilités XXXI, LNM 1655, Springer, pp. 113-125.
3. Feller, W. (1970). An Introduction to Probability and its Applications. Vol. 2. Wiley.
4. Galtchouk, L. and Novikov A.A. (1997). On Wald's equation. Discrete time case. Séminaire de Probabilités XXXI, LNM 1655, Springer, pp. 126-135.
5. Jacod, J. and Shiryaev A.N. (1987). Limit theorems for stochastic processes. Springer.
6. Kallsen, J. and Shiryaev, A.N. (2002). The cumulant process and Esscher's change of measure. Finance and Stochastics 6, pp. 397-428.
7. Liptser, R.S. and Novikov A.A. (2006) On tail distributions of supremum and quadratic variation of local martingales. Stochastic Calculus to Mathematical Finance. The Shiryaev Festschrift. Springer, pp. 421-432.
8. Madan, D.B. and Yor, M. (2005). Ito's integrated formula for strict local martingales. Séminaire de Probabilités XXXIX, LNM 1874, Springer, pp. 157–170.
9. Novikov, A.A. (1996). Martingales, Tauberian theorem and gambling. Theory of Prob. Appl. 41 No 4, pp. 716-729.
10. Takaoka, K. (1999). Some remarks on the uniform integrability of continuous martingales. Séminaire de Probabilités XXXIII, LNM 1709, Springer, pp. 327-333.

The Burkholder-Davis-Gundy Inequality for Enhanced Martingales

Peter Friz and **Nicolas Victoir**

Department of Pure Mathematics and Mathematical Statistics
University of Cambridge
e-mail: P.K.Friz@statslab.cam.ac.uk

Summary. Multi-dimensional continuous local martingales, enhanced with their stochastic area process, give rise to geometric rough paths with a.s. finite homogenous p-variation, $p \in (2,3)$. The purpose of this paper is to establish quantitative bounds of the homogenous p-variation norm in the form of a BDG inequality, combining old ides by Lépingle with recent geometric insights to rough path theory. Such BDG inequalities appear to be a useful addition to the toolbox available for the study of stochastic processes via rough path. Some applications are discussed, in particular Wong-Zakai type approximations.

1 Introduction

The theory of rough paths provides a new and robust way to drive differential equations by multi-dimensional stochastic processes in a *deterministic way*. In most cases, this is achieved by taking into account a certain stochastic area process and by establishing fine regularity properties of the resulting *enhanced* process. The object of study in this paper is a d-dimensional continuous local martingale M null at 0 for which the area is defined by iterated stochastic integration; the area process A_t is simply the anti-symmetric part of the iterated Stratonovich integral,

$$\mathbf{M}_t^2 \equiv \int_0^t \int_0^s dM_r \otimes \circ dM_s \in \mathbb{R}^d \otimes \mathbb{R}^d.$$

Note that the symmetric part of $\mathbf{M}_t^2$ is given by $\frac{1}{2} M_t \otimes M_t$ and hence redundant if one knows $\mathbf{M}_t^1 \equiv M_t$. It follows that the enhanced process $\mathbf{M} \equiv \left(1, \mathbf{M}^1, \mathbf{M}^2\right) \in \mathbb{R} \oplus \mathbb{R}^d \oplus \mathbb{R}^d \otimes \mathbb{R}^d$ lives in submanifold, namely in $G^2\left(\mathbb{R}^d\right) \equiv \exp\left(\mathbb{R}^d \oplus so\left(d\right)\right)$, where $\exp : (x,a) \mapsto \left(1, x, a + \frac{1}{2} x \otimes x\right)$. The space $\mathbb{R}^d \oplus so\left(d\right)$ carries a Lie algebra structure and induces a (Lie-)group

---
[*] Partially supported by a Leverhulme Research Fellowship.

structure on $G^2\left(\mathbb{R}^d\right)$. The interest in this algebraic exercise is that the result-ing product operation on $G^2\left(\mathbb{R}^d\right)$ is exactly what one needs to patch together "iterated integral increments" over adjacent intervals. $G^2\left(\mathbb{R}^d\right)$ is also a metric (in fact, Polish) space under the Carnot-Caratheodory metric d. Intuitively, the distance of two points under this metric is the length of the shortest path in $\mathbb{R}^d$ which wipes out a prescribed area. When $d = 2$, geodesics are seen to be parts of circles. $G^2\left(\mathbb{R}^d\right)$ carries a dilation induced by $(x, a) \mapsto \left(\lambda x, \lambda^2 a\right)$ for real λ. In fact, the CC-metric is induced by a sub-additive norm, homogenous w.r.t. dilation. Since all continuous homogenous norms are Lipschitz equiv-alent, computations are often carried out w.r.t. $|||(x,a)||| = |x| + |a|^{1/2}$. We refer to [6, 7] for background on rough paths, [2] contains a more detailed discussion of the relevant geometry and algebra. The notion of a (weak) geo-metric p-rough path [3] becomes quite elegant: by definition, one requires that the $G^2\left(\mathbb{R}^d\right)$-valued path $\mathbf{M}$ has finite p-variation in the sense

$$\|\mathbf{M}\|_{p\text{-var};[0,T]} = \left(\sup_{0 \le t_1 \le \ldots \le t_n \le T} \sum d\left(\mathbf{M}_{t_i}, \mathbf{M}_{t_{i+1}}\right)^p\right)^{1/p} < \infty$$

for $[p] = 2$ i.e. for $p \in [2, 3)$. Is is known [1] that this holds for a.e. $\mathbf{M} = \mathbf{M}\left(\omega\right)$ when $p > 2$ but not for $p = 2$, exactly as for semi-martingales [5, 11]. The first contribution of this paper is to establish quantitative bounds of the p-variation norm in the form of a two-sided BDG inequality: for any moderate function F such as $x \mapsto x^r$ for $r > 0$,

$$\mathbb{E}\left(F\left(\|\mathbf{M}\|_{p\text{-var};[0,T]}\right)\right) \sim \mathbb{E}\left(F\left(|\langle M\rangle_T|^{1/2}\right)\right).$$

The algebraic and geometric preparations above, prompted by recent progress in rough theory and novel in the context of martingale inequalities, allow for a successful combination with the classical arguments given in Lépingle's seminal paper [5] from 1976. This BDG inequality, and also one for $\|\mathbf{M}\|_{\infty;[0,T]}$ which is much easier to obtain, appear to be a generally useful addition to the toolbox available for the study of continuous martingales, and their integration theory, via rough path.

Secondly, we discuss approximations and show L^q-convergence of lifted piecewise linear approximations of a continuous L^q-martingale w.r.t. homoge-nous p-variation topology.

The authors would like to thank D. Lépingle for a helpful email exchange.

2 Preliminaries

We write $\mathcal{M}^c_{0,\mathrm{loc}}\left([0,\infty), \mathbb{R}^d\right)$ or $\mathcal{M}^c_{0,\mathrm{loc}}\left(\mathbb{R}^d\right)$ for the class of $\mathbb{R}^d$-valued con-tinuous local martingales $M : [0,\infty) \to \mathbb{R}^d$ null at 0. The bracket process $\langle M\rangle : [0,\infty) \to \mathbb{R}^d$ is defined component-wise, the ith component is given by the usual bracket $\langle M^i\rangle = \langle M^i, M^i\rangle$.

The area-process $A : [0, \infty) \to so\,(d)$ is defined by Itô- or Stratonovich stochastic integration. As the matrix $\langle M^i, M^j \rangle$ is symmetric both lead to the *same* area,

$$A_t^{i,j} = \frac{1}{2} \left(\int_0^t M^i dM^j - \int_0^t M^j dM^i \right)$$
$$= \frac{1}{2} \left(\int_0^t M^i \circ dM^j - \int_0^t M^j \circ dM^i \right).$$

We note that the area-process is a vector-valued continuous martingale. By disregarding a null-set we can and will assume that M and A are continuous.

Definition 1. *Set* $S_2\,(M) := \mathbf{M} := \exp\,(M + A)$ *so that* $\mathbf{M} \in C\big([0, \infty)$, $G^2(\mathbb{R}^d)\big)$. *The resulting class of enhanced (continuous, local) martingales is denoted by* $\mathcal{M}_{0,loc}^c\big(G^2\,(\mathbb{R}^d)\big)$. *We refer to the operation* $S_2 : M \mapsto \mathbf{M}$ *as lift.*

The lift is compatible with the stopping and time-changes.

Lemma 2. *(i) Let* τ *be a stopping time. Then* $\mathbf{M}^\tau = S_2\,(M^\tau)$. *(ii) Let* ϕ *be a time-change, that is, a family* ϕ_s, $s \geq 0$, *of stopping times such that the maps* $s \mapsto \phi_s$ *are a.s. increasing and right-continuous. Assume that* M *is constant on each interval* $[\phi_{t-}, \phi_t]$. *Then* $M \circ \phi$ *is a continuous local martingale and*

$$\mathbf{M} \circ \phi = S_2\,(M \circ \phi).$$

Proof. Stopped processes are special cases of time-changed processes (take $\phi_t = t \wedge \tau$) so it suffices to show the second statement. To this end, recall the compatibility of a time change ϕ and stochastic integration w.r.t. a continuous local martingale, constant on each interval $[\phi_{t-}, \phi_t]$, Proposition V.1.5. (ii) of [9]. The lift is a special case of stochastic integration.   □

The lift is also compatible with respect to scaling and concatenation of (local martingale) paths.

Lemma 3. *(i) If* $\delta_c : G^2\,(\mathbb{R}^d) \to G^2\,(\mathbb{R}^d)$ *is the dilation operator given by* $\delta_c \exp\,(x + a) = \exp\big(cx + c^2 a\big)$ *then* $\delta_c \mathbf{M} = S_2\,(cM)$. *(ii) We have*

$$S_2\,(M)_{0,t} = S_2\left(M|_{[0,s]} * M|_{[s,t]} \right)_{0,t} = S_2\,(M)_{0,s} \otimes S_2\,(M)_{s,t}, \quad 0 \leq s \leq t < \infty$$

Proof. (i) is trivial consequence of linearity of stochastic integrals. (ii) follows from the additivity of the stochastic integral.   □

Definition 4. $F : \mathbb{R}^+ \to \mathbb{R}^+$ *is moderate if (i)* F *is continuous and increasing, (ii)* $F\,(x) = 0$ *if and only if* $x = 0$ *and (iii) for some (and then for every)* $\alpha > 1$,

$$\sup_{x > 0} \frac{F\,(\alpha x)}{F\,(x)} < \infty.$$

We now recall the Burkholder-Davis-Gundy inequality for martingales, as found in [10, p93] for instance.

Theorem 5 (Burkholder-Davis-Gundy). *Let F be a moderate function, $M \in \mathcal{M}_{0,loc}^c (\mathbb{R})$. Then there exists a constant $C = C (F, |\cdot|)$ so that*

$$C^{-1}\mathbb{E} \left(F \left(|\langle M \rangle_\infty|^{1/2} \right) \right) \leq \mathbb{E} \left(F \left(\sup_{s \geq 0} |M_s| \right) \right) \leq C\mathbb{E} \left(F \left(|\langle M \rangle_\infty|^{1/2} \right) \right).$$

A few properties of moderate functions that we shall use are collected in

Lemma 6. *(i) $x \mapsto F (x)$ moderate iff $\mapsto F \left(x^{1/2} \right)$ moderate.*
(ii) Given $c, A, B > 0 : c^{-1}A \leq B \leq cA \implies \exists C = C (c, F) :$

$$C^{-1}F (A) \leq F (B) \leq CF (A).$$

(iii) $\exists C : \forall x, y > 0 : F (x + y) \leq C [F (x) + F (y)].$

Proof. (i),(ii) are left to the reader. Ad (iii): W.l.o.g. $x < y$, then $F (x + y) \leq F (2y) \leq CF (y)$ by moderate growth of F. □

A BDG inequality for $\mathbb{R}^d$-valued martingales (even Hilbert-space valued martingales) is well-known; we include it for the reader's convenience. (If $\mathbb{R}^d$ is equipped with the Euclidean norm, the constant below can be seen to be independent of d.)

Corollary 7. *Let F be a moderate function, $M \in \mathcal{M}_{0,loc}^c \left(\mathbb{R}^d \right)$ and $|\cdot|$ a norm on $\mathbb{R}^d$. Then there exists a constant $C = C (F, d, |\cdot|)$ so that*

$$C^{-1}\mathbb{E} \left(F \left(|\langle M \rangle_\infty|^{1/2} \right) \right) \leq \mathbb{E} \left(F \left(\sup_{s \geq 0} |M_s| \right) \right) \leq C\mathbb{E} \left(F \left(|\langle M \rangle_\infty|^{1/2} \right) \right).$$

Proof. When $|a| = \max \left\{ |a^1|, ..., |a^d| \right\}$ this is a simple consequence of BDG for $\mathcal{M}_{0,loc}^c (\mathbb{R})$, applied componentwise. The lemma above shows that one can switch to Lipschitz equivalent norms. □

From Lépingle [5], $\sup_{s \geq 0} |M_s|$ above can be replaced by the p-variation norm[2]. Noting that the p-variation of a discrete-time martingale (Y_n) is naturally defined as

$$|Y|_{p\text{-var}} \equiv \left[\sup_{(n_k) \nearrow} \sum_k |Y_{n_{k+1}} - Y_{n_k}|^p \right]^{1/p},$$

the following lemma is best viewed as a BDG-type upper bound for discrete-time martingales.

---

[2] In the next section, we will see a more general version of this.

Lemma 8. *Let F be moderate. If $1 < q < p \leq 2$ or $1 = q = p$ then there exists a constant c such that for all, possibly $\mathbb{R}^d$-valued, discrete-time martingales $(Y_n : n \in \mathbb{Z}^+)$*

$$\mathbb{E}\left(F\left(|Y|_{p\text{-}var}\right)\right) \leq c\mathbb{E}\left[F\left(\left[\sum_n |Y_{n+1} - Y_n|^q\right]^{1/q}\right)\right].$$

Proof. For $d = 1$ this follows from Proposition 2.b in [5] with the remark that a discrete-time martingale can be viewed as a particular case of a continuous-time martingale with purely discontinuous sample paths. As above, the extension to $d > 1$ does not pose any difficulty.   □

Remark 9 *Pisier and Xu [8, Theorem 2.1. (ii)] have shown that the preceding lemma holds true for $1 \leq p = q < 2$, at least for $F(x) = x^r$ with $1 \leq r < \infty$.*

3 BDG on the group

Theorem 10 (BDG inequality on the group). *Let F be a moderate function, $\mathbf{M} \in \mathcal{M}_{0,loc}^c\left(G^2\left(\mathbb{R}^d\right)\right)$, and $|\cdot|$, $\|\cdot\|$ continuous homogenous norm on $\mathbb{R}^d, G^2\left(\mathbb{R}^d\right)$ respectively. Then there exists a constant $C = C\left(F, d, |\cdot|, \|\cdot\|\right)$ so that*

$$C^{-1}\mathbb{E}\left(F\left(|\langle M \rangle_\infty|^{1/2}\right)\right) \leq \mathbb{E}\left(F\left(\sup_{s,t\geq 0} \|\mathbf{M}_{s,t}\|\right)\right) \leq C\mathbb{E}\left(F\left(|\langle M \rangle_\infty|^{1/2}\right)\right).$$

Proof. The lower bound comes from $\|\mathbf{M}_{s,t}\| \geq |M_{s,t}|$, monotonicity of F and the classical BDG lower bound. For the upper bound we note that $\sup_{u,v\geq 0} \|\mathbf{M}_{u,v}\| \leq 2\sup_{t\geq 0} \|\mathbf{M}_t\|$. By equivalence of homogeneous norm,

$$\|\mathbf{M}_t\| \leq C\left(|M_t| + |A_t|^{1/2}\right)$$

and using "$F(x+y) \lesssim F(x) + F(y)$", combined with the classical BDG upper bound, it suffices to show that,

$$\mathbb{E}\left(F\left(\sup_{t\geq 0} |A_t|^{1/2}\right)\right) \leq C\mathbb{E}\left(F\left(|\langle M \rangle_\infty|^{1/2}\right)\right).$$

But this is easy using the fact that $F\left((\cdot)^{1/2}\right)$ is moderate and A itself is a martingale to which we can apply BGD. Absorbing changing constants into $\lesssim$ we have,

$$\mathbb{E}\left(F\left(\sup_{t\geq 0}|A_t|^{1/2}\right)\right) \lesssim \mathbb{E}\left(F\left(|\langle A\rangle_\infty|^{1/4}\right)\right)$$

$$\lesssim \mathbb{E}\left(F\left(\sup_{t\geq 0}|M_t|^{1/2}\times|\langle M\rangle_\infty|^{1/4}\right)\right)$$

$$\lesssim \mathbb{E}\left(F\left(\sup_{t\geq 0}|M_t|+|\langle M\rangle_\infty|^{1/2}\right)\right)$$

$$\lesssim \mathbb{E}\left(F\left(\sup_{t\geq 0}|M_t|\right)\right)+\mathbb{E}\left(F\left(|\langle M\rangle_\infty|^{1/2}\right)\right)$$

$$\lesssim \mathbb{E}\left(F\left(|\langle M\rangle_\infty|^{1/2}\right)\right).$$

Here, we used "$F(xy) \leq F(x^2+y^2) \lesssim F(x^2)+F(y^2)$" and, of course, the classical BDG upper bound in the last step.   □

4 Enhanced martingale p-variation regularity

Let $p \in (2,3)$. From [1] it is known that every $\mathbf{M} \in \mathcal{M}^c_{0,\mathrm{loc}}\left(G^2\left(\mathbb{R}^d\right)\right)$ is a geometric rough path. In other words, for every $T > 0$

$$\|\mathbf{M}\|_{p\text{-var};[0,T]} < \infty \text{ a.s.} \tag{1}$$

The BDG inequality on the group allows for a quick and simplified proof of this.

Proposition 11 (Enhanced martingale p-variation regularity). *Let* $\mathbf{M} \in \mathcal{M}^c_{0,loc}\left(G^2\left(\mathbb{R}^d\right)\right)$. *Then, for every* $T > 0$,

$$\|\mathbf{M}\|_{p\text{-}var;[0,T]} < \infty \text{ a.s.}$$

Proof. There exists a sequence of stopping times $\tau_n \to \infty$ a.s. such that M^{τ_n} and $\langle M^{\tau_n}\rangle$ are bounded (for instance, $\tau_n = \inf\{t : |M_t| > n \text{ or } |\langle M\rangle_t| > n\}$ will do.) Since

$$\mathbb{P}\left(\|\mathbf{M}\|_{p\text{-var};[0,T]} \neq \|\mathbf{M}\|_{p\text{-var};[0,T\wedge\tau_n]}\right) \leq \mathbb{P}\left(\tau_n < T\right) \to 0 \text{ as } n \to \infty$$

it suffices to consider the lift of a bounded continuous martingale with bounded quadratic variation. We can work with the l^1-norm on $\mathbb{R}^d$, $|a| = \sum_{i=1}^d |a_i|$. The time change $\phi(t) := \inf\{s : |\langle M\rangle_s| > t\}$ may have jumps but continuity of $|\langle M\rangle|$ ensures that $|\langle M\rangle_{\phi(t)}| = t$. From definition of ϕ and the BDG inequality on the group, both $|\langle M\rangle|$ and $\mathbf{M}$ are constant on the intervals $[\phi_{t-}, \phi_t]$. It follows that $\mathbf{X}_t = \mathbf{M}_{\phi(t)}$ defines a continuous[3] path from $[0, |\langle M\rangle_T|]$ to

---

[3] From Lemma 2, $\mathbf{X} = S_2(M \circ \phi)$, the lift of a continuous local martingale. In particular, this is another way to see continuity of $\mathbf{X}$.

$G^2\left(\mathbb{R}^d\right)$ and it is easy to see that

$$\|\mathbf{X}\|_{p\text{-var},[0,|\langle M\rangle_T|]} = \|\mathbf{M}\|_{p\text{-var},[0,T]}\,.$$

As argued in the beginning of the proof, we may assume that $|\langle M\rangle_T| \leq R$ for some deterministic R large enough. Therefore,

$$\mathbb{P}\left(\|\mathbf{M}\|_{p\text{-var},[0,T]} > K\right) = \mathbb{P}\left(\|\mathbf{X}\|_{p\text{-var},[0,|\langle M\rangle_T|]}, |\langle M\rangle_T| \leq R\right) \qquad (2)$$

$$\leq \mathbb{P}\left(\|\mathbf{X}\|_{p\text{-var},[0,R]} > K\right).$$

We go on to show that $\mathbf{X}$ is in fact Hölder continuous. For $0 \leq s \leq t \leq R$, we can use the BDG inequality on the group, theorem 10, to obtain

$$\mathbb{E}\left(\|\mathbf{X}_{s,t}\|^{2q}\right) = \mathbb{E}\left(\|\mathbf{M}_{\phi(s),\phi(t)}\|^{2q}\right) \leq C_q \mathbb{E}\left(\left|\langle M\rangle_{\phi(t)} - \langle M\rangle_{\phi(s)}\right|^q\right).$$

Observe that

$$\left|\langle M\rangle_{\phi(t)} - \langle M\rangle_{\phi(s)}\right| = \sum_i \left(\langle M^i\rangle_{\phi(t)} - \langle M^i\rangle_{\phi(s)}\right)$$

$$= \left|\langle M\rangle_{\phi(t)}\right| - \left|\langle M\rangle_{\phi(s)}\right| = t - s.$$

Thus, for all $q < \infty$ there exists a constant C_q s.t.

$$\mathbb{E}\left(\|\mathbf{X}_{s,t}\|^{2q}\right) \leq C_q |t - s|^q\,.$$

Knowing that $\mathbf{X}$ is continuous, we can apply the Garsia-Rodemich-Rumsey lemma[4] for paths in $\left(G^2\left(\mathbb{R}^d\right), d\right)$ to see that $\|\mathbf{X}\|_{1/p\text{-Hölder},[0,R]} \in L^q$ for all $q \in [1, \infty)$ and

$$\mathbb{P}\left(\|\mathbf{X}\|_{p\text{-var},[0,R]} > K\right) \leq \frac{\mathbb{E}\left(\|\mathbf{X}\|_{p\text{-var},[0,R]}\right)}{K} \leq \frac{\mathbb{E}\left(\|\mathbf{X}\|_{1/p\text{-Hölder},[0,R]}\right)}{K}$$

tends to zero as $K \to \infty$. Together with (2) we see that $\|\mathbf{M}\|_{p\text{-var},[0,T]} < \infty$ with probability 1 as claimed.   □

5 BDG on the group in p-variation norm

Following a classical pattern to prove BDG inequalities [4,5], we first prove a Chebyshev-type estimate, carefully adapting the arguments of Lépingle [5] to our setting.

---

[4] There is no modification of $\mathbf{X}$ needed.

Lemma 12. *There exists a constant A such that for all continous local martingales M, for all $\lambda > 0$,*

$$\mathbb{P}\left(\|\mathbf{M}\|_{p\text{-}var;[0,\infty)} > \lambda\right) \leq A \frac{\mathbb{E}\left(|\langle M\rangle_\infty|\right)}{\lambda^2}. \tag{3}$$

Proof. If suffices to prove the statement when $\lambda = 1$ (the general case follows by considering M/λ with lift $\delta_{1/\lambda}\mathbf{M}$). The statement then reduces to

$$\exists A : \forall M : \mathbb{P}\left[\|\mathbf{M}\|_{p\text{-}var;[0,\infty)} > 1\right] \leq A\,\mathbb{E}\left(|\langle M\rangle_\infty|\right).$$

Assume this is false. Then for every A, and in particular for $A(k) \equiv k^2$, there exists $M \equiv M^{(k)}$ with lift $\mathbf{M}^{(k)}$ s.t. the condition is violated,

$$k^2 \mathbb{E}\left[\left|\left\langle M^{(k)}\right\rangle_\infty\right|\right] < \mathbb{P}\left[\left\|\mathbf{M}^{(k)}\right\|_{p\text{-}var;[0,\infty)} > 1\right] \leq 1.$$

Set $u_k = \mathbb{P}\left[\left\|\mathbf{M}^{(k)}\right\|_{p\text{-}var;[0,\infty)} > 1\right]$, $n_k = [1/u_k + 1] \in \mathbb{N}$ and note that $1 \leq n_k u_k \leq 2$. Take n_k copies of each $M^{(k)}$ and get a sequence of martingales of form

$$\left(\tilde{M}\right) \equiv (\underbrace{M^{(1)}, ..., M^{(1)}}_{n_1}; \underbrace{M^{(2)}, ..., M^{(2)}}_{n_2}; M^{(3)}, ...).$$

Then

$$n_k k^2 \mathbb{E}\left[\left|\left\langle M^{(k)}\right\rangle_\infty\right|\right] \leq n_k \mathbb{P}\left[\left\|\mathbf{M}^{(k)}\right\|_{p\text{-}var;[0,\infty)} > 1\right] = n_k u_k \leq 2.$$

and

$$\sum_k \mathbb{P}\left[\left\|\tilde{\mathbf{M}}^{(k)}\right\|_{p\text{-}var;[0,\infty)} > 1\right] = \sum_k n_k u_k = +\infty$$

while

$$\sum_k \mathbb{E}\left[\left|\left\langle \tilde{M}^{(k)}\right\rangle_\infty\right|\right] = \sum_k n_k \mathbb{E}\left[\left|\left\langle M^{(k)}\right\rangle_\infty\right|\right] \leq \sum_k \frac{2}{k^2} < \infty.$$

Thus, if the claimed statement is false, there exists a sequence of martingales, we now revert to write $M^{(k)}, \mathbf{M}^{(k)}$ instead of $\tilde{M}^{(k)}, \tilde{\mathbf{M}}^{(k)}$ respectively, each defined on some filtered probability space $\left(\Omega^k, \left(\mathcal{F}_t^k\right), \mathbb{P}^k\right)$ with the two properties

$$\sum_k \mathbb{P}^k\left[\left\|\mathbf{M}^{(k)}\right\|_{p\text{-}var;[0,\infty)} > 1\right] = +\infty \text{ and } \sum_k \mathbb{E}^k\left[\left|\left\langle M^{(k)}\right\rangle_\infty\right|\right] < \infty.$$

Define the probability space $\Omega = \bigotimes_{k=1}^\infty \Omega^k$, the probability $\mathbb{P} = \bigotimes_{k=1}^\infty \mathbb{P}^k$, and the filtration $(\mathcal{F}_t)$ on Ω given by

$$\mathcal{F}_t = \left(\bigotimes_{i=1}^{k-1} \mathcal{F}_\infty^i\right) \otimes \mathcal{F}_{g(k-t)}^k \otimes \left(\bigotimes_{j=k+1}^{\infty} \mathcal{F}_0^k\right) \quad \text{for } k-1 \leq t < k.$$

where $g(u) = 1/u - 1$ maps $[0,1] \to [0,\infty]$. Then, a continous martingale on $(\Omega, (\mathcal{F}_t), \mathbb{P})$ is defined by concatenation,

$$M_t = \sum_{i=1}^{k-1} M_\infty^{(i)} + M_{g(k-t)}^{(k)} \quad \text{for} \quad k-1 \leq t < k.$$

which implies

$$\mathbf{M}_t = \left(\bigotimes_{i=1}^{k-1} \mathbf{M}_\infty^{(i)}\right) \otimes \mathbf{M}_{g(k-t)}^{(k)}.$$

We also observe that, again for $k-1 \leq t < k$,

$$\langle M \rangle_t = \sum_{i=1}^{k-1} \left\langle M^{(i)} \right\rangle_\infty + \left\langle M^{(k)} \right\rangle_{g(k-t)}.$$

In particular, $\langle M \rangle_\infty = \sum_k \left\langle M^{(k)} \right\rangle_\infty$ and, using the second property of the martingale sequence, $\mathbb{E}\left(|\langle M \rangle_\infty|\right) < \infty$. Define the events

$$A_k = \left\{ \|\mathbf{M}\|_{p\text{-var};[k-1,k]} > 1 \right\}.$$

Then, using the first property of the martingale sequence,

$$\sum_k \mathbb{P}(A_k) = \sum_k \mathbb{P}^k \left(\|\mathbf{M}^k\|_{p\text{-var};[0,\infty)} > 1 \right) = \infty.$$

Since the events $\{A_k : k \geq 1\}$ are independent, the Borel-Cantelli lemma implies that $\mathbb{P}(A_k \text{ i.o.}) = 1$. Thus, almost surely, for all $K > 0$ there exists a finite number of increasing times $t_0, \cdots, t_n \in [0,\infty)$ so that

$$\sum_{i=1}^{n} \|\mathbf{M}_{t_{i-1},t_i}\| > K$$

and $\|\mathbf{M}\|_{p\text{-var};[0,\infty)}$ must be equal to $+\infty$ with probability one. We now define a martingale N by time-change, namely via $f(t) = t/(1-t)$ for $0 \leq t < 1$ and $f(t) = \infty$ for $t \geq 1$,

$$N : t \mapsto M_{f(t)}.$$

Note that $\mathbb{E}\left(|\langle M \rangle_\infty|\right) < \infty$ so that M can be extended to a (continuous) martingale indexed by $[0,\infty]$ and N is indeed a continuous martingale with lift $\mathbf{N}$. Since lifts interchange with time changes, $\|\mathbf{N}\|_{p\text{-var};[0,1]} = \|\mathbf{M}\|_{p\text{-var};[0,\infty)} = +\infty$ with probability one. But this contradicts the p-variation regularity of enhanced martingales. $\square$

The passage from the above Chebyshev-type estimate to the full BDG inequality is made possible by

Lemma 13 (Good λ inequality, [10, p.94]). *Let X, Y be nonnegative random variables, and suppose there exists $\beta > 1$ such that for all $\lambda > 0, \delta > 0$,*

$$\mathbb{P}(X > \beta\lambda, Y < \delta\lambda) \leq \psi(\delta)\,\mathbb{P}(X > \lambda)$$

where $\psi(\delta) \searrow 0$ when $\delta \searrow 0$. There, for each moderate function F, there exists a constant C depending only on β, ψ, F such that

$$\mathbb{E}(F(X)) \leq C\mathbb{E}(F(Y)).$$

Theorem 14 (BDG inequality on the group and in homogenous p-variation norm). *Let F be a moderate function, $\mathbf{M} \in \mathcal{M}_{0,loc}^c\left(G^2\left(\mathbb{R}^d\right)\right)$, and $|\cdot|, \|\cdot\|$ continuous homogenous norm on $\mathbb{R}^d, G^2\left(\mathbb{R}^d\right)$ respectively and $p > 2$. Then there exists a constant $C = C\left(p, F, d, |\cdot|, \|\cdot\|\right)$ so that*

$$C^{-1}\mathbb{E}\left(F\left(|\langle M\rangle_\infty|^{1/2}\right)\right) \leq \mathbb{E}\left(F\left(\|\mathbf{M}\|_{p\text{-}var;[0,\infty)}\right)\right) \leq C\mathbb{E}\left(F\left(|\langle M\rangle_\infty|^{1/2}\right)\right).$$

Proof. Only the upper bound requires a proof. Fixing $\lambda, \delta > 0$ and $\beta > 1$, we define the stopping times

$$S_1 = \inf\left\{t > 0, \|\mathbf{M}\|_{p\text{-}var;[0,t]} > \beta\lambda\right\},$$
$$S_2 = \inf\left\{t > 0, \|\mathbf{M}\|_{p\text{-}var;[0,t]} > \lambda\right\},$$
$$S_3 = \inf\left\{t > 0, |\langle M\rangle_t|^{1/2} > \delta\lambda\right\},$$

with the convention that the infimum of the empty set if ∞. Define the local martingale $N_t = M_{S_3 \wedge S_2, (t+S_2)\wedge S_3}$ noting that $N_t \equiv 0$ on $\{S_2 = \infty\}$. It is easy to see that

$$\|\mathbf{M}\|_{p\text{-}var;[0,S_3]} \leq \|\mathbf{M}\|_{p\text{-}var;[0,S_3 \wedge S_2]} + \|\mathbf{N}\|_{p\text{-}var}. \tag{4}$$

where $\|\mathbf{N}\|_{p\text{-}var} \equiv \|\mathbf{N}\|_{p\text{-}var;[0,\infty)}$. By definition of the relevant stopping times,

$$\mathbb{P}\left(\|\mathbf{M}\|_{p\text{-}var} > \beta\lambda, |\langle M\rangle_\infty|^{1/2} \leq \delta\lambda\right) = \mathbb{P}\left(S_1 < \infty, S_3 = \infty\right).$$

On the event $\{S_1 < \infty, S_3 = \infty\}$ one has

$$\|\mathbf{M}\|_{p\text{-}var;[0,S_3]} > \beta\lambda$$

and, since $S_2 \leq S_1$, one also has $\|\mathbf{M}\|_{p\text{-}var;[0,S_3 \wedge S_2]}$. Hence, on $\{S_1 < \infty, S_3 = \infty\}$,

$$\|\mathbf{N}\|_{p\text{-}var} \geq \|\mathbf{M}\|_{p\text{-}var;[0,S_3]} - \|\mathbf{M}\|_{p\text{-}var;[0,S_3 \wedge S_2]} \geq (\beta - 1)\lambda.$$

Therefore, using (3),

$$\mathbb{P}\left(\|\mathbf{M}\|_{p\text{-var}} > \beta\lambda, |\langle M\rangle_\infty|^{1/2} \le \delta\lambda\right) \le \mathbb{P}\left(\|\mathbf{N}\|_{p\text{-var}} \ge (\beta - 1)\lambda\right)$$

$$\le \frac{A}{(\beta - 1)^2 \lambda^2}\mathbb{E}\left(|\langle N\rangle_\infty|\right).$$

From the definition of N, for every $t \in [0, \infty]$,

$$\langle N\rangle_t = \langle M\rangle_{S_3 \wedge S_2, (t+S_2)\wedge S_3}.$$

On $\{S_2 = \infty\}$ we have $\langle N\rangle_\infty = 0$ while on $\{S_2 < \infty\}$ we have, from definition of S_3,

$$|\langle N\rangle_\infty| = \left|\langle M\rangle_{S_3 \wedge S_2, S_3}\right| = \left|\langle M\rangle_{S_3} - \langle M\rangle_{S_3 \wedge S_2}\right| \le 2\left|\langle M\rangle_{S_3}\right| = 2\delta^2\lambda^2.$$

It follows that

$$\mathbb{E}\left(|\langle N\rangle_\infty|\right) \le 2\delta^2\lambda^2\mathbb{P}\left(S_2 < \infty\right) = 2\delta^2\lambda^2\mathbb{P}\left(\|\mathbf{M}\|_{p\text{-var}} > \lambda\right)$$

and we have the estimate

$$\mathbb{P}\left(\|\mathbf{M}\|_{p\text{-var}} > \beta\lambda, |\langle M\rangle_\infty|^{1/2} \le \delta\lambda\right) \le \frac{2A\delta^2}{(\beta - 1)^2}\mathbb{P}\left(\|\mathbf{M}\|_{p\text{-var}} > \lambda\right).$$

An application of the good λ-inequality finishes the proof.   □

It is crucial that one can choose $p \in (2, 3)$ above since only then $\mathbf{M} = \exp(M + A)$ is a geometric p-rough path for which rough path results apply. Here is a typical application.

Corollary 15. *Let M be a continuous, $\mathbb{R}^d$-valued local martingale and $S_N(M)$ the collection of iterated Stratonovich integrals up to level N, i.e.*

$$M_t, \cdots, \int_{\{0 < s_1 < \ldots s_N < t\}} \circ dM_{s_1} \circ \cdots \circ dM_{s_N}$$

viewed as path in $G^N(\mathbb{R}^d)$. Then there exists $C = C(N, F, d, |\cdot|)$ such that for all stopping times τ

$$\mathbb{E}\left(F\left(\left|\int_{\{0 < s_1 < \ldots s_N < \tau\}} \circ dM_{s_1} \circ \cdots \circ dM_{s_N}\right|^{1/N}\right)\right)$$

$$\le C\mathbb{E}\left(F\left(\|S_N(M)\|_{p\text{-var};[0,\tau]}\right)\right)$$

$$\le C\mathbb{E}\left(F\left(|\langle M\rangle_\tau|^{1/2}\right)\right)$$

Proof. By considering M stopped at τ it suffices to consider M on $[0, \infty)$. The first inequality is an immediate consequence of equivalence of homogenous norms on $G^N(\mathbb{R}^d)$ and we only have to show

$$\mathbb{E}\left(F\left(\|S_N(M)\|_{p\text{-var};[0,\infty)}\right)\right) \lesssim \mathbb{E}\left(F\left(|\langle M\rangle_\infty|^{1/2}\right)\right).$$

Clearly, $S_2(M) = \mathbf{M}$ a.s. and, by a basic theorem of Lyons [6], there is a deterministic lift of $\mathbf{M}$ to a $G^N(\mathbb{R}^d)$-valued path with finite homogenous $p \in (2, 3)$-variation, denoted by $S_N(\mathbf{M})$, such that $\|S_N(\mathbf{M})\|_{p\text{-var}} \leq C(N)\|\mathbf{M}\|_{p\text{-var}}$. On the other hand,

$$S_N(M) = S_N(\mathbf{M}) \quad \text{a.s.}$$

which can be seen, for instance, from our approximation result Theorem 21, the classical Wong-Zakai theorem for stochastic integrals and continuity of the lift $\mathbf{M} \mapsto S_N(\mathbf{M})$ in p-variation distance [6]. It follows from Theorem 14 that

$$\mathbb{E}\left(F\left(\|S_N(M)\|_{p\text{-var}}\right)\right) = \mathbb{E}\left(F\left(\|S_N(\mathbf{M})\|_{p\text{-var}}\right)\right) \leq C\mathbb{E}\left(F\left(|\langle M\rangle_\infty|^{1/2}\right)\right)$$

as required. □

6 Approximations

We now only consider (lifted) local martingales on $[0, T]$, defined or identified with local martingales stopped at $T > 0$.

6.1 Geodesic approxiations

The p-variation norm of geodesics approximations [3] is uniformly controlled by the original p-variation norm. Therefore

$$\mathbb{E}\left(F\left(\sup_D \left\|\mathbf{M}^{[D]}\right\|_{p\text{-var};[0,T]}\right)\right) \leq C\mathbb{E}\left(F\left(|\langle M\rangle_T|^{1/2}\right)\right)$$

where $\mathbf{M}^{[D]}$ denotes the geodesics approxiation to $\mathbf{M}$ based on some dissection D of $[0, T]$. Note that this is stronger than

$$\sup_D \mathbb{E}\left(F\left(\left\|\mathbf{M}^{[D]}\right\|_{p\text{-var};[0,T]}\right)\right) \leq C\mathbb{E}\left(F\left(|\langle M\rangle_T|^{1/2}\right)\right)$$

which is what we are going to show for piecewise linear approximations.

6.2 Piecewise linear approximations

Let $D = (t_i)$ be a subdivision of $[0, T]$. Given $x \in C([0, T], \mathbb{R}^d)$ we define x^D to be the piecewise linear approximation of x which coincides with x on D. Since x^D is of bounded variation, it admits a canonical lift to a $G^2(\mathbb{R}^d)$-valued path, denoted by $\mathbf{x}^D$. This notation applies path-by-path to $M \in \mathcal{M}_{0,\text{loc}}^c(\mathbb{R}^d)$, we write $\mathbf{M}^D = \mathbf{M}^D(\omega)$ for the lifted piecewise linear approximation to $M(\omega)$. The next lemma involves no probabilty.

Lemma 16. *Set $\mathbf{x}^D = S_2(x^D)$ where x^D is linear between the points of D. Let $p \geq 1$. Then there exists a constant $C = C(p)$ such that*

$$\left\| \mathbf{x}^D \right\|_{p\text{-}var;[0,T]} \leq C \left\| \mathbf{x} \right\|_{p\text{-}var;[0,T]} + C \left(\max_{(s_k) \subset D} \sum_k d\left(\mathbf{x}_{s_k, s_{k+1}}, \mathbf{x}^D_{s_k, s_{k+1}} \right)^p \right)^{1/p}.$$

Proof. We first note that $\left\| \mathbf{x}^D_{s,t} \right\|^p \leq 3^{p-1} \left[\left| x^D_{s,s^D} \right|^p + \left\| \mathbf{x}^D_{s^D, t_D} \right\|^p + \left| x^D_{t_D, t} \right|^p \right]$. Now let (u_k) be a dissection of $[0, T]$, unrelated to D. Recall that u^D resp. u_D refers to the right- resp. left-neighbours of u in D.

$$3^{1-p} \sum_k \left\| \mathbf{x}^D_{u_k, u_{k+1}} \right\|^p \leq \sum_k \left[\left| x^D_{u_k, u_k^D} \right|^p + \left| x^D_{u_{k+1,D}, u_k} \right|^p \right] + \sum_k \left\| \mathbf{x}^D_{u_k^D, u_{k+1,D}} \right\|^p$$

$$\leq 2 \left| x^D \right|^p_{p\text{-}var;[0,T]} + \max_{(s_k) \subset D} \sum_k \left\| \mathbf{x}^D_{s_k, s_{k+1}} \right\|^p$$

$$\leq 2C \left| x \right|^p_{p\text{-}var;[0,T]} + \max_{(s_k) \subset D} \sum_k \left\| \mathbf{x}^D_{s_k, s_{k+1}} \right\|^p.$$

Trivially, $|x|_{p\text{-}var;[0,T]} \leq \|\mathbf{x}\|_{p\text{-}var;[0,T]}$. On the other hand, using $(a+b)^p \leq 2^{p-1}(a^p + b^p)$ when $a, b > 0$, the triangle inequality gives

$$2^{1-p} \max_{(s_k) \subset D} \sum_k \left\| \mathbf{x}^D_{s_k, s_{k+1}} \right\|^p \leq \max_{(s_k) \subset D} \sum_k d\left(\mathbf{x}_{s_k, s_{k+1}}, \mathbf{x}^D_{s_k, s_{k+1}} \right)^p + \|\mathbf{x}\|^p_{p\text{-}var;[0,T]}.$$

$\square$

Lemma 17. *Let F be a moderate function, $\mathbf{M} \in \mathcal{M}_{0,\text{loc}}^c(G^2(\mathbb{R}^d))$, and $|\cdot|, \|\cdot\|$ continuous homogenous norm on $\mathbb{R}^d, G^2(\mathbb{R}^d)$ respectively. Assume $2 < p' < p \leq 4$. Then there exists a constant $C = C(p, p', F, d, |\cdot|, \|\cdot\|)$ so that for all dissections $D = \{t_l\}$ of $[0, T]$,*

$$\mathbb{E} \left[F \left(\left(\max_{(s_k) \subset D} \sum_k d\left(\mathbf{M}_{s_k, s_{k+1}}, \mathbf{M}^D_{s_k, s_{k+1}} \right)^p \right)^{1/p} \right) \right]$$

$$\leq C \mathbb{E} \left[F \left(\left(\sum_l \left\| \mathbf{M}_{t_l, t_{l+1}} \right\|^{p'} \right)^{1/p'} \right) \right].$$

Remark 18 *By using a stronger version of Lemma 8, see Remark 9, this estimate holds for $p' = p$, but this does not improve any of our results below.*

Proof. For fixed k, there are $i < j$ so that $s_k = t_i$ and $s_{k+1} = t_j$. Then

$$\mathbf{M}_{s_k,s_{k+1}} = \bigotimes_{l=i}^{j-1} \exp\left(M_{t_l,t_{l+1}} + A_{t_l,t_{l+1}}\right), \quad \mathbf{M}^D_{s_k,s_{k+1}} = \bigotimes_{l=i}^{j-1} \exp\left(M_{t_l,t_{l+1}}\right).$$

From equivalence of homogenous norms we have

$$d\left(\mathbf{M}_{s_k,s_{k+1}}, \mathbf{M}^D_{s_k,s_{k+1}}\right) = \left\|\mathbf{M}^{-1}_{s_k,s_{k+1}} \otimes \mathbf{M}^D_{s_k,s_{k+1}}\right\|$$

$$= \left\|\exp\left(\sum_{l=i}^{j-1} A_{t_l,t_{l+1}}\right)\right\| \le C \left|\sum_{l=i}^{j-1} A_{t_l,t_{l+1}}\right|^{1/2}. \quad (5)$$

The idea is to introduce the (vector-valued) discrete-time martingale

$$Y_j = \sum_{l=0}^{j-1} A_{t_l,t_{l+1}} \in so\,(d)$$

so that

$$\max_{(s_k) \subset D} \sum_k d\left(\mathbf{M}_{s_k,s_{k+1}}, \mathbf{M}^D_{s_k,s_{k+1}}\right)^p \le C \max_{\{i_1,\ldots,i_n\} \subset \{1,\ldots,\#D\}} \sum_k \left|Y_{i_{k+1}} - Y_{i_k}\right|^{p/2}$$

which can be rewritten as

$$\left(\max_{(s_k) \subset D} \sum_k d\left(\mathbf{M}_{s_k,s_{k+1}}, \mathbf{M}^D_{s_k,s_{k+1}}\right)^p\right)^{1/p} \le C\sqrt{|Y|_{p/2\text{-var}}}.$$

Noting that $F \circ \sqrt{\cdot}$ is moderate and that $1 < p'/2 < p/2 \le 2$, Lemma 8 yields

$$\mathbb{E}\left[F \circ \sqrt{\cdot}\left(|Y|_{p/2\text{-var}}\right)\right] \le \mathbb{E}\left[F \circ \sqrt{\cdot}\left(\left(\sum_l |Y_{l+1} - Y_l|^{p'/2}\right)^{2/p'}\right)\right]$$

$$= \mathbb{E}\left[F \circ \sqrt{\cdot}\left(\left(\sum_l |A_{t_l,t_{l+1}}|^{p'/2}\right)^{2/p'}\right)\right]$$

$$\le \mathbb{E}\left[F\left(\left(\sum_l \|\mathbf{M}_{t_l,t_{l+1}}\|^{p'}\right)^{1/p'}\right)\right].$$

$\square$

Theorem 19. *Let F be a moderate function, $\mathbf{M} \in \mathcal{M}^c_{0,loc}\left(G^2\left(\mathbb{R}^d\right)\right)$, and $|\cdot|, \|\cdot\|$ continuous homogenous norm on $\mathbb{R}^d, G^2\left(\mathbb{R}^d\right)$ respectively. Then there exists a constant $C = C\left(p, F, d, |\cdot|, \|\cdot\|\right)$ so that for all dissections D of $[0,T]$,*

$$\mathbb{E}\left(F\left(\left\|\mathbf{M}^D\right\|_{p\text{-}var;[0,T]}\right)\right) \leq C\mathbb{E}\left(F\left(\left|\langle M\rangle_T\right|^{1/2}\right)\right).$$

Proof. From Lemma 16,

$$\left\|\mathbf{M}^D\right\|_{p\text{-}var;[0,T]} \leq C\left\|\mathbf{M}\right\|_{p\text{-}var;[0,T]} + C\left(\max_{(s_k)\subset D}\sum_k d\left(\mathbf{M}_{s_k,s_{k+1}}, \mathbf{M}^D_{s_k,s_{k+1}}\right)^p\right)^{1/p}.$$

Using "$F\left(x+y\right) \lesssim F\left(x\right) + F\left(y\right)$" and the above lemma, with $p' = 1 + p/2$ for instance, we obtain

$$\mathbb{E}\left[F\left(\left\|\mathbf{M}^D\right\|_{p\text{-}var}\right)\right] \leq C\mathbb{E}\left[F\left(\left\|\mathbf{M}\right\|_{p\text{-}var}\right)\right] + C\mathbb{E}\left[F\left(\left(\sum_l \left\|\mathbf{M}_{t_l,t_{l+1}}\right\|^{p'}\right)^{1/p'}\right)\right]$$

$$\leq C\mathbb{E}\left[F\left(\left\|\mathbf{M}\right\|_{p\text{-}var}\right)\right] + C\mathbb{E}\left[F\left(\left\|\mathbf{M}\right\|_{p'\text{-}var}\right)\right].$$

The proof is now finished with the BDG inequality on the group in p- (resp. p')-variation norm. □

Remark 20 *We don't expect a lower BDG bound uniformly over all dissections D of $[0,T]$. For instance,*

$$C^{-1}\mathbb{E}\left(F\left(\left|\langle M\rangle_T\right|^{1/2}\right)\right) \leq \mathbb{E}\left(F\left(\left|M^D\right|_{\infty;[0,T]}\right)\right)$$

can't hold since $D = \{0,T\}$ implies $M^D_{\infty;[0,T]} = |M_T|$ and for $F\left(x\right) = x$ we would control

$$\mathbb{E}\left(\left|M\right|_{\infty;[0,T]}\right) \sim \mathbb{E}\left(\left|\langle M\rangle^{1/2}_T\right|\right)$$

in terms of $\mathbb{E}\left(|M_T|\right)$ which is Doob's L^q maximal inequality with $q = 1$. But, as is well known, one needs $q > 1$ for Doob's L^q-inequality to hold true.

Theorem 21. *Assume that M is a continuous martingale such that*

$$\left|M\right|_{\infty;[0,T]} \in L^q\left(\Omega\right) \text{ for some } q \geq 1. \tag{6}$$

Then, $d_{p\text{-}var;[0,T]}\left(\mathbf{M}^D, \mathbf{M}\right)$ converges to 0 in L^q. If M is a continuous local martingale, then convergence holds in probability.

Remark 22 *If $q > 1$, Doob's maximal inequality implies that (6) holds for any L^q-martingale.*

Proof. When $t = t_j \in D$, as in the last lemma,

$$d\left(\mathbf{M}_t, \mathbf{M}_t^D\right) \leq C \left| \sum_{l=0}^{j-1} A_{t_l,t_{l+1}} \right|^{1/2}.$$

Next, consider $t \in [t_i, t_{i+1}]$ for some i. The path $M_\cdot^D$ restricted to $[t_i, t_{i+1}]$ is a straight line with no area, hence

$$\mathbf{M}_{t_i,t}^D = \exp\left(\frac{t-s}{t_{i+1} - t_i} M_{t_i,t_{i+1}} \right).$$

and

$$
\begin{aligned}
d\left(\mathbf{M}_{t_\cdot}, \mathbf{M}_t^D\right) &= d\left(\mathbf{M}_{t_i} \otimes \mathbf{M}_{t_i,t}, \mathbf{M}_{t_i}^D \otimes \mathbf{M}_{t_i,t}^D\right) \\
&= \left\| \left(\mathbf{M}_{t_i,t}^D\right)^{-1} \otimes \left(\mathbf{M}_{t_i}^D\right)^{-1} \otimes \mathbf{M}_{t_i} \otimes \mathbf{M}_{t_i,t} \right\| \\
&\leq \left\| \left(\mathbf{M}_{t_i,t}^D\right) \right\| + \left\| \left(\mathbf{M}_{t_i}^D\right)^{-1} \otimes \mathbf{M}_{t_i} \right\| + \left\| \mathbf{M}_{t_i,t} \right\| \\
&\leq 2 \sup_{0 < v - u \leq |D|} \|\mathbf{M}_{u,v}\| + C \max_{i,j} \left| \sum_{l=i}^{j-1} A_{t_l,t_{l+1}} \right|^{1/2}.
\end{aligned}
$$

Since $\mathbf{M}$ is almost surely continuous, and hence uniformly continuous $[0, T]$,

$$\sup_{0 < v - u \leq |D|} \|\mathbf{M}_{u,v}\| \to 0 \text{ a.s. with } |D| \to 0.$$

and, by dominated convergence (with $\|\mathbf{M}\|_\infty \in L^q$, seen by (6) and BDG on the group) this convergence holds in L^q. To deal with the second term, pick $2 < p' < p < 3$. By equivalence of homogenous norms,

$$
\begin{aligned}
\max_{i,j} \left| \sum_{l=i}^{j-1} A_{t_l,t_{l+1}} \right|^{\frac{1}{2}} &= \left| \max_{i,j} \left| \sum_{l=i}^{j-1} A_{t_l,t_{l+1}} \right|^{\frac{p}{2}} \right|^{\frac{1}{p}} \\
&\leq C \left| \max_{(s_k) \subset D} \sum_k d\left(\mathbf{M}_{s_k,s_{k+1}}, \mathbf{M}_{s_k,s_{k+1}}^D\right)^p \right|^{\frac{1}{p}},
\end{aligned}
$$

and Lemma 17, applied with $F(x) = x^q$, gives

$$\mathbb{E}\left(\max_{i,j} \left| \sum_{l=i}^{j-1} A_{t_l,t_{l+1}} \right|^{q/2} \right) \leq C\mathbb{E}\left[\left| \left(\left(\sum_{l:t_l \in D} \|\mathbf{M}_{t_l,t_{l+1}}\|^{p'} \right)^{1/p'} \right)^q \right| \right].$$

But the last expression tends to zero. Indeed, $\mathbf{M} \in C^{0;p'\text{-var}}$, the $d_{p'\text{-var}}$-closure of smooth lifted paths, and from [3],

$$\lim_{|D|\to 0}\left(\sum_{l:t_l\in D}\left\|\mathbf{M}_{t_l,t_{l+1}}\right\|^{p'}\right)^{1/p'}=0 \text{ a.s.}$$

and the converging sequence is dominated by $\|\mathbf{M}\|_{p'\text{-var};[0,T]}\in L^q$, this being a consequence of our BDG inequalities

$$\mathbb{E}\left[\|\mathbf{M}\|^q_{p'\text{-var};[0,T]}\right]\sim \mathbb{E}\left[|\langle M\rangle_T|^{q/2}\right]\sim \mathbb{E}\left[|M|^q_{\infty;[0,T]}\right]<\infty.$$

Recall the notations

$$d_{0;[0,T]}\left(\mathbf{x},\mathbf{y}\right)=\sup_{s,t\in[0,T]}d\left(\mathbf{x}_{s,t},\mathbf{y}_{s,t}\right),\ d_{\infty;[0,T]}=\sup_{t\in[0,T]}d\left(\mathbf{x}_t,\mathbf{y}_t\right)$$

We just showed that $d_\infty\left(\mathbf{M},\mathbf{M}^D\right)\equiv d_{\infty;[0,T]}\left(\mathbf{M},\mathbf{M}^D\right)\to 0$ in L^q. Writing $d_0 = d_{0;[0,T]}$, recall that

$$d_0\left(\mathbf{M},\mathbf{M}^D\right)\le d_\infty\left(\mathbf{M},\mathbf{M}^D\right)+C\left|\|\mathbf{M}\|_\infty\, d_\infty\left(\mathbf{M},\mathbf{M}^D\right)\right|^{1/2}.$$

It suffices to use Cauchy-Schwarz,

$$\mathbb{E}\left(\left|\|\mathbf{M}\|_\infty\, d_\infty\left(\mathbf{M},\mathbf{M}^D\right)\right|^{q/2}\right)$$

$$\le \left(\mathbb{E}\left(|\|\mathbf{M}\|_\infty|^q\right)\right)^{1/2}\left(\mathbb{E}\left(\left|d_\infty\left(\mathbf{M},\mathbf{M}^D\right)\right|^q\right)\right)^{1/2}$$

to see that $d_0\left(\mathbf{M},\mathbf{M}^D\right)\to 0$ in L^q. Writing $d_{p\text{-var}}\equiv d_{p\text{-var};[0,T]}$, we then use the interpolation formula [3]

$$d_{p\text{-var}}\left(\mathbf{M},\mathbf{M}^D\right)\le Cd_0\left(\mathbf{M},\mathbf{M}^D\right)^{1-\frac{p'}{p}}\left(\|\mathbf{M}\|^{\frac{p'}{p}}_{p'-var}+\|\mathbf{M}^D\|^{\frac{p'}{p}}_{p'-var}\right),\ 2<p'<p.$$

Hence,

$$\mathbb{E}\left(\left|d_{p\text{-var}}\left(\mathbf{M}^D,\mathbf{M}\right)\right|^q\right)$$

$$\le C\mathbb{E}\left(\left(\|\mathbf{M}\|^{q\frac{p'}{p}}_{p'\text{-var}}+\|\mathbf{M}^D\|^{q\frac{p'}{p}}_{p'\text{-var}}\right)d_0\left(\mathbf{M},\mathbf{M}^D\right)^{q\left(1-\frac{p'}{p}\right)}\right)$$

Using Hölder with conjugate exponents $1/\left(p'/p\right)$ and $1/\left(1-p'/p\right)$ gives

$$\mathbb{E}\left(\left|d_{p\text{-var}}\left(\mathbf{M}^D,\mathbf{M}\right)\right|^q\right)$$

$$\le C\mathbb{E}\left(\|\mathbf{M}\|^q_{p'\text{-var}}+\|\mathbf{M}^D\|^q_{p'\text{-var}}\right)^{p'/p}\left[\mathbb{E}\left(d_0\left(\mathbf{M},\mathbf{M}^D\right)^q\right)\right]^{1-p'/p}.$$

But now it suffices to remark, using our BDG estimates, that

$$\mathbb{E}\left(\|\mathbf{M}\|_{p'-var;[0,T]}^{q}\right), \mathbb{E}\left(\|\mathbf{M}^{D}\|_{p'-var;[0,T]}^{q}\right)$$
$$\leq C\mathbb{E}\left(|\langle M\rangle_{T}|^{q/2}\right) \leq C\mathbb{E}\left(\left||M|_{\infty;[0,T]}\right|^{q}\right)$$

and the last term is finite by assumption. We proved that $d_{p\text{-var}}\left(\mathbf{M}^{D},\mathbf{M}\right) \to 0$ in L^{q} for any martingale M s.t. $|M|_{\infty;[0,T]} \in L^{q}$. If M is a local martingale one obtains convergence in probability by a simple localization argument.   □

References

1. Coutin Laure, Lejay Antoine: Semi-martingales and rough paths theory, Electronic Journal of Probability, Vol. 10, Paper 23, 2005.
2. Friz, Peter; Victoir, Nicolas: Approximations of the Brownian Rough Path with Applications to Stochastic Analysis, Annales de l'Institut Henri Poincare (B), Probability and Statistics, Volume 41, Issue 4, 703-724, 2005.
3. Friz, Peter; Victoir, Nicolas: On the notion of Geometric Rough Paths, Probab. Theory Relat. Fields, Vol 136 Nr 3, 395-416, 2006.
4. Lenglart Érik, Lépingle Dominique, Pratelli Maurizio: Présentation unifiée de certaines inégalités de la théorie des martingales. LNM 1404, 1980.
5. Lépingle Dominique: La variation d'ordre p des semi-martingales, Z. Wahrscheinlichkeitstheorie und Verw. Gebiete, Volume 36, Issue 4,1976.
6. Lyons, Terry: Differential equations driven by rough signals. Rev. Mat. Iberoamericana 14, no. 2, 215–310, 1998.
7. Lyons, Terry; Qian, Zhongmin: System Control and Rough Paths, OUP, 2002.
8. Pisier, Gilles; Xu, Quanhua: The Strong p -Variation of Martingales and Orthogonal Series, Probab. Theory Relat. Fields, Vol 77, 497-514, 1988.
9. Revuz Daniel, Yor Marc: Continuous Martingales and Brownian Motion, 3rd edition, Springer, 1999.
10. Rogers LCG, Williams David: Diffusions, Markov Processes, and Martingales : Itô Calculus, CUP, 2000.
11. Stricker, Christophe: Sur la p-variation des surmartingales. Séminaire de probabilités de Strasbourg, 13 (1979), p. 233-237

On Martingale Selectors of Cone-Valued Processes

Yuri Kabanov[1,2] and Christophe Stricker[1]

[1] Laboratoire de Mathématiques, Université de Franche-Comté
16 Route de Gray, 25030 Besançon, cedex, France
[2] Central Economics and Mathematics Institute, Moscow, Russia
e-mails: youri.kabanov@univ-fcomte.fr, christophe.stricker@univ-fcomte.fr

Summary. We discuss a result of Guasoni, Rásonyi, and Schachermayer on the existence of martingale selectors for a class of continuous cone-valued processes. The setting includes that arising in models of financial markets with transaction costs.

Keywords: Cone-valued process, Martingale selector, Transaction costs, Dalang–Morton–Willinger theorem, Consistent price system.

MSC 2000: 60G44

1 Introduction

Let C be a cone in $\mathbf{R}^d$ containing the vector $\mathbf{1} = (1, ..., 1)$ in its interior. Let $S = (S_t)_{0 \le t \le 1}$ be a $\mathbf{R}^d$-valued continuous adapted process with strictly positive components defined on a stochastic basis $(\Omega, \mathcal{F}, \mathbf{F}, P)$, and consider the diagonal matrix $\Sigma_t := \operatorname{diag} S_t$. The question is: *when is the set* $\mathcal{M}_0^1(\Sigma C \setminus \{0\})$ *non-empty?* That is, when does there exist an $\mathbf{R}^d$-valued martingale M such that the process $\Sigma_t^{-1} M_t$ takes values in $C \setminus \{0\}$?

This type of martingale selection problem arises in models of financial markets with constant proportional transaction costs where S is the price process and $C = K^*$, the dual of the solvency cone K (the investor positions are measured in units of a numéraire). In "canonical" notations $\Sigma_t C$ is just $\widehat{K}_t^*$ where $\widehat{K}_t$ is the solvency cone (random because of price movements) when the investor positions are measured in "physical" units. In the theory of markets with transaction costs the martingales evolving in $\widehat{K}^* \setminus \{0\}$ play the role of (densities of) martingale measures, see [4], [5], [6] etc. They are called consistent price systems, [8].

To formulate the result we introduce the following hypotheses.

If τ and σ are two stopping times with values in $[0,1]$ such that $\sigma \geq \tau$, let $A_{\tau,\sigma}$ denote the (random) topological support of the regular conditional distribution $P_{\tau,\sigma}(dx,\omega)$ of $S_\sigma - S_\tau$ with respect to $\mathcal{F}_\tau$.

H$_1$: $0 \in \mathrm{ri\,conv}\, A_{\tau,\sigma}$ a.s. on $\{\tau < 1\}$ for all stopping times τ and σ such that $\sigma \geq \tau$ (ri means: relative interior).

H$_2$: $0 < P(\sup_{\tau \leq r \leq 1} |S_r - S_\tau| \leq \varepsilon | \mathcal{F}_\tau) < 1$ a.s. on $\{\tau < 1\}$ for all $\varepsilon > 0$ and all stopping times τ.

Theorem 1. *Assume that* **H$_1$** *and* **H$_2$** *hold. Then* $\mathcal{M}_0^1(\Sigma C \setminus \{0\}) \neq \emptyset$.

This note can be viewed as a seminar comment to the interesting recent paper [3], where the authors suggested a sufficient condition for the non-emptiness of $\mathcal{M}_0^1(\Sigma C)$. Though our formulation sounds slightly more general (as we prefer the Levental–Skorohod type condition, [7]), the arguments follow the same lines. We only take a shortcut, in the proof of the key lemma (interesting on its own), by directly using the Dalang–Morton–Willinger (DMW) theorem, [1], [2], instead of repeating a part of its proof (cf. Lemma 3.3 in [3]).

2 Key Lemma

Let $X = (X_n)_{n \geq 0}$ be an $\mathbf{R}^d$-valued discrete-time adapted process on a stochastic basis $(\Omega, \mathcal{F}, \mathbf{G} = (\mathcal{G}_n), P)$. Put $\xi_n = \Delta X_n$.

Lemma 1. *Suppose that the following conditions hold:*
(i) for each finite N the process $(X_n)_{n \leq N}$ has the NA-property;
(ii) the sets $\Gamma_n := \{\xi_n = 0\}$ are increasing and $P(\cup_n \Gamma_n) = 0$;
(iii) $0 < P(\Gamma_n | \mathcal{G}_{n-1}) < 1$ a.s. on Γ_{n-1}^c for each $n \geq 1$.
Then there exists $Q \sim P$ such that X is a Q-martingale bounded in $L^2(Q)$.

Proof. Recall that according to the DMW theorem condition (i) is equivalent to the NA-property for each one-step model: the relation $\gamma \xi_n \geq 0$ with $\gamma \in L^0(\mathbf{R}^d, \mathcal{G}_{h-1})$ may hold only if $\gamma \xi_n = 0$. Suppose that $P(\Gamma_{n-1}) < 1$. In virtue of (iii) the set $\Omega_n' := \{\xi_n \neq 0\} = \Gamma_n^c$ has positive probability. Consider the space $(\Omega_n', \mathcal{G}_n', P_n')$ where $\mathcal{G}_n'$ is the trace of $\mathcal{G}_n$ on Γ_n^c and P_n' is the restriction of P on $\mathcal{G}_n'$ normalized to be a probability measure. The restriction ξ_n' of ξ_n to Ω_n' retains the NA-property. Thus, by another part of the DMW theorem, there is a bounded random variable $\alpha_n' > 0$ on Ω_n' such that $E_n'(\alpha_n' | \mathcal{G}_{n-1}') = 1$, $E_n' \alpha_n' |\xi_n'| < \infty$, $E_n'(\alpha_n' \xi_n' | \mathcal{G}_{n-1}') = 0$ and, moreover, $E_n'(\alpha_n' \xi_n'^2 | \mathcal{G}_{n-1}') \leq c_n$ where c_n is a constant. Define on Ω a $\mathcal{G}_n$-measurable random variable $\alpha_n > 0$ by the formula

$$\alpha_n = I_{\Gamma_{n-1}} + (1 - \delta_n) I_{\Gamma_{n-1}^c} \frac{1}{P(\Gamma_n | \mathcal{G}_{n-1})} I_{\Gamma_n} + \delta_n I_{\Gamma_{n-1}^c} \tilde{\alpha}_n I_{\Gamma_n^c}.$$

where $\delta_n := 2^{-n}/(1 + c_n)$ and $\tilde{\alpha}_n := \alpha_n'/P(\Gamma_n^c)$ on Γ_n^c. In the case $P(\Gamma_{n-1}) = 1$ we put consistently $\alpha_n = 1$.

Clearly, $E(\alpha_n|\mathcal{G}_{n-1}) = 1$, $E(\alpha_n\xi_n^2|\mathcal{G}_{n-1}) \le 2^{-n}$, $E(\alpha_n\xi_n|\mathcal{G}_{n-1}) = 0$.

The process $Z_n := \alpha_1...\alpha_n$ is a martingale. It converges stationarily a.s. to a random variable $Z_\infty > 0$ with $EZ_\infty \le 1$. Since $I_{\Gamma_n} \uparrow 1$ (a.s.) and $Z_\infty I_{\Gamma_n} = Z_n I_{\Gamma_n}$,

$$EZ_\infty = E \lim_n Z_\infty I_{\Gamma_n} = \lim_n EZ_\infty I_{\Gamma_n} = \lim_n EZ_n I_{\Gamma_n} = 1 - \lim_n EZ_n I_{\Gamma_n^c}.$$

It follows that $EZ_\infty = 1$ (i.e. (Z_n) is uniformly integrable martingale):

$$EI_{\Gamma_n^c} Z_n = E \prod_{k \le n} \delta_k \tilde{\alpha}_k I_{\Gamma_k^c} \le \prod_{k \le n} \delta_k \to 0.$$

Thus, $Q := Z_\infty P$ is a probability measure under which X is a martingale. At last,

$$E_Q X_n^2 = \sum_{k \le n} EZ_n \xi_k^2 \le \sum_{k \le n} \delta_n c_n \le 1,$$

i.e. X_n belongs to the unit ball of $L^2(Q)$. □

3 Martingale Selection Theorem: Proof

Fix $\theta > 1$. Define the sequence of stopping times, $\tau_0 = 0$,

$$\tau_n := \inf\{t \ge \tau_{n-1} : \max_{i \le d} |\ln S_t^i - \ln S_{\tau_{n-1}}^i| \ge \ln \theta\} \wedge 1, \qquad n \ge 1,$$

and the stopping time $\tau_t := \min\{\tau_n : \tau_n > t\}$ for $t \in [0,1[$. Put also $\sigma_t := \max\{\tau_n : \tau_n \le t\}$ and $\nu := \max\{n : \tau_n < 1\}$. Since the ratios $S_t^i/S_{\sigma_t}^i$ and $S_{\tau_t}^i/S_{\sigma_t}^i$ take values in the interval $[\theta^{-1}, \theta]$, we have the bounds

$$\theta^{-2} \le S_{\tau_t}^i/S_t^i \le \theta^2, \qquad i \le d. \tag{1}$$

Set $X_n := S_{\tau_n} I_{\{\tau_n < 1\}} + S_{\tau_\nu} I_{\{\tau_n = 1\}}$, $\mathcal{G}_n := \mathcal{F}_{\tau_n}$. Suppose that the discrete-time process $X = (X_n)$ satisfies the conditions of the lemma. Then X is a uniformly integrable Q-martingale with respect to some probability measure $Q = Z_\infty P$ equivalent to P. Consider the continuous-time martingale $\tilde{S}_t := E_Q(X_\infty|\mathcal{F}_t)$, $t \in [0,1]$. Since $S_{\tau_n} = X_n$ we have the inequalities

$$\theta^{-1} \le \tilde{S}_{\tau_n}^i/S_{\tau_n}^i \le \theta$$

where τ_n can be replaced by τ_t. Using this and the bounds (1) we get

$$\theta^{-3} \le \tilde{S}_{\tau_t}^i/S_t^i \le \theta^3.$$

But $\tilde{S}_t^i/S_t^i = E_Q(\tilde{S}_{\tau_t}^i/S_t^i|\mathcal{F}_t)$ and, therefore, the ratios $\tilde{S}_t^i/S_t^i$ take values in the interval $[\theta^{-3}, \theta^3]$. Thus, for θ sufficiently close to unit, the Q-martingale $\tilde{S}$ evolves in $\Sigma C \setminus \{0\}$ and so does also the P-martingale $M := Z\tilde{S}$.

It remains to note that properties (i) and (iii) hold by virtue of $\mathbf{H}_1$ and $\mathbf{H}_2$ while (ii) is always fulfilled for continuous S. □

Remark. An important part of the paper [3] is devoted to the property of S called "conditional full support", implying $\mathbf{H}_1$ and $\mathbf{H}_2$. This property is shown to hold for a wide class of continuous processes.

References

1. Dalang R.C., Morton A., Willinger W. Equivalent martingale measures and no-arbitrage in stochastic securities market model. *Stochastics and Stochastic Reports*, **29** (1990), 185–201.
2. Jacod J., Shiryaev A.N. Local martingales and the fundamental asset pricing theorem in the discrete-time case. *Finance and Stochastics*, **2** (1998), 3, 259–273.
3. Guasoni P., Rásonyi M., Schachermayer W. Consistent price systems and face-lifting pricing under transaction costs. Preprint, 2007.
4. Kabanov Yu.M. Hedging and liquidation under transaction costs in currency markets. *Finance and Stochastics*, **3** (1999), 2, 237–248.
5. Kabanov, Yu.M., Stricker, Ch. The Harrison–Pliska arbitrage pricing theorem under transaction costs. *J. Math. Economics*, **35**, 2001, 2, 185-196.
6. Kabanov, Yu.M., Rásonyi M., Stricker, Ch. On a closedness of sums of convex cones in L^0 and the robust no-arbitrage property. *Finance and Stochastics*, **7** (2003), 3, 403–411.
7. Levental S., Skorohod A.V. On the possibility of hedging options in the presence of transaction costs. *The Annals of Applied Probability*, **7** (1997), 410–443.
8. Schachermayer, W.: The Fundamental Theorem of Asset Pricing under proportional transaction costs in finite discrete time. *Mathematical Finance*, **14**, 1 (2004), 19-48.

No asymptotic free lunch
reviewed in the light of Orlicz spaces

Irene Klein

Dept. of Statistics and Decision Support Systems, University of Vienna
Brünnerstr. 72, A-1210 Vienna, Austria
e-mail: Irene.Klein@univie.ac.at

Summary. No asymptotic free lunch (NAFL) was introduced in [11] and led to a general version of the Fundamental Theorem of Asset Pricing (FTAP) for large financial markets. The present note observes that NAFL can be defined in a natural way using Orlicz spaces. This gives a transparent proof of the FTAP–result.

MSC 2000: 46A20, 46A22, 46N10, 60G44, 60H05

Key words: fundamental theorem of asset pricing, equivalent martingale measure, free lunch, large financial market, asymptotic free lunch, contiguity of measures, Orlicz space

1 Introduction

In [11] the notion of no asymptotic free lunch (NAFL) was introduced and gave a general version of the Fundamental Theorem of Asset Pricing (FTAP) for large financial markets. The present note is a revision showing that the appropriate theoretical background of the rather 'hand–knitted' proof there is the theory of Orlicz spaces. This new sight adds substantially to the transparency of the FTAP for large financial markets and its proof.

Speaking mathematically, the classical continuous–time model of a financial market is just a stochastic basis $\mathbf{B} = (\Omega, \mathcal{F}, (\mathcal{F}_t), P)$ with a vector–valued semimartingale $S = (S_t)$ interpreted as the price process. The theory initiated by Kabanov and Kramkov in [9] suggests to describe a *large financial market* by a sequence of classical models $(\mathbf{B}^n, S^n)$. In a certain sense this is a modern version of the Huberman–Ross arbitrage pricing theory, see for comparison [20] and [5] and the paper [8] which gives a clear overview of the theory. In [9] it was assumed that for each classical model the set of equivalent martingale measures M^n for S^n consists of a single measure Q^n. The authors of [9] introduced the notions of asymptotic arbitrage of first and second kind and showed that their absence is equivalent to the contiguity $(P^n) \triangleleft (Q^n)$ and $(Q^n) \triangleleft (P^n)$, respectively. These results can be understood as one–sided versions of the FTAP for large financial markets as contiguity corresponds to the property of absolute continuity of measures in the classical model. Criteria for the general situation (where M^n is not a singleton) look more involved, see [13], [14] and in a different formulation [10]. E.g., in [10] it was shown that the above contiguity conditions can be replaced by $(P^n) \triangleleft (\bar{Q}^n)$ and $(\underline{Q}^n) \triangleleft (P^n)$, where $\bar{Q}^n(A) = \sup_{Q \in M^n} Q(A)$, $\underline{Q}^n(A) = \inf_{Q \in M^n} Q(A)$.

However, a FTAP in the context of a large financial market might rather be the equivalence of an asymptotic no arbitrage condition and the existence of an *equivalent* martingale measure for the large financial market, that is, a sequence $Q^n \in M^n$ such that simultaneously $(P^n) \triangleleft (Q^n)$ and $(Q^n) \triangleleft (P^n)$. In [11] this property of *bicontiguity* was related to the condition NAFL (no asymptotic free lunch). The idea of NAFL is to generalize the classical no free lunch condition of Kreps [16] for a sequence of market models. The definition of NAFL uses a special description of the zero–neighbourhoods for the Mackey–topology of L^∞, which was done in a self–made way that conceals the connection to the theory of Orlicz spaces. The aim of the present note is to reveal this connection and thus present a better looking candidate for NAFL which is formulated using polars of balls of Orlicz spaces. The proof of the FTAP–result is recalled now using this NAFL; it is more transparent as definitions are more natural and some technicalities become superfluous being straightforward consequences of properties of N–functions. Moreover in Chapter 5 another version of the NAFL condition is presented, which uses the norm–topology of Orlicz spaces instead of the Mackey–topology. This definition turns out to be equivalent to the original one and is of interest as the approximation in the non–metrizable Mackey–topology is replaced by an approximation with respect to norms. The connection to the no free lunch condition of Kreps becomes clear and is based on a characterization of the weak–star–closure of a convex subset of L^∞ in terms of Orlicz spaces, see also [17].

2 Definitions and notations

Let $(\Omega, \mathcal{F}, (\mathcal{F}_t)_{t \in [0,T]}, P)$ be a filtered probability space where the filtration satisfies the usual assumptions. (We choose a finite time horizon $T > 0$ to avoid technical subtleties, but note that $T = \infty$ works as well.) Whenever it is clear which P is meant, the notation L^p is used for $L^p(\Omega, \mathcal{F}, P)$; whenever the dependence on a certain measure R is stressed we use the notation $L^p(R)$. Let $(S_t)_{t \in [0,T]}$ be an $(\mathcal{F}_t)$–adapted semimartingale with values in $\mathbb{R}^d$, describing the price processes of d tradeable assets. Let H be a predictable S–integrable process and $(H \cdot S)_t$ the stochastic integral of H with respect to S. The process

H is an admissible trading strategy if there is $a > 0$ such that $(H \cdot S) \geq -a$. Define

$$\mathbf{K} = \{(H \cdot S)_T : H \text{ admissible}\} \text{ and } \mathbf{C} = (\mathbf{K} - L_+^0) \cap L^\infty.$$

$\mathbf{K}$ can be interpreted as the cone of all replicable claims, and $\mathbf{C}$ is the cone of all claims in L^∞ that can be superreplicated. Define the set M_a of absolutely continuous and the set M of equivalent separating measures

$$M_a = \{Q \ll P : E_Q[f] \leq 0 \text{ for all } f \in \mathbf{K}\} \text{ and } M = \{Q \in M_a : Q \sim P\}.$$

If S is bounded (locally bounded) then M_a (M) consists of all P–absolutely continuous (P–equivalent) probability measures such that S is a martingale (local martingale). In general, for unbounded S, M_a (M) is the set of P–absolutely continuous (P–equivalent) probabilities such that the admissible stochastic integrals are supermartingales.

Recall two well-known generalizations of the no arbitrage condition, namely, no free lunch with vanishing risk (NFLVR) and no free lunch (NFL):

$$\begin{aligned} \text{(NFLVR)} &\quad \bar{\mathbf{C}} \cap L_+^\infty = \{0\} \\ \text{(NFL)} &\quad \bar{\mathbf{C}}^* \cap L_+^\infty = \{0\}, \end{aligned} \tag{2.1}$$

where $\bar{\mathbf{C}}$ is the L^∞–norm closure and $\bar{\mathbf{C}}^*$ the weak–star–closure of $\mathbf{C}$.

The Fundamental Theorem of Asset Pricing (FTAP) says that an appropriate 'no-arbitrage'–condition is equivalent to $M \neq \emptyset$. Kreps [16] proved that NFL is equivalent to $M \neq \emptyset$. Delbaen and Schachermayer [1] introduced NFLVR and proved the deep theorem that under NFLVR we have $\mathbf{C} = \bar{\mathbf{C}}^*$. Thus, the seemingly weaker condition NFLVR is, in fact, equivalent to NFL. Moreover, in [2] they showed that, under NFLVR, the set M^σ of all equivalent sigma–martingale measures is dense in M. Therefore NFLVR is equivalent to the existence of an equivalent σ–martingale measure, compare also [7].

In a *large financial market* a sequence of market models is considered, that is, a sequence of semimartingales S^n based on $(\Omega^n, \mathcal{F}^n, (\mathcal{F}_t^n), P^n)$. The interpretation of the superscript n in expressions such as $\mathbf{K}^n$, $\mathbf{C}^n$, M_a^n, M^n etc. is then obvious. Throughout the paper we assume

$$M^n \neq \emptyset, \qquad \text{for all } n \in \mathbb{N}. \tag{2.2}$$

So, any no arbitrage condition (such as NFLVR and NFL) holds for each model.

However, there is still the possibility of various approximations of an arbitrage profit by trading on the sequence of small markets, compare for example [9], [10], [13], [14]. The present note is focused on the condition no asymptotic free lunch (NAFL) which is the large financial market analogue of NFL, see [11].

The following object plays the role of an equivalent (sigma–, local–) martingale measure for the large financial market: it is a sequence of measures

$Q^n \in M^n$ such that $(Q^n) \triangleleft \triangleright (P^n)$; here the notation $(Q^n) \triangleleft \triangleright (P^n)$ means that the sequence of probability measures (Q^n) is contiguous with respect to the sequence of probability measures (P^n) and vice versa.

Definition 2.1 *A sequence of measures (Q^n) is called contiguous with respect to (P^n), denoted by $(Q^n) \triangleleft (P^n)$, if and only if for any sequence $(A^n)_{n=1}^{\infty}$, $A^n \in \mathcal{F}^n$, $P^n(A^n) \to 0$ implies that $Q^n(A^n) \to 0$.*

Let now $Q^n \ll P^n$, for all n. Throughout the paper we will use the notation Z^n for $\frac{dQ^n}{dP^n}$. Then an alternative criterion to $(Q^n) \triangleleft (P^n)$ is that $(Z^n|Q^n)$ is $\mathbb{R}$–tight which is equivalent to the condition that $(Z^n|P^n)$ is uniformly integrable. Compare [6], Lemma V.1.6 and V.1.10, for these criteria in a more general form. Another obvious formulation of contiguity shows that it is a concept of absolute continuity in a uniform way for sequences of probability measures: indeed, $(Q^n) \triangleleft (P^n)$ is equivalent to the following condition: for all $\varepsilon > 0$ there is $\delta > 0$ such that, for all $n \in \mathbb{N}$ and $A^n \in \mathcal{F}^n$, $P^n(A^n) < \delta$ implies $Q^n(A^n) < \varepsilon$.

3 NAFL and the FTAP for large financial markets

A FTAP for large financial markets is a theorem that shows the equivalence between a condition of no asymptotic arbitrage type and the existence of a sequence of measures $Q^n \in M^n$ such that $(Q^n) \triangleleft \triangleright (P^n)$. In general, one has to use the condition of no asymptotic free lunch (NAFL) of [11] to get this equivalence. A recent result shows that one gets the equivalence with an asymptotic condition of no market free lunch type as well, but this is a slightly different approach using preferences of investors, see [12] and [3]. In the present note we will see that NAFL can be defined in a more elegant way using polars of balls of Orlicz spaces as Mackey–neighbourhoods of 0 in L^{∞}. In the original definition this connection of the Mackey–topology to the theory of Orlicz spaces is concealed as there a rather 'hand–knitted' description of the Mackey–neighbourhoods of 0 was used.

First recall the definition of NAFL. We will define the sets $V^{F,n} \subseteq L^{\infty}(P^n)$ below. In Section 4 we will see that, when F runs through all N–functions (see Definition 4.1), these sets form a fundamental system for all Mackey–neighborhoods of 0 of $L^{\infty}(P^n)$. Compare [11] for a definition of $V^{F,n}$ that works but is artificial and complicated. For each $\varepsilon > 0$ let

$$D^{\varepsilon,n} = \{w \in L^{\infty}(P^n) : 0 \leq w \leq 1 \text{ and } E_{P^n}[w] \geq \varepsilon\}. \tag{3.1}$$

Remark 3.1 For a sequence $w^n \in D^{\alpha,n}$ the following holds: if $(P^n) \triangleleft (Q^n)$ then there is $\beta > 0$ such that $w^n \in D^{\beta}(Q^n)$ for all $n \in \mathbb{N}$.

Indeed, it is clear that $P^n(w^n \geq \frac{\alpha}{2}) \geq \frac{\alpha}{2}$ for each $w^n \in D^{\alpha,n}$. By contiguity there is $\delta > 0$ such that $Q^n(w^n \geq \frac{\alpha}{2}) \geq \delta$, for all n, and so $E_{Q^n}[w^n] \geq \frac{\alpha}{2}\delta =: \beta$.

□

Let F be an N–function (see Definition 4.1 below). We put

$$B^F(P^n) = \{f \in L^1(P^n) : E_{P^n}[F(|f|)] \le 1\}. \tag{3.2}$$

Let $V^F(P^n)$ be the polar of $B^F(P^n)$, that is,

$$V^F(P^n) = (B^F(P^n))^\circ \tag{3.3}$$
$$= \{g \in L^\infty(P^n) : |E_{P^n}[gh]| \le 1 \text{ for all } h \in B^F(P^n)\}.$$

We use the notations $B^{F,n}$ and $V^{F,n}$ for $B^F(P^n)$ and $V^F(P^n)$, respectively, if it is clear which measure P^n is meant. In Section 5 we will see the connection to Orlicz spaces and, in particular, that the set $B^{F,n}$ is the closed unit ball of the Orlicz space $L_F(P^n)$.

Definition 3.2 *We say that the large financial market satisfies the NAFL condition if for any $\varepsilon > 0$ there exists an N–function F such that, for all $n \in \mathbb{N}$, $\mathbf{C}^n \cap (D^{\varepsilon,n} + V^{F,n}) = \emptyset$.*

This means that $\mathbf{C}^n$ is, for each $\varepsilon > 0$, separated from $D^{\varepsilon,n}$ by some Mackey–neighbourhood $V^{F,n}$ of 0 (where the F does not depend on n). NAFL is the analogue of NFL for a sequence of L^∞ spaces as it is not possible to approximate a strictly positive gain by elements of the sequence of sets $(\mathbf{C}^n)_{n\in\mathbb{N}}$ in a Mackey sense (or, equivalently, as the sets $\mathbf{C}^n$ are convex, in a weak star sense). The following version of the FTAP for large financial markets holds, see [11].

Theorem 3.3 *NAFL $\Leftrightarrow$ there is $Q^n \in M^n$ such that $(Q^n) \lhd\rhd (P^n)$.*

If S^n is (locally) bounded, for all n, then (Q^n) is a sequence of (local) martingale measures. For unbounded S^n, Theorem 3.3 implies the existence of a bicontiguous sequence of sigma–martingale–measures. This is an easy consequence of the fact that $M^{\sigma,n}$ is dense in M^n for the variation topology.

4 N–functions, NAFL and the proof of the FTAP

One of the two crucial properties of the sets V^F, for all $F \in \mathcal{N}$, in Definition 3.2 is that they form a fundamental system for all Mackey–neighbourhoods of 0. The Mackey–topology of L^∞ is the topology of uniform convergence on all weakly compact subsets of L^1, see [4]. A fundamental system for all Mackey–neighbourhoods of 0 is given by the polars of all weakly compact subsets of L^1. So one should look for an appropriate way to describe all weakly compact subsets of L^1. To this end we introduce a class of Young functions, the so-called N–functions, see [19].

Definition 4.1 $F : [0,\infty) \to [0,\infty)$ *is an N–function if F is convex, continuous, $F(0) = 0$, $\frac{F(t)}{t} \uparrow \infty$ as $t \uparrow \infty$ and $\frac{F(t)}{t} \downarrow 0$ as $t \downarrow 0$. The set of all N–functions is denoted by $\mathcal{N}$.*

Remark 4.2 Note that in [12] we used a more restrictive (that is *differentiable*) class of Young functions to define NAFL. However, all results of [12] hold as well when we use the class $\mathcal{N}$ above, as all relevant results of Kusuoka [17], which were referred to, hold for this class as well.

Lemma 4.3 *The set B^F of (3.2) is closed with respect to L^1–norm and uniformly integrable. In particular,* $\sup_{h \in B^F} E[|h|\mathbb{1}_{\{|h|\geq\kappa\}}] \leq \frac{\kappa}{F(\kappa)}$. *On the other hand, for each uniformly integrable $A \subseteq L^1$ there is $F \in \mathcal{N}$ such that $A \subseteq B^F$.*

Proof. The proof of uniform integrability of B^F follows by the definition of B^F and the properties of the function F (criterion of De La Vallée-Poussin). To get the inequality note that for $h \in B^F$ and $\kappa > 0$

$$E[|h|\mathbb{1}_{\{|h|\geq\kappa\}}] = E\left[\frac{|h|}{F(|h|)}F(|h|)\mathbb{1}_{\{|h|\geq\kappa\}}\right] \leq \frac{\kappa}{F(\kappa)}E[F(|h|)] \leq \frac{\kappa}{F(\kappa)},$$

as $\frac{F(y)}{y}$ is increasing. To show closedness in L^1 take $h^n \in B^F$ with $h^n \to h$ in L^1, then a subsequence of $F(|h^n|)$ (still denoted by n) converges to $F(|h|)$ a.s. By Fatou

$$E[F(|h|)] = E[\lim F(|h^n|)] \leq \liminf E[F(|h^n|)] \leq 1,$$

as $F(|h^n|) \geq 0$. This shows that $h \in B^F$. The second part of the statement follows again by De La Vallée-Poussin. $\square$

Remark 4.4 B^F is a weakly compact convex balanced subset of L^1. Therefore, by the Bipolar Theorem, $B^F = (B^F)^{\circ\circ} = (V^F)^\circ$, see [4]. Indeed, it is clear that B^F is balanced and convex. So L^1–closedness implies closedness for the topology $\sigma(L^1, L^\infty)$. By the Dunford–Pettis criterion (see for example [21]) a subset of L^1 is relatively weakly compact if and only if it is uniformly integrable. So, by Lemma 4.3 we get weak compactness of B^F.

Lemma 4.3 and Remark 4.4 show that the weakly compact subsets of L^1 can indeed be described completely with the help of the sets B^F. Hence the polars V^F of all B^F form a fundamental system of the Mackey 0–neighbourhoods. Moreover, by Remark 4.4 the sets V^F fulfill the second crucial property which was used in the proof in [11]. Namely, $(V^F)^\circ = B^F$ which is uniformly integrable. Later on this will give that $(Q^n) \triangleleft (P^n)$ because of the following relation of contiguity of sequences of measures to the sets B^F.

Lemma 4.5 *Let $Q^n \ll P^n$, for all n. Then $(Q^n) \triangleleft (P^n)$ if and only if there is $F \in \mathcal{N}$ such that, for all n, $Z^n \in B^{F,n}$.*

Proof. $(Q^n) \triangleleft (P^n)$ if and only if $(Z^n|P^n)$ is uniformly integrable. The rest follows by Lemma 4.3, details in [12]. $\square$

Proof (Proof of Theorem 3.3). ($\Rightarrow$) The proof works exactly as in [11] but for the new sets $V^{F,n}$. For the convenience of the reader we provide the details. By NAFL, for any $\varepsilon > 0$, there is $F_\varepsilon \in \mathcal{N}$ such that, for all n, $\mathbf{C}^n \cap (D^{\varepsilon,n} + V^{F_\varepsilon,n}) = \emptyset$. For notational simplicity we suppress the indices n and ε for the moment, but keep in mind that everything holds for any n and that $F = F_\varepsilon$ and $D = D^\varepsilon$. As $V^F = -V^F$ we have that $(\mathbf{C} + V^F) \cap D = \emptyset$. Replace for the moment V^F by its non–empty Mackey–interior $\hat{V}^F$, then, by Hahn–Banach, we can separate the disjoint convex sets D and $A = \mathbf{C} + \hat{V}^F$ (which is Mackey–open). This gives $g \in L^1$, which is not identical to 0 (so we can choose $\|g\|_{L^1} = 1$), such that for the original V^F we still have that

$$\sup_{f \in \mathbf{C}+V^F} E[fg] \leq \inf_{h \in D} E[hg]. \tag{4.1}$$

As $0 \in V^F$ and $-L^\infty_+ \subseteq \mathbf{C}$ we have that $g \geq 0$. As $\mathbf{C}$ is a cone we have that $\sup_{f \in \mathbf{C}} E[fg] \leq 0$. This gives in particular that g is the density of a measure $Q \in M_a$. Moreover, $0 \in \mathbf{C}$ and for all $h \in D$ we have that $E[hg] \leq \|h\|_{L^\infty} \|g\|_{L^1} \leq 1$. Therefore by (4.1) and as $V^F = -V^F$ we get

$$\sup_{f \in V^F} |E[fg]| \leq 1,$$

hence $g \in (V^F)^\circ = B^F$.

We claim that there is δ depending only on ε (but not on n) such that

$$\inf_{h \in D} E[gh] \geq \delta.$$

Indeed, as $g \in B^F$, by Lemma 4.3 we have that $E[g\mathbb{1}_{\{g \geq \kappa\}}] \leq \frac{\kappa}{F(\kappa)}$. Choose κ large enough such that $\frac{\kappa}{F(\kappa)} < \frac{1}{3}$. Then it is easy to see (use $E[g] = 1$) that

$$P\left(g \geq \frac{1}{3}\right) \geq \frac{1}{3\kappa}. \tag{4.2}$$

Define now $\delta = \frac{1}{9\kappa\gamma}$, where $\gamma > 0$ is a uniform L^1–bound of the uniformly integrable set B^F. Note that γ and κ depend only on ε and so does δ. Clearly, we have that $\bar{f} := \frac{1}{\gamma}\mathbb{1}_{\{g \geq \frac{1}{3}\}} \in V^F = (B^F)^\circ$. Hence we get

$$\inf_{h \in D} E[hg] \geq \sup_{f \in V^F} E[fg] \geq E[\bar{f}g] \geq \frac{1}{3\gamma}P\left(g \geq \frac{1}{3}\right) \geq \delta,$$

where the first inequality holds by (4.1) and the last one by (4.2).

Let us summarize what we proved. For all $\varepsilon > 0$ there is $F_\varepsilon \in \mathcal{N}$, $Q^{n,\varepsilon} \in M_a^n$ and $\delta > 0$ such that, for all n,

(i) $Z^{n,\varepsilon} \in B^{F_\varepsilon,n}$ (and so by Lemma 4.4, $(Q^{n,\varepsilon}) \lhd (P^n)$, for all ε),
(ii) $P^n(A^n) \geq \varepsilon$ implies that $Q^{n,\varepsilon}(A^n) \geq \delta$ (as $\mathbb{1}_{A^n} \in D^{\varepsilon,n}$).

Let now $\varepsilon = 2^{-j}$, $j \geq 1$, and define $Q^n = \sum_{j=1}^{\infty} 2^{-j} Q^{n,2^{-j}}$. For this sequence it holds that $Q^n \in M^n$ and $(Q^n) \triangleleft \triangleright (P^n)$, see Lemma 4.6 below.

($\Leftarrow$) Suppose that NAFL does not hold. As $(Q^n) \triangleleft (P^n)$ there is $\varphi \in \mathcal{N}$ such that, for all n, $Z^n \in B^{\varphi,n}$. Define, for each $\varepsilon > 0$, $F_\varepsilon \in \mathcal{N}$ by $F_\varepsilon(x) := \varphi(\varepsilon x)$. Observe that $V^{F_\varepsilon,n} = \varepsilon V^{\varphi,n}$. (Indeed, it is easy to see that $\frac{1}{\varepsilon} B^{\varphi,n} = B^{F_\varepsilon,n}$ and so $\varepsilon (B^{\varphi,n})^\circ = V^{F_\varepsilon,n}$ by the properties of the polar.) By assumption there is $\alpha > 0$ such that for F_ε there is $n = n(F_\varepsilon)$ and $f^\varepsilon \in \mathbf{C}^n$ with $f^\varepsilon = h^\varepsilon + g^\varepsilon$, where $h^\varepsilon \in D^{\alpha,n}$ and $g^\varepsilon \in V^{F_\varepsilon,n}$. As $Q^n \in M^n$ we have that $E_{Q^n}[f^\varepsilon] \leq 0$. Moreover $|E_{Q^n}[g^\varepsilon]| = |E_{P^n}[Z^n g^\varepsilon]| \leq \varepsilon$, as $Z^n \in B^{\varphi,n}$ and $g^\varepsilon \in \varepsilon V^{\varphi,n}$. Hence

$$E_{Q^n}[h^\varepsilon] = E_{Q^n}[f^\varepsilon] + E_{Q^n}[-g^\varepsilon] \leq \varepsilon.$$

As $(P^n) \triangleleft (Q^n)$, by Remark 3.1 there is $\beta > 0$ such that $\inf_{h \in D^{\alpha,n}} E_{Q^n}[h] \geq \beta$ for any n. This is a contradiction for small ε. $\qquad\square$

Lemma 4.6 *Suppose that for each $\varepsilon > 0$ there is a sequence $Q^{n,\varepsilon} \in M_a^n$ such that*

i) there is $\delta > 0$ such that for all n and $A^n \in \mathcal{F}^n$, $P^n(A^n) \geq \varepsilon$ implies that $Q^{n,\varepsilon}(A^n) \geq \delta$, and
ii) $(Q^{n,\varepsilon}) \triangleleft (P^n)$.

Let $Q^n = \sum_{j=1}^{\infty} 2^{-j} Q^{n,2^{-j}}$. Then $Q^n \in M^n$ and, moreover, $(Q^n) \triangleleft \triangleright (P^n)$.

Proof. $\{Z^n = 0\} \subseteq \bigcap_{j=1}^{\infty} \{Z^{n,2^{-j}} < \delta_j\}$, so i) implies that $P^n(Z^n = 0) = 0$. Hence $Q^n \in M^n$. Let us now prove that $(Q^n) \triangleleft (P^n)$. Let $\gamma > 0$ be arbitrary but fixed. We have to show that there is $\mu > 0$ such that $P^n(A^n) < \mu$ implies $Q^n(A^n) < \gamma$ for all $A^n \in \mathcal{F}^n$. Let $N \in \mathbb{N}$ be large enough such that $\sum_{j=N+1}^{\infty} 2^{-j} < \frac{\gamma}{2}$. By ii) $(Q^{n,2^{-j}})_{n \geq 1} \triangleleft (P^n)_{n \geq 1}$ for $j = 1, 2, \ldots, N$, whence for each j there exists $\mu_j > 0$ such that, for all n, $P^n(A^n) < \mu_j$ implies that $Q^{n,2^{-j}}(A^n) < \frac{\gamma}{2}$. Let now $\mu = \min_{j \leq N} \mu_j$ and $A^n \in \mathcal{F}^n$ be such that $P^n(A^n) < \mu$. Then

$$Q^n(A^n) = \sum_{j=1}^{N} 2^{-j} Q^{n,2^{-j}}(A^n) + \sum_{j=N+1}^{\infty} 2^{-j} Q^{n,2^{-j}}(A^n) < \frac{\gamma}{2} + \frac{\gamma}{2} = \gamma.$$

To prove that $(P^n) \triangleleft (Q^n)$ observe that i) implies the following: for all $j \in \mathbb{N}$, there is μ_j such that, for any n, $Q^{n,2^{-j}}(A^n) < \mu_j$ implies that $P^n(A^n) < 2^{-j}$. Let $\gamma > 0$ be fixed and choose $N \in \mathbb{N}$ such that $2^{-(N-1)} < \gamma$. Define $\mu = 2^{-2N} \mu_N$. Let now $A^n \in \mathcal{F}^n$ such that $Q^n(A^n) < \mu$. Then

$$P^n(A^n) = P^n \left(A^n \cap \left\{ Z^{n,2^{-N}} < \mu_N \right\} \right) + P^n \left(A^n \cap \left\{ Z^{n,2^{-N}} \geq \mu_N \right\} \right)$$

$$< 2^{-N} + P^n \left(A^n \cap \{ Z^n \geq 2^{-N} \mu_N \} \right) < 2^{-N} + \frac{2^N}{\mu_N} Q^n(A^n) < \gamma.$$

$\qquad\square$

5 NAFL, NFL and Orlicz spaces

In Section 3 we mentioned that NAFL is a generalization of the concept NFL of Kreps [16] which was recalled in (2.1). We introduced Mackey–neighbourhoods of 0 to be able to describe an approximation with respect to the Mackey–topology for a sequence of L^∞ spaces. Let us now take another look at NFL using the theory of Orlicz spaces. Although we have not yet mentioned the connection to Orlicz spaces, we already implicitly used them above by introducing the sets B^F. We will present now an alternative sight that enables us to replace the Mackey–neighborhoods by an approximation using *norm–topologies* (of the Orlicz spaces L_F). For each $F \in \mathcal{N}$ let

$$L_F(P) = \{f \in L^0(P) : E[F(a|f|)] < \infty \text{ for some } a > 0\}.$$

We use the notation L_F if it is clear which measure P is meant. The so-called gauge norm on L_F is given by

$$\|f\|_F = \inf\{a > 0 : E[F(\tfrac{1}{a}|f|)] \le 1\}.$$

The space L_F is called Orlicz space and it is well known that it is a Banach space with respect to $\|.\|_F$, see for example [19]. The set $B^F(P)$ defined in (3.2) and used in the proof of Theorem 3.3 is in fact the closed unit ball of $L^F(P)$. Indeed, it is easy to see that

$$B^F(P) = \{f \in L_F(P) : \|f\|_F \le 1\}.$$

We will need a few facts on Orlicz spaces with respect to N–functions. We refer to [19] for all facts on Orlicz spaces. For $F \in \mathcal{N}$ there is a complementary N–function given by

$$G(y) = \sup_{x \ge 0}(xy - F(x)). \qquad (5.1)$$

The complementary of G is again F. For each $F \in \mathcal{N}$ let

$$L_F^0(P) = \{f \in L^0(P) : E_P[F(a|f|)] < \infty \text{ for all } a > 0\}.$$

For $F \in \mathcal{N}$ the space L_F^0 is the closed linear span of L^∞ in $(L^F, \|.\|_F)$. Moreover the following holds.

Proposition 5.1 *Let (F, G) be a complementary pair of N–functions.*

i) $E[|fg|] \le 2\|f\|_F\|g\|_G$ for $f \in L_F$ and $g \in L_G$.
ii) If $g \in L_G$, then $\Phi : L_F^0 \to \mathbb{R}$ given by $\Phi(f) = E[fg]$ is a continuous linear functional and

$$\|g\|_G \le \|\Phi\|_{(L_F^0)^*} \le 2\|g\|_G.$$

iii) $L^\infty = \bigcap_{F \in \mathcal{N}} L_F$ and $L^1 = \bigcup_{F \in \mathcal{N}} L_F$.

Recall that $V^F(P) = (B^F(P))^\circ$. The following relation between the sets V^F and B^G holds.

Lemma 5.2 *Let (F, G) be a complementary pair of N–functions. Then*

$$\frac{1}{2} B^G \cap L^\infty \subseteq V^F \subseteq B^G \cap L^\infty.$$

Proof. By Proposition 5.1 i) the first inclusion is clear. Let now $g \in V^F$. As $g \in L^\infty$ it is clear that g defines a continuous linear functional Φ on L_F^0. As $|E[gh]| \leq 1$ for all $h \in B^F$, we have that $\|\Phi\|_{(L_F^0)^*} \leq 1$. Proposition 5.1 ii) implies that $\|g\|_G \leq \|\Phi\|_{(L_F^0)^*} \leq 1$. Hence $g \in B^G \cap L^\infty$. $\square$

We will use the following characterization of the closure with respect to the Mackey–topology of a convex subset A of L^∞. A very similar result can be found in Kusuoka [17]. We use the notation $\overline{A}^F$ for the closure of A in $(L^F, \|.\|_F)$.

Lemma 5.3 *Let A be a convex subset of L^∞. Then $\bigcap_{F \in \mathcal{N}} \overline{A}^F = \overline{A}^\tau$, where $\overline{A}^\tau$ is the closure of A in L^∞ with respect to the Mackey topology. (Note that, by convexity, $\overline{A}^\tau = \overline{A}^*$.)*

Proof. It is clear that $\mathbf{A} := \bigcap_{F \in \mathcal{N}} \overline{A}^F$ is a subset of L^∞ by Proposition 5.1, (iii). We show that $\bigcap_{F \in \mathcal{N}} \overline{A}^F$ is closed with respect to the Mackey topology. Indeed, let $f \in \overline{\mathbf{A}}^\tau$. We will show that $f \in \mathbf{A}$. Fix an arbitrary N–function G. We have to show that $f \in \overline{A}^G$. Define, for any $\varepsilon > 0$, $G_\varepsilon(x) = G(\frac{1}{\varepsilon}x)$ and let F_ε be the complementary N–function of G_ε. As $f \in \overline{\mathbf{A}}^\tau$ and the sets V^F are Mackey–neighborhoods we have that $f \in \mathbf{A} + V^F$ for all $F \in \mathcal{N}$, and so, in particular, $f \in \mathbf{A} + V^{F_\varepsilon}$ for any $\varepsilon > 0$. Hence there exist $f_\varepsilon \in \mathbf{A}$ and $h_\varepsilon \in V^{F_\varepsilon}$ such that $f = f_\varepsilon + h_\varepsilon$. By Lemma 5.2 we have that $V^{F_\varepsilon} \subseteq B^{G_\varepsilon} \cap L^\infty$. By the definition of G_ε moreover $B^{G_\varepsilon} = \varepsilon B^G$, and so $\|h_\varepsilon\|_G \leq \varepsilon$. As $f_\varepsilon \in \mathbf{A}$ we have that, in particular, $f_\varepsilon \in \overline{A}^G$ and so there exists $\tilde{f}_\varepsilon \in A$ such that $\|f_\varepsilon - \tilde{f}_\varepsilon\|_G \leq \varepsilon$. Hence we get that

$$\|f - \tilde{f}_\varepsilon\|_G \leq \|f_\varepsilon - \tilde{f}_\varepsilon\|_G + \|h_\varepsilon\|_G \leq 2\varepsilon.$$

As $\tilde{f}_\varepsilon \in A$ and as this can be done for any $\varepsilon > 0$ we get that $f \in \overline{A}^G$. This works for any $G \in \mathcal{N}$ and so $f \in \mathbf{A}$, hence $\mathbf{A}$ is Mackey–closed.

Kusuoka [17] showed that $\overline{A}^\tau = \left(\bigcap_{F \in \mathcal{N}} Cl_F(A) \right) \cap L^\infty$, where $Cl_F(A)$ consists of all $f \in L_F^0$ such that there is a sequence $f^n \in A$ with $f^n \to f$ in probability and $\sup_n E[F(a|f^n|)] < \infty$ for all $a > 0$. It is straightforward that $\overline{A}^F \subseteq Cl_F(A)$. This concludes the proof. $\square$

A formulation of NAFL in terms of Orlicz norms

First we will reformulate the NFL–condition of (2.1).

Lemma 5.4 *NFL holds if and only if for any $\varepsilon > 0$*

$$\left(\bigcap_{F \in \mathcal{N}} \overline{\mathbf{C}}^F\right) \cap D^\varepsilon = \emptyset.$$

Proof. Suppose there is NFL. Assume that there is $\alpha > 0$ such that there exists $w \in \left(\bigcap_{F \in \mathcal{N}} \overline{\mathbf{C}}^F\right) \cap D^\alpha$. By Lemma 5.3 this is a contradiction, as $\bigcap_{F \in \mathcal{N}} \overline{\mathbf{C}}^F = \overline{\mathbf{C}}^*$. Now assume that the condition of the Lemma holds and that there is $w \in \overline{\mathbf{C}}^* \cap L_+^\infty$ and $w \neq 0$. As $\mathbf{C}$ is a cone we have that $\tilde{w} := \frac{w}{\|w\|_\infty} \in \overline{\mathbf{C}}^*$ as well. It is clear that there is $\alpha > 0$ such that $\tilde{w} \in D^\alpha$. By Lemma 5.3 this is a contradiction. $\qquad\square$

The formulation of NFL as in Lemma 5.4 gives rise to the following alternative definition of no asymptotic free lunch using Orlicz norms. We denote the distance in $L^F(P)$ with respect to the norm by dist_F.

Definition 5.5 *We say that the large financial market satistfies the NAFL' condition if for any $\varepsilon > 0$ there exist $F \in \mathcal{N}$ and $\delta > 0$ such that for all $n \in \mathbb{N}$*

$$\mathrm{dist}_F(\mathbf{C}^n, D^{\varepsilon,n}) \geq \delta.$$

The notion NAFL' says that it is not possible to approximate a strictly positive gain in all Orlicz norms. The advantage to Definition 3.2 is that we do not use the Mackey–topology which is not metrizable and therefore more technical than norm topologies. It turns out that the condition NAFL' is indeed equivalent to the condition NAFL. The proof is straightforward with the help of Lemma 5.2.

Proposition 5.6 *NAFL' $\Longleftrightarrow$ NAFL.*

Proof. Assume that NAFL does not hold. Fix an arbitrary $F \in \mathcal{N}$ and $\varepsilon > 0$. Let G be the complementary N–function of F and define $G_\varepsilon(x) = G(\varepsilon x)$. This implies that $V^{G_\varepsilon,n} = \varepsilon V^{G,n}$ and so, by Lemma 5.2, $V^{G_\varepsilon,n} \subseteq \varepsilon(B^{F,n} \cap L^\infty(P^n))$. By assumption there is $\alpha > 0$ (which does not depend on F or ε) such that for G_ε there exist n_ε and $f^\varepsilon \in \mathbf{C}^{n_\varepsilon}$ with $f^\varepsilon = w^\varepsilon + g^\varepsilon$, where $w^\varepsilon \in D^{\alpha,n_\varepsilon}$ and $g^\varepsilon \in V^{G_\varepsilon,n_\varepsilon}$. Therefore we have that

$$\mathrm{dist}_F(D^{\alpha,n_\varepsilon}, \mathbf{C}^{n_\varepsilon}) \leq \|g^\varepsilon\|_F \leq \varepsilon,$$

hence NAFL' does not hold.

Assume now that NAFL' does not hold. Let $F \in \mathcal{N}$ and G be the complementary function of F. By assumption there is $\alpha > 0$ (which does not depend on F), such that for any $\varepsilon > 0$ there is n_ε, $f \in \mathbf{C}^{n_\varepsilon}$ and $w \in D^{\alpha,n_\varepsilon}$ such that $\|f - w\|_G \leq \varepsilon$. For $\varepsilon = 1$ this gives that $g \in B^{G,n_1} \cap L^\infty(P^{n_1}) \subseteq V^{F,n_1}$, where $g := f - w$. Hence NAFL does not hold. $\qquad\square$

References

1. F. Delbaen and W. Schachermayer. A general version of the fundamental theorem of asset pricing. *Math. Ann.*, 300:463-520, 1994.
2. F. Delbaen and W. Schachermayer. The fundamental theorem of asset pricing for unbounded stochastic processes. *Math. Ann.*, 312:215-250, 1998.
3. M. Frittelli. Some remarks on arbitrage and preferences in securities market models. *Math. Finance*,14:351-357, 2004
4. J. Horvath. *Topological Vector Spaces and Distributions*. Addison–Wesley, 1966.
5. G. Huberman. A simple approach to arbitrage pricing theory. *J. Econom. Theory*, 28:183–191, 1982.
6. J. Jacod and A. N. Shiryaev. *Limit Theorems for Stochastic Processes*. Springer, 1987.
7. Y. Kabanov. On the FTAP of Kreps–Delbaen–Schachermayer. In Statistics and Control of Stochastic Processes, the Lipster Festschrift (Y. Kabanov, B-L. Rozovskii and A.N. Shiryaev, eds.), *World Scientific, Singapore*, pages 191–203, 1997.
8. Y. Kabanov Arbitrage Theory In Handbooks in Mathematical Finance. Option Procing: Theory and Practice, *Cambridge University Press*, pages 3-42, 2001.
9. Y. Kabanov and D. Kramkov. Large financial markets: asymptotic arbitrage and contiguity. *Theory Probab. Appl.*, 39: 222–228, 1994.
10. Y. Kabanov and D. Kramkov. Asymptotic arbitrage in large financial markets. *Finance and Stochastics*, 2:143-172, 1998.
11. I. Klein. A fundamental theorem of asset pricing for large financial markets. *Math. Finance*, 10:443-458, 2000.
12. I. Klein. Market free lunch and large financial markets. to appear in Ann. Appl. Probab.
13. I. Klein and W. Schachermayer Asymptotic arbitrage in non-complete large financial markets. *Theory Probab. Appl.*, 41:927-934, 1996.
14. I. Klein and W. Schachermayer A quantitative and a dual version of the Halmos–Savage theorem with applications to mathematical finance. *Ann. Probab.*, 24:867-881, 1996.
15. M. A. Krasnosel'skii and Y. B. Rutickii. *Convex Functions and Orlicz Spaces*. Gordon and Breach, 1961.
16. D. M. Kreps. Arbitrage and equilibrium in economies with infinitely many commodities. *J. Math. Econom.*, 8:15-35, 1981.
17. S. Kusuoka. A remark on arbitrage and martingale measure. *Publ. Res. Inst. Math. Sci.*, 29:833-840, 1993.
18. J. Mémin. Espaces de semimartingales et changement de probabilité. *Z. Wahrsch. Verw. Gebiete.*, 52:9-39, 1980.
19. M. M. Rao and Z. D. Ren. *Theory of Orlicz Spaces*. Dekker, 1991.
20. S. A. Ross. The arbitrage theory of asset pricing. *J. Econom. Theory*, 13:341-360, 1976.
21. P. Wojtaszczyk. *Banach spaces for analysts*. Cambridge University Press, 1991.

New methods in the arbitrage theory of financial markets with transaction costs*

Miklós Rásonyi

Computer and Automation Institute of the Hungarian Academy of Sciences
1518 Budapest, P. O. Box 63., Hungary
e-mail: rasonyi@sztaki.hu

Summary. Using entirely new methods, we reprove the main result of [6]: strict absence of arbitrage is equivalent to the existence of a strictly consistent price system in markets with efficient proportional transaction costs in finite discrete time. We also improve on that result by considering a more general class of models.

1 Introduction

In this note we propose a new way of proving various versions of the fundamental theorem of asset pricing in discrete-time financial market models with proportional transaction costs. We work in the geometric framework of [7] and [6].

This new method is based on the notion of conditional expectation for random sets (see [8]) and applies an inductive argument inspired by that of [3], see also [5]. Further tools are finite-dimensional separation theorems combined with measurable selection. Similar results have been obtained by D. B. Rokhlin in the recent paper [11], see also [12] and [13].

In the usual approach ([6], [14]) one proceeds by proving that the set of attainable claims is closed in some topological vector space. There are cases where this approach is bound to fail as the set of attainable claims is *not* closed, see Example 1.3 of [3] or Example 5.3.2 of [9]. The method presented here is hoped to apply in such situations, too.

---

* I would like to thank Fabian Astic, Bruno Bouchard, Pavel G. Grigoriev and Nizar Touzi for helpful discussions on topics related to this paper. Special thanks go to Yuri M. Kabanov for his suggestions concerning the presentation and to an anonymous referee for spotting several errors. The author was supported by the Hungarian State Eötvös Fellowship and by Hungarian National Science Foundation (OTKA) grants F 049094 and T 047193. Part of this research was carried out while staying at Université Paris 7 and CREST. I am grateful to Walter Schachermayer for an invitation to Vienna University of Technology where I could find important ingredients for these results.

2 Model

Scalar product in $\mathbb{R}^d$ will be denoted by $\langle \cdot, \cdot \rangle$, we will also need the unit ball $U := \{x \in \mathbb{R}^d : |x| \leq 1\}$. The positive dual cone of a closed cone $K \subset \mathbb{R}^d$ is defined as

$$K^* := \{x \in \mathbb{R}^d : \langle x, c \rangle \geq 0, \text{ for all } c \in K\},$$

this is also a closed cone. We say that K is proper if int $K^* \neq \emptyset$.

Let $(\Omega, \mathcal{F}, (\mathcal{F}_t)_{0 \leq t \leq T}, P)$ be a discrete-time stochastic basis with finite time horizon. In this paper we suppose that each σ-algebra we are dealing with contains all P-zero sets. If $\mathcal{G}$ is a σ-algebra and $\mathcal{B}(\mathbb{R}^d)$ denotes the Borel-sets of $\mathbb{R}^d$, an element $A \in \mathcal{G} \otimes \mathcal{B}(\mathbb{R}^d)$ is called a $\mathcal{G}$-measurable random set. If the sections $A(\omega)$ of this random set are cones for almost all ω then we call it a random cone and similarly for random closed cones, random convex sets, etc ...

If A is some (not necessarily $\mathcal{G}$-measurable) random set which is nonempty a.s. then $L^0(A, \mathcal{G})$ denotes the family of all $\mathcal{G}$-measurable $\mathbb{R}^d$-valued functions f such that $f(\omega) \in A(\omega)$ for almost all ω. Such an f is called a $\mathcal{G}$-*measurable selector* of A.

We pursue an abstract geometric approach here, see [7], [6] or [14] for detailed descriptions of a model with proportional transaction costs which fits into the present framework.

We assume that G_t is an $\mathcal{F}_t$-measurable random closed cone in $\mathbb{R}^d$ containing $\mathbb{R}^d_+$ a.s., for $t = 0, \ldots, T$. In the financial context one may think of $G_t(\omega)$ as the set of "nonnegative" positions in d assets at time t and in the state of the world $\omega \in \Omega$. This model class is more general than that of [6] where the G_t are assumed to be random *polyhedral* cones.

We define

$$A_t := \sum_{s=0}^{t} L^0(-G_s, \mathcal{F}_s).$$

Remark 2.1. In the financial setting traders in the market are assumed to have the information structure $(\mathcal{F}_t)_{0 \leq t \leq T}$ and to act in a self-financing way hence A_t corresponds to the set of attainable positions at time t from 0 initial endowment.

Definition 2.2 *We say that there is* efficient friction *(EF) if G_t is proper a.s. for $t = 0, \ldots, T$.*

In the model of [6] condition (EF) means that there are no freely exchangeable assets (even allowing indirect transfers).

Definition 2.3 *We say that there is* strict absence of arbitrage *(NA^s), if for all $0 \leq t \leq T$,*

$$A_t \cap L^0(G_t, \mathcal{F}_t) = \{0\}.$$

Now we introduce the dual variables of this model.

Definition 2.4 *The set of martingales Z such that $Z_t \in$ int G_t^*, $0 \leq t \leq T$ is denoted by $\mathcal{M}_0^T(\text{int } G^*)$.*

Following the terminology of [14], these martingales are called *strictly consistent price systems* in the context of financial modelling.

Remark 2.5. By standard arguments, if K is a $\mathcal{G}$-measurable random closed cone, then K^* is also a $\mathcal{G}$-measurable random closed cone, see e.g. Appendix C of [9]. We also remark that if C is a random convex set then int C is also a random convex set, see e.g. the argument of Proposition 4.13 in [10].

3 Result

With our method it is possible to reprove the main result of [6] in a slightly stronger form: in the cited paper it is assumed that the cones G_t are polyhedral a.s., the proof presented here does not need this assumption.

Theorem 3.1 *Under (EF), (NA^s) holds iff $\mathcal{M}_0^T(\text{int } G^*) \neq \emptyset$.*

Unlike in the original proof, we do not rely on the closedness of the set A_T of attainable claims. For the frictionless Dalang-Morton-Willinger theorem (see [1]) the existing proofs either use this closedness property or apply measurable selection. Now our results create a similar alternative for markets with efficient transaction costs.

4 Proof

Let us remember some basic facts about convex sets.

Proposition 4.1 *If C is a convex set, $p \in C$, $q \in$ int C then for all $\alpha \in [0, 1)$ we have $\alpha p + (1 - \alpha)q \in$ int C.*

Proof. See Lemma 2.16 on p. 104 of [4]. □

This implies, in particular, that if int $C \neq \emptyset$ then $\overline{C} = \overline{\text{int } C}$. The next statement is an immediate corollary.

Fact 4.2 *If the convex sets K, G have nonempty interior and $K \cap (\text{int } G) \neq \emptyset$ then $(\text{int } K) \cap (\text{int } G) = \text{int}(K \cap G) \neq \emptyset$.* □

The following "projections" play a key rôle in our arguments.

Lemma 4.3 *Let $\mathcal{G} \subset \mathcal{H}$ be σ-algebras. Let $C \subset U$ be an $\mathcal{H}$-measurable random convex compact set. Then there exists a $\mathcal{G}$-measurable random convex compact set $E(C|\mathcal{G}) \subset U$ satisfying*

$$L^0(E(C|\mathcal{G}), \mathcal{G}) = \{E(\vartheta|\mathcal{G}): \ \vartheta \in L^0(C, \mathcal{H})\}.$$

If $0 \in C$ a.s. then $0 \in E(C|\mathcal{G})$ a.s., too.

Proof. Everything follows from Theorem 1.49 on p. 173 of [8] except for $E(C|\mathcal{G}) \subset U$ and the last statement, which are trivial.   □

Define recursively the following sets, using Lemma 4.3:

$$C_T := G_T^* \cap U, \quad C_t = E(C_{t+1}|\mathcal{F}_t) \cap G_t^*, \ 0 \le t \le T - 1.$$

Each C_t is an $\mathcal{F}_t$-measurable random convex closed set containing 0 and satisfying $C_t \subset U$ a.s., by Lemma 4.3.

The next Lemma describes a further property of $E(C|\mathcal{G})$ which we need in the sequel.

Lemma 4.4 *Let $\mathcal{G} \subset \mathcal{H}$ be σ-algebras and let $C \subset U$ be a $\mathcal{H}$-measurable random compact convex set containing 0 such that $\operatorname{int} C \ne \emptyset$ a.s. Then $\operatorname{int} E(C|\mathcal{G}) \ne \emptyset$ a.s. and*

$$\{E(\vartheta|\mathcal{G}) : \ \vartheta \in L^0(\operatorname{int} C, \mathcal{H})\} \subset L^0(\operatorname{int} E(C|\mathcal{G}), \mathcal{G})$$
$$\subset \{E(\vartheta|\mathcal{G}) : \ \vartheta \in L^0(2 \operatorname{int} C, \mathcal{H})\}.$$

It is clear from the proof that $2 \operatorname{int} C$ can be replaced by $(1 + \delta)\operatorname{int} C$ for arbitrary $\delta > 0$.

Proof. Take any $\vartheta \in L^0(\operatorname{int} C, \mathcal{H})$, then for some $\varepsilon \in L^0((0,1), \mathcal{H})$ and for a sequence $(q_n)_{n \in \mathbb{N}}$ everywhere dense in U we have, almost surely, $\vartheta + \varepsilon q_n \in C$ for all n. Then a.s. $E(\vartheta|\mathcal{G}) + E(\varepsilon|\mathcal{G})q_n \in E(C|\mathcal{G})$ for all n. As $E(C|\mathcal{G})$ is closed, we necessarily have

$$E(\vartheta|\mathcal{G}) + E(\varepsilon|\mathcal{G})U \subset E(C|\mathcal{G}),$$

so $\operatorname{int} E(C|\mathcal{G}) \ne \emptyset$ a.s. and the first inclusion holds.

To show the second inclusion, fix some $\vartheta_0 \in L^0(\operatorname{int} C, \mathcal{H})$, and define $\kappa := E(\vartheta_0|\mathcal{G})$. Now take any $\lambda \in L^0(\operatorname{int} E(C|\mathcal{G}), \mathcal{G})$. For some $\varepsilon \in L^0((0,1), \mathcal{G})$, we have $\lambda - \varepsilon\kappa \in E(C|\mathcal{G})$ a.s., so by Lemma 4.3 we have $E(\vartheta_1|\mathcal{G}) = \lambda - \varepsilon\kappa$ for some $\vartheta_1 \in L^0(C, \mathcal{H})$. Defining $\vartheta := \vartheta_1 + \varepsilon\vartheta_0$, one gets $\lambda = E(\vartheta|\mathcal{G})$ and $\vartheta/(1 + \varepsilon) \in \operatorname{int} C$ by Proposition 4.1, so $\vartheta \in 2 \operatorname{int} C$ indeed, by $0 \in C$.   □

We now prove a measurability lemma which will be used in Lemma 4.6 below.

Lemma 4.5 *Suppose that $W \subset \Omega \times \mathbb{R}^d$ is defined as*

$$W := \{(\omega, x) : \langle x, k \rangle \le 0 \text{ for all } k \in K(\omega), \ \langle x, g \rangle > 0 \text{ for all } g \in \operatorname{int} C(\omega)\}.$$

where K, C are $\mathcal{G}$-measurable closed convex random sets such that $\operatorname{int} C \ne \emptyset$ a.s. Then $W \in \mathcal{G} \otimes \mathcal{B}(\mathbb{R}^d)$.

Proof. Theorem 2.3 on p. 26 of [8] implies that there are sequences (σ_n), (τ_n) of $\mathcal{G}$-measurable d-dimensional random variables such that

$$K(\omega) = \overline{\{\sigma_n(\omega),\ n \in \mathbb{N}\}},$$
$$C(\omega) = \overline{\{\tau_n(\omega),\ n \in \mathbb{N}\}},$$

for almost all ω. Under the given conditions we have

$$W = \bigcap_{n=1}^{\infty} \{(\omega, x) :\ \langle x, \sigma_n(\omega) \rangle \leq 0,\ \langle x, \tau_n(\omega) \rangle \geq 0\},$$

which is easily seen to be in $\mathcal{G} \otimes \mathcal{B}(\mathbb{R}^d)$.   □

We assume that the reader has some familiarity with the measurable selection theorem (see e.g. III.44 of [2]), which will be crucial in what follows.

Lemma 4.6 *Suppose that (EF) holds and* int $C_s \neq \emptyset$ *almost surely for $s \geq k + 1$. If $\xi \in A_k$ is such that for almost all $\omega \in \Omega$,*

$$\langle \xi(\omega), g \rangle \geq 0 \text{ for all } g \in E(C_{k+1} | \mathcal{F}_k)(\omega), \tag{1}$$

and on a set of positive probability

$$\langle \xi(\omega), g \rangle > 0 \text{ for all } g \in int\, E(C_{k+1} | \mathcal{F}_k)(\omega), \tag{2}$$

then (NA^s) fails.

Proof. The proof is by induction on $T - k$. Consider the case $T - k = 1$ first, now $C_T = G_T^* \cap U$. We claim that $\xi \in G_T$ a.s. Indeed, if we had $P(B) > 0$ for $B := \{\xi \notin G_T\}$ then by the measurable selection theorem there would exist $\kappa \in L^0(G_T^*, \mathcal{F}_1)$ such that

$$|\kappa| \leq 1, \quad \langle \kappa, \xi \rangle < 0 \text{ a.s.}$$

on B, hence for $\zeta := E(\kappa I_B | \mathcal{F}_{T-1})$ one would get $\langle \zeta, \xi \rangle < 0$ with positive probability, which is absurd: by Lemma 4.3 we have $\zeta \in E(C_T | \mathcal{F}_{T-1})$ a.s. and (1) implies that $\langle \zeta, \xi \rangle \geq 0$. So $\xi \in G_T$ a.s. holds. Since int $C_T \neq \emptyset$ a.s. by (EF), we have int $E(C_T | \mathcal{F}_{T-1}) \neq \emptyset$ a.s. from Lemma 4.4, hence ξ can't be a.s. zero because of (2).

Now supppose that the statement is true for $T - k = l$, let us proceed to show its validity for $T - k = l + 1$.

Step 0 Define the random halfspace $X(\omega) = \{x \in \mathbb{R}^d :\ \langle x, \xi(\omega) \rangle \leq 0\}$. Consider the following $\mathcal{F}_{T-l}$-measurable partition of Ω:

$$B_1 := \{\text{int } E(C_{T-l+1} | \mathcal{F}_{T-l}) \cap (\text{int } G_{T-l}^*) \cap X \neq \emptyset\},$$
$$B_2 := \{(\text{int } E(C_{T-l+1} | \mathcal{F}_{T-l})) \cap G_{T-l}^* \cap X \neq \emptyset,$$
$$(\text{int } E(C_{T-l+1} | \mathcal{F}_{T-l})) \cap G_{T-l}^* \cap X \subset \partial G_{T-l}^*\},$$
$$B_3 := \{(\text{int } E(C_{T-l+1} | \mathcal{F}_{T-l})) \cap G_{T-l}^* \cap X = \emptyset\}.$$

Remember that int $E(C_{T-l+1} | \mathcal{F}_{T-l}) \cap$ int $G_{T-l}^* \supset$ int $C_{T-l} \neq \emptyset$ by hypothesis.

Step 1 We first claim that $P(B_1) = 0$. Indeed, by Fact 4.2, on B_1 we have
int $E(C_{T-l+1}|\mathcal{F}_{T-l}) \cap (\text{int } G^*_{T-l}) \cap X = \text{int}(E(C_{T-l+1}|\mathcal{F}_{T-l}) \cap G^*_{T-l}) \cap X \neq \emptyset$
and, again by Fact 4.2, we also have int$(E(C_{T-l+1}|\mathcal{F}_{T-l}) \cap G^*_{T-l}) \cap$ int $X =$
(int $C_{T-l}) \cap \{x \in \mathbb{R}^d : \langle \xi, x \rangle < 0\} \neq \emptyset$. Take a (bounded) measurable selector
κ of this latter set on B_1 and define $\kappa = 0$ on B_1^C.

Then, by Lemma 4.3, we have $\zeta := E(\kappa|\mathcal{F}_{T-l-1}) \in E(C_{T-l}|\mathcal{F}_{T-l-1})$ and
$\langle \zeta, \xi \rangle < 0$ with positive probability if $P(B_1) > 0$: a contradiction with (1), so
$P(B_1) = 0$ must hold.

Step 2 Obviously, $E(C_{T-l+1}|\mathcal{F}_{T-l}) \cap G^*_{T-l} \cap X = C_{T-l} \cap X \subset H := \{x \in \mathbb{R}^d : \langle \xi, x \rangle = 0\}$ a.s. by (1). Thus we can write

$$B_2 = \{(\text{int } E(C_{T-l+1}|\mathcal{F}_{T-l})) \cap G^*_{T-l} \cap H \neq \emptyset,$$
$$(\text{int } E(C_{T-l+1}|\mathcal{F}_{T-l})) \cap G^*_{T-l} \cap H \subset \partial G^*_{T-l}\}.$$

Step 3 Using Fact 4.2 in the relative topology of H it is easy to see that

$$\{(\text{int } E(C_{T-l+1}|\mathcal{F}_{T-l})) \cap G^*_{T-l} \cap X \neq \emptyset\} \cap \{(\text{int } G^*_{T-l}) \cap H \neq \emptyset\}$$
$$\subset \{((\text{int } E(C_{T-l+1}|\mathcal{F}_{T-l})) \cap H) \cap (G^*_{T-l} \cap H) \neq \emptyset, \ (\text{int } G^*_{T-l}) \cap H \neq \emptyset\}$$
$$\subset \{\text{int } E(C_{T-l+1}|\mathcal{F}_{T-l}) \cap (\text{int } G^*_{T-l}) \cap H \neq \emptyset\} \subset B_1.$$

It follows that $B_2 \cap \{(\text{int } G^*_{T-l}) \cap H \neq \emptyset\} = \emptyset$. Hence for almost all $\omega \in B_2$,
either int $G^*_{T-l}(\omega) \subset \{x : \langle \xi(\omega), x \rangle < 0\}$ or int $G^*_{T-l}(\omega) \subset \{x : \langle x, \xi(\omega) \rangle > 0\}$.
The first possibility can only occur on a P-zero set as int $C_{T-l} \subset$ int G^*_{T-l}
and (1) holds. So we conclude that $\xi \in G_{T-l}$ a.s. on B_2.

Step 4 We next treat the case $P(B_2) = 1$. Since ξ is not a.s. 0 by (2), ξ
violates (NA^s) at time $T - l$ and we are done.

Step 5 Finally, if $P(B_2) < 1$ then $P(B_3) > 0$. Fixing $\omega \in B_3$, the finite-
dimensional version of the Hahn-Banach theorem shows that for some $\gamma(\omega) \in \mathbb{R}^d$ and $\alpha(\omega) \in \mathbb{R}$,

$$\langle \gamma(\omega), k \rangle \leq \alpha(\omega) \text{ for all } k \in G^*_{T-l}(\omega) \cap X(\omega),$$
$$\langle \gamma(\omega), g \rangle > \alpha(\omega) \text{ for all } g \in \text{int } E(C_{T-l+1}|\mathcal{F}_{T-l})(\omega). \quad (3)$$

Since $G^*_{T-l} \cap X$ is a cone and $0 \in E(C_{T-l+1}|\mathcal{F}_{T-l})$ we necessarily have
$\alpha(\omega) = 0$. Consider the set

$$W := \{(\omega, x) : \langle x, k \rangle \leq 0 \text{ for all } k \in G^*_{T-l}(\omega) \cap X(\omega),$$
$$\langle x, g \rangle > 0 \text{ for all } g \in \text{int } E(C_{T-l+1}|\mathcal{F}_{T-l})(\omega)\}.$$

By Lemma 4.5 we have $W \in \mathcal{F}_{T-l} \otimes \mathcal{B}(\mathbb{R}^d)$ and the projection of W on Ω
contains B_3 by the considerations above. Hence, by the measurable selection
theorem, there exists $\gamma \in L^0(\mathbb{R}^d, \mathcal{F}_{T-l})$ such that (3) holds with $\alpha(\omega) = 0$ for
almost all $\omega \in B_3$. As $(G^*_{T-l} \cap X)^* = G_{T-l} + X^*$ we get that $\gamma = \beta\xi + \eta$ with

$\beta \in L^0(\mathbb{R}_+, \mathcal{F}_{T-l})$ and $\eta \in L^0(-G_{T-l}, \mathcal{F}_{T-l})$. If $P(B) > 0$ for $B := \{\beta = 0\}$ then, by (3), the induction hypothesis applies to ηI_B and the proof is finished.

If $P(B) = 0$ then set $\eta' = -\xi$ on B_2 (remember that $\xi \in G_{T-l}$ on B_2 by Step 3) and $\eta' := \eta/\beta$ on B_3. Again by (3), the induction hypothesis is applicable to $\xi + \eta' \in A_{k+1}$ and (NA^s) fails. $\qquad\square$

Proof of Theorem 3.1. If $\mathcal{M}_0^T(\text{int } G^*) \neq \emptyset$ then it is standard to show that (NA^s) holds, see [6].

To show the converse implication notice that if int $C_s \neq \emptyset$ for all $0 \leq s \leq T$ then there exists $Z \in \mathcal{M}_0^T(\text{int } G^*)$. As int $C_s \subset$ int G_s^* for all s, it is enough to show $\mathcal{M}_0^T(\text{int } C) \neq \emptyset$ (i.e. there is a martingale Z with $Z_s \in$ int C_s, $0 \leq s \leq T$). To see this, we will show by induction on t that $\mathcal{M}_0^t(\text{int } C) \neq \emptyset$ for $0 \leq t \leq T$. First take an $\mathcal{F}_0$-measurable selector Z_0 of int C_0. Now suppose that $\tilde{Z} \in \mathcal{M}_0^t(\text{int } C)$ has been defined. Use Lemma 4.4 to get an $\mathcal{F}_{t+1}$-measurable $\tilde{Z}_{t+1} \in 2\,\text{int } C_{t+1}$ such that $E(\tilde{Z}_{t+1}|\mathcal{F}_t) = \tilde{Z}_t$. Then, by $0 \in C_s$, $s = 0, \ldots, t$, we have $Z := \tilde{Z}/2 \in \mathcal{M}_0^{t+1}(\text{int } C)$ and we are done.

From now on we may and will suppose that for some k, $P(\text{int } C_s \neq \emptyset) = 1$ for $s \geq k+1$ and $P(B) > 0$ for $B := \{\text{int } C_k = \emptyset\}$ and we will show the failure of (NA^s). We claim that on B one also has $(\text{int } E(C_{k+1}|\mathcal{F}_k)) \cap G_k^* = \emptyset$. Indeed, as both $E(C_{k+1}|\mathcal{F}_k)$ and G_k^* have nonempty interior by Lemma 4.4 and by (EF), respectively, $(\text{int } E(C_{k+1}|\mathcal{F}_k)) \cap G_k^* \neq \emptyset$ and Fact 4.2 would imply $\text{int}(E(C_{k+1}|\mathcal{F}_k) \cap G_k^*) = \text{int } C_k \neq \emptyset$, but this fails by the definition of B.

Thus by a separation argument (similar to the one in Step 5 of Lemma 4.6) we obtain $\xi \in L^0(\mathbb{R}^d, \mathcal{F}_k)$ such that $\xi = 0$ on B^c and for $\omega \in B$

$$\langle \xi(\omega), l \rangle > 0 \text{ for all } l \in \text{int } E(C_{k+1}|\mathcal{F}_k)(\omega),$$
$$\langle \xi(\omega), g \rangle \leq 0 \text{ for all } g \in G_k^*(\omega).$$

Clearly, $\xi \in -G_k$ a.s. and using Lemma 4.6 finishes the proof of this Theorem. $\qquad\square$

References

1. Dalang, R.C., Morton, A. and Willinger, W. (1990) Equivalent martingale measures and no-arbitrage in stochastic securities market models. *Stochastics Stochastics Rep.*, **29**, 185–201.
2. Dellacherie, C. and Meyer, P. A. (1978) *Probabilities and Potential.* North-Holland, Amsterdam.
3. Grigoriev, P. G. (2005) On low dimensional case in the fundamental theorem of asset pricing theorem with transaction costs. *Statist. Decisions*, **23**, 33–48.
4. Hirriart-Urruty, J.-B. and Lemaréchal, C. (1993) *Convex analysis and minimization algorithms, vol. I.* Springer-Verlag, Berlin.
5. Kabanov, Yu. M. (2005) On the Grigoriev theorem. *manuscript.*

6. Kabanov, Yu. M. , Rásonyi, M. and Stricker, Ch. (2002) No-arbitrage criteria for financial markets with efficient friction. *Finance Stoch.*, **6**, 371–382.

7. Kabanov, Yu. M. and Stricker, Ch. (2001) The Harrison-Pliska arbitrage pricing theorem under transaction costs. *J. Math. Econom.*, **35**, 185–196.

8. Molchanov, I. (2005) *Theory of random sets.* Springer-Verlag, Berlin.

9. Rásonyi, M. (2002) *On certain problems of arbitrage theory in discrete-time financial market models.* PhD thesis, Université de Franche-Comté, Besançon. `http://www.sztaki.hu/~rasonyi`

10. Rásonyi, M. and Stettner, L. (2006) On the existence of optimal portfolios for the utility maximization problem in discrete time financial market models. *In: From Stochastic Calculus to Mathematical Finance, The Shiryaev Festschrift,* ed. Kabanov, Yu., Liptser, R., Stoyanov, J., 589–608.

11. Rokhlin, D. B. (2006) Martingale selection problem and asset pricing in finite discrete time. *preprint.* `arXiv:math.PR/0602594`, 6 pages.

12. Rokhlin, D. B. (2006) Martingale selection theorem for a stochastic sequence with relatively open convex values. *preprint.* `arXiv:math.PR/0602587`, 7 pages.

13. Rokhlin, D. B. (2006) Constructive no-arbitrage criterion under transaction costs in the case of finite discrete time. *preprint.* `arXiv:math.PR/0603284`, 18 pages.

14. Schachermayer, W. (2004) The fundamental theorem of asset pricing under proportional transaction costs in finite discrete time. *Math. Finance,* **14**, 19–48.

NOTE TO CONTRIBUTORS

Contributors to the Séminaire are reminded that their articles should be formatted for the Springer Lecture Notes series.

The manuscripts should be prepared with LaTeX version 2e, using the `svmult.cls` environment provided by Springer-Verlag for multi-authored LNM volumes. The style files can be downloaded from

ftp://ftp.springer.de/pub/tex/latex/mathegl/mult.zip

Authors addresses in the contributions: Each author's address should be written in two lines, and include all information about: institute or laboratory, then university, then road, then post code or PO BOX, and then country. The first line should include information about institute etc, and university; the second line should include postal information such as road, postal code and city, country. The e-mail (in italics) should always be on a separate third line under the postal address.

Lecture Notes in Mathematics

For information about earlier volumes
please contact your bookseller or Springer
LNM Online archive: springerlink.com

Vol. 1905: C. Prévôt, M. Röckner, A Concise Course on Stochastic Partial Differential Equations (2007)

Vol. 1906: T. Schuster, The Method of Approximate Inverse: Theory and Applications (2007)

Vol. 1907: M. Rasmussen, Attractivity and Bifurcation for Nonautonomous Dynamical Systems (2007)

Vol. 1908: T.J. Lyons, M. Caruana, T. Lévy, Differential Equations Driven by Rough Paths, Ecole d'Été de Probabilités de Saint-Flour XXXIV 2004 (2007)

Vol. 1909: H. Akiyoshi, M. Sakuma, M. Wada, Y. Yamashita, Punctured Torus Groups and 2-Bridge Knot Groups (I) (2007)

Vol. 1910: V.D. Milman. G. Schechtman (Eds.), Geometric Aspects of Functional Analysis. Israel Seminar 2004 2005 (2007)

Vol. 1911: A. Bressan, D. Serre, M. Williams, K. Zumbrun, Hyperbolic Systems of Balance Laws. Cetraro, Italy 2003. Editor: P. Marcati (2007)

Vol. 1912: V. Berinde, Iterative Approximation of Fixed Points (2007)

Vol. 1913: J.E. Marsden, G. Misiołek, J.-P. Ortega, M. Perlmutter, T.S. Ratiu, Hamiltonian Reduction by Stages (2007)

Vol. 1914: G. Kutyniok, Affine Density in Wavelet Analysis (2007)

Vol. 1915: T. Bıyıkoğlu, J. Leydold, P.F. Stadler, Laplacian Eigenvectors of Graphs. Perron-Frobenius and Faber-Krahn Type Theorems (2007)

Vol. 1916: C. Villani, F. Rezakhanlou, Entropy Methods for the Boltzmann Equation. Editors: F. Golse, S. Olla (2008)

Vol. 1917: I. Veselić, Existence and Regularity Properties of the Integrated Density of States of Random Schrödinger (2008)

Vol. 1918: B. Roberts, R. Schmidt, Local Newforms for GSp(4) (2007)

Vol. 1919: R.A. Carmona, I. Ekeland, A. Kohatsu-Higa, J.-M. Lasry, P.-L. Lions, H. Pham, E. Taflin, Paris-Princeton Lectures on Mathematical Finance 2004. Editors: R.A. Carmona, E. Çinlar, I. Ekeland, E. Jouini, J.A. Scheinkman, N. Touzi (2007)

Vol. 1920: S.N. Evans, Probability and Real Trees. Ecole d'Été de Probabilités de Saint-Flour XXXV-2005 (2008)

Vol. 1921: J.P. Tian, Evolution Algebras and their Applications (2008)

Vol. 1922: A. Friedman (Ed.), Tutorials in Mathematical BioSciences IV. Evolution and Ecology (2008)

Vol. 1923: J.P.N. Bishwal, Parameter Estimation in Stochastic Differential Equations (2008)

Vol. 1924: M. Wilson, Littlewood-Paley Theory and Exponential-Square Integrability (2008)

Vol. 1925: M. du Sautoy, L. Woodward, Zeta Functions of Groups and Rings (2008)

Vol. 1926: L. Barreira, V. Claudia, Stability of Nonautonomous Differential Equations (2008)

Vol. 1927: L. Ambrosio, L. Caffarelli, M.G. Crandall, L.C. Evans, N. Fusco, Calculus of Variations and Non-Linear Partial Differential Equations. Cetraro, Italy 2005. Editors: B. Dacorogna, P. Marcellini (2008)

Vol. 1928: J. Jonsson, Simplicial Complexes of Graphs (2008)

Vol. 1929: Y. Mishura, Stochastic Calculus for Fractional Brownian Motion and Related Processes (2008)

Vol. 1930: J.M. Urbano, The Method of Intrinsic Scaling. A Systematic Approach to Regularity for Degenerate and Singular PDEs (2008)

Vol. 1931: M. Cowling, E. Frenkel, M. Kashiwara, A. Valette, D.A. Vogan, Jr., N.R. Wallach, Representation Theory and Complex Analysis. Venice, Italy 2004. Editors: E.C. Tarabusi, A. D'Agnolo, M. Picardello (2008)

Vol. 1932: A.A. Agrachev, A.S. Morse, E.D. Sontag, H.J. Sussmann. V.I. Utkin, Nonlinear and Optimal Control Theory. Cetraro, Italy 2004. Editors: P. Nistri, G. Stefani (2008)

Vol. 1933: M. Petkovic, Point Estimation of Root Finding Methods (2008)

Vol. 1934: C. Donati-Martin, M. Émery, A. Rouault, C. Stricker (Eds.), Séminaire de Probabilités XLI (2008)

Vol. 1935: A. Unterberger, Alternative Pseudodifferential Analysis (2008)

Vol. 1936: P. Magal, S. Ruan (Eds.), Structured Population Models in Biology and Epidemiology (2008)

Vol. 1937: G. Capriz, P. Giovine, P.M. Mariano (Eds.), Mathematical Models of Granular Matter (2008)

Vol. 1938: D. Auroux, F. Catanese, M. Manetti, P. Seidel, B. Siebert, I. Smith, G. Tian, Symplectic 4-Manifolds and Algebraic Surfaces. Cetraro, Italy 2003. Editors: F. Catanese, G. Tian (2008)

Vol. 1939: D. Boffi, F. Brezzi, L. Demkowicz, R.G. Durán, R.S. Falk, M. Fortin, Mixed Finite Elements, Compatibility Conditions, and Applications. Cetraro, Italy 2006. Editors: D. Boffi, L. Gastaldi (2008)

Vol. 1940: J. Banasiak, V. Capasso, M.A.J. Chaplain, M. Lachowicz, J. Miękisz, Multiscale Problems in the Life Sciences. From Microscopic to Macroscopic. Będlewo, Poland 2006. Editors: V. Capasso, M. Lachowicz (2008)

Vol. 1941: S.M.J. Haran, Arithmetical Investigations. Representation Theory, Orthogonal Polynomials, and Quantum Interpolations (2008)

Vol. 1942: S. Albeverio, F. Flandoli, Y.G. Sinai, SPDE in Hydrodynamic. Recent Progress and Prospects. Cetraro, Italy 2005. Editors: G. Da Prato, M. Röckner (2008)

Vol. 1943: L.L. Bonilla (Ed.), Inverse Problems and Imaging. Martina Franca, Italy 2002 (2008)

Vol. 1944: A. Di Bartolo, G. Falcone, P. Plaumann, K. Strambach, Algebraic Groups and Lie Groups with Few Factors (2008)

Recent Reprints and New Editions

Vol. 1702: J. Ma, J. Yong, Forward-Backward Stochastic Differential Equations and their Applications. 1999 – Corr. 3rd printing (2007)

Vol. 830: J.A. Green, Polynomial Representations of GL_n, with an Appendix on Schensted Correspondence and Littelmann Paths by K. Erdmann, J.A. Green and M. Schoker 1980 – 2nd corr. and augmented edition (2007)

Vol. 1693: S. Simons, From Hahn-Banach to Monotonicity (Minimax and Monotonicity 1998) – 2nd exp. edition (2008)

Vol. 470: R.E. Bowen, Equilibrium States and the Ergodic Theory of Anosov Diffeomorphisms. With a preface by D. Ruelle. Edited by J.-R. Chazottes. 1975 – 2nd rev. edition (2008)

Vol. 523: S.A. Albeverio, R.J. Høegh-Krohn, S. Mazzucchi, Mathematical Theory of Feynman Path Integral. 1976 – 2nd corr. and enlarged edition (2008)